Tracking Environmental Change Using Lake Sediments.
Volume 1: Basin Analysis, Coring, and Chronological Techniques

Developments in Paleoenvironmental Research

VOLUME 1

Tracking Environmental Change Using Lake Sediments Volume 1: Basin Analysis, Coring, and Chronological Techniques

Edited by

William M. Last
Department of Geological Sciences,
University of Manitoba

and

John P. Smol
Department of Biology,
Queen's University

KLUWER ACADEMIC PUBLISHERS
DORDRECHT / BOSTON / LONDON

A C.I.P. Catalogue record for this book is available from the Library of Congress.

ISBN 0-7923-6482-1

Published by Kluwer Academic Publishers,
P.O. Box 17, 3300 AA Dordrecht, The Netherlands.

Sold and distributed in North, Central and South America
by Kluwer Academic Publishers,
101 Philip Drive, Norwell, MA 02061, U.S.A.

In all other countries, sold and distributed
by Kluwer Academic Publishers,
P.O. Box 322, 3300 AH Dordrecht, The Netherlands.

Cover Photos: East Basin Lake, Victoria, Australia.
Inset: Varved lake sediment
(from Saarinen & Petterson, Volume 2).

Printed on acid-free paper

02-0603-200ts
First printed 2001, Reprinted with corrections 2004

Printed in the Netherlands.

DEDICATION

Dedicated to Prof. B. E. Berglund, whose edited volume *Handbook of Holocene Palaeoecology and Palaeohydrology* has guided researchers for over 15 years.

CONTENTS

Part I: Basin Analysis Techniques

Introduction
Seismic sequence stratigraphy applied to lake basins
Theoretical considerations of the seismic reflection method
Practical considerations, instrumentation and scales of observation
Sequence stratigraphic interpretation of lacustrine seismic reflection data
Summary
Acknowledgments
References

Introduction
Ground-Penetrating radar theory
Data acquisition
Data processing and interpretation
Applications
Summary
Acknowledgments
References

Sample collection and preparation
What types of depositional environments are suitable for luminescence dating?
What can lead to an inaccurate optical age?
Summary
Acknowledgements
References

Introduction
Principles of ESR analysis
Sample collection
ESR analysis
ESR microscopy and other new techniques
Applications and datable materials in limnological settings
Summary
Acknowledgements
References

Introduction
Recording fidelity of geomagnetic behavior by sediments
Field and laboratory methods
Holocene SV records
Magnetostratigraphic studies of Neogene lake sediments
Excursions, short events and relative paleointensity
Conclusions
Summary
References

Introduction
Racemization and epimerization
Sampling protocols
AAR analysis
AAR dating
Dating and aminostratigraphic correlation in lacustrine environments
Paleoenvironmental analyses in lacustrine environments
Summary
Acknowledgements
References

PREFACE

The explosive growth of paleolimnology over the past two decades has provided impetus for the publication of this series of monographs detailing the numerous advances and new techniques being applied to the interpretation of lake histories. This is the first volume in the series and deals mainly with the acquisition and archiving of cores, chronological techniques, and large-scale basin analysis methods. Volume 2 (Last & Smol, 2001) examines the physical and geochemical parameters and methods; Volumes 3 and 4 (Smol et al., 2001a, b) provide a comprehensive overview of the many biological techniques that are used in paleolimnology. A fifth volume that is currently being prepared (Birks et al., in preparation) examines statistical and data handling methods. It is our hope that these monographs will provide sufficient detail and breadth to be useful handbooks for both seasoned practitioners as well as newcomers to the area of paleolimnology. These books should also be useful to non-paleolimnologists (e.g., limnologists, environmental scientists, archeologists, palynologists, geographers, geologists, etc.) who continue to hear and read about paleolimnology, but have little chance to explore the vast and sometimes difficult to access journal-based reference material for this rapidly expanding field. Although the chapters in these volumes target mainly lacustrine settings, many of the techniques described can also be readily applied to fluvial, glacial, marine, estuarine, and peatland environments.

The 16 chapters in this volume are organized into three major parts. The four chapters in Part I provide an overview of the most common, large-scale basin analysis methods. Part II summarizes the suite of sample acquisition, archiving and logging techniques routinely used in paleolimnology. The third and largest part of this book includes eight chapters summarizing chronostratigraphic techniques. Following this is a comprehensive glossary and list of acronyms and abbreviations.

Many people have helped with the planning, development, and final production of this volume. In addition to the hard work provided by the authors of these contributions, this publication benefitted from the technical reviews furnished by our scientific colleagues, many of whom remain anonymous. Each chapter was critically examined by two external referees as well as the editors. In order to assure readability for the major target audience, we asked many of our graduate students to also examine selected chapters; their insight and questioning during the reviewing and editorial process are most gratefully acknowledged. The staff of the Environmental, Earth and Aquatic Sciences Division of Kluwer Academic Publishers are commended for their diligence in production of the final presentation. In particular, we would also like to thank Ad Plaizier, Anna Besse-Lototskaya (Publishing Editor, Aquatic Science Division), and Rene Mijs (former Publishing Editor, Biosciences Division) for their long-term support of this new series of monographs and their interest in paleoenvironmental research. Finally, we would like to thank our respective universities and colleagues for support and encouragement during this project.

THE EDITORS

William M. Last is a professor in the Department of Geological Sciences at University of Manitoba (Winnipeg, Manitoba, Canada) and is co-editor of the *Journal of Paleolimnology*.

John P. Smol is a professor in the Biology Department at Queen's University (Kingston, Ontario, Canada), with a cross-appointment at the School of Environmental Studies. He co-directs the Paleoecological Environmental Assessment and Research Lab (PEARL). Professor Smol is co-editor of the *Journal of Paleolimnology* and holds the *Canada Research Chair in Environmental Change*.

AIMS AND SCOPE OF *DEVELOPMENTS*
IN PALEOENVIRONMENTAL RESEARCH SERIES

Paleoenvironmental research continues to enjoy tremendous interest and progress in the scientific community. The overall aims and scope of the *Developments in Paleoenvironmental Research* book series is to capture this excitement and document these developments. Volumes related to any aspect of paleoenvironmental research, encompassing any time period, are within the scope of the series. For example, relevant topics include studies focused on terrestrial, peatland, lacustrine, riverine, estuarine, and marine systems, ice cores, cave deposits, palynology, isotopes, geochemistry, sedimentology, paleontology, etc. Methodological and taxonomic volumes relevant to paleoenvironmental research are also encouraged. The series will include edited volumes on a particular subject, geographic region, or time period, conference and workshop proceedings, as well as monographs. Prospective authors and/or editors should consult the series editors for more details. The series editors also welcome any comments or suggestions for future volumes.

CONTENTS OF VOLUMES 1 TO 4 OF THE SERIES

Contents of Volume 3: *Tracking Environmental Change Using Lake Sediments: Terrestrial, Algal, and Siliceous Indicators.*

Contents of Volume 4: *Tracking Environmental Change Using Lake Sediments: Zoological Indicators.*

SAFETY CONSIDERATIONS AND CAUTION

Paleolimnology has grown into a vast scientific pursuit with many branches and subdivisions. It should not be surprising, therefore, that the tools used by paleolimnologists are equally diverse. Virtually every one of the techniques described in this book requires some familiarity with standard laboratory or field safety procedures. In some of the chapters, the authors have made specific reference to appropriate safety precautions; others have not. The responsibility for safe and careful application of these methods is yours. Never underestimate the personal risk factor when undertaking either field or laboratory investigations. Researchers are strongly advised to obtain all safety information available for the techniques they will be using and to explicitly follow appropriate safety procedures. This is particularly important when using strong acids, alkalies, or oxidizing reagents in the laboratory or many of the analytical and sample collection/preparation instruments described in this volume. Most manufacturers of laboratory equipment and chemical supply companies provide this safety information, and many Internet and other library resources contain additional safety protocols. Researchers are also advised to discuss their procedures with colleagues who are familiar with these approaches, and so obtain further advice on safety and other considerations.

The editors and publisher do not necessarily endorse or recommend any specific product, procedure, or commercial service that may be cited in this publication.

LIST OF CONTRIBUTORS

P. G. Appleby
Department of Mathematical Sciences
University of Liverpool
P. O. Box 147
Liverpool, L69 3BX, UK
e-mail: appleby@liverpool.ac.uk

Svante Björck
GeoBiosphere Science Centre, Quaternary Sciences
Lund University
Sölveg. 12
SE-223 62 Lund, Sweden
e-mail: svante.bjorck@geol.lu.se

Bonnie A. B. Blackwell
Department of Chemistry
Williams College
Williamstown, MA 01267 USA
e-mail: Bonnie.A.B.Blackwell@williams.edu

Steve M. Colman
U.S. Geological Survey
384 Woods Hole Rd
Woods Hole, MA 02543, USA
e-mail: scolman@usgs.gov

John R. Glew
Paleoecological Environmental Assessment and Research Lab (PEARL)
Department of Biology
Queen's University
Kingston, Ontario, K7L 3N6, Canada
e-mail: glewj@biology.queensu.ca

D. J. Huntley
Department of Physics
Simon Fraser University
Burnaby, B.C., V5A 1S6, Canada
e-mail: huntley@sfu.ca

J. H. Fred Jansen
Netherlands Institute for Sea Research
P.O. Box 59
NL-1790 AB Den Burg (Texel)
The Netherlands
e-mail: jansen@nioz.nl

John King
Graduate School of Oceanography
University of Rhode Island
Narragansett, RI 02882-1197, USA
e-mail: jking@gso.uri.edu

Scott Lamoureux
Department of Geography
Queen's University
Kingston, Ontario, K7L 3N6, Canada
e-mail: lamoureux@lake.geog.queensu.ca

William M. Last
Department of Geological Sciences
University of Manitoba
Winnipeg, Manitoba, R3T 2N2, Canada
e-mail: WM_Last@UManitoba.ca

Suzanne A. G. Leroy
Department of Geography and Earth Sciences
Brunel University
Uxbridge UB8 3PH, Middlesex, UK
e-mail: suzanne.leroy@brunel.ac.uk

Olav B. Lian
Department of Geography
Royal Holloway, University of London
Egham, Surrey, U.K. TW20 0EX
e-mail: olav.lian@rhul.ac.uk

J. J. Lowe
Centre for Quaternary Research
Geography Department
Royal Holloway, University of London
Egham, Surrey, TW20 0EX, UK
e-mail: J.Lowe@rhbnc.ac.uk

Jens Mingram
GeoForschungsZentrum Potsdam
PB 3.3 – Sedimentation and Basin Analysis
Telegrafenberg, D-14473 Potsdam, Germany
e-mail: ojemi@gfz-potsdam.de

Brian Moorman
Earth Science Program
University of Calgary
2500 University Drive N.W.
Calgary, AB, T2N 1N4, Canada
e-mail: moorman@ucalgary.ca

Rudolf Naumann
GeoForschungsZentrum Potsdam
PB 4.2 - Material Properties and Transport Processes
Telegrafenberg, D-14473 Potsdam, Germany
e-mail: rudolf@gfz-potsdam.de

Mats B. Nilsson
Department of Forest Ecology
Swedish University of Agricultural Sciences
SE-901 83 Umeå, Sweden
e-mail: Mats.B.Nilsson@sek.slu.se

Norbert R. Nowaczyk
GeoForschungsZentrum Potsdam
Projektbereich 3.3
"Sedimente und Beckenbildung"
Telegrafenberg, 14473 Potsdam, Germany
e-mail:nowa@gfz-potsdam.de

John Peck
Department of Geology
University of Akron
Akron, OH, 44325-4101, USA
e-mail: jpeck@uakron.edu

Dorothy Sack
Department of Geography
122 Clippinger Labs
Ohio University
Athens, OH 45701 USA
e-mail: sack@oak.cats.ohiou.edu

Christopher A. Scholz
Department of Earth Sciences
Syracuse University
Syracuse, New York, 13152, USA
e-mail: cascholz@syr.edu

John P. Smol
Paleoecological Environmental Assessment and Research Lab (PEARL)
Department of Biology
Queen's University
Kingston, Ontario, K7L 3N6, Canada
e-mail: SmolJ@BIOLOGY.QueensU.Ca

C. S. M. Turney
School of Archaeology and Palaeoecology
Queen's University
Belfast, BT7 1NN UK
e-mail: c.turney@qub.ac.uk

Sjerry van der Gaast
NIOZ, P.O. Box 59
NL-1790 AB Den Burg (Texel), The Netherlands
e-mail: gaast@nioz.nl

Barbara Wohlfarth
Department of Physical Geography & Quaternary Geology
Stockholm University
SE-106 91 Stockholm
e-mail: Barbara@geo.su.se

Bernd Zolitschka
Geomorphologie und Polarforschung (GEOPOL)
Institut für Geographie
Universitaet Bremen
Celsiusstr. FVG-M
D-28359 Bremen, Germany
e-mail: zoli@uni-bremen.de

1. AN INTRODUCTION TO BASIN ANALYSIS, CORING, AND CHRONOLOGICAL TECHNIQUES USED IN PALEOLIMNOLOGY

WILLIAM M. LAST (WM_Last@UManitoba.ca)
Department of Geological Sciences
University of Manitoba
Winnipeg, Manitoba
R3T 2N2, Canada

JOHN P. SMOL (SmolJ@Biology.QueensU.Ca)
Paleoecological Environmental Assessment
and Research Lab (PEARL)
Department of Biology
Queen's University
Kingston, Ontario
K7L 3N6, Canada

Paleolimnology, the interpretation of past conditions and processes in lake basins, is a multidisciplinary science whose roots extend back nearly two centuries. Despite this long history of development, the science has seen a surge of interest and application over the past decade. Today paleolimnology assumes a pivotal role in paleoclimatic and global change investigations, many fields of environmental science, and hydrocarbon and mineral resource exploration and exploitation. Associated with this dramatic increase in research activity involving lake sediments, there has been an equally rapid advance in the techniques and methods used by paleolimnologists.

The objective of this volume is to provide a state-of-the-art summary of the major field methods, chronological techniques, and concepts used in the study of large-scale lacustrine basin analysis. This and the other techniques volumes in this series build on the foundation provided by previous compilations of paleoenvironmental techniques, such as Kummel & Raup (1965), Bouma (1969), Carver (1971), Berglund (1986), Gray (1988), Tucker (1988), Warner (1990), and Rutter & Catto (1995), many of which continue to serve as essential handbooks. However, the development of new and different methods for studying lake sediments over the past decade, as well as advancements and modifications to old methods, have provided impetus for a new series of monographs for this rapidly expanding topic.

Three additional books from this series deal with other components of paleolimnology. Volume 2 (Last & Smol, 2001) focuses on the vast array of physical, mineralogical, and geochemical parameters that can be applied to the interpretation of lake histories. Volumes 3 and 4 (Smol et al., 2001a, b) address the great range of biological techniques that have been

1

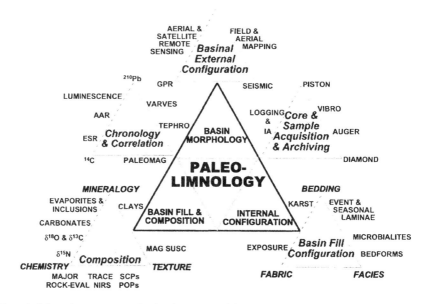

Figure 1. Schematic representation showing the spectrum of physical and chemical techniques and coring methods used in paleolimnology. This volume deals with basin configuration, chronology and correlation, and core and sample acquisition, whereas Volume 2 (Last & Smol, 2001) covers the composition of basin fill and its internal configuration.

and continue to be such an important aspect of many paleolimnological efforts. Although chapters in each of these books discuss the quantitative aspects of lake sediment interpretations, a separate volume on statistical and data handling approaches in paleolimmnology is currently in preparation (Birks et al., in preparation). Our intent with this series of volumes is to provide sufficient methodological and technical detail to allow both practitioners and newcomers to the area of paleolimnology to use them as handbooks/laboratory manuals and as primary research texts. Although we recognize that the study of pre-Quaternary lakes is a very rapidly growing component in the vast arena of paleolimnology, this volume is directed primarily towards those involved in Quaternary lacustrine deposits and Quaternary environmental change. Nonetheless, many of the techniques and approaches are applicable to other time frames. Finally, we anticipate that many of the techniques discussed in these volumes apply equally well to marine, estuarine, and other depositional environments, although we have not specifically targeted non-lacustrine settings.

This volume contains 16 chapters divided into three parts. As illustrated in Figure 1, the three main components of paleolimnology addressed in Volume 1 are: (i) methods used in the field collection and archiving of sediment cores, (ii) techniques for establishing chronology of the recovered stratigraphic sequence, and (iii) concepts and techniques that can be applied to large-scale lake basin analysis. Following this short introduction, Part I outlines several of the most commonly used basin analysis methods. Often these large-scale basin configuration and subsurface remote sensing techniques are a requisite first step for the paleolimnologist, applied prior to sample collection or core site selection. In the 1970's, application of seismic reflection technology to marine settings instigated a major paradigm shift in the traditional science of stratigraphy. These seismic sequence stratigraphy concepts and techniques are now routinely applied in lacustrine basins as summarized by Scholz. In the second chapter of Part I, Moorman discusses ground-penetrating radar (GPR), another geophysical method for lake basin sub-bottom profiling, which offers great promise for paleolimnologists working within freshwater systems. Lake basin geomorphology has traditionally formed the cornerstone for many paleolake investigations extending back to the mid-19th century. In the final chapter of this section, Sack provides an overview of geomorphological techniques that have been and continue to be of critical importance to many aspects of lake basin research.

The four chapters in Part II of this volume summarize the suite of sample acquisition and archiving techniques routinely used in Quaternary paleolimnology. Collecting continuous, undisturbed sediment cores from the upper, highly-fluid portion of a lacustrine sequence presents many technical difficulties. Glew et al. review the wide variety of coring devices that have been implemented to overcome these problems. Paleolimnologists interested in retrieving longer records from extant lakes or in acquiring cores from very deep lacustrine basins similarly face a considerable range of drilling and multiple-entry coring systems as compiled and described by Leroy & Colman. Once a suitable core or series of cores has been collected from a lake, it is desirable to gain as much information about the sediment as possible before actually subsampling and physically altering the recovered sequence. Zolitschka et al. consider the many new logging instruments now available to paleolimnologists that are capable of carrying out rapid, quasi-continuous, non-destructive, high-resolution analyses of the sediment's physical properties. In the final chapter of Part II, Nowaczyk summarizes the different methods of logging magnetic susceptibility, a sediment parameter that provides not only a valuable tool for core-to-core correlation, but also gives the researcher an indication of sediment composition, provenance (source), and depositional setting.

The third and final part of this book deals with techniques that can be used to establish the chronostratigraphic framework of a recovered lacustrine sequence. Age dating of Quaternary deposits is a large and diverse field, which has resulted in many publications. In the first chapter of Part III, Appleby reviews the models and methodologies involved in dating the recent ($\sim$150 year) record in lakes: ^{210}Pb, ^{137}Cs and ^{241}Am techniques. The single most widely used chronostratigraphic tool for paleolimnologists, ^{14}C dating, is summarized in the chapter by Björck & Wohlfarth. Because annually laminated lake deposits are particularly important for high-resolution paleolimnological studies, a considerable effort has been made to establish varve-based chronologies in many lake regions. Lamoureux gives a succinct overview of the construction of varve chronologies, and the methods for assessing the accuracy and consistency of varved records. Optical and thermoluminescence dating techniques, discussed by Lian & Huntley, have been applied extensively to aeolian deposits, but may also provide important chronological control in glacial lake sequences and in nonglacial lakes in which suitable organic matter for ^{14}C analysis is not present. Electron spin resonance (ESR) dating, reviewed by Blackwell, can provide absolute dating over a substantial time range ($\sim$0.5 ka to 5–10 Ma). Because of the wide variety of materials that can be dated with this technique, ESR holds considerable promise for establishing the chronology of older Quaternary and late Tertiary lacustrine deposits. King & Peck summarize the chronostratigraphic use of temporal variations in the Earth's magnetic field (geomagnetic secular variation or SV) and magnetostratigraphy. Blackwell provides a comprehensive review of amino acid racemization (AAR) dating and aminostratigraphy, which have been shown to be powerful tools for Quaternary chronostratigraphy and correlation. The geologically instantaneous deposition of atmospheric particles following a volcanic eruption allows tephrochronology to achieve a degree of precision unmatched by nearly any other chronostratigraphic method. In the final chapter of this section, Turney & Lowe discuss some of the recent advances in this technique and outline the methods of identification and correlation of tephra deposits and the establishment of a regional tephrochronological framework.

Although we have assumed readers will have a basic knowledge of earth and environmental science jargon, we have included a comprehensive glossary and list of acronyms and abbreviations to assist with the large and perhaps unfamiliar vocabulary. This is followed by a subject index.

Some related aspects are not covered in this volume. As we write this, the "lakes of Mars" are common headlines in the popular scientific press. The arsenal of remote sensing tools and techniques that are required for this type of extraterrestrial paleolimnology are not included in this volume (e.g., Warton et al., 1995). Due to space limitations, we have not included summary chapters on traditional field mapping and outcrop description techniques (e.g., Daniel, 1986), which are often part of a Quaternary scientist's bailiwick. Electrical resistivity methods and several other geophysical mapping and delineation techniques are not specifically discussed, nor do we address any of the large number of "down-hole" petrophysical logging devises that have been occasionally used in Quaternary and pre-Quaternary paleolimnology, although selected aspects of these topics are covered in several chapters and this topic is also dealt with in Volume 2 (Last & Smol, 2001).

As in many other areas of environmental earth science, the continued development of our understanding of the histories of lake basins will depend on the interaction of specialists in a wide range of disciplines. The tools and investigative techniques these specialists bring to bear on the task of characterizing the sediments and fluids within these basins are continuing to advance at a rapid pace. The applications and research avenues fronting paleolimnology are immense, but so are the potential analytical and logistical problems. One of the greatest challenges facing the next generation of paleolimnologists is to maintain the integrated multidisciplinary approach to the study of lake sediments that has evolved over the past decade. We see no evidence that the accelerated rate of progress that this field has enjoyed will be slowing down in the near future. Many new approaches and applications are certainly soon to be discovered, and will hopefully be covered in subsequent volumes of this series.

References

Berglund, B. E. (ed.), 1986. Handbook of Holocene Palaeoecology and Palaeohydrology. John Wiley & Sons, New York, 869 pp.

Birks, H. J. B., S. Juggins, A. Lotter & J. P. Smol, In preparation. Tracking environmental change using lake sediments. Volume 5: Data Handling and Statistical Techniques. Kluwer Academic Publishers, Dordrecht, The Netherlands.

Bouma, A. H., 1969. Methods for the Study of Sedimentary Structures. John Wiley & Sons, New York, 458 pp.

Carver, R. E. (ed.), 1971. Procedures in Sedimentary Petrology. Wiley Interscience, New York, 653 pp.

Daniel, E., 1986. Geological survey mapping. In Berglund, B. E. (ed.) Handbook of Holocene Palaeoecology and Palaeohydrology. John Wiley & Sons, New York: 195–201.

Gray, J. (ed.), 1988. Paleolimnology: Aspects of Freshwater Paleoecology and Biogeography. Elsevier, Amsterdam, 678 pp.

Kummel, B. & D. Raup (eds.), 1965. Handbook of Paleontological Techniques. W. H. Freeman and Company, San Francisco, 862 pp.

Last, W. M. & J. P. Smol (eds.), 2001. Tracking environmental change using lake sediments. Volume 2: Physical and Geochemical Methods. Kluwer Academic Publishers, Dordrecht, The Netherlands, 504 pp.

Rutter, N. W. & N. R. Catto (eds.), 1995. Dating Methods for Quaternary Deposits. GEOtext 2, Geological Association of Canada, St. John's, NF., 308 pp.

Smol, J. P., H. J. B. Birks & W. M. Last (eds.), 2001a. Tracking environmental change using lake sediments. Volume 3: Terrestrial, Algal, and Siliceous Indicators, Kluwer Academic Publishers, Dordrecht, The Netherlands, 371 pp.

Smol, J. P., H. J. B. Birks & W. M. Last (eds.), 2001b. Tracking environmental change using lake sediments. Volume 4: Zoological Indicators, Kluwer Academic Publishers, Dordrecht, The Netherlands.

Tucker, M. E. (ed.), 1988. Techniques in Sedimentology. Blackwell Scientific Publications Ltd., Osney Mead, Oxford, UK, 394 pp.

Warner, B. G. (ed.), 1990. Methods in Quaternary Ecology. Geoscience Canada Reprint Series 5, St. John's, NF., 170 pp.

Wharton, R. A., J. M. Crosby, C. P. McKay & J. W. Rice, 1995. Paleolakes on Mars. J. Paleolim. 13: 267–283.

2. APPLICATIONS OF SEISMIC SEQUENCE STRATIGRAPHY IN LACUSTRINE BASINS

CHRISTOPHER A. SCHOLZ (cascholz@syr.edu)
Department of Earth Sciences
Syracuse University, Syracuse
NY 13152, USA

Keywords: lakes, lacustrine basins, sequence stratigraphy, reflection seismology, seismic stratigraphy, basin analysis, seismic profiling, lacustrine sedimentation

Introduction

Sedimentary basin studies in lakes routinely require the imaging or remote characterization of the subsurface, prior to the collection of sediment samples by sediment coring or drilling. Acoustic techniques such as echosounding, sub-bottom seismic reflection profiling, and multibeam or swath bathymetry mapping are commonly employed for lake-bottom and sub-bottom characterization. Over the past several decades, the proliferation of digital seismic reflection technology is in part responsible for modernizing the discipline of stratigraphy and subsurface analysis, including greatly expanding our understanding of lacustrine basin evolution. Whereas the stratigraphic sciences were formerly dominated by compilations of one-dimensional data sets such as outcrop stratigraphic columns, well-logs or sediment cores, abundant two-dimensional seismic profiles and 3-D seismic volumes now permit the rendering of complex structural and depositional features in the subsurface. This technological transformation has enabled the refinement of stratigraphic concepts initially introduced by Sloss and coworkers (e.g., Sloss, 1963), and which culminated in publications by Vail and colleagues from the oil and gas industry that present the fundamental concepts of seismic sequence stratigraphy (e.g., Payton, 1977; AAPG Memoir 26 and subsequent publications). Their efforts were based largely on basin-scale industry seismic reflection profiles acquired on marine passive margins, but now many researchers have demonstrated that the main concepts of sequence stratigraphy are largely independent of scale and environment. These concepts are now routinely applied in non-marine settings (e.g., Shanley & McCabe, 1998), including lacustrine basins. This chapter presents a brief overview of principles of the seismic reflection technique, a discussion of practical applications on different scales, and concludes with a summary of the unique aspects of seismic sequence stratigraphy in lacustrine systems.

Prior to the 1980s digital seismic technology was mainly the domain of large corporations that could afford expensive computer processing power. The proliferation of inexpensive desktop computer technology now makes digital acquisition, processing and

W. M. Last & J. P. Smol (eds.), 2001. *Tracking Environmental Change Using Lake Sediments. Volume 1: Basin Analysis, Coring, and Chronological Techniques.* Kluwer Academic Publishers, Dordrecht, The Netherlands.

sophisticated interpretation procedures accessible to many in academic settings. Whereas analysis of seismic reflection data recorded on analog paper profiles was previously the norm, rapid interpretation and extensive subsurface mapping can now be carried out entirely on computer workstations, with little need for paper reproduction.

Seismic sequence stratigraphy applied to lake basins

Basin evolution of lacustrine systems encompasses several spatial scales of observation, and a wide range of geological time. Whereas investigations of the Earth's oldest tectonic lakes consider millions of years of geologic time, many current studies are focused on lakes in existence only since the late Pleistocene or early Holocene (e.g., Tiercelin, 1999). The small size of many lake systems and the lack of energetic geostrophic currents in most lake basins are the most obvious differences between marine and lacustrine sedimentary basins. Additionally, most lakes are inherently closed sedimentary systems, whereas some marine depositional systems may recover or lose sediments from adjacent basins. Because basin physiography is often compressed in lake basins relative to ocean basins, sedimentary facies changes may be abrupt, both laterally and vertically.

The hydrologic controls on lake levels are often in a delicate state of balance in many lacustrine basins, particularly those in the tropics or in arid settings. For instance in Lake Malawi, in the southern part of the east African rift, nearly 90% of the water lost from the lake annually is evaporated off its surface, whereas only about 10% of the total water loss is through the lake's outlet. This produces seasonal fluctuations of 1–2 m in this enormous lake, and long-term changes in water level are measured in the hundreds of meters over time frames of hundreds to thousands of years (Owen et al., 1990). Consequently, the stratigraphic response to environmental or geological change is often highly amplified in these lacustrine basins compared to marine systems. This sensitivity to external forcing produces depositional systems that can be superb high-resolution archives of environmental change, yet in some instances the stratigraphic record from lake basins may be highly punctuated, when considered over long periods of geological time.

Theoretical considerations of the seismic reflection method

Seismic methods

Reflection seismic techniques are routinely employed in the search for oil and gas and were initially introduced more than 80 years ago (Dobrin & Savit, 1988). Past refinements of the methodology in the early years were prompted by opportunities associated with the energy industry and most recent innovations in 3-D seismic technology stem from efforts related to hydrocarbon exploration. Many research volumes and scientific journals are dedicated to reviews of the techniques, and many handbooks (e.g., Sheriff & Geldart, 1995) present a comprehensive treatment of the principles and practice of the seismic method. Marine seismic methods evolved rapidly following World War II, and are now routinely used in coastal and continental margin studies, as well as in lake surveys.

The seismic reflection method uses acoustic waves to remotely image the stratigraphic contrasts in sediment physical properties, manifested in sediment lithology. Reflectivity across a lithologic interface is quantified as the contrast in acoustic impedance, the product of saturated bulk density and P-wave velocity of the adjacent sediments (e.g., Sheriff & Geldart, 1995; Dobrin & Savit, 1988). The seismic response to the layered earth reflectivity is produced when a percussive source signal is imparted into the earth, and energy reflected from lithologic boundaries is recorded on the Earth's surface. In the case of lake surveys, source energy is generated by a towed source array, that may consist of a transducer, boomer plate, sparker, airgun or similar sound-generating device. In the case of land surveys, receiving signals are recorded by ground accelerations, and in the marine or lacustrine case by passage of a pressure wave past a hydrophone. Swept frequency pulse or CHIRP systems (e.g., Schock & LeBlanc, 1990) are now commonly used in many marine and lacustrine studies. Rather than employ a percussive source, the swept frequency pulse system generates a prolonged pulse over a range of frequencies that is later correlated to replicate an impulsive spike. Advantages of CHIRP systems are their lower electrical power demands compared to impulsive sound sources, and their ability to return data of variable resolution using a broad bandwidth pulse. Repetitive firing of the seismic source along a given vessel transect yields profiles that are composed of hundreds or thousands of discrete seismic traces.

Single channel seismic profiling methods

Single-channel seismic (SCS) reflection techniques assume that source and receiver positions are coincident or nearly so, such that the seismic rays travel along a normal incidence pathway from source to reflecting horizon and back to the receiving array. The operational aspects of this technique are simple and straightforward, and usually involve a towed source and receiver array close to one another, and both towed close to the stern of the research vessel. Whereas SCS surveys are operationally and logistically simple to deploy, they provide only the most fundamental subsurface information. Sources of coherent noise that develop from water bottom or interbed multiples may be difficult to extract from primary reflections, and overall signal-to-noise ratio may be modest compared with data acquired using multichannel methods. No seismic velocity information is derived from SCS data.

Multichannel seismic profiling methods

Multichannel (MCS) or common midpoint (CMP) seismic reflection profiling involves the deployment of long receiver arrays to concentrate raypaths and produce an "acoustic lens" over successive reflecting points in the subsurface (Fig. 1). This requires considerable post-acquisition seismic processing, including resorting numerous returning signals or traces from the shot domain to the common midpoint domain (e.g., Fig. 1). Following the application of the "normal moveout" correction of traces that are offset from the source in order to bring arrivals from the same horizon to similar arrival times, traces are summed or "stacked", resulting in significantly enhanced signal/noise over the common reflecting point. Additionally, MCS data yield information on the seismic velocity structure of the subsurface, which permits time-to-depth conversions of seismic reflection profiles. Desktop computers now have the processing power necessary for completing processing on all but the largest of pre-stack MCS data sets. The procedures for acquiring and processing marine-type MCS are highly refined, and are identical for marine and lake surveys (see Sheriff &

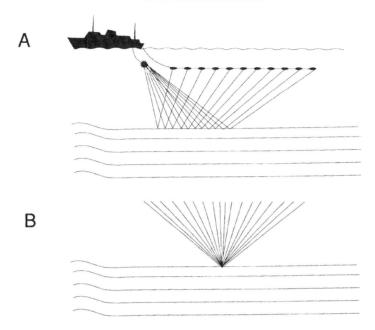

Figure 1. A) Deployment of a multichannel seismic receiver array behind a research vessel, and hypothetical raypaths from source. B) Arrangement of common midpoint traces (CMP gather) assembled after sorting from different shot gathers.

Geldart, 1995; Dobrin & Savit 1988 for additional discussions). In general, the limiting factor in the deployment of MCS operations on lakes is the size of the air compressor systems required for delivering high-pressure air to large airgun arrays. Consequently, MCS operations generally require deployments from larger research vessels.

Seismic stratigraphic interpretation methods

The application of seismic methods to stratigraphic problems greatly expanded following the proliferation of digital seismic methods and the resulting improvements in the fidelity of subsurface records. The stratigraphic analysis of seismic reflection profiles uses the characterization of stratal relationships, in the form of reflection terminations, to identify breaks and changes in the sedimentary record of a basin. Principal stratal relationships and their importance in deciphering basin histories are described in Table I, and can also be applied to basin studies in lacustrine basins, as is demonstrated in Figures 2–4.

Several assumptions are made in seismic stratigraphic analysis. Foremost among these is that seismic reflections have chronostratigraphic significance, and that discrete reflections represent time lines from which relative stratigraphic and structural histories within the basin can be derived. Early sequence stratigraphers reached this conclusion empirically, following the examination of many thousands of kilometers of marine seismic profiles. We adhere to a similar assumption in our sequence stratigraphic analyses of lacustrine basins.

Table I. Reflection terminations and stratal geometries.

Termination Type	Geometry	Significance	
EROSIONAL TRUNCATION	overlying flat strata truncating dipping strata	indicates erosion and removal of material; fall in lake level	
TOPLAP	low-angle dipping strata with overlying flat strata	caused by non-deposition, minor erosion, sedimentary bypassing; often taken to indicate a still-stand in lake level	
ONLAP	horizontal strata terminating against an underlying inclined surface	indicates a transgression after a lake level fall	
DOWNLAP	inclined strata terminating onto a flat surface	indicates sediment starvation deep in the basin; highstand conditions	

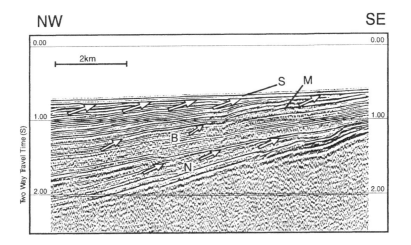

Figure 2. Multichannel seismic profile from Lake Malawi showing four discrete depositional sequences or sequence sets, bounded by erosional unconformities (S,M,B,N) (after Scholz et al., 1998).

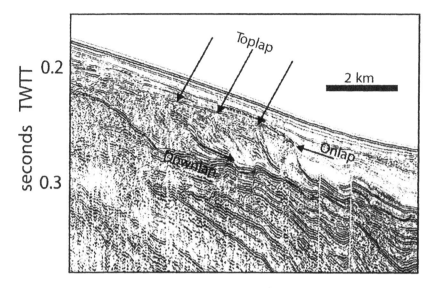

Figure 3. Example of toplapping strata and corresponding downlap on upper surface of progradational lacustrine lowstand delta lobe. Note also onlap of younger strata onto delta lobe. Example from offshore of the Songwe delta, northern Lake Malawi.

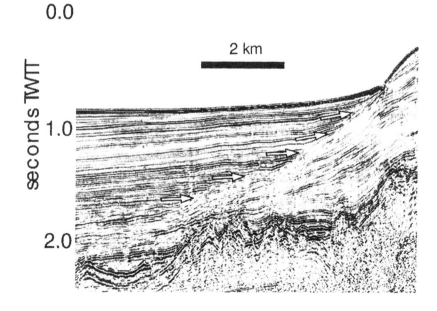

Figure 4. Example of onlap of strata in upper sequence onto lower sequence. MCS profile from northern Lake Baikal.

Practical considerations, instrumentation and scales of observation

The ages of modern lake basins span several orders of magnitude. In many areas that were inundated by continental ice sheets during the last glaciation, lakes have only been in existence for ~10 ka, whereas deep tectonic lakes such as Lakes Baikal and Tanganyika are likely more than 10 ma in age (Cohen et al., 1993; Mats 1993). Consequently, the sediment fills of lakes of different sizes and ages will differ dramatically, and thus demand markedly different tools for thorough subsurface investigations. Presented here are discussions of three different survey approaches useful for: 1) "basin-scale" surveys of the largest of the world's tectonic lake basins; for 2) sedimentary facies scale studies requiring "intermediate resolution" seismic methods; and 3) high-resolution seismic tools for detailed site surveys and analyses of the late-Quaternary.

Deep basin-scale studies

The world's oldest and most deeply subsided lakes are typically of tectonic origin (Herdendorf, 1990; Johnson, 1984) and may contain sediment fills that are measured in thicknesses of many kilometers. For instance the maximum sediment fill of the Lake Baikal Rift is estimated to be more than 9 km (Scholz et al., 1998) and the basin is thought to be more than 25 M.Y. in age (Mats, 1993). Similarly, Lake Tanganyika in the east African rift is also a lacustrine rift basin, and has a sediment fill of more than 5 km and is thought to be 9–12 M.Y. in age (Cohen et al., 1993). In order to image through several kilometers of lacustrine sediment, seismic imaging tools that are routinely employed in marine basins using large research vessels are necessary. Although small, specialized lake seismic vessels are used under some circumstances, (e.g., Rosendahl & Livingstone, 1983), larger "ships of opportunity" are commonly the most practical approach for such surveys.

Imaging to depths of several kilometers in the subsurface requires large acoustic sources and long receiver arrays. In general, large airgun arrays (totaling more than several hundred cubic inches), such as are used in marine surveys, provide the best subsurface penetration. Maximum receiver offsets (or total length of hydrophone streamer arrays) are used that are of the same order of the desired or anticipated maximum depth of penetration. For instance, Figure 5 shows a seismic image from Lake Baikal that used a 3000 m-long streamer with 48 channels to recover the crystalline basement reflection at a two-way travel time of about 5 seconds. Lower frequency sources (4–50 Hz) commonly yield much improved penetration of the sediment column, but with a significantly diminished vertical resolution than higher frequency sources.

Multichannel seismic acquisition programs will commonly yield images of the lake subsurface through soft sediment as well as through great thicknesses of lithified material down to bedrock or crystalline basement rocks. First order depositional sequences can be observed in such data sets (e.g., Fig. 4), representing major changes in tectonic conditions or profound drainage basin alterations. The ability to penetrate considerable thicknesses of section comes with a major trade-off in seismic resolution. A rule-of-thumb for estimating the thickness of the minimum resolvable bed (vertical resolution) is 1/4 of the wave length of the signal (e.g., Sheriff & Geldart, 1995; Dobrin & Savit, 1988). For instance, an industry "standard" seismic source array, with a bandwidth of ~8–50 Hz, will typically yield a maximum vertical resolution of ~25 m with normal seismic velocities at mid-

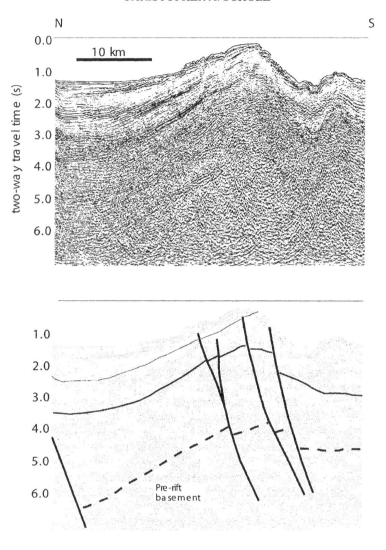

Figure 5. Basin-scale multichannel seismic profile from southern Lake Baikal, showing thick sedimentary section and pre-rift crystalline basement reflections.

basinal depths. Whereas major unconformities can be observed on such profiles, it is rare for detailed information about depositional facies to be easily extracted from such low-resolution images (e.g., Fig. 5). Sedimentary basins that have total sediment thicknesses of less than 500 m do not necessarily require such low frequency source arrays to image the complete sedimentary section.

Multichannel seismic techniques offer enhanced signal-to-noise ratios compared to single-channel seismic systems, and in particular permit the suppression of coherent noise on seismic records. Pre-stack processing routines offer a variety of approaches to suppression of unwanted noise, and for enhancing the geologic signals in the data (e.g., Yilmaz, 1987). Information on the basin seismic velocity structure is also obtained from MCS surveys, and can be extremely useful for constraining true basin geometries and lithofacies characteristics. Velocity information is commonly required for siting deep scientific boreholes and for producing accurate reconstructions of complex structures in the subsurface. The operational deployment of MCS systems can be complex, however, and such programs are typically undertaken only on large lakes with the support of well-equipped research vessels.

The calibration of deep basin scale MCS data requires the collection of data from deep penetrating scientific boreholes or oil wells. Such boreholes are usually drilled only in conjunction with major scientific drilling programs or during hydrocarbon exploration. In the case of oil field or deep scientific drilling, down-hole geophysical tools are used to help characterize subsurface lithologies, and these data can be used to tightly calibrate or constrain coincident seismic profiles.

Many of the world's largest lakes have been surveyed using MCS techniques, including the East African Rift Valley Lakes (e.g., Rosendahl, 1987), Lake Baikal (e.g., Hutchinson et al., 1992), and Lake Superior (Cannon et al., 1989). The focus of these studies has been on basin evolution, long time-frame studies of paleoclimates, and on deciphering long-term tectonic histories.

Sedimentary facies scale studies using intermediate resolution seismic reflection tools

Projects focused on the characterization of sedimentary facies variations within lake basins are carried out using seismic sources with a higher frequency range than the large airgun source arrays described above. For instance instruments with a bandwidth of ~50–~1000 Hz, which include small airguns ($<\sim20\,in^3$) and sparkers, can penetrate through all types of unconsolidated materials including mud, coarse sands and gravel. Sparkers often produce the best compromise between depth of penetration and resolution for facies evaluation (e.g., Back et al., 1998). However sparkers can be hazardous to handle, and require a conductive medium for shooting. Thus when used in freshwater lakes, sparker sources must be deployed in a container or bag containing salt water. Under ideal conditions, small airgun source arrays can also penetrate as much as 1–2 km into the subsurface (Fig. 6). Such sources, when combined with a digital single-channel seismic systems, are very effective at constraining facies geometries in most modern sedimentary environments.

At a peak frequency of ~250 Hz and a mean seismic velocity of <2000 m/sec, the vertical resolution of intermediate-resolution seismic systems is ~1–2 m. Surveys using this instrumentation are of significant utility on a wide variety of lake basins, and are useful for characterizing depositional architecture, for neotectonic studies and for Plio-Pleistocene paleoclimate studies. Ground-truth calibration of seismic data sets of this type involves the shallow drilling of scientific boreholes to depths of hundreds of meters into the subsurface. Whereas both single-channel and multichannel receiver geometries can be employed in

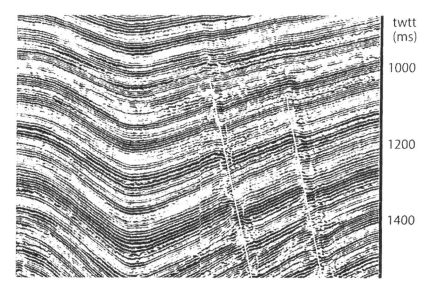

Figure 6. Small airgun, single-channel seismic reflection profile illustrating thick (700 m) thick section of hemipelagic lacustrine sediments. Profile from central Lake Malawi. Note normal faults interpreted on right side of section.

these surveys, the ease of deployment of SCS systems makes this type of approach most common in lake environments.

High-resolution seismic tools

High-resolution seismic tools use high-frequency transducers, electromagnetic plates (boomers) or swept-frequency sources (CHIRP systems) to generate signals that penetrate from a few 10's of meters to a few hundred meters into the subsurface under ideal circumstances. Such systems are the tool of choice for siting soft-sediment cores intended to sample late-Quaternary sediments. Because the source bandwidth of such systems may range from ~1 kHz to ~6–12 kHz depending on the specific instrumentation, it is possible to resolve sedimentary features and horizons with a vertical resolution as fine as 10–20 cm (Fig. 7). Although very effective at imaging fine-grained and water-saturated muds, they are generally unable to penetrate or image coarse sand or gravel deposits to any significant degree.

Operationally, such sources are almost always used in SCS mode, and are commonly deployed from relatively small research vessels, or even from inflatable boats. Power requirements vary with the specific source, but small portable generators generally provide satisfactory electrical power to drive high-resolution seismic systems.

Calibration of high-resolution seismic data can be accomplished with standard wire-line deployable oceanographic or limnological core samplers. Measurements of bulk sediment density and seismic velocity (see chapter on physical properties studies) can be used to

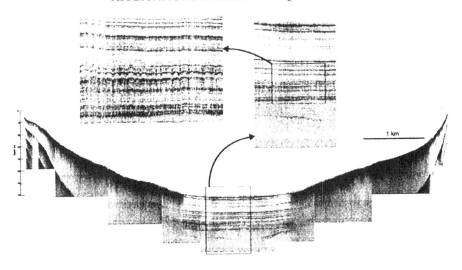

Figure 7. Example of high-resolution seismic profile from Lake Bosumtwi, shot using a 4–24 kHz CHIRP seismic system. Note high-quality images of the stratigraphic section observed in the center of the profile, yet limited penetration on margins. This latter effect is on account of biogenic gas in the sediment.

create synthetic seismograms, and in many instances precisely link core information directly to seismic data.

Sequence stratigraphic interpretation of lacustrine seismic reflection data

Marine versus lacustrine controls on depositional sequences

Depositional sequences are thought to be the stratigraphic building blocks of all sedimentary basins (Vail, 1987). Sequences differ from geological formations, in that they are chronostratigraphically constrained packages of sediments or sedimentary rocks, that are bounded at the top and the bottom by unconformities, whereas formations are lithologically distinctive groupings of strata that are, in some cases, time-transgressive. Seismic reflection data are a tremendous benefit to stratigraphic interpretation, because these data permit observations of distinct subsurface stratal geometries in two dimensions, which is the key to establishing the sequence-based stratigraphic framework of the basin.

The main controls on the formation of depositional sequences in sedimentary basins are eustasy (sea- or lake-level change), subsidence, and sediment supply (e.g., Vail, 1987). In addition to these three main controls, other factors such as pre-existing slope or basin morphology (Sarg & Pratson, 2000), and sediment compaction and flexural isostasy (Reynolds et al., 1991) may also significantly modify the structure and geometry of depositional sequences. Whereas the three primary controls or variables are essentially independent in marine systems, they are connected to a high degree in lacustrine basins, which are effectively closed depositional systems. For instance in marine basins, sediment supply and sea level are largely unrelated over most periods of the Phanerozoic. However in lake basins, lake levels and sediment influx will likely be tightly coupled through regional climate and hydrology, although not necessarily in a direct, linear relationship.

Global water level fluctuations in the oceans are controlled by glacio-eustasy, and local sea level changes in marine basins are controlled by subsidence, crustal rebound, sedimentation (e.g., delta progradation) or one of the other controlling factors listed above. In lacustrine basins, maximum potential lake levels are ultimately controlled by basin sill depth, which can be altered by tectonic or erosive events (Scholz et al., 1998). Where the outflow point or sill depth of a lake is situated in a geologically dynamic location, catastrophic changes in lake level can result.

Lake hydrology, which is dominantly controlled by the balance between precipitation and evapotranspiration/evaporation in the catchment/lake, will control lake levels on short-duration time frames (e.g., Scholz et al., 1998). When hydrology and sediment input are factored together, lake basins can be classified between either underfilled or overfilled end-member models (Carroll & Bohacs, 1999). Lake basins that are underfilled by a combination of water and sediment may produce rapid lake level responses, whereas overfilled basins, where most of the water leaves the system through the lake outlet, may experience long-term lake level stability.

Because the sedimentation rates of lake systems tend to be considerably higher than marine systems, high-frequency cyclical processes are far easier to observe directly in seismic records from lake basins than in records from most marine settings. For example, sedimentation rates of 1–10 cm/ka are common in marine settings, yet the combination of the muted response for the marine system to the climatic cycles, with the slow sedimentation rates, yields seismic images that generally do not resolve climate cycles that develop at orbital (Milankovitch) frequencies. However the stratigraphic response to climate forcing is commonly amplified in many lacustrine systems, inducing enormous fluctuations in lake level or in sediment input. Given that overall lacustrine sedimentation rates are an order of magnitude (or more) higher than in marine systems, it is possible to directly observe orbital scale climate cyclicity in intermediate- or high-resolution seismic reflection records (e.g., Scholz, 1995; Lezzar et al., 1996). Figure 8 illustrates a series of stacked sublacustrine fans from Lake Malawi where ~40–100 m-thick packages of sublacustrine fan deposits alternate with deep-water hemipelagic sediment. This geometry is thought to be generated by 100 ka climate cycles in the Pleistocene, in response to fluctuating lake levels in Lake Malawi (Scholz, 1995).

Indicators of lake level change in seismic reflection records

Seismic reflection records commonly record strong evidence for changing lake level conditions. Examples of specific indicators of past lake levels include erosional truncation surfaces, beach berms, and low stand progradational delta deposits.

Unconformities in the form of erosional truncation surfaces may indicate areas and episodes of subaerial exposure within a sedimentary section. Whereas erosion can develop subaqueously in large marine basins because of contour currents or through major down-slope disturbances, such erosional areas are limited to localized slides and slumps lacustrine basins. In localities where erosional truncation surfaces can be mapped around the rim of a basin with a dense set of seismic reflection records, a valid case can be made for an extended period of low lake conditions (e.g., Fig. 9).

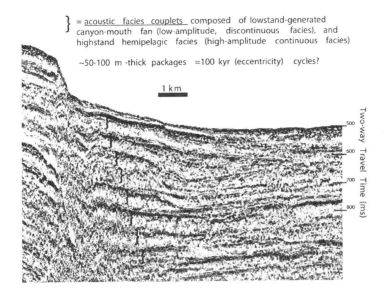

Figure 8. Acoustic facies couplets (}) composed of lake-level lowstand-generated canyon-mouth fan (low-amplitude, discontinuous facies), and highstand hemipelagic facies (high-amplitude continuous facies). These ~50–100 m-thick packages are thought to be deposited over 100 kyr (eccentricity) cycles.

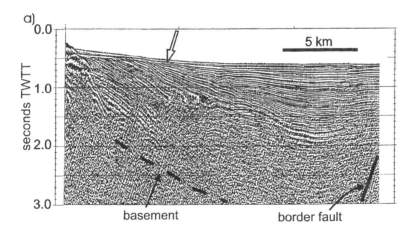

Figure 9. Example of a broad zone of erosional truncation, observed on the flexural margin side of the northern half-graben basin in Lake Malawi. Note pronounced reflection terminations on the left side of profile, just beneath upper depositional sequence. Down-dip limit of erosional truncation may indicate approximate location of paleoshoreline (white arrow).

Perhaps the best indicator of instantaneous low lake or paleoshoreline conditions are progradational low stand delta deposits (Fig. 3). The "roll-over" transitions from topset to foreset strata on a delta clinoform packages are commonly interpreted to indicate the "depositional shelf break", which generally forms just lakeward of the frontal distributary mouth bar in most lacustrine deltas. The classic clinoform geometry can thus be used to establish a record of low lake levels, especially when several deltas advancing into a lake yield comparable geometries in similar water depths and stratigraphic levels in the subsurface.

The several tropical lakes mentioned above are among the largest and oldest lakes in the world, and it is these ancient lakes at the lower latitudes that seem prone to dramatic shifts in lake level. Many Alpine and glacial lakes of Europe and North America have been investigated using seismic reflection methods (e.g., Mullins et al., 1996; Giovanoli et al., 1982; Houbolt & Jonker, 1968). In general, these lakes existed only since the terminal Pleistocene deglaciation. Whereas lake levels in these systems rose and fell dramatically during deglaciation, they are generally overfilled systems that have had relatively stable water levels for much of the Holocene, at least compared to the highly reactive lakes of the lower latitudes.

Summary

Seismic sequence stratigraphy provides an approach to lacustrine basin analysis that is useful at many scales of observation. As has been observed in the case of marine systems, hierarchies of depositional sequences can be observed using different types of imaging tools, each of which offers different types of geological information. Deep-penetrating multichannel seismic imaging provides extensive information on ancient basins, and the sequences observed in these records provide histories of basin evolution and tectonism. Intermediate-resolution seismic systems and mid-range seismic sources, such as small airguns and sparkers, permit observations and analyses of depositional facies architecture in lake basins. High-resolution seismic tools are ideal for late-Quaternary studies, for imaging fine-grained and unlithified sediments, and for locating ideal coring sites for Holocene and late-Pleistocene sediment sampling in lakes.

Seismic sequence analysis in lacustrine basins is similar to the approach taken in marine basins; the stratigraphic framework is established by mapping unconformities using reflection terminations and stratal relationships observed on two-dimensional seismic reflection profiles. Because stratigraphic forcing mechanisms, however, behave somewhat differently in lake basins, care must be taken to avoid model-driven interpretations based on principles developed wholly from the classic marine examples. For instance, changes in available accommodation space in marine systems are measured on scales of less than 100–200 m in marine basins, and the frequency of glacio-eustatic sea level change is typically measured on a time frame of tens of thousands of years. In lake basins however, the amplitude of lake level change may be measured in hundreds of meters in large basins, and shifts of this magnitude occur on century-to-millennial time frames. Depositional slopes may be steep in some tectonic lake basins, and the resulting facies transitions are commonly abrupt. Classic components of the marine-based sequence model, such as transgressive systems tracts, may thus be exceedingly thin and difficult to image seismically in lacustrine systems. Whereas glacio-eustasy is a common controlling process in the evolution of marine basins, even adjacent lake basins may be controlled by radically different climate regimes and hydrologies. Consequently the development of lake basin sequences may be highly variable in space and time.

Acknowledgments

Over the past several years my ideas on lacustrine sequence stratigraphy have matured through discussions with numerous colleagues, including, M. de Batist, M. Soreghan, H. T. Mullins, D. J. Reynolds, J. Neal, F. Anselmetti, K. Lezzar, A. Cohen, J. J. Tiercelin, K. Bohacs, and A. Carroll.

References

Back, S., M. De Batist, P. Kirillov, M. R. Strecker & P. Vanhauwaert, 1998. The Frolikha Fan; a large Pleistocene glaciolacustrine outwash fan in northern Lake Baikal, Siberia. J. Sed. Res. 68: 841–849.

Cannon, W. F., A. G. Green, D. R. Hutchinson, M. W. Lee, B. Milkereit, J. C. Behrendt, H. C. Halls, J. C. Green, A. B. Dickas, G. B. Morey, R. H. Sutcliffe & C. Spencer, 1989. The North American Midcontinent Rift beneath Lake Superior from GLIMPCE seismic reflection profiling. Tectonics 8: 305–332.

Carroll, A. R. & K. M. Bohacs, 1999. Stratigraphic classification of ancient lakes; balancing tectonic and climatic controls. Geology 27: 99–102.

Cohen, A. S., M. J. Soreghan & C. A. Scholz, 1993. Estimating the age of lakes: An example from Lake Tanganyika, East African Rift system. Geology 21: 511–514.

Dobrin, M. B. & C. H. Savit, 1988. Introduction to Geophysical Prospecting. McGraw-Hill Book Co., New York, NY, United States, 867 pp.

Giovanoli, F., G. Eberli, P. Finckh, W. Finger, Q. He, C. Heim, K. Kelts, G. Lister, C. Peng, C. Sidler, X. Zhao & K. Hsu, 1982. Zubo 80, a 200 m core in the deepest part of the alpine Lake Zuerich, Switzerland; geophysical investigations. Eleventh international congress on sedimentology. Hamilton, ON, Canada: Aug. 22–27, 1982, International Congress on Sedimentology 11: 129.

Johnson, T. C., 1984. Sedimentation in large lakes. Ann. Rev. Earth Planet. Sci. 12: 179–204.

Herdendorf, C. E., 1990. Distribution of the world's large lakes. In Tilzer, M. M. & C. Serruya (eds.) Large Lakes. Springer-Verlag, Berlin, 3–38.

Houbolt, J. J. H. C. & J. B. M. Jonker, 1968. Recent sediments in the eastern part of the Lake of Geneva (Lac Leman). Geologie en Mijnbouw 47: 131–148.

Hutchinson D. R., A. J. Golmstok, L. P. Zonenshain, T. C. Moore, C. A. Scholz & K. D. Klitgord, 1992. Depositional and tectonic framework of the rift basins of Lake Baikal from multichannel seismic data. Geology 20: 589–592.

Lezzar, K. E., J.-J. Tiercelin, M. De Baptist, A. S. Cohen, T. Bandora, P. Van Rensbergen, C. Le Turdu, W. Mifundi & J. Klerx, 1996. New seismic stratigraphy and late-tertiary history of the north Tanganyika basin, East African Rift system, deduced from multichannel and high-resolution reflection seismic data and piston core evidence. Basin Research 8: 1–28.

Mats, V. D., 1993. The Structure and development of the Baikal rift depression. Earth-Science Reviews 34: 81–118.

Moore, T. C., Jr., K. D. Klitgord, A. J. Golmshtok & E. Weber, 1997. Sedimentation and subsidence patterns in the central and north basins of Lake Baikal from seismic stratigraphy. Geological Society of America Bulletin 109: 746–766.

Mullins, H. T., E. J. Hinchey, R. W. Wellner, D. B. Stephens, W. T. Anderson Jr., T. R. Dwyer & A. C. Hine, 1996. Seismic stratigraphy of the Finger Lakes; a continental record of Heinrich event H-1 and Laurentide ice sheet instability, in Subsurface geologic investigations of New York Finger Lakes; implications for late Quaternary deglaciation and environmental change. Special Paper, Geological Society of America 311: 1–35.

Owen, R. B., R. Crossley, T. C. Johnson, D. Tweddle, I. Kornfield, S. Davison & D. H. Eccles, 1990. Major low levels of Lake Malawi and their implications for speciation rates in cichlid fishes. Proc. R. Soc. Lond. B Biol. Sci. B 240: 519–553.

Payton, C. E., 1977. Seismic stratigraphy; applications to hydrocarbon exploration. American Association of Petroleum Geologists Memoir 26: 477–502.

Reynolds, D. J., M. S. Steckler & B. J. Coakley, 1991. The role of the sediment load in sequence stratigraphy; the influence of flexural isostasy and compaction. J. Geophys. Res. B 96: 6931–6949.

Rosendahl, B. R., 1987. Architecture of continental rifts with special reference to East Africa. Ann. Rev. Earth Planet. Sci. 15: 445–503.

Rosendahl, B. R. & D. A. Livingstone, 1983. Rift lakes of East Africa, new seismic data and implications for future research. Episodes 1983: 14–19.

Sarg, J. F. & L. Pratson, 2000. Erosional surfaces in sedimentary sequences and the impact of basin slope and relief on lowstand depositional geometry. (Abstract) AAPG 2000 Annual Convention. p. A131.

Schock, S. & L. R. Leblanc, 1990. CHIRP sonar; new technology for sub-bottom profiling. Sea Technology 31: 35–43.

Shanley, K. W. & P. J. McCabe, 1998. Relative role of eustasy, climate, and tectonism in continental rocks. Special Publication, SEPM (Society for Sedimentary Geology), no. 59.

Scholz, C. A., 1995. Deltas of the Lake Malawi Rift, East Africa, Seismic Expression and Exploration Implications. AAPG Bulletin 79: 1679–1697.

Scholz, C. A, T. C. Moore, D. R. Hutchinson, K. D. Klitgord & A. Ja. Golmshtok, 1998. Comparative Sequence Stratigraphy of Low-Latitude versus High-Latitude Lacustrine Rift Basins: Seismic Data Examples from the East African and Baikal Rifts. Palaeogeogr, Palaeoclim, Palaeoecol 140: 401–420.

Sheriff, R. E. & L. P. Geldart, 1995. Exploration seismology. Cambridge University Press, GBR, 592 pp.

Sloss, L. L., 1963. Sequences in the cratonic interior of North America. Geological Society of America Bulletin 74: 93–113.

Tiercelin, J.-J. (ed.), 1999. Proceedings of the second International Congress of Limnogeology, Brest France, May 1999.

Vail, P. R., 1987. Seismic stratigraphy interpretation using sequence stratigraphy, in atlas of seismic stratigraphy. Bally, A. W. (ed.) AAPG Studies in Geology #27, vol. 1, pp. 1–10.

Yilmaz, O., 1987. Seismic data processing. Soc. Explor. Geophys., Tulsa, OK, United States, 526 pp.

3. GROUND-PENETRATING RADAR APPLICATIONS IN PALEOLIMNOLOGY

BRIAN J. MOORMAN (moorman@ucalgary.ca)
Earth Science Program
University of Calgary
2500 University Drive
Calgary, Alberta
T2N 1N4, Canada

Keywords: ground-penetrating radar, GPR, geophysics, sub-bottom profiling, bathymetric mapping

Introduction

Application of ground-penetrating radar to mapping
lake basins

One of the initial requirements for many paleolimnological studies is a bathymetric map and an estimation of the distribution, thickness, and stratigraphy of the sediments underlying the lake. This aids in locating coring sites and conducting the overall basin analysis. Of the two main geophysical methods suitable for sub-bottom profiling, namely ground-penetrating radar (GPR) and sonar, GPR offers the ability to survey in the winter through an ice cover, simplifying surveying and coring. Sonar systems require the transponder to be in direct contact with the water. Ground-penetrating radar also offers the advantage of providing fully digital data that makes the process of data manipulation and interpretation simpler and easier. Modern GPR systems are compact and light-weight enough to be easily operated by one or two persons on foot or from a vehicle. The biggest limitation of GPR for surveying lakes is the large impact salinity has on reducing the depth of penetration of the signal. Ground-penetrating radar is not suitable for imaging through saline water.

This section of the book will include a discussion of the theoretical concepts underlying GPR operation and the application of GPR for paleolimnological research. Ground-penetrating radar is widely utilized for a wide variety of geoscience applications from imaging permafrost and glaciers, to delineating groundwater contamination. Accordingly there are many configurations of GPR systems and survey designs. The GPR concepts and system configurations applicable to paleolimnology and lake basin analysis will be discussed here.

W. M. Last & J. P. Smol (eds.), 2001. *Tracking Environmental Change Using Lake Sediments. Volume 1: Basin Analysis, Coring, and Chronological Techniques*. Kluwer Academic Publishers, Dordrecht, The Netherlands.

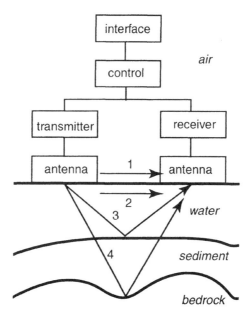

Figure 1. A block diagram of a GPR system. The interface module enables the user to enter the system parameters, and displays and records the data. The control unit generates the timing signals so that all of the components operate in unison. This unit also does some preliminary data processing. The pulse travel paths in order of arrival are 1) direct air wave, 2) direct ground wave, 3 and 4) reflections.

Instrumentation

There are currently a number of different GPR systems on the market that are suitable for surveying lakes. They all have slightly different configurations, however, as shown in Figure 1, they are all comprised of five main components 1) control unit, 2) transmitter, 3) receiver, 4) antennas, and 5) interface, data storage, and display module.

The control unit receives the survey parameters from the interface and generates the timing signals for the transmitter and receiver. It also receives the data from the receiver and does the initial processing before sending it to the storage device. In some systems the interface, data storage, display and control unit are all incorporated into one unit. In other systems, the control unit is separate and a laptop or palmtop computer is employed to enter the survey parameters, store the data, and provide real-time data display.

On the command of the control unit, the transmitter generates the electromagnetic pulse that is emitted through an antenna connected to it. The transmitter/antenna pair determine the centre frequency and bandwidth of the signal that is sent into the ground. An antenna identical to that attached to the transmitter is attached to the receiver. This antenna intercepts reflected energy and sends it to the receiver where it is amplified, digitized and sent to the control unit. The time versus received energy graph is called a trace. Data are collected at frequent intervals along a profile, such that the traces can be plotted side by side, creating a pseudo-section through the ground.

In higher frequency system (e.g., greater than approximately 300 MHz) the antennas can be shielded from unwanted externally generated electromagnetic (EM) energy. If the antennas are shielded, standard wire data cables can be used to connect the transmitter and receiver to the control unit. At lower frequencies the antennas are larger and it is difficult to effectively shield them, thus they are susceptible to the EM noise produced by the electrical signals being sent down a wire communications cable. Thus, fibre optic cables are frequently employed when the antennas are not shielded.

Ground-penetrating radar systems generally have resistively loaded dipole antennas with fixed frequencies. As such, GPR systems are designed such that antennas can be easily switched and multiple sets of antennas can be tested at the study site to determine the optimal frequency to utilize in a given setting. In general, GPR antenna frequencies vary from 12.5 MHz up to 1000 MHz.

History

The history of GPR is a relatively short one, and can be divided into four stages of development. The first stage was characterized by initial equipment experimentation. The first application of EM energy to finding metal objects is attributed to Hulsmeyer (1904). Lowy (1911) developed a technique that would give an indication of the depth of subsurface interfaces, but the use of radar for mapping through geologic materials was first carried out by Stern (1930). Over the next three decades there was very little activity in the field, mainly due to lack of application identification.

It was not until the 1960's that interest in radar technology for subsurface imaging applications started to take hold. The second stage of development saw the custom construction of systems built for specific purposes. Early systems were designed for applications such as measuring ice thickness (Behrendt et al., 1979; Bently et al., 1979; Bryan, 1974; Evans, 1963; Steenson, 1951), mapping bedrock structures in underground mines (Cook, 1973; 1975; 1977; Dellwig & Bare, 1978), and finding the location of hidden pipes and utilities (Caldecott et al., 1988; Morey, 1974; 1976; Osumi & Ueno, 1988). Some work was done on measuring the depth of shallow water (Austin & Austin, 1974; Ulriksen, 1982); however, the study of glaciers and ice caps, salt, coal, and hard rock mines dominated this early work with GPR.

The third stage of development was issued in with the development of the first commercial analog GPR system (Lerner, 1974). In the 1970's the activity level dramatically increased, with a variety of applications identified. However, the moderate system performance of early commercial systems and the lack of digital data processing capabilities limited their utility. These systems tended to be large and heavy creating logistical problems for their use.

In the mid 1980's the first digital GPR system was introduced by Sensors and Software of Mississauga, Ontario, beginning the fourth and most recent stage of development. This system had a higher system performance, enabling deeper depth of penetration, and it generated digital data which was suitable for computer processing and manipulating. During the 1990's GPR has become a standard tool for a variety of geologic, geotechnical, and

Figure 2. An example of how a GPR can be towed across ice on foot. The person in front is also carrying a differential GPS receiver so that the location of the profiles can be precisely mapped.

environmental applications. The current systems have fast data processors and data transfer systems, robust designs, and are small enough to be carried in a boat, on a sled or in a backpack (Fig. 2).

Ground-penetrating radar theory

Signal propagation

The strong relationship between the physical properties of geologic materials (including water) and their electromagnetic properties enables the identification of physical structures in the subsurface using electrical methods (Davis & Annan, 1989; Dallimore & Davis, 1987; Delaney & Arcone, 1982; Scott et al., 1978). The large contrast between the EM properties of water and those of sediment makes GPR a particularly effective method for mapping lake basins.

Most geologic materials (in bulk form) are considered as semi-conductors, or dielectrics, thus they can be characterized by three electromagnetic properties: electrical conductivity, electric permittivity, and magnetic permeability. The electrical conductivity of a material is a measure of its ability to transmit a DC current and is inversely proportional to the voltage drop experienced across a given distance for a given DC current. Magnetic permeability is defined as the ratio of the magnetic flux induction to the magnetizing force. The magnetic susceptibility of a material is a function of the permeability. Electrical permittivity is the ratio of the capacitance of an electrical condenser filled with a dielectric to the capacitance of the same condenser when evacuated.

The movement of electromagnetic energy within the subsurface is governed by the propagation constant of the material it travels through. The basic propagation constant of electromagnetic waves in free space (κ_0) is defined as:

$$\kappa_0 = \left(\omega^2 \mu_0 \, \varepsilon_0\right)^{0.5}, \tag{1}$$

where ω is the frequency in radians, μ_0 is the magnetic permeability of free space and ε_0 is the electric permittivity of free space. In the ground, electromagnetic propagation is complicated by the electrical conductivity of the material it travels through. The calculation of the propagation constant becomes:

$$\kappa = \left(\omega^2 \mu \varepsilon + i\omega \, \sigma_{DC}\right)^{0.5}, \tag{2}$$

where μ, ε, and σ_{DC} are the magnetic permeability, the electric permittivity, and the DC electrical conductivity of the material, respectively, and i is $(-1)^{0.5}$. As the magnetic permeability of most sediment and water is of minor consequence in GPR applications (Telford et al., 1976), the value for free space, $4\pi \times 10^{-7}\,\mathrm{Hm^{-1}}$, is usually used. The electric permittivity is of great importance in GPR applications and is usually expressed in the form of the relative permittivity or dielectric constant:

$$\varepsilon/\varepsilon_0 = \kappa^* = \kappa'\varepsilon_0 + i\kappa''\varepsilon_0, \tag{3}$$

where κ^* is the complex dielectric constant and κ' is termed the dielectric constant and κ'' is the loss factor. Thus the electromagnetic propagation factor in the subsurface can be expressed as:

$$\kappa = \left[\omega\mu_0\varepsilon_0\left(\omega\kappa' + i\omega\kappa'' + i\sigma_{DC}\right)\right]^{0.5}, \tag{4}$$

where $\omega\kappa'$ is the dielectric constant factor, $i\omega\kappa''$ is the dielectric loss factor, and $i\sigma_{DC}$ is the DC conductivity loss factor. This equation demonstrates how the DC conductivity and the complex dielectric constant are both important in determining the amount of energy which will be dispersed in a given material. This is especially important if the lake you are profiling has a salinity greater than about $10\,\mathrm{mS/m}$ (see section on Depth of Penetration). The dielectric constants for a variety materials are given in Table I. The rate at which electromagnetic energy is dissipated in the ground is measured by the attenuation coefficient, which is defined as:

$$\alpha = \omega c^{-1}\left\{\left[\kappa'\left((1 + \tan^2\delta)^{0.5} + 1\right)\right]/2\right\}^{0.5}, \tag{5}$$

where $\tan^2\delta = \sigma_{DC}(\omega\kappa'\varepsilon_0)^{-1}$ and c is the speed of light in free space, $2.998 \times 10^8\,\mathrm{m\,s^{-1}}$. With the conductivity and the dielectric constant being the dominant influence on the attenuation coefficient A-CUBED (1983) gives an estimation of the attenuation using:

$$\alpha \approx (60\pi\,\sigma_{DC})/(\kappa')^{0.5} \quad \left(\mathrm{m^{-1}}\right), \tag{6}$$

$$\text{or} \quad \alpha \approx 1.64\delta/(\kappa')^{0.5} \quad \left(\mathrm{dB\,m^{-1}}\right), \tag{7}$$

Table 1. Dielectric constant, DC electrical conductivity, propagation velocity and attenuation properties of common geologic materials. (Sources: Davis & Annan, 1989; A-CUBED, 1983; Ulriksen, 1982).

Material	Dielectric constant κ'	DC electrical conductivity σ_{DC} (mS m^{-1})	Typical velocity v (m ns^{-1})	Attenuation α (dB m^{-1})
air	1	0	0.3	0
distilled water	80	0.01	0.033	0.002
fresh water	80	0.5	0.033	0.1
sea water	80	30000	0.01	1000
pure ice	3–4	0.01	0.16	0.01
fresh water saturated sand	20–30	0.1–1.0	0.06	0.03–0.3
fresh water saturated silt	10	50	0.09	26
silts (various)	5–30	1–100	0.07	1–100
fresh water saturated clay	10	500	0.09	260
clays (various)	5–40	2–1000	0.06	1–300
limestone	4–8	0.5–2	0.12	0.4–1
shales	5–15	1–100	0.09	1–100
granite	4–6	0.01–1	0.13	0.01–1

where σ_{DC} is measured in S m^{-1}. The rate of attenuation in a variety of materials is given in Table I.

Signal reflection

Ground-penetrating radar principles are similar to those of reflection seismic, in that a pulse of energy is sent into the subsurface, and a portion of it is reflected back to the surface by interfaces between different materials. In contrast to seismic, GPR energy is electromagnetic, not vibrational, and the interfaces are dielectric, not sonic velocity interfaces. Most ground-penetrating radar antennae are poorly focused. In water, the energy is transmitted with a near radial geometry (Davis & Annan, 1989). This results in several simple ray paths which the energy tends to follow, travelling between the transmitter and the receiver (Fig. 1). After the transmitter has emitted a signal, the first energy to arrive at the receiver is the direct air wave. This is the first because it travels directly from the transmitter to the receiver, through the air (at near the speed of light). As the travel time of the direct air wave is easily calculated and stays relatively constant, its arrival time is often used as a marker for static correction. The next return is the direct ground wave. It travels directly from the transmitter to the receiver through the top skin of the subsurface. The next returns are the reflections from the dielectric interfaces. They arrive in order of their depth (top

Table II. Vertical incidence reflection coefficients for some lacustrine settings (Source: A-CUBED, 1983).

From	To	Reflection coefficient
water ($\kappa' = 80$)	gyttja ($\kappa' = 50$)	0.12
gyttja ($\kappa' = 50$)	wet sediment ($\kappa' = 25$)	0.17
water ($\kappa' = 80$)	wet sediment ($\kappa' = 25$)	0.28
wet sediment ($\kappa' = 25$)	rock ($\kappa' = 8$)	0.28
water ($\kappa' = 80$)	rock ($\kappa' = 8$)	0.52
ice ($\kappa' = 3.2$)	water ($\kappa' = 80$)	-0.67

first). The radar waves may also be refracted, however wide-angle reflection and refraction soundings have shown that refracted returns are not usually generated in normal profiling mode (Arcone & Delaney 1982). Due to the complexity of refracted radar waves, detailed analysis has yet to be done on refracted waves as it has in refraction seismic.

The three variables that affect the strength and arrival time of returns are: the strength of the reflections, the propagation velocity of the radar waves, and the rate of attenuation of the signal. Just as light is reflected from an interface between two substances with different refractive indexes, a portion of the energy in a radar energy pulse is reflected from an interface between two materials with different dielectric constants. The percentage of the incident energy that is reflected is proportional to the difference of the dielectric constants. Reflection coefficient (R) for vertical incidence between two perfectly dielectric materials is given by:

$$R = \left(\kappa'^{0.5}_1 - \kappa'^{0.5}_2\right) \Big/ \left(\kappa'^{0.5}_1 + \kappa'^{0.5}_2\right), \tag{8}$$

where κ'_1 and κ'_2 are the dielectric constants of the two media (A-CUBED, 1983). The reflection coefficient for several common interfaces is given in Table II. Due to the length of radar waves, the reflection coefficient from thin layers is also proportional to the frequency of the energy and the thickness of the layer. In materials that are very heterogeneous, small objects (which act like point reflectors) and thin layers are the cause of a great deal of backscatter. This can be used to characterize the material, however, it is often considered as geologic noise. Annan & Davis (1977) and A-CUBED Inc. (1983) discuss this extensively in relation to the range of GPR systems.

Lake-bottom multiples (also known as secondary reflections) were observed on some profiles beneath the lake bottom (Fig. 3). They are easy to distinguish from sub-bottom reflectors because they have the exact same shape as the lake bottom reflector, and arrive after double the travel time of the bottom reflector. In the example shown in Figure 3, some of the sub-bottom features are also displayed in the lake bottom multiple. Generally, the lake bottom multiples arrived substantially after the deepest reflections, such that there was no interference between primary and secondary reflections. There is only potential for interference in lakes where there is extremely good sub-bottom penetration or the water depth is very shallow. For example, with silt ($V = 0.09 \, \text{m ns}^{-1}$) underlying a lake, the ratio

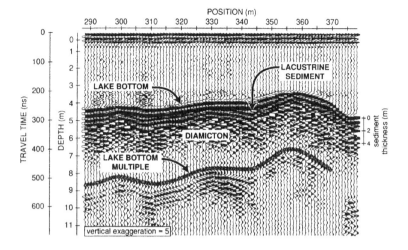

Figure 3. Example of lake-bottom multiple reflection.

of depth of penetration to water depth would have to be greater than 3:1 before interference between the primary and secondary reflections would occur. In a 7 m deep lake the secondary bottom reflection will arrive at the same time as primary sub-bottom reflections from 14 m beneath the lake bottom. In shallow portions of lakes (i.e. less than 2 m deep) the potential for interference from lake bottom multiples increases, however lacustrine sediments tend to be much thinner near the lake edges.

The three main types of GPR reflections generated from the bottom of a lake or from within the sub-bottom sediments are: 1) continuous line, 2) hyperbolic curve, and 3) chaotic. A continuous line reflection is found to result from a continuous, horizontal to sloping interface. A-CUBED Inc. (1983) describe in detail the reflections expected from simple single interfaces and closely spaced multiple interfaces. Hyperbolic reflections are produced by small bodies (i.e. around the size of the radar wavelength around 10^1–10^3 cm in diameter) which have a considerably different dielectric constant than the surrounding material. These act as point reflectors. Meckel & Nath (1985) discuss the character of hyperbolic returns from targets of different size and related horizontal resolution. Boulders and bedrock pinnacles on a lake bottom have been found to generate hyperbolic returns. Chaotic reflections are produced by thin discontinuous layers and very small point reflectors (i.e., considerably less than the wavelength of the radar e.g., in the order of 10 cm for 100 MHz antennas). Some of the causes of chaotic reflections have been found to be coarser clasts in diamicton, joints in bedrock, the disrupted layering in disturbed sediments.

Signal velocity and depth determination

The velocity at which electromagnetic energy travels in the ground is important in determining the depth of reflectors. In free space electromagnetic energy travels at the speed of light, $0.3 \, \mathrm{m \, ns^{-1}}$. In the subsurface it travels at a fraction of the speed of light, usually in the range $0.01-0.16\,\mathrm{m\,ns^{-1}}$. Both the dielectric constant and the DC conductivity strongly influence the propagation velocity (v) of a medium:

$$v = c\left\{\kappa'\left((1 + \tan^2 \delta)^{0.5} + 1\right)/2\right\}^{-1}.$$ (9)

An estimate of the propagation velocity was also determined by A-Cubed (1983) to be:

$$v = 0.3(\kappa')^{-0.5} \qquad (\mathrm{m\,ns^{-1}}).$$ (10)

A more detailed analysis of the specific effects that the dielectric constant and the electrical conductivity have on the attenuation and velocity of electromagnetic energy at a given frequency is given by Olhoeft (1978).

The typical propagation velocity of radar energy in a variety of materials is given in Table I. The propagation velocity in water tends to be fairly constant and the typical values shown in the table are generally employed to calculate water depth. When it comes to measuring the thickness of lake-bottom sediment layers, velocity determination becomes more important as the propagation velocity in sediment can vary considerably. The propagation velocity can be determined in three ways: 1) direct water depth measurements and core depth logging, 2) common-mid point (CMP) velocity surveys (Fig. 4), where the velocity is calculated from the direct arrivals and reflections off horizontal interfaces (Fig. 5a), and 3) point-source reflection analysis, where the velocity is determined using the shape of diffraction patterns produced when profiling over point-source reflectors (Fig. 5b) (Moorman & Michel, 2000). The same geometry applies to the reflection patterns generated by a point-source reflector in a profile, and a horizontal interface in a velocity survey. The travel time equation is:

$$t^2 = \frac{x^2}{V^2} + t_0^2,$$ (11)

where x is the antenna separation in a velocity survey or the lateral offset from the point-source reflector in a profile, V is the propagation velocity, t_0 is the one-way travel time at zero separation/offset (Telford et al., 1976), can be rearranged to calculate velocity:

$$V = \sqrt{\frac{x^2}{t^2 - t_0^2}}.$$ (12)

The later two methods are generally more effective for determining velocities for the upper layers in settings with relatively strong and simple reflections.

As the travel time to a certain depth is dependent on the velocity of all of the layers above, the spatial variability of sediments can influence the shape of underlying reflection patterns. Variability in the propagation velocity throughout a profile often need to be taken into consideration when interpreting the shape of reflections. An example of this is schematically illustrated in Figure 6.

Once the velocity is determined, a depth scale can be constructed. Generally GPR software will automatically generate a single depth scale for a profile. In settings where the velocity changes dramatically with depth or along the profile, it may be required to manually construct a depth scale for different units within the profile.

BRIAN J. MOORMAN

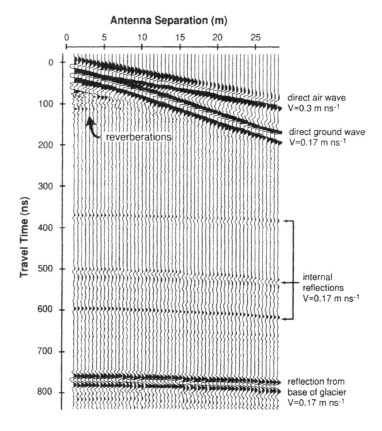

Figure 4. Example of a CMP velocity survey profile from a glacier.

Where the subsurface structure is simple, it is currently possible to automatically correct for velocity changes as well as geometric errors by employing migration software. However, this is labour intensive, imperfect and, in all but the simplest situations, requires far more velocity information that can be easily acquired in a GPR survey. When the velocity varies in two dimensions (i.e., laterally and with depth) migration is much more complicated. Thus, the minor errors in depth and shape of reflections in GPR profiles are usually manually corrected for in the interpretation process or left in the data.

Depth of penetration

The depth to which GPR can image below the surface is dependant on three main factors: 1) the number of interfaces that generate reflections and the dielectric contrast at each interface, 2) the rate at which the signal is attenuated as it travels through the subsurface, and 3) the centre frequency of the antennas. As the GPR pulse arrives at each interface, a portion of it is returned to the surface and the rest continues into the next layer. As

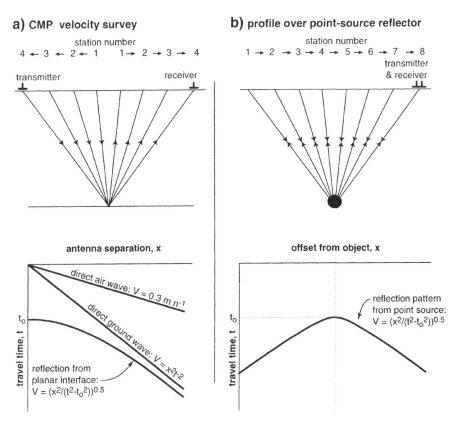

a) CMP velocity survey

station number

4 ← 3 ← 2 ← 1   1 → 2 → 3 → 4

transmitter   receiver

antenna separation, x

direct air wave: V = 0.3 m n⁻¹

direct ground wave: V = x2t-2

reflection from planar interface: V = $(x^2/(t^2-t_o^2))^{0.5}$

travel time, t

t_o

b) profile over point-source reflector

station number

1 → 2 → 3 → 4 → 5 → 6 → 7 → 8

transmitter & receiver

offset from object, x

reflection pattern from point source: V = $(x^2/(t^2-t_o^2))^{0.5}$

travel time, t

t_o

Figure 5. Subsurface velocities can be determined from CMP surveys where traces are recorded as the transmitter and receiver are separated about a common mid-point a), producing a separation vs. travel time plot as shown, or by profiling over a point-source reflector b), and using the geometry of the diffraction tails generated on the profile.

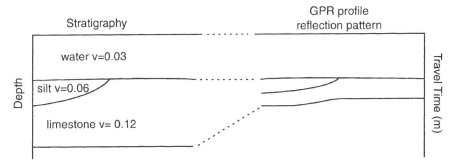

Stratigraphy

GPR profile reflection pattern

water v=0.03

silt v=0.06

limestone v= 0.12

Depth

Travel Time (m)

Figure 6. Since GPR profiles are a graph of reflection strength versus travel time along the survey line, the apparent depth of reflectors is a function of propagation velocity. Where the signal moves faster, the reflections will arrive sooner.

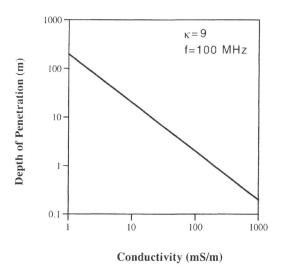

Conductivity (mS/m)

Figure 7. As the electrical conductivity of the propagation medium (water in the case of lakes) increases, the GPR signal is attenuated faster, thus decreasing the depth of penetration.

the number of interfaces increases, the proportion of energy that propagates to depth is reduced. In addition, the greater proportion of energy that is reflected back to the surface at each interface, the less energy that is available to propagate deeper into the ground. In sediment, localized dielectric contrasts can create chaotic reflections. This limits the depth of investigation because the reflections of interest get masked by the clutter of the chaotic returns.

The conductivity of the material that a GPR signal is travelling through has a major influence on the depth to which the signal will penetrate. As the conductivity increases, the material acts more like a conductor than a semi-conductor. The conductive currents in a material are an energy dissipating mechanism for an EM field. In this case, energy is irreversibly extracted from the EM field and transferred to the medium it is in. Figure 7 illustrates the impact conductivity has on the potential depth of penetration.

The frequency used is also of importance since the resolution of the system and the rate of signal attenuation is proportional to the frequency (Fig. 8) of the GPR system. With lower frequencies, the wavelength is longer and as a result there is less attenuation due to conductive losses and less scattering from the chaotic reflections from small clutter. The main disadvantage of using very low frequencies is that the resolution decreases, such that the thickness of small layers can no longer be measured and small objects are not detected. A practical consideration is that, as the frequency decreases, the length of the antennas increase in size and become more difficult to work with.

In fresh-water lakes it is reasonable to expect a depth of penetration of 20–30 m and then a propagation of several metres into the sub-bottom, depending on the material type (Sellman et al., 1992; Moorman & Michel 1997).

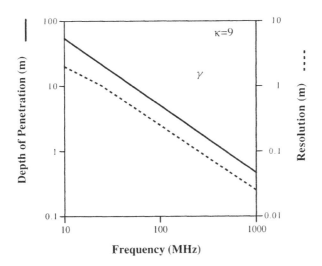

Figure 8. When selecting the frequency to use for a survey there is a trade off between depth of penetration and resolution.

Feature detection and resolution

The three main factors that determine where an object or a thin layer of material will be detected are the size of the object or layer thickness, the frequency of the GPR system, and the propagation velocity of the medium. Higher frequency antennas generate shorter waves and thus have finer resolution and can detect smaller objects. The propagation velocity of the medium is important because the size of the wavelet in the subsurface is not only determined by the frequency of the antenna that created it, but also the velocity at which it propagates through the subsurface. If a pulse is forced to slow down in a medium with a slower propagation velocity (e.g., water), the pulse is compressed to accommodate the slower velocity.

The precise vertical resolution can be determined by surveying two reflectors that intersect at a gentle angle (Sheriff, 1985), however wave theory suggests that the greatest vertical resolution that can be expected is 1/4 of the size of a wavelet (Sheriff & Geldart, 1982). The size of the wavelets that are recorded on a GPR profile is a function of the pulse width of the original transmitted pulse. The pulse width produced by the 100 MHz and 50 MHz antennae and the resultant maximum theoretical resolution in various geologic materials is displayed in Table III.

The horizontal resolution is a function of the spacing between traces and the footprint of the radar pulse. When surveying lakes, the footprint is the area over which the GPR pulse is reflected from the lake bottom. The wavelength, radiation pattern and, water depth determine the size of the footprint. For common dipole antennas, Annan (1992) provides a

Table III. Pulse width and theoretical resolution of 100 MHz and 50 MHz GPR antennas that have a bandwidth to frequency ratio of 1. Sources: Annan, 1992; Davis & Annan, 1989; Ulriksen, 1982.

Material	100 MHz		50 MHz	
	Pulse width	Theoretical resolution	Pulse width	Theoretical resolution
	(m)	(m)	(m)	(m)
water	0.33	0.08	0.66	0.16
ice	1.6	0.4	3.2	0.8
saturated sand	0.6	0.15	1.2	0.3
saturated clay	1.0	0.25	2.0	0.5
limestone	1.2	0.3	2.4	0.6
shale	1.0	0.25	2.0	0.5
granite	1.3	0.33	2.6	0.66

way to estimate the footprint using:

$$A = \frac{\lambda}{2} + \frac{d}{\sqrt{\kappa - 1}}. \tag{13}$$

Where A is the long axis diameter of the oval footprint, d is the distance (or depth), and κ is the dielectric constant of the medium (80 for water). The short axis of the oval footprint is roughly half the length of the long axis. For example, at a depth of 20 m, in water, the effective footprint of a 50 MHz pulse has a mean diameter of approximately 10 m. This is a little smaller than the footprint of an acoustic subbottom profiler (Sellmann et al., 1992).

Data acquisition

Effective data acquisition is dependant on a well designed survey and effective implementation. There are two main elements in designing a GPR survey, the GPR system parameters and the profile parameters. Both of these sets of parameters are adjusted depending on the site conditions and the information desired.

Survey design-system parameters

Most of the GPR systems currently on the market allow for the adjustment of many of the system parameters enabling the optimization of the system for the specific survey environment. The main system parameters that can be adjusted include: operating frequency, time window, sampling interval, stacking, antenna spacing, and antenna orientation.

The frequency of antennas chosen influence the exploration depth that is possible, the resolution of the data, and the amount of clutter that is present on the profiles. As the frequency is lowered, the depth of penetration is increased and the amount of clutter present in the profiles decreases; however, the ability to detect and resolve smaller objects decreases.

In limnological surveys, the maximum depth of the lake and the electrical conductivity of the water are the major influences on the required frequency. Generally, 12.5 MHz to 100 MHz antennas are employed to survey lakes.

The time window is the duration over which the system records the returns at each position. This and the propagation velocity determine the depth to which data will be recorded from. It is generally recommended that it should be set appreciably larger than the maximum possible depth of interest. It is better to collect too much data than to find that a reflector of interest runs off the bottom of the plot.

The sampling interval determines how well the received wave form is digitally represented. The sampling interval required to effectively represent the shape of the wave form is a function of the frequency and the propagation velocity. Setting the interval too large degrades the vertical resolution of the data. Over sampling results in excess data file sizes and slower computing and plotting time. However, with the current computer technology, over sampling is not a significant issue.

Trace stacking involves the GPR automatically taking repeat measurements at each station, and averaging the traces. This increases the signal to noise ratio by suppressing the unwanted, random, non-geologic noise. Because each trace takes so little time to acquire, having larger numbers of stacks only slightly effects the length of time it takes to complete data acquisition at each station. However, depending on the amount of noise present, there is a limit to the number of stacks required before no further improvement in the data is achieved.

The separation between the transmitter antenna and receiver antenna controls the geometry of the travel path of the radar pulse, the power reaching the target of interest, and the amount of energy travelling directly from the transmitter to the receiver (referred to as the direct air wave and the direct ground wave). A large antenna separation results in a time compression of the near surface reflections. This is generally not a significant issue when profiling lakes, since the water column takes up the upper portion of the profile. If the two antennas are too close together, the large pulse of energy contained in the direct air wave and direct ground wave can saturate the receiver electronics and decrease its ability to detect the full range of data received within a few metres of the surface. To attain the optimal amount of energy reaching the target of interest, the antenna spacing can be refined using the relationship from Annan (1992):

$$S = \frac{2d}{\sqrt{(\kappa - 1)}}, \tag{14}$$

where d is the depth of the target and κ is the dielectric constant of the medium. However, when profiling lakes, the spacing is not as crucial since the water column simplifies the radiation patterns around the antenna and slightly focusses the energy directly beneath the antenna. As a result, the antenna spacing is often kept within the 1–3 m range with the specific separation determined by the logistical constrains of the survey.

The final system parameter that can be adjusted is the orientation of the antennas. Most GPR antennas currently being produced commercialy are polarized and thus the relative orientation of the antennas can influence the polarity of returns. The elongation of the EM field radiated out from the transmitter antenna means that, depending on the antenna orientation, there is more spatial averaging in one direction over the other. However, the impact of antenna orientation on the overall data quality is relatively small, thus survey logistical requirments often determine antenna orientation.

Survey design-profile parameters

The profile parameters that need to be taken into consideration when designing a GPR survey include station spacing along the survey profile, spacing between profiles, and the type of survey grid desired.

Ensuring that the station spacing is close to enable interpretation of the profile is very important when designing a survey. Station spacing is a function of the frequency used, the dielectric constant of the material, and the complexity of the subsurface. If the lake bottom and sub-bottom structure are horizontal, it is not as crucial to ensure that the traces are close together; however, sloping reflectors are much more difficult to interpret if the station spacing is too large (Fig. 9). The relationship between frequency and station spacing in GPR surveying of lakes is illustrated in Figure 10.

The layout of profile lines can take two forms: a rectilinear grid where the distance between profiles stays constant, or a series of vectors from the shore of the lake. The rectilinear grid takes longer to complete but since all of the profiles intercept at right angles, it tends to be easier to interpret the three dimensional structure from the profiles. The vectors can be completed faster since the end of one profile (where it reaches the shore) is the start of another profile. However, the interpretation tends to be trickier. When there are dipping reflectors in the lake sediments (e.g., deltaic foreset beds), it is best to profile parallel to their dip direction. This produces the most realistic representation of the sloping horizons. The profiles should also be close enough that the footprint of the GPR overlaps from one profile to the next.

Along with the size of the footprint, the complexity of the lake bottom will influence the profiling detail required. The acquisition mode (e.g., on an ice platform or from a boat) strongly influences the details of how that detail is achieved. If the GPR survey is conducted on an ice platform, where the system is towed by hand or behind a vehicle, positioning and velocity control are relatively simple. When surveying from a boat on open water, position control is more complicated.

Acquisition mode

There are two main ways to conduct GPR surveys of lakes, towing the system over an ice cover or surveying through the bottom of a boat. Thus, this enables GPR to be used for surveying lakes for a large portion of the year.

When surveying on an ice cover, the GPR system can be towed behind a slow moving vehicle or a person walking. Since motorized vehicles generate a certain amount of EM noise, and are a large radar target, the radar system must be at least 5 m behind the vehicle, depending on the type of vehicle. Depending on the GPR system employed, and the system parameters chosen, data can be collected at up to several kilometres per hour. The data can be collected at a constant rate and thus variations in the speed at which it is towed will affect the trace (i.e., station) spacing, or recent systems offer odometer wheels that trigger the GPR to collect a trace at regular distance intervals.

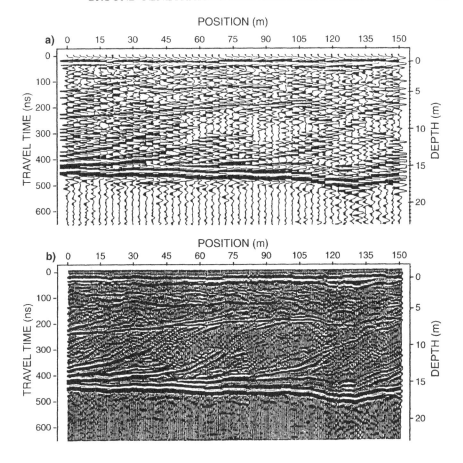

Figure 9. The effect of station spacing is illustrated when trying to interpret the sloping reflectors in this 50 MHz profile across deltaic deposits. In a) the station spacing is 3 m, while much more detail can be observed in b) where the same section is displayed with a station spacing of 0.5 m (after Annan, 1992).

The stratigraphy of GPR profiles collected on an ice platform are more complicated due to the additional layer of ice being imaged. However, the very fast propagation velocity of the GPR pulse through ice generally results in the ice layer having little influence on the character of the GPR profile. This is especially true if the ice thickness is relatively constant. When the thickness of ice varies dramatically, such as on rivers, GPR can be used to measure the thickness of the ice (Annan & Davis, 1977; Delaney et al., 1990).

When surveying on open water, the antennas are placed in a non-metallic flat bottom boat, such as an inflatable rubber boat. This allows the antennas to be in close contact with the water without experiencing interference from the boat. Experience has shown that the antennas work fine through a one-inch wood floor. A metal floor on a boat will completely

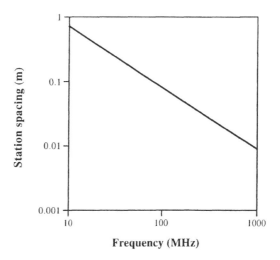

Figure 10. The optimal station spacing (distance between traces) decreases as the frequency increases.

attenuate the signal before it reaches the water. To avoid interference from the boat's motor, sometimes it is necessary to tow the GPR system in one boat 5 m or so behind another boat.

Some initial attempts have been made to perform GPR surveys by slinging the antennas beneath a helicopter and flying the profile route of the lake (e.g., O'Neill & Arcone, 1991). However, there are still a number of technological issues that must be addressed. Since the antennas are not in contact with the surface of the lake, a large proportion of the energy in the pulse is lost in the reflection from the lake surface. This limits the depth of penetration through and beneath the lake. Data acquisition speed is generally too slow, as it is difficult to fly helicopters at a very slow speed. As well, the considerable interference from the helicopter must be eliminated.

No matter what mode of data acquisition is used, survey positions must be acquired if the data are to be transferred onto a map. When working on an ice platform, standard survey techniques can be easily employed. Mapping the position of profiles is trickier when you are in constant motion in a boat or helicopter. However, with the advent of differential global positioning systems (DGPS) and other rapid acquisition high precision techniques, this problem is becoming inconsequential.

Data processing and interpretation

Processing techniques

Compared to seismic data, relatively little processing is generally done to GPR data. A gain function is usually applied to the data to emphasize the weaker returns. There are a number of gain functions that can be applied depending of the data quality and the profile elements that you want to emphasize (e.g., automatic gain control (AGC), spherical and exponentially compensation (SEC) gain, user defined gain function). An example of the effects of different gain functions are displayed in Moorman & Michel (1997).

Temporal filtering involves filtering along the time axis of each trace. This can involve high-pass, low-pass filters or frequency filters such as Fourier analysis. These filters involve reduction or elimination of certain unwanted returns along each trace. For example by running mean along each trace, high frequency noise can be reduced (Moorman & Michel, 1997).

As with temporal filtering, spatial filtering is employed to remove unwanted spatial variations. One technique is to perform a running mean of data points at the same time across traces. This type of trace to trace running mean is used to emphasize horizontal reflections while de-emphasizing sloping returns. Other types of spatial filters can be applied to emphasize sloping reflections such as trace-to-trace differential filters.

There are also a number of more advanced filtering techniques such as migration that are occasionally employed to solve a specific problem with the data set. However, these more complicated filtering techniques are generally too time consuming to be regularly applied to all data sets.

Feature identification

Feature identification is primarily accomplished by examining the reflection characteristics (e.g., continuous line, hyperbola, or multiple discontinuous chaotic reflections). By comparing the reflection character to known examples, one can construct a radar stratigraphic facies interpretation of the data (Fig. 11). Radar stratigraphic examples have been published from a few environments to date (e.g., Beres & Haeni, 1991; van Heteren et al., 1994; Smith & Jol, 1997).

In some applications, by studying the pulse polarity, you can determine the relative magnitude of the dielectric constant on either side of an interface (Arcone et al., 1995; Moorman & Michel, 2000). However, this probably has limited use in studying lake basin sediments.

Along with identifying features of interest, the interpreter must be alert to the potential for the incorporation of unwanted returns into the GPR profile. Because the energy is being radially emitted from the antenna, in areas of very uneven topography there is the potential of recording a reflection from a feature that is not beneath the antennas, but off to the side or even nearby on the surface. These are called offline reflections. This is most evident when the profile is parallel to a very steep lake bottom slope (i.e., greater than 45°). When the slope is greater than 45°, lake bottom reflections will not be directed back to the receiver, but off to the side. Similarly, if the profile runs near to a steep slope, unwanted offline reflections may be recorded from the slope off to the side. An example of the chaotic noise that can be generated from offline reflectors is illustrated in Figure 12.

The refraction of a propagated signal can clutter a profile with misleading returns. Fortunately, with the antenna separation generally being so close, this is seldom an issue. Channeling of the emitted pulse in the ice cover can also lead to anomalous data. Since the dielectric constant of ice is much smaller that water, if the ice is thick enough reflections from within the ice off to the side of the profile may be recorded. As the propagation velocity of ice is about 5 times faster than that in water, a reflection 5 m off to the side

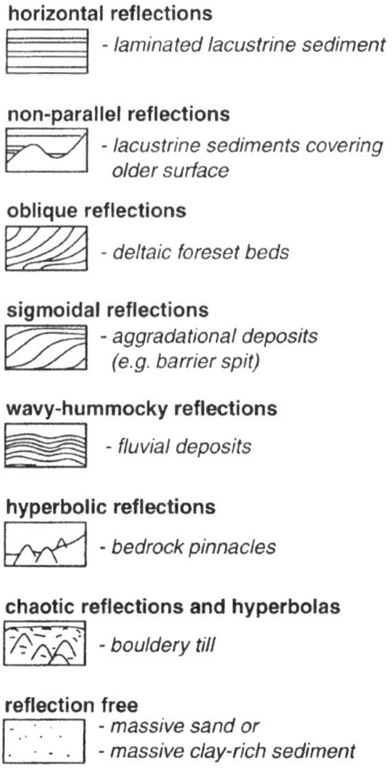

horizontal reflections
- laminated lacustrine sediment

non-parallel reflections
- lacustrine sediments covering older surface

oblique reflections
- deltaic foreset beds

sigmoidal reflections
- aggradational deposits (e.g. barrier spit)

wavy-hummocky reflections
- fluvial deposits

hyperbolic reflections
- bedrock pinnacles

chaotic reflections and hyperbolas
- bouldery till

reflection free
- massive sand or
- massive clay-rich sediment

Figure 11. Examples of radar stratigraphic reflection patterns from lacustrine environments.

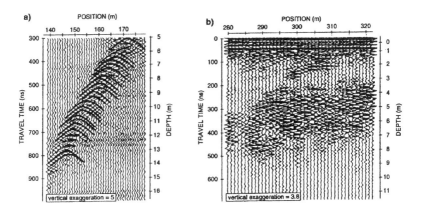

Figure 12. Steeply dipping slopes can create difficulties in determining the exact depth of a lake. In a) the GPR profile was run perpendicular to a steeply sloping lake edge. The hyperbolas were generated by boulders. b) GPR profile was oriented parallel to steep edge of the lake. Reflections from "off-line" targets cause chaotic noise.

within the ice could technically arrive at the same time as a reflection 1 m down the water column. Multiple reflections from the lake bottom are frequently observed on GPR profiles and have the potential to mask some of the sub-bottom data. However, this is usually not a problem unless there is extremely good penetration into the sub-bottom (Moorman & Michel, 1997).

In general, GPR data are relatively straight forward to process and interpret. The lake bottom produces a continuous line reflection or hyperbolic reflections. The biggest problem arises when not enough energy reaches the lake bottom to produce a reflection that is detectable at the surface. Depending on the thickness of the lake sediments and the scale of the structures within it, sedimentary structures may be able to be identified on a GPR profile.

As with any geophysical investigation, the geophysical data should be verified with direct measurements. This is especially true for the identification and thickness of lake bottom sediments. Water depth measurements are less likely to be in error since the dielectric structure of lakes remains constant and the propagation velocity of water varies little.

Applications

The application of GPR to paleolimnology is still in its infancy. The bulk of the research on investigating the application of GPR to paleolimnological problems has been initiated in the last ten years. To date, the research has focused on four areas: bathymetry, sediment thickness and structure, peat stratigraphy, and floating ice thickness.

Since the early work of Annan & Davis (1977) and Kovacs (1978), it has been apparent that GPR is an effective tool for mapping the depth of fresh water lakes. Using a 120 MHz system, Trumen et al. (1991) found that water depth in a reservoir could be mapped with confidence. When the GPR profiles were correlated to measured depths, an R^2 value was calculated to be 0.989. Kovacs (1991) showed that GPR was more effective at mapping bathymetry than sonar when the water depth was shallow and when the weed concentration was high. Sellman et al. (1992) & Delaney et al. (1992) demonstrated that 100 MHz GPR has a resolution similar to a 7 kHz acoustic survey for measuring water depth, and that the GPR was far superior for imaging the bedding structure. Using 50 MHz antennas, up to 30 m of water could be mapped, with up to 7 m of penetration into the bed (Sellman et al., 1992).

A number of authors have reported mapping the distribution and thickness of lacustrine sediments (Delaney et al. 1992; Moorman & Michel, 1997). Smith & Jol have under taken extensive work on mapping the structure of modern and ancient deltaic deposits (e.g., Jol & Smith, 1991, 1992; Smith & Jol, 1997). Lowe (1985) reported being able to discern multiple layers of lacustrine sediments and tephra up to 10 m depth.

Various trials have been undertaken showing the capabilities of GPR for mapping the thickness and stratigraphy of organic deposits in ponds and bogs (Worsfold et al., 1986; Warner et al., 1990; Hanninen, 1992; Doolittle et al., 1992; Mellett, 1995). Depending on how much the physical properties vary with depth, GPR results varied from just detecting the bottom of the peat to differentiating the actual peat stratigraphy.

As ice enables radar pulses to propagate with little attenuation, GPR has been shown effective for imaging the thickness of lake, sea and river ice (Arcone et al., 1997; Kovacs, 1978; Kovacs & Morey, 1979, 1985, 1992). As the dielectric contrast between ice and water is much greater than that of ice and air, O'Niell & Arcone (1991) found that lake ice thickness could be detected from a helicopter. A number of technical issues complicate the collection of good quality data, however, in some situations, it is feasible.

Summary

From the research conducted to date, it appears that GPR has the potential to become a valuable tool in paleolimnological investigations. By carrying out gridded surveys over a lake, the bathymetry and the lake bottom sediment distribution, thickness and structure can be mapped with relative ease. Operating on foot, on lake ice several kilometres of data can be collected per day. Towing the GPR behind a vehicle or boat enables even faster data collection.

Currently, the bathymetry of lakes up to 30 m in depth can be mapped with GPR. Up to seven metres of sediment have been imaged beneath lakes. In the future it is anticipated that these values will improve, as GPR systems performances continue to improve.

The new GPR systems currently being released are pointing the way towards very simplified GPR mapping. Whether this will include simplified lake surveying will depend on market forces. Since the simpler a system is made use, the less flexible it is. Simple lake profiling will require a market demand for such a system.

Further development of airborne GPR surveying is required before a final determination can be made on how effective airborne systems can be for mapping bathymetry and lake bottom sediments.

Acknowledgments

The author gratefully acknowledges the advice and guidance of Dr. Fred Michel, and the field assistance of Deborah Kliza, Lynn Moorman, Mark Elver, and Amy Lyttle while working in the Arctic. Much of the field work was supported by The Northern Science Training Program and the Polar Continental Shelf Project. These two organizations are gratefully acknowledged. Valuable discussions with Les Davis of Sensors and Software Inc. are also greatly appreciated.

References

A-CUBED, 1983. General State of the Art Review of Ground Probing Radar. A-CUBED, Mississauga, Ontario, 89 pp.

Annan, A. P., 1992. Ground Penetrating Radar Workshop Notes. Sensors and Software Inc., Mississauga, Ontario, 56 pp.

Annan, A. P. & J. L. Davis, 1977. Radar range analysis for geological materials. GSC Paper 77–1B. Current Research Part B: 117–124.

Arcone, S. A. & A. J. Delaney, 1982. Measurement of Ground Dielectric Properties Using Wide-angle Reflection and Refraction. U.S. Army Cold Regions Research and Engineering Laboratories, Hanover, New Hampshire, 11 pp.

Arcone, S. A., D. E. Lawson & A. J. Delaney, 1995. Short-pulse radar wavelet recovery and resolution of dielectric contrasts within englacial and basal ice of Matanuska Glacier, Alaska, USA. J. Glaciology 41: 68–86.

Arcone, S. A., N. E. Yankielun & E. F. Chacho, 1997. Reflection profiling of arctic lake ice using microwave FM-CW radar. IEEE Transactions on Geoscience and Remote Sensing 35: 436–443.

Austin, G. L. & L. B. Austin, 1974. The use of radar in urban hydrology. J. Hydrology 22: 131–142.

Behrendt, J. C., D. Drewry, E. Jankowski & A. W. England, 1979. Aeromagnetic and radar ice sounding indicate substantially greater area for dufek intrusion in Antarctica. American Geophysical Union Transactions 60: 245.

Bentley, C. R., J. W. Clough, K. C. Jezek & S. Shabtaie, 1979. Ice thickness patterns and the dynamics of the Ross Ice Shelf, Antarctica. J. Glaciology 24: 287–294.

Beres, M., Jr. & F. P. Haeni, 1991. Application of Ground-penetrating-radar methods in hydrogeologic studies. Ground Water 29: 375–386.

Bryan, M. L., 1974. Ice Thickness Variability on Silver Lake, Genesee County, Michigan: A Radar Approach, Advanced Concepts in the Study of Snow and Ice Resources. United States Contribution to the International Hydrological Decade: 213–223.

Caldecott, R., M. Poirier, D. Scofea, D. E. Svoboda & A. J. Terzuoli, 1988. Underground mapping of utility lines using impulse radar. Institute of Electrical Engineers Proceedings F. Communications, Radar and Signal Processing 135 part F: 343–361.

Cook, J. C., 1973. Radar exploration through rock in advance of mining. Transactions of A.I.M.E., Society of Mining Engineers 254: 140–146.

Cook, J. C., 1975. Radar transparencies of mine and tunnel rocks. Geophysics 40: 865–885.

Cook, J. C., 1977. Borehole-radar exploration in a coal seam. Geophysics 42: 1254–1257.

Dallimore, S. R. & J. L. Davis, 1987. Ground probing radar investigations of massive ground ice and near surface geology in continuous permafrost. GSC Paper 87–1A. Current Research Part A: 913–918.

Davis, J. L. & A. P. Annan, 1989. Ground-penetrating radar for high-resolution mapping of soil and rock stratigraphy. Geophysical Prospecting 37: 531–551.

Delaney, A. J. & S. A. Arcone, 1982. Laboratory Measurements of Soil Electric Properties between 0.1 and 5 GHz. U.S. Army Cold Regions Research and Engineering Laboratories, Hanover, New Hampshire, 7 pp.

Delaney, A. J., S. A. Arcone & E. F. Chacho, Jr., 1990. Winter short-pulse radar studies on the Tanana River, Alaska. Arctic 43: 244–250.

Delaney, A. J., P. V. Sellmann & S. A. Arcone, 1992. Sub-bottom profiling: a comparison of short-pulse radar and acoustic data. In Hanninen, P. & S. Autio (eds.) Fourth International Conference on Ground Penetrating Radar. Geological Survey of Findland Special Paper 16: Rovaniemi, Finland: 149–157.

Dellwig, L. F. & J. E. Bare, 1978. A radar investigation of North Louisiana salt domes. Photogrametric Engineering and Remote Sensing 44: 1411–1419.

Doolittle, J. A., M. A. Hardisky & S. Black, 1992. A ground-penetrating radar study of Goodream palsas, Newfoundland, Canada. Arctic and Alpine Research 24: 173–178.

Evans, S., 1963. Radio techniques for the measurement of ice thickness. Polar Record 11: 406–410.

Hänninen, P., 1992. Application of Ground Penetrating Radar and Radio Wave Moisture Probe Techniques to Peatland Investigations. Geological Survey of Finland, Rovaniemi, Finland, 71 pp.

Hulsmeyer, C., 1904. German Patent Number 165546.

Jol, H. M. & D. G. Smith, 1991. Ground penetrating radar of northern lacustrine deltas. Canadian J. Earth Sci. 28: 1939–1947.

Jol, H. M. & D. G. Smith, 1992. Geometry and structure of deltas in large lakes: a ground penetrating radar overview. In Hanninen, P. & S. Autio (eds.) Fourth International Conference on Ground Penetrating Radar. Geological Survey of Finland Special Paper 16, Geological Survey of Finland, Rovaniemi, Finland: 159–168.

Kovacs, A., 1978. Remote detection of water under ice-covered lakes on the north slope of Alaska. Arctic 31: 448–458.

Kovacs, A., 1991. Impulse Radar Bathymetric Profiling in Weed-infested Fresh Water. U.S. Army Corps of Engineers, Cold Regions Research & Engineering Laboratory, Hanover, New Hampshire, 19 pp.

Kovacs, A. & R. M. Morey, 1979. Remote detection of a freshwater pool off the Sagavanirktok River Delta, Alaska. Arctic 32: 161–164.

Kovacs, A. & R. M. Morey, 1985. Electromagnetic Measurements of Multi-year Sea Ice Using Impulse Radar. U.S. Army Cold Regions Research and Engineering Laboratory, Hanover, New Hampshire, 26 pp.

Kovacs, A. & R. M. Morey, 1992. Estimating sea ice thickness from impulse radar sounding time of flight data. In Pilon, J. A. (ed.) Ground Penetrating Radar. Geological Survey of Canada Paper 90–4, Geological Survey of Canada, Ottawa: 117–124.

Lerner, R. M., 1974. Ground Radar System. United States Patent #3831173. Assignee: Massachusetts Institute of Technology.

Lowe, D. J., 1985. Application of impulse radar to continuous profiling of tephra-bearing lake sediments and peats: an initial evaluation. New Zealand of Geology and Geophysics 28: 667–674.

Lowy, H., 1911. German Patent Number 254517.

Meckel, L. D., Jr. & A. K. Nath, 1985. Geologic considerations for stratigraphic modeling and interpretation. Seismic Stratigraphy-Applications to Hydrocarbon Exploration. AAPG Memoir 26. American Association of Petroleum Geologists, Tulsa, Oklahoma: 417–438.

Mellett, J. S., 1995. Profiling of ponds and bogs using ground-penetrating radar. J. Paleolim. 14: 233–240.

Moorman, B. J. & F. A. Michel, 1997. Bathymetric mapping and sub-bottom profiling through lake ice with ground-penetrating radar. J. Paleolim. 18: 61–73.

Moorman, B. J. & F. A. Michel, 2000. Glacial hydrological system characterization using ground-penetrating radar. Hydrological Processes 14: 2645–2667.

Morey, R. M., 1974. Continuous sub-surface profiling by impulse radar. Conference on Subsurface Exploration for Underground Excavation and Heavy Construction. American Society of Civil Engineers, 213–232.

Morey, R. M., 1976. Detection of subsurface cavities by ground penetrating radar. Highway Geological Symposium 27: 28–30.

Olhoeft, G. R., 1978. Electrical Properties of Permafrost. Third International Conference on Permafrost: Edmonton, Alberta. National Research Council of Canada, Ottawa: 127–131.

O'Neill, K. & S. A. Arcone, 1991. Investigations of Freshwater and Ice Surveying Using Short-pulse Radar. U.S. Army Cold Regions Research and Engineering Laboratory, Hanover, New Hampshire, 22 pp.

Osumi, N. & K. Ueno, 1988. Detection of buried plant. Institute of Electrical Engineers Proceedings F. Communications, Radar and Signal Processing 135 part F: 330–342.

Scott, W. J., P. V. Sellmann & J. A. Hunter, 1978. Geophysics in the study of permafrost. Third International Conference on Permafrost: Edmonton, Alberta. National Research Council of Canada, Ottawa: 93–115.

Sellmann, P. V., A. J. Delaney & S. A. Arcone, 1992. Sub-bottom Surveying in Lakes with Ground-penetrating Radar. U.S. Army Cold Regions Research and Engineering Laboratories, Hanover, New Hampshire, 18 pp.

Sheriff, R. E., 1985. Limitations on resolution of seismic reflections and geologic detail derivable from them. Seismic Stratigraphy-Applications to Hydrocarbon Exploration. American Association of Petroleum Geologists, Tulsa Oklahoma: 3–14.

Sheriff, R. E. & R. E. Geldart, 1982. Exploration Seismology, Volume 1: History, Theory, and Data Acquisition, Cambridge University Press, New York, 253 pp.

Smith, D. G. & H. M. Jol, 1997. Radar structure of a Gilbert-type delta, Peyto Lake, Banff National Park, Canada. Sed. Geol. 113: 195–209.

Steenson, B. O., 1951. Radar Methods for the Exploration of Glaciers. Pasadena, California. California Institute of Technology, Pasadena, California.

Stern, W., 1930. Principles, methods and results of electrodynamic thickness measurement of glacier ice. Zeitschrift fur Gletscherkunde 18: 24.

Telford, W. M., L. P. Geldart, R. E. Sheriff & D.A. Keys, 1976. Applied Geophysics: Cambridge. Cambridge University Press, New York, 455 pp.

Truman, C. C., L. E. Asmussen & H. D. Allison, 1991. Ground-penetrating radar: a tool for mapping reservoirs and lakes. J. Soil and Water Conservation 46: 370–373.

Ulriksen, C. P., 1982. Application of impulse radar to civil engineering. Published Ph.D. Thesis, Lund University of Technology, Lund, Geophysical Survey Systems, Inc. Hudson, New Hampshire, 179 pp.

van Heteren, S., D. M. Fitzgerald & P. A. McKinlay, 1994. Application of ground penetrating radar in coastal stratigraphic studies. GPR '94, Volume Proceeding of the Fifth International Conference on Ground Penetrating Radar. University of Waterloo, Kitchener, Ontario: 869–881.

Warner, B. G., D. C. Nobes & B. D. Theimer, 1990. An application of ground penetrating radar to peat stratigraphy of Ellice Swamp, southwestern Ontario. Can. J. Earth Sci. 27: 932–938.

Worsfold, R. D., S. K. Parashar & T. Perrott, 1986. Depth profiling of peat deposits with impulse radar. Can. Geotech. J. 23: 142–154.

4. SHORELINE AND BASIN CONFIGURATION TECHNIQUES IN PALEOLIMNOLOGY

DOROTHY SACK (sack@ohio.edu)
Department of Geography
Ohio University
122 Clippinger Labs
Athens, Ohio
USA 45701

Keywords: coastal geomorphology, paleolake geomorphology, lacustrine landforms, closed-basin lakes, paleolake shorelines, shoreline mapping, paleolake level, lake morphometry, late Pleistocene paleolakes, the Great Basin

Introduction

Background and rationale

This chapter considers methods and techniques by which data from the geomorphic evidence of paleolakes are collected, analyzed, and utilized. From a historical perspective, landform evidence can be viewed as one of the foundations of paleolake studies because (1) the identification of abandoned shoreline landforms is often the first indication noted for the existence of a former expanded water body (e.g., Whittlesey, 1850; Stansbury, 1852; Russell, 1885; Taylor, 1897; Schewennesen, 1918; Meinzer, 1922), and (2) many of the early researchers emphasized the geomorphic evidence of paleolakes (Beckwith, 1855; Gilbert, 1871, 1890; Russell, 1885; Leverett, 1897; Goldthwait, 1906; Leverett & Taylor, 1915; Street-Perrott & Harrison, 1985). Nevertheless, because landforms are either deposits of sediment (broadly defined) or remnant features from which sediments have been eroded, the disciplines of geomorphology and sedimentology are closely related and those early paleolake workers typically augmented their primarily geomorphic work with sedimentologic and stratigraphic observations (Smith & Street-Perrott, 1983).

The presence of geomorphology in physical paleolimnological research has varied through time. Following the initial dominance of landform evidence in the paleolake literature of the mid-19th to the early 20th century (e.g., Russell, 1885; Gilbert, 1890), the role of sedimentology and stratigraphy grew in the first half of the 20th century becoming at least as important as the geomorphic aspect (e.g., Thompson, 1929; Blackwelder & Ellsworth, 1936). Starting about 1950 the geomorphic component of paleo-

 W. M. Last & J. P. Smol (eds.), 2001. *Tracking Environmental Change Using Lake Sediments. Volume 1: Basin Analysis, Coring, and Chronological Techniques.* Kluwer Academic Publishers, Dordrecht, The Netherlands.

lake research began to dwindle for two major reasons, one from outside and the other from inside the discipline of geomorphology. From sedimentology and stratigraphy came new developments in lithostratigraphy, soil stratigraphy, and chronostratigraphy, and a surge in their application to paleolake research (Hunt et al., 1953; Broecker & Orr, 1958; Bissell, 1963; Morrison, 1964; Sack, 1989). Internally, many geomorphologists began turning away from topics involving longer term landscape evolution as they became increasingly disillusioned with the Davisian historical-evolutionary approach, which had been dominant in much of geomorphology for decades (Davis, 1899; Strahler, 1950, 1952; Sack, 1992a). At least in North America the use of geomorphic methods in paleolake research reached its nadir in the 1970s when process, as opposed to historical, geomorphology dominated that field. Since 1979, geomorphic analysis in paleolake studies has been rebounding, accompanied by advances in surveying and geographic positioning technologies as well as renewed interest among geomorphologists in Quaternary science.

Perhaps because of these historical trends within 20th century paleolake studies, geomorphology has occasionally been portrayed as an unimportant (Morrison, 1965a, 1966), less important (Street-Perrott & Harrison, 1985), or even nonexistent (Frey, 1988) subfield of paleolimnology. Whether done consciously or unconsciously, such portrayals do paleolimnologists a disservice because in addition to providing present evidence of an ancient lake, geomorphology is one of the past environmental factors that influenced the existence, extent, and changing size of the paleolake. In other words, geomorphology is an essential aspect of paleolake studies because it is involved as both a cause and an effect of the water bodies.

Compared to other evidence of past water levels, such as that available from core stratigraphy, relict shorelines give a much more direct and precise indication of paleolake extent because they were created at the interface between the water surface and the land. The preservation of abandoned shorelines as distinguishable landforms at the earth's surface is crucial for documenting the actual size and shape of a paleolake, that is, for reconstructing its hydrography, and for deciphering its hydrographic changes through time. These changes have traditionally been portrayed diagrammatically with hydrographs, which show reconstructed values of water level, surface area, or volume on the ordinate plotted against time on the abscissa (Fig. 1) (Gilbert, 1890; Antevs, 1945; Crittenden, 1963; Oviatt et al., 1992; Currey, 1994b). Hypsographs illustrate the variation in surface area with changing water levels.

The predominant rationale for many paleolake studies conducted in the last decades of the 20th century is paleoclimatic reconstruction. The existence, extent, and changing extent of a lake, however, are influenced by geomorphic (including structural) as well as climatic factors (Eardley et al., 1957; Currey, 1994a, 1994b; Mason et al., 1994; Sack, 1994). For example, at the most basic level, adequate climatic conditions will not result in a lake if there is no basin of topographic closure. Through orographic and other effects, drainage basin topography influences the amount of precipitation falling and potentially reaching the lake as surface runoff (Snyder & Langbein, 1962; Wilkins, 1997). Lake level changes are caused by variations in climate but also by such geomorphic factors as stream capture, outlet incision, or spillover from adjacent basins (Lewis & Anderson, 1989; King, 1993; Pengelly et al., 1997; Sack, in press). Even the amount of lake level rise resulting from an increase in hydrologic input depends on lake basin configuration and whether or not the lake has fluvial

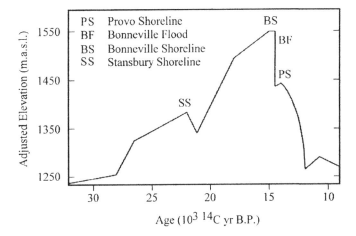

Figure 1. Generalized hydrograph of late Pleistocene Lake Bonneville (after Burr & Currey, 1988; Oviatt et al., 1992). Elevations are adjusted to pre-isostatic rebound values.

outflow (Street-Perrott & Harrison, 1985). Because the lacustrine response documented in a reconstructed paleolake hydrograph reflects both the geomorphic and the climatic history of the drainage basin, an accurate reconstruction of the paleoclimate depends on properly distinguishing the climatically influenced and geomorphically influenced aspects of the hydrographic response.

Open- versus closed-basin lakes

Whether a lake basin is hydrographically closed (closed-basin lake) or open (open-basin lake) is a fundamental and useful distinction in paleolake studies. Because a closed-basin lake has no outflowing stream, changes in its water input-to-output ratio are expressed as lake level and especially lake surface area fluctuations (Benson & Paillet, 1989). A discrete shoreline is potentially created each time the water level restabilizes at a new position. In contrast, considerably less paleohydrologic data are generally available from the shoreline record of open-basin lakes, which respond to changes in the water balance by varying the amount of discharge over the topographic threshold. Variations in this essentially fluvial outlet discharge can be accomplished by adjustments in flow velocity and width with very little change in depth (McCarthy & McAndrews, 1988; Tinkler et al., 1992). As a result, compared to closed-basin lakes, open-basin lakes typically experience much less drastic changes in water level and surface area, making it much less likely that a water-balance change will be recorded and preserved as a distinctive element in the paleoshoreline record. In general then, closed-basin paleolakes represent significantly better sources of paleohydrologic data than do lakes which were hydrographically open during most of their existence.

Closed-basin lakes are more common in arid than in humid climates because the water surplus in humid regions causes most closed-basin lakes in that environment to reach their thresholds, overflow, and become open-basin lakes (Langbein, 1961). With additional time those outlets will be lowered by fluvial erosion, and the overflowing lakes will eventually be replaced by through-flowing streams. Water can also pond in the closed basins of arid lands, but because arid climates have higher evaporation and lower precipitation values than humid climates, lakes in desert basins are much less likely to reach the point of overflow and self-destruct. As a result, closed basins and closed-basin lakes are more numerous in arid than in humid climates (Langbein, 1961).

Despite high evaporation losses on the desert floor, ephemeral and intermittent lakes (i.e., playas or playa lakes) can form there because there are at least intervals during which desert-basin precipitation exceeds evaporation loss, assuming negligible groundwater flux. Paleolimnological evidence of these shallow lakes is overwhelmingly sedimentologic rather than geomorphic. More persistent lakes exist in arid basins where the modest annual input from direct precipitation is augmented by considerable surface runoff from the surrounding drainage basin. Although basin floor evaporation rates remain high, having more water move *through* the lake-basin system allows a larger and more permanent lake to be maintained. In order to supply this surface runoff the catchment must experience, over the long term, an excess of precipitation over evaporation. In other words, although the lake basin lies in an arid climate, part of its drainage basin must have a humid climate. This situation occasionally exists in very large catchments, such as that of Lake Eyre, where arid-region lakes are maintained by surface inflow from distant humid regions (Bowler, 1986). In many other cases, humid enclaves are found within arid regions because of elevation and orographic effects (Fig. 2). These part-arid, part-humid, or "hemiarid" (Currey, 1994b), closed-basin-lake-promoting settings of high relief result from hypertectonism in otherwise arid environments (Currey, 1994a).

Because of its numerous hemiarid basins and generally more arid climate subsequent to the most recent (very late Pleistocene) deep-lake cycle, the North American Great Basin provides some excellent geomorphic evidence of past lakes, and this chapter draws heavily from work done in that region. Nevertheless, preserved segments of paleolake shorelines exist on every continent and in many different circumstances, and methods of collecting and analyzing data from the geomorphic evidence of paleolakes have been derived, tested, and applied in a variety of settings.

Coastal landform nomenclature

Coastal geomorphology is the geomorphic subfield that considers, at least in principal, landforms and landforming processes found along both marine and lacustrine coasts. Most lacustrine coastal, that is, lacustrine shorezone, landforms have an equivalent or approximately equivalent marine counterpart referred to by the same name. More than any other subfield of geomorphology, landform terms from coastal geomorphology seem especially prone to incorrect usage. The coastal bar, for example, a submerged depositional feature, is probably the most misused term in all of modern geomorphology. It is rarely found among relict paleolake landforms. Because of this terminology problem the more basic and common lacustrine coastal depositional landforms are distinguished in Table I. The modern problem may exist partly for historical reasons since coastal terms were sometimes used differently in the 19th century literature than they are today.

Figure 2. Aerial view of part of the Great Basin of western North America illustrating the precipitation-accumulating role of mountains in desert regions.

Table I. Major coastal depositional landforms[a]

Landform	Description
Beach	Attached to land along entire planimetric length (i.e., fringing). Length to width ratio is greater than 1.0. Common types: pocket, bayhead, bayside.
Foreland	Attached to land along entire planimetric length. Length to width ratio is less than 1.0. Common types: cuspate, arcuate.
Beach Ridges	Approximately parallel, alternating ridges and swales, comprising many beaches, forelands, or barriers.
Spit	Attached to land only at proximal end, with the rest projecting freely into the water body.
Tombolo	Connected at both ends to islands or to an island at one end and the mainland at the other.
Barrier	Depositional barricade protecting the land from direct attack of open-water waves. Separated from the land by a lagoon. Common types: bayhead, baymouth.

[a]All features listed project above the associated mean formative lake level except during rare storm events.

Table II. Some principal factors affecting shoreline obliteration and preservation (after Sack, 1995, Table I)

Obliteration processes:

Erosion, reworking, or burial by coastal processes and by saturation-induced gravitational processes during subsequent oscillations and lake cycles

Burial by subsequent pelagic deposits

Subaerial weathering

Erosion, reworking, or burial by fluvial, mass wasting, eolian, volcanic, glacial, and periglacial processes

Preservation factors:

Recency of shoreline formation

Strong geomorphic development (large size) of coastal landforms

Isolation from subsequent subaqueous obliterating factors via tectonic uplift, outlet lowering, or rapid lake recession

Low magnitude, frequency, and spatial intensity of subaerial obliterating processes

Protection from subaerial obliterating processes by topographic barriers, e.g., higher shorelines

Shoreline identification

Finding relict shorelines preserved at the earth's surface in a nonmarine setting constitutes indisputable evidence of a previously expanded lake (Allison, 1945; Reeves, 1969; Bowler, 1971). Indeed the lack of relict shorelines has been used to argue against the existence of hypothesized deep lakes (Quade, 1986). In many cases preserved geomorphic evidence of shorelines is easily recognized even by casual observers (Russell, 1876; Currey, 1980); at other sites it is more difficult to discern due to either weak construction or poor preservation of the landforms. The construction of well-developed coastal landforms requires water depths greater than about 2 to 4 m (Cowardin et al., 1979), shorezone characteristics amenable to sediment erosion, transportation, and deposition by lacustrine waves and currents (Currey, 1994b), and an interval without large and rapid fluctuations in lake level. Table II lists some of the principal factors that contribute to the obliteration and preservation of paleolake shorezone landforms.

Subaerially exposed paleolake shorelines are most commonly identified from ground observations and the stereoscopic study of aerial photographs (Petty et al., 1996), but other useful means include direct aerial observations (Grove et al., 1975) and topographic map interpretation (Thomas, 1963). Because of its small scale, satellite imagery has only rarely played a role in shoreline identification (Löffler & Sullivan, 1979). In some cases, submerged wave-cut terraces may be discernible on seismic profiles (Lister et al., 1991).

The base of wave-eroded cliffs or bluffs and the crest of wave-deposited landforms approximately mark the elevation of the water plane that formed those features. Preserved shoreline segments, therefore, display a lateral continuity in elevation that gives them a distinctively horizontal appearance (Fig. 3). This lateral horizontality is heightened by the tendency of coastal landforms to have a smoother, that is, less crenulate, appearance in

Figure 3. This bayhead barrier, located in west-central Utah, displays the lateral horizontality characteristic of many relict lacustrine coastal landforms.

planimetric view than many other landscape elements. Largely by virtue of this lateral horizontality, well-developed and well-preserved paleolake shorelines often stand in stark visual contrast to the more vertical, contour-crossing trends which tend to be created by endogenic geomorphic processes and the gravity-driven subaerial geomorphic processes, such as stream erosion (Fig. 4). Where lacustrine landforms are small and less well preserved this horizontal nature can often still be discerned in the field, from the air, or on air photos under favorable, low-angle light conditions especially if the landforms consist of a vertical series of erosional bluffs (Gale, 1915; Blackwelder, 1933). Evidence from nonlacustrine landforms, such as an alignment of either small eolian dunes or the apexes of segments of adjacent alluvial fans, also help direct the eye to the position of a relict shoreline. Geomorphic differences between coastal landforms and noncoastal landforms with which they might be confused were well described by Gilbert (1885, 1890) over a century ago.

Where morphologic evidence alone is not sufficient to identify the remnant of a shoreline with confidence, sedimentologic evidence is consulted. In some cases black and white, color, or color infrared air photos reveal grain size variations, or consequent permeability and vegetation differences, between a shoreline and the surrounding landscape through the photographic recognition elements of texture, tone, and color (Avery & Berlin, 1992). Most typically, however, a possible shoreline postulated in the field or from air photos on the basis of form is tested with field-derived information on grain size, sorting, and sedimentary structures (Coventry, 1976; Gassé & Street, 1978). In addition, the combined use of geomorphic, sedimentologic, and stratigraphic methods enables researchers to identify buried or partially buried shorelines. This, of course, requires sufficient stratigraphic

Figure 4. The horizontal nature of this Lake Bonneville shoreline stands in stark contrast to the more vertical trends created by mountain building and stream erosion.

exposure to reveal the distinguishing form and internal composition of coastal depositional landforms buried by subsequent coastal, subaqueous, or subaerial sediments (Blackwelder & Ellsworth, 1936; Currey et al., 1983; Sack, 1999).

Shoreline reconstruction

Once identified, individual preserved shoreline segments are characterized in order to facilitate correlation with other isolated segments in the basin that may represent the same lake level. Physical attributes commonly used for correlating such preserved pieces include elevation, topographic profiles, and distinctive deposits (Russell, 1885; Tackman et al., 1998). Relative and numeric age determinations, which are also employed in intrabasin segment correlation, are discussed in the context of hydrograph construction in a subsequent section. Chronostratigraphic techniques are the specific topics of other chapters in this volume.

    Early paleolake investigators clearly realized the importance of shoreline elevation measurements and collected these data relative to contemporary lake levels, railroad lines, and other shorelines of interest using barometers, transits, and spirit levels (Russell, 1885; Gilbert, 1890; Lawson, 1893; Upham, 1895; Goldthwait, 1910; Gale, 1915). The same principles remain in use today (Cowan, 1985; Lewis & Anderson, 1989; Tackman et al., 1998), but measurement accuracy, precision, speed, and ease have been enhanced through technical improvements in barometers and the development of electronic optical surveying instruments. An increase in coverage by topographic maps and in the number of vertical geodetic control points have also helped to improve elevation estimates and surveyed measurements. In addition, global positioning system (GPS) technology has become available for acquiring elevation data, and quite accurate results can be achieved through differential GPS, provided the site has clear access to satellites. Depths to submerged shoreline features may be obtained through echo sounding (Lewis & Anderson, 1989).

Using elevation to correlate shoreline segments requires an understanding of possible height variations in a single shoreline around a basin. Although the base of erosional cliffs or bluffs and the crest of depositional shoreline landforms mark the *approximate* level of the formative water plane, some inherent differences in the height occur along a single shoreline because of variations in such factors as wave environment, sediment supply, shoreline configuration, and whether the segment is erosional or depositional (Gilbert, 1890; Eardley et al., 1957; Dulhunty, 1975; Coventry, 1976; Currey, 1982; Tackman et al., 1998). Lateral variation in shoreline height also stems from postlacustrine processes, such as erosion of or deposition onto relict shorelines, or through neotectonism, which includes faulting and hydro- or glacio-isostatic rebound. Measured topographic profiles and cross sections may reveal the break or change in slope that marks the mean position of the paleowater plane, and have been helpful in sorting out water-level changes responsible for complex vertical suites of several closely spaced shorelines (Gilbert, 1890; Goldthwait, 1906; Allison, 1952, 1954; Bowman, 1971; Burr & Currey, 1988; Stine, 1990; Avouac et al., 1996).

Water body delineation

An accurate depiction of water body size and configuration derives from shoreline mapping, which relies on the techniques of shoreline segment identification and correlation discussed above. In fact, most mapping proceeds with shoreline identification, correlation, and delineation occurring interactively. By necessity, early paleolake mappers relied exclusively on field observations (Russell, 1885; Gilbert, 1890; Upham, 1895; Leverett, 1908; Meinzer, 1922), often making their own topographic base maps along the way. The value of vertical aerial photographs in paleoshoreline mapping was realized virtually as soon as they became available (Allison, 1940; Dennis, 1944), and, together with accurate topographic base maps on which to trace the shorelines (Gale, 1915; Leverett & Taylor, 1915; Davis, 1983), they remain the most important tool in shoreline mapping (Fig. 5) (Löffler & Sullivan, 1979; Mifflin & Wheat, 1979; Currey, 1982). For most environments shorelines are easier to delineate while viewed stereoscopically on air photos than from field observations alone (Crittenden, 1963), but to ensure the highest accuracy, wherever possible shoreline mapping is accomplished with the combined use of air photos, topographic maps, and fieldwork (Sack, 1995).

Geologic mapping of areas previously inundated by expanded water bodies aids in reconstructing the existence and extent of paleolakes (Feth, 1961). Geologic maps are especially valuable sources of information regarding the nature and history of paleolakes

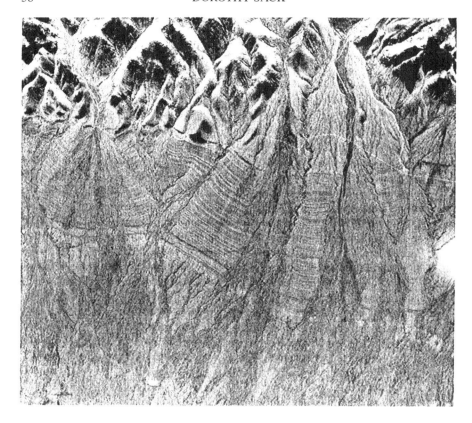

Figure 5. Relict paleolake shorelines are often most easily mapped from vertical aerial photographs. This U.S. government photo (GS-VCSW 3-43) of an alluvial fan piedmont is one of many that were used in mapping shorelines and surficial materials in the Tule Valley subbasin of Lake Bonneville (Sack, 1990). North is to the left. The figure covers about 5.5 km north to south and 4.75 km east to west. Note the north-south trending piedmont fault scarp which offsets the shorelines near the photo center.

when they depict relict shorelines and portray Quaternary units to the same level of detail as bedrock (e.g., Oviatt, 1989, 1991a, 1991b; Sack, 1990, 1993).

Shoreline mapping provides more than just a visual indication of the paleolake perimeter at different stillstands; it supplies information on neotectonism, the hydrologic status of the former water body, and details of its hydrography. Elevation changes found along the length of a once approximately level, isochronous shoreline enable researchers to assess the amount and pattern of isostatic rebound that has affected the shoreline since its formation. In most hemiarid lake basins the rebound signature is primarily hydroisostatic, especially generated from the larger paleolakes, such as Lakes Bonneville and Lahontan in the Great Basin (Gilbert, 1890; Crittenden, 1963; Mifflin & Wheat, 1971; Currey, 1990; Bills et al., 1994). Differential glacioisostatic rebound, however, is the stronger influence on shorelines of paleolakes that were located closer to the margin of Quaternary ice sheets, such as the

Laurentian Great Lakes or the lakes of southern Manitoba (e.g., Farrand & Drexler, 1985; Larsen, 1985b; Tackman et al., 1998). Compared to isostatic deflection, faulting causes more abrupt changes in the elevation of an isochronous shoreline (Currey, 1982).

Whether a given paleolake overflowed its initial basin or not is often discernible through shoreline mapping. By studying shoreline position with respect to passes and the nature of the adjoining topography, reconstructed shorelines help researchers determine if adjacent closed-basin lakes coalesced into a single large closed-basin lake, if one closed-basin lake overflowed through a connecting stream channel into another closed-basin lake, or if a lake experienced an interval of overflow to fluvial systems reaching the sea (Gale, 1915; Allison, 1954; Eardley et al., 1957; Snyder & Langbein, 1962; Morrison, 1965b; Mifflin & Wheat, 1979).

Mapping the reconstructed perimeter of a former standing water body onto a topographic base creates a valuable resource from which can be derived qualitative and quantitative data regarding the three-dimensional configuration of the paleolake. It must be remembered, however, that the modern topography of a former lake basin will differ to some extent from the topography and bathymetry present during the paleolake interval because of intervening erosion, deposition, and neotectonism. This spatial reconstruction of the paleolake and the morphometric variables derived from it, therefore, are actually best estimates of the paleolake and its attributes.

Some kinds of lake basin configuration information, especially when derived for multiple shorelines within the same basin, lead to hypotheses regarding the nature and distribution of various lacustrine sediments and subenvironments (Currey, 1990). For example, a reconstructed shoreline map will show if the lake displayed a relatively simple planimetric outline or if it had multiple arms connected by narrow straits which might have restricted the circulation of water and sediments (Currey, 1990). Likewise, the map may indicate whether the lake was composed of a single basin or of multiple subbasins that were separated from each other by various sills, implying a complicated history of spillover and subbasin integration (Benson, 1978, 1994; Davis, 1983; Last & Slezak, 1986). The location of major inflowing streams should also be evident from the map.

Many quantitative variables describing the size and shape of a paleolake and its drainage basin can be measured from topographic maps of reconstructed shorelines (Table III). These are the same morphometric variables used to describe modern lakes and their drainage basins (Hutchinson, 1957; Håkanson, 1981), thus quantitative comparisons can be made among ancient and modern lakes as well as among various paleolakes, or different levels of a single paleolake. Not surprisingly, such fundamental variables as maximum lake depth, lake surface area, and drainage basin area are historically among the first data to be reported for a given paleolake, and they remain the most frequently cited attributes. As discussed in more detail below, the ratio of a closed-basin paleolake's surface area to the area of the rest of its drainage basin provides a very important means for linking the physical paleolake record with paleoclimatic attributes (Snyder & Langbein, 1962; Mifflin & Wheat, 1979).

Chronologic considerations

In addition to the three-dimensional spatial data needed to reconstruct a lake's size and shape, relict paleolake shorelines contain some evidence relevant to when they were occupied by the water plane. Thus, from a suite of relict shorelines it is in some cases possible

Table III. Lacustrine morphometric variables (Håkanson, 1981)

maximum length
maximum effective length
effective length
effective fetch
maximum width
maximum effective width
mean width
maximum depth
mean depth
median depth
first quartile depth
third quartile depth
relative depth
direction of major axis
shoreline length
total lake area
lake area
volume
mean slope
median slope
shore development
lake bottom roughness
form roughness
volume development
insulosity
lake form

to determine not only when a lake existed and how large it became, but also how it changed in size and shape over time.

Inferences regarding the relative age of paleolake shorelines have been made from various kinds of shoreline geomorphic evidence at least since the work of Gilbert (1885, 1890). By virtue of longer exposure to postformational subaerial and/or subaqueous processes, older shorelines commonly experience greater amounts of weathering, erosion, and burial than younger shorelines in the same physical environment. As a result, an older age is inferred for shorelines displaying greater soil development, thicker rock varnish, lower slopes, more rounded profiles, greater dissection by streams, and less lateral continuity, which can be quantified by the shoreline preservation index (Gilbert, 1890; Gale, 1915; Blackwelder, 1933, 1954; Thomas, 1963; Dulhunty, 1975; Petersen, 1984; Sack, 1995; Avouac et al., 1996; Wilkins & Currey, 1997). Conclusions concerning the relative age of paleolake shorelines also derive from cross-cutting relationships with other landforms (e.g., King, 1993), stratigraphic relationships (e.g., Coventry & Walker, 1977), and archaeological evidence (e.g., Bowler, 1971; Giraudi, 1989).

A slightly different temporal variable, the relative duration of a lake at a given level, is sometimes inferred from the size of a shoreline's coastal landforms (Gilbert, 1890; Blackwelder & Ellsworth, 1936; Street-Perrott & Harrison, 1985; Teller & Last, 1990). Everything else being equal, shorelines composed of wide terraces and massive coastal depositional landforms, which are often referred to as well-developed shorelines, have been interpreted as reflecting a long duration of water plane occupation, with more poorly developed shorelines reflecting shorter occupations by the water plane (Gilbert, 1890; Dulhunty, 1975). The presence of very well developed coastal landforms may suggest shoreline formation during an open-basin interval of threshold control (Gale, 1915). Relative duration deductions made on the basis of comparative landform size, however, are made very cautiously since several factors besides water-level duration may contribute to the total amount of geomorphic work performed at a shoreline. In other words, "everything else" is typically not equal. These factors include a host of climatic, sedimentologic, and geomorphic variables that influence the size, frequency, and effectiveness of the waves which strike a given shoreline segment as well as its preservation potential (Gilbert, 1890; Dulhunty, 1975; Komar, 1976; Pethick, 1984; Bowler, 1986; Stine, 1990; Petty et al., 1996; Wilkins & Currey, 1997).

Numeric age estimates are made from various materials found directly on or within relict coastal landforms, including wave-cut caves (Broecker & Orr, 1958; Thompson et al., 1986; Benson & Thompson, 1987). Conventional and accelerator mass spectrometer (AMS) radiocarbon ages of shorezone materials are most typically obtained from wood fragments, shells, and tufa (Feth & Rubin, 1957; Broecker & Orr, 1958; Bedwell, 1973; Scott, 1988; Oviatt & Nash, 1989; Oviatt et al., 1992), but age estimates of postlake organic matter, including varnish on shoreline clasts (Dorn et al., 1990) and dung from wave-cut caves (Jennings, 1957), have also been radiocarbon dated. Uranium-series dating techniques have been applied to shoreline tufa and shells (Broecker & Kaufman, 1965; Szabo et al., 1996). Optically stimulated luminescence and cosmogenic isotope dating methods have expanded the roster of datable materials to include coastal clastic sediments and erosional terraces.

Numeric age estimates from multiple relict shorelines within a paleolake basin reveal considerable detail about how the lake fluctuated over time. For those basins that have them, dated shorelines constitute the principal benchmarks for reconstructing the paleolake hydrograph. Age estimates derived from nonshoreline lacustrine deposits, including deeper water sediments collected by coring, provide additional useful data but, except for desiccated intervals, less specific information regarding the elevation of the paleowater plane.

Although a hydrograph may show changes in either lake surface elevation or area versus time, elevation remains the more commonly used variable (Fig. 1) (Currey & Madsen, 1974; Farrand & Drexler, 1985; Larsen, 1985a; Oviatt et al., 1992; Oviatt, 1997). One reason for this is that geomorphic preservation is not always sufficient to allow complete basinwide delineation of shorelines, and therefore measurement of surface area, with confidence. Ideally, hydrographs portray original shoreline elevations, which might differ from modern values because of neotectonism, and which might vary substantially around the basin due to differential isostatic deformation. Correcting the modern elevation of a preserved shoreline segment for postlacustrine isostatic deformation may be relatively simple or very complicated depending on how much is known about the neotectonic and outlet history of the lake (Currey & Oviatt, 1985; Hansel et al., 1985; Lewis & Anderson, 1989; Sun & Teller, 1997; Tackman et al., 1998).

In diagram form, many paleolake hydrographs give the impression of precision, certainty, and authority beyond what the data warrant (Currey, 1990). Because of the estimated nature of the temporal and reconstructed spatial data, however, hydrographs are best regarded as sets of working hypotheses which are subject to corrections and further refinement.

Paleoclimatic reconstruction

Once reconstructed, the history of a paleolake's oscillations can be used to help interpret aspects of the regional paleoclimatic history. Lacustrine hydrographs, however, depict water-level variations over time regardless of their origin. Most major water-level changes in open basins are determined by threshold dynamics rather than by climate change. Threshold dynamics include outlet incision, changes in outlet elevation due to blockage or neotectonism, and outlet switching, which can be caused by blockage of the old outlet, blockage removal at the new outlet, or differential isostatic rebound between the two (Petty et al., 1996; Pengelly et al., 1997; Sun & Teller, 1997; Tackman et al., 1998). In addition, when an outlet channel is broad or when a small rise in lake level opens additional outlets, threshold control dampens the impact of non-outlet hydrologic factors, including climate change, drainage basin change by stream capture, and receipt of outburst overflow from upstream lakes and glacial meltwater channels (McCarthy & McAndrews, 1988; Lewis & Anderson, 1989; Tinkler & Pengelly, 1995). Reconstructed hydrographs of open-basin paleolakes, therefore, usually indicate the paleoclimatic regime in only a general way.

The absence of external threshold control, on the other hand, leaves closed-basin lakes free to fluctuate in response to hydrologic changes of any origin. These include climatic as well as any non-outlet geomorphically induced hydrologic events that might occur, such as spillin from adjacent or upstream lakes, spillover to adjacent closed basins, and drainage basin changes by stream capture (Eardley et al., 1957; Mifflin & Wheat, 1979; Smith & Street-Perrott, 1983; Sack, in press). Not all closed-basin lakes experience these geomorphic events, which are sometimes minor when they do take place, and spill to or from adjacent closed basins is ultimately climate-driven. Hydrographs of closed-basin lakes, therefore, predominantly reflect the climatic signal, with geomorphic events affecting the magnitude and duration of the climatically driven trends or appearing as short-term water-level rises, falls, or stillstands (Gassé & Street, 1978; Cerling, 1986). Paleoclimatic reconstruction is more fruitfully accomplished using data from closed- rather than open-basin paleolakes because the former have fuller records of water-level fluctuations and stillstands, and those are more likely to be climate-driven.

"Existing, as they do, in balance between factors of input and output, the fluctuations in water level of closed lakes are often viewed as indexes of climatic variation" (Langbein, 1961:7). Although lake level data are convenient to use when constructing hydrographs, they are not the most appropriate data for making climatic inferences from the paleolake shoreline record. This is the case because the fundamental variable that actually connects a change in the hydrologic input to the response of the closed-lake system is lake surface area. Assuming no outflow to the groundwater system, closed-basin lakes respond directly to variations in hydrologic input by adjusting their *surface area* to the size at which evaporation losses from the lake surface just balance the hydrologic input (Gale, 1915; Eardley et al., 1957; Langbein, 1961; Benson & Paillet, 1989). The elevation at which a given surface area is attained is determined by the three-dimensional shape of the lake basin, and the hydrographic responsiveness of a basin can vary substantially with changes in basin slope along the perimeter of the lake (Langbein, 1961; Street-Perrott & Harrison, 1985; Benson & Thompson, 1987; Benson & Paillet, 1989; Currey, 1994b; Mason et al., 1994).

Paleovalues of precipitation, evaporation, and runoff have been estimated from closed paleolake basins using the z ratio, or pluvial hydrologic index,

$$z = AL/AB, \tag{1}$$

where AL is the reconstructed paleolake surface area and AB is the area of the rest of the lake's drainage basin (Broecker & Orr, 1958; Snyder & Langbein, 1962; Mifflin & Wheat, 1979). Early in the twentieth century Meinzer (1922) had used paleolake surface area as a percentage of drainage basin area to compare the hydrologic productivity of three western North American late Pleistocene lake drainage basins. Broecker & Orr (1958) connected the ratio of closed-basin lake area to drainage basin area more directly to climatic variables by considering the equilibrium situation of constant surface area. Steady surface area results when annual evaporative loss from the closed-basin lake surface equals the annual input of direct precipitation onto the lake plus runoff from the drainage basin. This water balance is represented by the equation

$$R + AL(PL) = AL(EL), \tag{2}$$

where R is runoff into the lake from the tributary drainage basin, PL is precipitation onto the lake, and EL is evaporation from the lake. By expressing R in terms of precipitation and evaporation over the tributary drainage basin, PB and EB, respectively, equation (2) may be rewritten

$$AB(PB - EB) + AL(PL) = AL(EL) \tag{3}$$

or

$$z = AL/AB = (PB - EB)/(EL - PL). \tag{4}$$

Reconstructed paleolake shorelines are necessary for measuring both AL and AB, and thus provide a critical link to the paleoclimatic variables. Although these equations do not generate unique values of the paleoclimatic variables, they have been used to estimate the most probable combinations of precipitation, temperature, evaporation, and runoff needed to maintain various late Pleistocene lakes, and reasonable estimates have been made on a regional basis (Broecker & Orr, 1958; Snyder & Langbein, 1962; Mifflin & Wheat, 1979; Street-Perrott & Harrison, 1985; Sack, 1994).

Other applications

Relict shorelines of known age are useful for scientific endeavors besides paleoclimatic reconstruction. Because of cross-cutting relationships with other Quaternary landforms or sediments, dated shorelines furnish maximum- and minimum-limiting age control on a variety of geomorphic and geologic events, including faulting (Fig. 5) (Petersen, 1984), basaltic eruptions (Hoover, 1974; Oviatt & Nash, 1989), slope failures (Van Horn, 1975; Currey, 1982), and others. They have figured prominently in studies of landscape evolution (Sack, 1992b, 1995). Because of their original horizontality, geophysicists employ now-deformed paleolake shorelines to quantify isostatic rebound, reconstruct deflection histories, and constrain estimates of upper mantle viscosity and lithospheric thickness (Crittenden, 1963; Passey, 1981; Nakiboglu & Lambeck, 1983; Currey, 1990; Bills et al., 1994; Lambert et al., 1998; Seppa & Tikkanen, 1998; Tackman et al., 1998). Paleoecological applications of the shoreline record can result from understanding the overflow history of a paleolake (Hubbs & Miller, 1948; Mifflin & Wheat, 1979). In addition, lake size and shape influence the accumulation of pollen and other biological and geochemical material, therefore paleoenvironmental interpretations derived from those deposits must consider the impact of paleolake morphometry (Punning & Koff, 1997).

Conclusions and future outlook

Geomorphology plays an essential role in paleolimnological research because it is involved as both a cause and an effect of the ancient water bodies. Geomorphic research, moreover, is not conducted in a disciplinary vacuum. Geomorphologists continue to combine landform, sediment, and stratigraphic evidence in their analyses of the paleolake shoreline record. Study of the relict landform evidence of Quaternary lakes is necessary for reconstructing their paleoshorelines, which reveal actual positions of the water plane. Many morphometric attributes of a paleolake can be determined once the paleoshorelines are delineated onto a topographic base map (Table III). Thus far, the most important of these variables is paleolake surface area, which provides a direct tie to the paleoclimate once geomorphic factors affecting lake level are accounted for.

During the second half of the 20th century, the landform evidence of paleolakes has generally received less attention than the evidence available from sediment cores. A growing interest among geomorphologists in Quaternary science, advances in surveying and geographic positioning technologies, and the development of optically stimulated luminescence and cosmogenic isotope dating methods, however, are generating renewed interest in paleolake geomorphology. Attention to the geomorphic evidence should continue to increase with further technological advances. A future surge in paleolake geomorphic research is expected once detailed chronologies have been worked out for more and more paleolakes, and researchers find new ways to analyze the landform evidence. Dated shorelines have already proven themselves to be valuable in a variety of applications, ranging from climate reconstruction through landscape evolution to geophysical modelling. With further refinement in shoreline dating techniques, such applications will undoubtedly become increasingly refined and increasingly useful.

Summary

In addition to providing relict shorezone landform evidence that documents the existence and changing extent of ancient lakes, geomorphology is one of the environmental factors that helped to determine the existence and changing extent of paleolakes. Because geomorphology acts as both a cause and an effect of lakes, it is truly a fundamental aspect of paleolimnological research. Many paleolake studies are conducted for the purpose of contributing to paleoclimatic reconstruction. Because changing water levels reflect both the geomorphic and the climatic history of the drainage basin, an accurate reconstruction of paleoclimate from paleolake data requires recognizing the geomorphically and climatically influenced aspects of the lacustrine hydrographic response.

Climatic factors alone are not responsible for the formation and evolution of a lake. The existence of a lake requires a topographic basin in addition to the appropriate mix of climatic variables. The topography of the drainage basin, moreover, affects precipitation values and the amount of surface runoff reaching a lake basin. Compared to lakes that have outflowing streams, closed-basin lakes fluctuate much more readily in size in response to changes in the hydrologic input-to-output ratio, and therefore represent significantly better sources of paleohydrologic data. The surface area of a closed-basin lake is controlled by precipitation onto the lake, evaporation from the lake, drainage basin precipitation, and drainage basin runoff. The three-dimensional shape of the closed lake basin determines the lake level at which an equilibrium lake area is attained. Paleovalues of climatic variables can be estimated by comparing paleolake area with drainage basin area. Thus, the ratio of a closed-basin paleolake's surface area to the area of the rest of its drainage basin provides a very important means for linking the physical paleolake record with paleoclimatic attributes.

Lake basin and drainage basin geomorphology influence lake formation and behavior, but the water body, in turn, has waves and currents that erode, transport, and deposit earth material into distinctive shorezone landforms, such as wave-cut terraces, wave-cut cliffs, beaches, barriers, spits, and tombolos. Preserved portions of relict shoreline landforms are the principal, surviving geomorphic effect of paleolakes, and they play a crucial role in paleolake studies. Because relict shorelines were created at the interface between the water surface and the land, they are direct evidence of paleolake level and of the location of the paleolake perimeter. Reconstructions of how closed-basin paleolake level and surface area changed through time require detailed knowledge of paleoshoreline position and age.

Relict shoreline segments are most commonly identified on aerial photographs and from geomorphic and sedimentologic observations in the field. Isolated shoreline segments within a single basin can be correlated using elevation measurements, topographic profiles, distinctive deposits, and relative and numeric dating techniques. Careful mapping of paleo-lake shorelines onto a topographic base relies on the techniques of shoreline identification and correlation, including field checking, and creates a valuable source of qualitative and quantitative data regarding the three-dimensional configuration of the former water body. Once reliable age estimates are obtained for them, delineated shorelines constitute the principal benchmarks for reconstructing a paleolake hydrograph. Paleolake hydrographs are the culmination of considerable research time and effort, and give the impression of precision, certainty, and authority, but they are best considered works in progress that are subject to correction and refinement as more field data are gathered and as further technological innovations occur.

Acknowledgments

I thank W. M. Last and J. P. Smol for inviting me to participate in this volume, and all reviewers of the manuscript for their time and thoughtful comments.

References

Allison, I. S., 1940. Study of Pleistocene lakes of south central Oregon. Carnegie Inst. Washington Yearbook 39: 299–300.

Allison, I. S., 1945. Pumice beds at Summer Lake, Oregon. Geol. Soc. Amer. Bull. 56: 789–808.

Allison, I. S., 1952. Dating of pluvial lakes in the Great Basin. Am. J. Sci. 250: 907–909.

Allison, I. S., 1954. Pluvial lake levels of south-central Oregon. Geol. Soc. Amer. Bull. 65: 1331.

Antevs, E., 1945. Correlation of Wisconsin glacial maxima. Am. J. Sci. 243A: 1–39.

Avery, T. E. & G. L. Berlin, 1992. Fundamentals of Remote Sensing and Airphoto Interpretation. Prentice Hall, Upper Saddle River (N.J.), 472 pp.

Avouac, J.-P., J.-F. Dobremez & L. Bourjot, 1996. Palaeoclimatic interpretation of a topographic profile across middle Holocene regressive shorelines of Longmu Co (western Tibet). Palaeogeogr. Palaeoclim. Palaeoecol. 120: 93–104.

Beckwith, E. G., 1855. Explorations for a route for the Pacific Railroad, of the line of the forty-first parallel of North Latitude. In Reports of Explorations and Surveys to Ascertain the Most Practicable and Economic Route for a Railroad from the Mississippi River to the Pacific Ocean 1853–4. Congr. Doc., 33rd Congress, 2nd Sess., Senate Exec. Doc. No. 78.

Bedwell, S. F., 1973. Fort Rock Basin: Prehistory and Environment. University of Oregon, Eugene, 159 pp.

Benson, L. V., 1978. Fluctuation in the level of pluvial Lake Lahontan during the last 40,000 years. Quat. Res. 9: 300–318.

Benson, L. V., 1994. Carbonate deposition, Pyramid Lake subbasin, Nevada, 1: Sequence of formation and elevational distribution of carbonate deposits (tufas). Palaeogeogr. Palaeoclim. Palaeoecol. 109: 55–87.

Benson, L. V. & F. L. Paillet, 1989. The use of total lake-surface area as an indicator of climatic changes: Examples from the Lahontan basin. Quat. Res. 32: 262–275.

Benson, L. V. & R. S. Thompson, 1987. Lake-level variation in the Lahontan basin for the past 50,000 years. Quat. Res. 28: 69–85.

Bills, B. G., D. R. Currey & G. A. Marshall, 1994. Viscosity estimates for the crust and upper mantle from patterns of lacustrine shoreline deformation in the eastern Great Basin. J. Geophys. Res. 99: 22, 059–22, 086.

Bissell, H. J., 1963. Lake Bonneville: Geology of southern Utah Valley, Utah. U.S. Geol. Surv. Prof. Paper 257-B.

Blackwelder, E., 1933. Lake Manly: An extinct lake of Death Valley. Geog. Rev. 23: 464–471.

Blackwelder, E., 1954. Pleistocene lakes and drainage in the Mojave region, southern California. Calif. Div. Mines & Geol. Bull. 170: 35–40.

Blackwelder, E. & E. E. Ellsworth, 1936. Pleistocene lakes of the Afton basin, California. Am. J. Sci. 231: 453–463.

Bowler, J. M., 1971. Pleistocene salinities and climatic change: Evidence from lakes and lunettes in southeastern Australia. In Mulvaney, D. J. & J. Golson (eds.) Aboriginal Man and Environment in Australia. Aust. National Univ. Press, Canberra: 47–65.

Bowler, J. M., 1986. Spatial variability and hydrologic evolution of Australian lake basins: Analogue for Pleistocene hydrologic change and evaporite formation. Palaeogeogr. Palaeoclim. Palaeoecol. 54: 21–41.

Bowman, D., 1971. Geomorphology of the shore terraces of the late Pleistocene Lisan Lake (Israel). Palaeogeogr. Palaeoclim. Palaeoecol. 9: 183–209.

Broecker, W. S. & A. Kaufman, 1965. Radiocarbon chronology of Lake Lahontan and Lake Bonneville II, Great Basin. Geol. Soc. Amer. Bull. 76: 537–566.

Broecker, W. S. & P. C. Orr, 1958. Radiocarbon chronology of Lake Lahontan and Lake Bonneville. Geol. Soc. Amer. Bull. 69: 1009–1032.

Burr, T. N. & D. R. Currey, 1988. The Stockton Bar. Utah Geol. & Min. Surv. Misc. Pub. 88-1: 66–73.

Cerling, T., 1986. A mass-balance approach to basin sedimentation: Constraints on the recent history of the Turkana basin. Palaeogeogr. Palaeoclim. Palaeoecol. 54: 63–86.

Coventry, R. J., 1976. Abandoned shorelines and the late Quaternary history of Lake George, New South Wales. J. Geol. Soc. Aust. 23: 249–273.

Coventry, R. J. & P. H. Walker, 1977. Geomorphological significance of late Quaternary deposits of the Lake George area, N.S.W. Aust. Geog. 13: 369–376.

Cowan, W. R., 1985. Deglacial Great Lakes shorelines at Sault Ste. Marie, Ontario. Geol. Assoc. Can. Sp. Paper 30: 33–37.

Cowardin, L. M., V. Carter, F. C. Golet & E. T. LaRoe, 1979. Classification of wetlands and deepwater habitats of the United States. U.S. Fish & Wildl. Serv. FWS/OBS-79/31, 103 pp.

Crittenden, M. D., Jr., 1963. New data on the isostatic deformation of Lake Bonneville. U.S. Geol. Surv. Prof. Paper 454E.

Currey, D. R., 1980. Coastal geomorphology of Great Salt Lake and vicinity. Utah Geol. & Min. Surv. Bull. 116: 69–82.

Currey, D. R., 1982. Lake Bonneville: Selected features of relevance to neotectonic analysis. U.S. Geol. Surv. Open-File Report 82–1070, 30 pp.

Currey, D. R., 1990. Quaternary palaeolakes in the evolution of semidesert basins, with special emphasis on Lake Bonneville and the Great Basin, USA. Palaeogeogr. Palaeoclim. Palaeoecol. 76: 189–214.

Currey, D. R., 1994a. Hemiarid lake basins: Geomorphic patterns. In Abrahams, A. D. & A. J. Parsons (eds.) Geomorphology of Desert Environments. Chapman & Hall, London: 422–444.

Currey, D. R., 1994b. Hemiarid lake basins: Hydrographic patterns. In Abrahams, A. D. & A. J. Parsons (eds.) Geomorphology of Desert Environments. Chapman & Hall, London: 405–421.

Currey, D. R. & D. B. Madsen, 1974. Holocene fluctuations of Great Salt Lake. Am. Quat. Assoc. Abstr. 3: 74.

Currey, D. R. & C. G. Oviatt, 1985. Durations, average rates, and probable causes of Lake Bonneville expansions, stillstands, and contractions during the last deep-lake cycle, 32,000 to 10,000 years ago. Geogr. J. Korea 10: 1085–1099.

Currey, D. R., C. G. Oviatt & G. B. Plyler, 1983. Lake Bonneville stratigraphy, geomorphology, and isostatic deformation in west-central Utah. Utah Geol. & Min. Surv. Sp. Study 62: 63–82.

Davis, J. O., 1983. Level of Lake Lahontan during deposition of the Trego Hot Springs tephra about 23,400 years ago. Quat. Res. 19: 312–324.

Davis, W. M., 1899. The geographical cycle. Geogr. J. 14: 681–504.

Dennis, P. E., 1944. Shore-lines of the Escalante Bay of Lake Bonneville. Proc. Utah Acad. Sci., Arts & Letters 19/20: 121–124.

Dorn, R. I., A. J. T. Jull, D. J. Donahue, T. W. Linick & L. J. Toolin, 1990. Latest Pleistocene lake shorelines and glacial chronology in the western Basin and Range province, U.S.A.: Insights from AMS radiocarbon dating of rock varnish and paleoclimatic implications. Palaeogeogr. Palaeoclim. Palaeoecol. 78: 315–331.

Dulhunty, J. A., 1975. Shoreline shingle terraces and prehistoric fillings of Lake Eyre. Trans. r. Soc. S. Aust. 99: 183–188.

Eardley, A. J., V. Gvosdetsky & R. E. Marsell, 1957. Hydrology of Lake Bonneville and sediments and soils of its basin. Geol. Soc. Amer. Bull. 68: 1141–1201.

Farrand, W. R. & C. W. Drexler, 1985. Late Wisconsinan and Holocene history of the Lake Superior basin. Geol. Assoc. Can. Sp. Paper 30: 17–32.

Feth, J. H., 1961. A new map of western conterminous United States showing the maximum known or inferred extent of Pleistocene lakes. U.S. Geol. Surv. Prof. Paper 424-B: 110–112.

Feth, J. H. & M. Rubin, 1957. Radiocarbon dating of wave-formed tufas from the Bonneville basin. Geol. Soc. Amer. Bull. 68: 1827.

Frey, D. G., 1988. What is paleolimnology? J. Paleolim. 1: 5–8.

Gale, H. S., 1915. Salines in the Owens, Searles, and Panamint basins, southeastern California. U.S. Geol. Surv. Bull. 580: 251–323.

Gassé, F. & F. A. Street, 1978. Late Quaternary lake-level fluctuations and environments of the northern Rift Valley and Afar region (Ethiopia and Djibouti). Palaeogeogr. Palaeoclim. Palaeoecol. 24: 279–325.

Gilbert, G. K., 1871. On certain glacial and postglacial phenomena of the Maumee Valley. Am. J. Sci. 1: 339–345.

Gilbert, G. K., 1885. Topographic features of lake shores. U.S. Geol. Surv. Fifth Annual Report: 69–123.

Gilbert, G. K., 1890. Lake Bonneville. U.S. Geol. Surv. Monograph 1, 438 pp.

Giraudi, C., 1989. Lake levels and climate for the last 30,000 years in the Fucino area (Abruzzo–central Italy) — A review. Palaeogeogr. Palaeoclim. Palaeoecol. 70: 249–260.

Goldthwait, J. W., 1906. Correlation of the raised beaches on the west side of Lake Michigan. J. Geol. 14: 411–424.

Goldthwait, J. W., 1910. Isobases of the Algonquin and Iroquois beaches and their significance. Geol. Soc. Amer. Bull. 21: 227–248.

Grove, A. T., F. A. Street & A. S. Goudie, 1975. Former lake levels and climatic change in the Rift Valley of southern Ethiopia. Geogr. J. 141: 177–202.

Håkanson, L., 1981. A Manual of Lake Morphometry. Springer-Verlag, N.Y., 78 pp.

Hansel, A. K., D. M. Mickelson, A. F. Schneider & C. E. Larsen, 1985. Late Wisconsinan and Holocene history of the Lake Michigan basin. Geol. Assoc. Can. Sp. Paper 30: 39–53.

Hoover, J. D., 1974. Periodic Quaternary volcanism in the Black Rock Desert, Utah. Brigham Young Univ. Geol. Studies 21: 3–72.

Hubbs, C. L. & R. R. Miller, 1948. The zoological evidence. Bull. Univ. of Utah 38: 17–166.

Hunt, C. B., H. D. Varnes & H. E. Thomas, 1953. Lake Bonneville: Geology of northern Utah Valley, Utah. U.S. Geol. Surv. Prof. Paper 257-A.

Hutchinson, G. E., 1957. A Treatise on Limnology, 1. J. Wiley & Sons, N.Y., 1015 pp.

Jennings, J. D., 1957. Danger Cave. Univ. Utah Anthropol. Papers 27, 328 pp.

King, G. Q., 1993. Late Quaternary history of the lower Walker River and its implications for the Lahontan paleolake system. Phys. Geog. 14: 81–96.

Komar, P. D., 1976. Beach Processes and Sedimentation. Prentice-Hall, Englewood Cliffs (N.J.), 429 pp.

Lambert, A., T. S. James & L. H. Thorleifson, 1998. Combining geomorphological and geodetic data to determine postglacial tilting in Manitoba. J. Paleolim. 19: 365–376.

Langbein, W. B., 1961. Salinity and hydrology of closed lakes. U.S. Geol. Surv. Prof. Paper 412, 20 pp.

Larsen, C. T., 1985a. A stratigraphic study of beach features on the southwestern shore of Lake Michigan: New evidence of Holocene lake level fluctuations. Illinois State Geol. Surv. Env. Geol. Notes 112, 31 pp.

Larsen, C. T., 1985b. Lake level, uplift, and outlet incision, the Nipissing and Algoma Great Lakes. Geol. Assoc. Can. Sp. Paper 30: 63–77.

Last, W. M. & L. A. Slezak, 1986. Paleohydrology, sedimentology, and geochemistry of two meromictic saline lakes in southern Saskatchewan. Geographie Phys. Quat. 40: 5–15.

Lawson, A. C., 1893. Sketch of the coastal topography of the north side of Lake Superior with special reference to the abandoned strands of Lake Warren. Minn. Geol. & Nat. Hist. Surv. Annual Report 20: 181–289.

Leverett, F., 1897. The Pleistocene features and deposits of the Chicago area. Chicago Acad. Sci., Geol. & Nat. Hist. Surv. Bull. 2, 86 pp.

Leverett, F., 1908. Ann Arbor folio. U.S. Geol. Surv. Folio No. 155.

Leverett, F. & F. B. Taylor, 1915. The Pleistocene of Indiana and Michigan, and the history of the Great Lakes. U.S. Geol. Surv. Monograph 53, 529 pp.

Lewis, C. F. M. & T. W. Anderson, 1989. Oscillations of levels and cool phases of the Laurentian Great Lakes caused by inflows from glacial Lakes Agassiz and Barlow-Ojibway. J. Paleolim. 2: 99–146.

Lister, G. S., K. Kelts, C. K. Zao, J. Yu & F. Niessen, 1991. Lake Qinghai, China: Closed-basin lake levels and the oxygen isotope record for ostracoda since the latest Pleistocene. Palaeogeogr. Palaeoclim. Palaeoecol. 84: 141–162.

Löffler, E. & M. E. Sullivan, 1979. Lake Dieri resurrected: An interpretation using satellite imagery. Z. Geomorph. 23: 233–242.

Mason, I. M., M. A. J. Guzkowska, C. G. Rapley & F. A. Street-Perrott, 1994. The response of lake levels and areas to climatic change. Clim. Change 27: 161–197.

McCarthy, F. M. G. & J. H. McAndrews, 1988. Water levels in Lake Ontario 4230–2000 years B.P.: Evidence from Grenadier Pond, Toronto, Canada. J. Paleolim. 1: 99–113.

Meinzer, O. E., 1922. Map of the Pleistocene lakes of the Basin-and-Range province and its significance. Geol. Soc. Amer. Bull. 33: 541–552.

Mifflin, M. D. & M. M. Wheat, 1971. Isostatic rebound in the Lahontan basin, northwestern Great Basin. Geol. Soc. Am. Abstr. 7: 647.

Mifflin, M. D. & M. M. Wheat, 1979. Pluvial lakes and estimated pluvial climates of Nevada. Nevada Bur. Mines & Geol. Bull. 94, 57 pp.

Morrison, R. B., 1964. Lake Lahontan: Geology of southern Carson Desert, Nevada. U.S. Geol. Surv. Prof. Paper 401, 156 pp.

Morrison, R. B., 1965a. New evidence on Lake Bonneville stratigraphy and history from southern Promontory Point, Utah. U.S. Geol. Surv. Prof. Paper 525C: 110–119.

Morrison, R. B., 1965b. Quaternary geology of the Great Basin. In Wright, H. E., Jr. & D. G. Frey (eds.) The Quaternary of the United States. Princeton University Press: 265–285.

Morrison, R. B., 1966. Predecessors of Great Salt Lake. Utah Geol. Soc. Guidebook Geol. of Utah 20: 77–104.

Nakiboglu, S. M. & K. Lambeck, 1983. A reevaluation of the isostatic rebound of Lake Bonneville. J. Geophys. Res. 88: 10, 439–10, 447.

Oviatt, C. G., 1989. Quaternary geology of part of the Sevier Desert, Millard County, Utah. Utah Geol. & Min. Surv. Sp. Study 70, 41 pp.

Oviatt, C. G., 1991a. Quaternary geology of the Black Rock Desert, Millard County, Utah. Utah Geol. & Min. Surv. Sp. Study 73, 23 pp.

Oviatt, C. G., 1991b. Quaternary geology of Fish Springs Flat, Juab County, Utah. Utah Geol. Surv. Sp. Study 77, 16 pp.

Oviatt, C. G., 1997. Lake Bonneville fluctuations and global climate change. Geology 25: 155–158.

Oviatt, C. G. & W. P. Nash, 1989. Late Pleistocene basaltic ash and volcanic eruptions in the Bonneville basin, Utah. Geol. Soc. Amer. Bull. 101: 292–303.

Oviatt, C. G., D. R. Currey & D. Sack, 1992. Radiocarbon chronology of Lake Bonneville, eastern Great Basin, USA. Palaeogeogr. Palaeoclim. Palaeoecol. 99: 225–241.

Passey, Q. R., 1981. Upper mantle viscosity derived from the difference in rebound of the Provo and Bonneville shorelines: Lake Bonneville basin, Utah. J. Geophys. Res. 86: 11, 701–11, 708.

Pengelly, J. W., K. J. Tinkler, W. G. Parkins & F. M. McCarthy, 1997. 12,600 years of lake level changes, changing sills, ephemeral lakes and Niagara Gorge erosion in the Niagara Peninsula and eastern Lake Erie basin. J. Paleolim. 17: 377–402.

Petersen, J. F., 1984. Equilibrium tendency in piedmont scarp denudation, Wasatch Front, Utah. In Morisawa, M. & J. T. Hack (eds.) Tectonic Geomorphology. Allen & Unwin, Boston: 209–233.

Pethick, J., 1984. An Introduction to Coastal Geomorphology. Edward Arnold, London, 260 pp.

Petty, W. J., P. A. Delcourt & H. R. Delcourt, 1996. Holocene lake-level fluctuations and beach-ridge development along the northern shore of Lake Michigan, USA. J. Paleolim. 15: 147–169.

Punning, J.-M. & T. Koff, 1997. The landscape factor in the formation of pollen records in lake sediments. J. Paleolim. 18: 33–44.

Quade, J., 1986. Late Quaternary environmental changes in the upper Las Vegas Valley, Nevada. Quat. Res. 26: 340–357.

Reeves, C. C., Jr., 1969. Pluvial Lake Palomas northwestern Chihuahua, Mexico. New Mex. Geol. Soc. Field Conf. Guidebook 20: 143–154.

Russell, I. C., 1876. On the formation of lakes. Pop. Sci. Monthly 9: 539–546.

Russell, I. C., 1885. Geological history of Lake Lahontan, a Quaternary lake of northwestern Nevada. U.S. Geol. Surv. Monograph 11, 288 pp.

Sack, D., 1989. Reconstructing the chronology of Lake Bonneville: An historical review. In Tinkler, K. J. (ed.) History of Geomorphology. Unwin-Hyman, London: 223–256.

Sack, D., 1990. Quaternary geology of Tule Valley, west-central Utah. Utah Geol. & Min. Surv. Map 124, 26 pp.

Sack, D., 1992a. New wine in old bottles: The historiography of a paradigm change. Geomorphology 5: 251–263.

Sack, D., 1992b. Obliteration of surficial paleolake evidence in the Tule Valley subbasin of Lake Bonneville. Soc. Sed. Geol. (SEPM) Sp. Pub. 48: 427–433.

Sack, D., 1993. Quaternary geologic map of Skull Valley, Tooele County, Utah. Utah Geol. Surv. Map 150, 16 pp.

Sack, D., 1994. Geomorphic evidence of climate change from desert-basin palaeolakes. In Abrahams, A. D. & A. J. Parsons (eds.) Geomorphology of Desert Environments. Chapman & Hall, London: 616–630.

Sack, D., 1995. The shoreline preservation index as a relative-age dating tool for late Pleistocene shorelines: An example from the Bonneville basin, USA. Earth Surface Process Landf. 20: 363–377.

Sack, D., 1999. The composite nature of the Provo level of Lake Bonneville, Great Basin, western North America. Quat. Res. 52: 316–327.

Sack, D. Fluvial linkages in Lake Bonneville subbasin integration. Smithson. Contr. Earth Sci. 33, in press.

Schwennesen, A. T., 1918. Ground water in the Animas, Playas, Hachita, and San Luis basins, New Mexico. U.S. Geol. Surv. Water-Supply Paper 422, 152 pp.

Scott, W. E., 1988. Transgressive and high-shore deposits of the Bonneville lake cycle near North Salt Lake, Utah. Utah Geol. Surv. Misc. Pub. 88-1: 38–42.

Seppa, H. & M. Tikkanen, 1998. The isolation of Kruunuvuorenlampi, southern Finland, and implications for Holocene shore displacement models of the Finnish coast. J. Paleolim. 19: 385–398.

Smith, G. I. & F. A. Street-Perrott, 1983. Pluvial lakes of the western United States. In Porter, S. C. (ed.) Late Quaternary Environments of the United States. Univ. Minnesota Press, Minneapolis: 190–212.

Snyder, C. T. & W. B. Langbein, 1962. The Pleistocene lake in Spring Valley, Nevada, and its climatic implications. J. Geophys. Res. 67: 2385–2394.

Stansbury, H., 1852. Exploration and Survey of the Valley of the Great Salt Lake of Utah. Congr. Doc., 32nd Congress, U.S. Senate Sp. Session, March 1851, Exec. Doc. 3. Lippincott, Grambo & Co., Philadelphia.

Stine, S., 1990. Late Holocene fluctuations of Mono Lake, eastern California. Palaeogeogr. Palaeoclim. Palaeoecol. 78: 333–381.

Strahler, A. N., 1950. Davis' concepts of slope development viewed in the light of recent quantitative investigations. Ann. Assoc. Am. Geog. 40: 209–213.

Strahler, A. N., 1952. Dynamic basis of geomorphology. Geol. Soc. Amer. Bull. 63: 923–938.

Street-Perrott, F. A. & S. P. Harrison, 1985. Lake levels and climate reconstructions. In Hecht, A. D. (ed.) Paleoclimate Analysis and Modeling. J. Wiley & Sons, N.Y.: 291–340.

Sun, C. S. & J. T. Teller, 1997. Reconstruction of glacial Lake Hind in southwestern Manitoba, Canada. J. Paleolim. 17: 9–21.

Szabo, B. J., C. A. Buch & L. V. Benson, 1996. Uranium-series dating of carbonate (tufa) deposits associated with Quaternary fluctuations of Pyramid Lake, Nevada. Quat. Res. 45: 271–281.

Tackman, G. E., D. R. Currey, B. G. Bills & T. S. James, 1998. Paleoshoreline evidence for postglacial tilting in southern Manitoba. J. Paleolim. 19: 343–363.

Taylor, F. B., 1897. Notes on the abandoned beaches of the north shore of Lake Superior. Am. Geol. 20: 111–127.

Teller, J. T. & W. M. Last, 1990. Paleohydrological indicators in playas and salt lakes, with examples from Canada, Australia, and Africa. Palaeogeogr. Palaeoclim. Palaeoecol. 76: 215–240.

Thomas, R. G., 1963. The late Pleistocene 150 foot fresh water beach line of the Salton Sea area. S. Calif. Acad. Sci. Bull. 62: 9–17.

Thompson, D. G., 1929. The Mohave Desert region, California. U.S. Geol. Surv. Water-Supply Paper 578.

Thompson, R. S., L. V. Benson & E. M. Hattori, 1986. A revised chronology for the last Pleistocene lake cycle in the central Lahontan basin. Quat. Res. 25: 1–9.

Tinkler, K. J. & J. W. Pengelly, 1995. Great Lakes response to catastrophic inflows from Lake Agassiz: Some simulations to a hydraulic geometry for chained lake systems. J. Paleolim. 13: 252–266.

Tinkler, K. J., J. W. Pengelly, W. G. Parkins & J. Terasmae, 1992. Evidence for high water levels in the Erie basin during the Younger Dryas chronozone. J. Paleolim. 7: 215–234.

Upham, W., 1895. The glacial Lake Agassiz. U.S. Geol. Surv. Monograph 25, 658 pp.

Van Horn, R., 1975. Largest known landslide of its type in the United States: A failure by lateral spreading in Davis County, Utah. Utah Geol. 2: 83–88.

Whittlesey, C., 1850. On the natural terraces and ridges of the country bordering Lake Erie. Am. J. Sci. 10: 31–39.

Wilkins, D. E., 1997. Hemiarid basin responses to abrupt climatic change: Paleolakes of the Trans-Pecos closed basin. Phys. Geog. 18: 460–477.

Wilkins, D. E. & D. R. Currey, 1997. Timing and extent of late Quaternary paleolakes in the Trans-Pecos closed basin, west Texas and south-central New Mexico. Quat. Res. 47: 306–315.

5. SEDIMENT CORE COLLECTION AND EXTRUSION

JOHN R. GLEW (GlewJ@Biology.QueensU.Ca)
& JOHN P. SMOL (SmolJ@Biology.QueensU.Ca)
Paleoecological Environmental Assessment
and Research Lab (PEARL)
Department of Biology
Queen's University, Kingston
Ontario K7L 3N6, Canada

WILLIAM M. LAST (WM_Last@UManitoba.CA)
Department of Geological Sciences
University of Manitoba
Winnipeg, Manitoba
R3T 2N2, Canada

Keywords: corers, sediments, piston corers, gravity corers, freeze corers, vibracorers, pneumatic corers, samplers, equipment, paleolimnology

Introduction

For many workers in the field of paleolimnology, the retrieval of an unmixed, continuous sediment core marks the first step in a lengthy process. Collecting the core is, in many ways, the most critical part of the paleolimnological process, as any errors or problems encountered during core collection can rarely be corrected "after-the-fact". The success of the entire project, and all the subsequent, time-consuming analyses, depends on this first step: the recovery of a good sample.

Retrieving good cores can be difficult and will always be considered by some as a rather unpredictable 'grey area' of the work. This is because coring operations often require skills and familiarity with equipment that are unlike those required for much of the other laboratory-based aspects of the work. Adding to this are the other factors associated with any kind of field work, such as incomplete knowledge of the sampling site, depth of water, nature of the sediment, as well as the effects of prevailing weather, the logistics of moving and setting up equipment in remote sites, and so forth. All these factors work together to make such operations seldom routine, even for the most experienced paleolimnologist.

"Paleolimnological coring" is a broad term and some of what is covered in this chapter would not necessarily be referred to as "coring" in other technical fields, including operations such as *in situ* freezing and sidewall sampling, etc. This is the result of the adoption

W. M. Last & J. P. Smol (eds.), 2001. *Tracking Environmental Change Using Lake Sediments. Volume 1: Basin Analysis, Coring, and Chronological Techniques.* Kluwer Academic Publishers, Dordrecht, The Netherlands.

of a variety of techniques from other disciplines. These techniques have been used by some paleolimnologists to augment or replace conventional coring, and the inconsistencies in naming the devices and techniques likely reflect the different origins of the equipment. Hence, the proliferation of paleolimnological coring equipment over the last five decades or so has resulted in a "mixed bag" of somewhat standardized devices that often have their origins in such fields as civil engineering, mining, soil science, ice drilling, groundwater investigations, and the marine sciences.

Our goal is to provide the reader with an overview of the major generic categories of equipment used for the retrieval of relatively short to moderate length (<10 m long) unconsolidated lake sediment cores, and to provide our opinions on the advantages and disadvantages of each type. The goal of this chapter is not to summarize all available apparatus and techniques used for lake sediment coring. Such a compilation would require enough pages to fill a large book, and is certainly beyond the scope of this chapter. Several reviews and monographs have been completed on lake sediment coring (e.g., Cushing & Wright, 1965; Wright et al., 1965; Bouma, 1969; ASTM, 1971; Acker, 1974; Kézdi, 1980; Wright, 1980, 1991; Håkanson & Jansson, 1983; Aaby, 1986; Murdoch & Azcue, 1995; Colman, 1996), and the reader is referred to these compilations for more details. Moreover, new coring devices, or more often modifications of existing coring apparatus, are constantly being proposed. A companion chapter in this volume deals with drilling equipment and the recovery of long sedimentary sequences (Leroy & Colman, this volume).

Our overall view is that no one type of sediment corer is applicable to all types of studies and conditions. We do, however, believe that virtually every lacustrine sediment sequence can be sampled in an effective manner, as long as the proper equipment and techniques are used. To do this often requires a number of different coring devices, depending on factors such as the type of sediment and the temporal resolution required by the investigators. In addition, a number of other conditions exist that determine the design and operation of the corer. Water depth is probably the most significant of these factors. On dry lake beds or in playa lake environments, the sediment surface may afford easy access and can be inspected and sampled with the simplest of equipment. For subaqueous limnological sampling, the sediments become more remote as water depth increases, and may ultimately require fully automated equipment to sample them.

Aim of good core recovery

The first requirement of sediment coring is to recover an undisturbed sample, typically including the sediment/water interface (i.e., a representative sample of the material that makes up the lake bed). The requirements of an undisturbed sample, or sediment core, were outlined by Hvorslev (1949) as meeting three criteria: (1) no disturbance of structure; (2) no change in water content or void ratio; and (3) no change in constituent or chemical composition.

Meeting these criteria is often difficult and many of the features that are key design elements of the equipment described here represent approaches to solving problems associated with recovering undisturbed samples.

Nature of lacustrine sediment

Lake sediments can vary greatly in density and consistency. The genesis of the sediment, the processes by which the material is transported to the lake, its pathway through the water body, and the diagenetic processes acting upon it as it is incorporated into the lake bed are all significant factors. From a sampling point of view, the most significant factor is the variation in mechanical shear strength that occurs with depth. Material at the sediment-water interface in the offshore areas of lakes often has such a high-water content and low strength that it is more liquid than solid (often more than 90% water content), whereas more deeply buried material or sediment in the nearshore and shoreline areas of a lake can be well compacted and consolidated.

Offshore sedimentation in lake basins can, under favourable conditions, approach a near-perfect example of the law of superposition. In such a situation, accumulating material from one period of time is systematically overlain by progressively younger material in a time-stratigraphic sequence. Often, the single largest change in the character of the accumulating sediment is that of increased bulk density due to a gradual decrease in porosity, a consequence of increasing lithostatic pressure as the sedimentary sequence thickens. This gradual change from material of high water content, with bulk densities approaching 1.0 g cm^{-3}, at the top of the sediment column to more compact material having higher bulk density (often $\geq 2.0 \text{ g cm}^{-3}$) lower in the sequence is so great that, in practice, no single technique or piece of equipment is able to sample both adequately. These differences have resulted in the development of a group of samplers that are able to recover the soft and high-water content surface materials, and a second group that has greater penetrating ability and can recover deeper, more compact, high strength sediments. In limnological sampling, these two groups are often referred to simply as "short" and "long" cores. In practice, short cores are considered to be anything that samples down to about one meter from the sediment-water interface, including the surface sediments and the base of the water column, and long cores are those usually recovered sequentially (e.g., 1 meter lengths) and may not necessarily sample the material at the sediment-water interface without significant disturbance.

Nature of the sampling process

When considering the removal of sediment from the lake bed in a way that lifts the sample intact to the surface, two main categories of physical forces have to be taken into account.

The first group of forces are such things as frictional forces that relate to the cohesion and density of the sediment (Fig. 1). The second group are those relating to the sampling mechanism or, in a broader context, the sampling system (Fig. 2). Into the latter category fall all the motions attributable to the environment in which the sampler is working (e.g., wave-coupled motion on a recovery line, dynamic imbalance as the device is lowered through the water, the weight distribution as the device is lowered and enters the sediment, as well as the effects of closing the device and withdrawing it). All these have some effect on the overall efficiency of the device, its ability to penetrate and recover sediment, and the degree to which the sample is disturbed.

Although the motion of a core tube driven into the sediment by any means is fairly simple, its actual motion and the forces that relate to it are worth reviewing, as they will help in understanding why particular features incorporated in many of the devices outlined

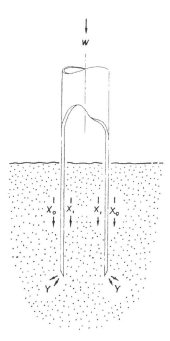

Figure 1. Forces acting on the core tube. X_o = force on the outside of the core tube; X_i = force on the inside of the core tube; Y = displacement forces concentrated at the leading edge of the core tube; W = driving force (in this case, the weight of the corer).

here are important. The following summary relates specifically to core tubes (cylinders) driven vertically into the sediment, although the principles relate equally to other types of apparatus.

Diagrammatically, the arrangement of forces outlined in Figure 1 are simplified and, in practice, not easy to separate. In the simple case of a gravity corer, the lowering of the core tube into the sediment transfers the weight from the suspended line to the lake sediment; in this situation the transfer is seldom uniform. First, there is a combination of motions illustrated in Figure 2 that affect the motion of the corer as it is suspended in the water column and as it enters the sediment. Secondly, a set of forces relating to the entry of the sampler into the sediment itself replaces those outlined in Figure 2. These forces are those related to the shear strength of the sediment and the hydrostatic pressure of water transmitted along the penetrating surfaces of the core tube. These can be considered as frictional forces acting on the side walls of the core tube 'X' in Figure 1 and are significant in terms of core penetration. If the sediment into which the core tube is being driven is of uniform consistency, the resisting forces acting on the tube increase proportionately to its surface area (tube length) in contact with the sediment. As the frictional force X approaches that of the applied driving force W, the descent of the core tube slows to a point where it may be considered stationary.

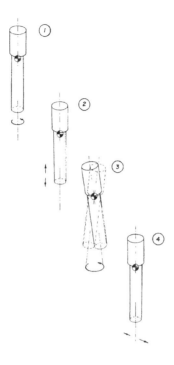

Figure 2. Basic motions relating to suspended core tubes which are likely to perturb initial entry of the core tube into the sediment surface. 1 = rotation about the vertical axis; 2 = oscillation along the vertical axis; 3 = nutation about centre of mass; 4 = lateral motion.

Effects of frictional and deformational forces on the sediment sample

Cores taken with open-barrel type equipment (see below) are frequently shorter than the penetrated depth. This is referred to as core shortening, or incorrectly as core compaction or core compression, and is defined as the length of the recovered core being shorter than the depth penetrated by the core tube. Analysis of many shortened cores reveal the geometry illustrated in Figure 3 where the stratigraphic elements of the core are usually thinned progressively down-core. This may occur more in some stratigraphic units than others (e.g., high water content units would be more affected). The process by which this thinning occurs has been the subject of speculation in the past, the most enigmatic aspect being the fate of the 'lost' material and the process involved in its removal (e.g., Crusius & Anderson, 1991; Cumming et al., 1993). Early work carried out by Emory & Dietze (1941) and Emory & Hulsemann (1964) found that, although the retrieved core was shorter than the penetrated sediment, the loss of water was insignificant. The cored section conformed to that illustrated in Figure 3, where the upper layers were represented in an undisturbed form and the lower strata were thinned progressively down the core (i.e., the boundaries between layers being closer together in the core than in the undisturbed sedimentary structure). This

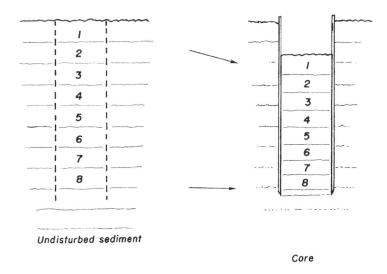

Undisturbed sediment

                                                                    Core

Figure 3. Effect of core shortening (core compression) in which layers in the sediment (numbered 1 to 8 in this hypothetical example) are progressively thinned down the core.

has sometimes been attributed to compression, although subsequent analysis (Piggot, 1941) of the test material determined that true compression (i.e., reduction in volume of a given mass with increased pressure) had not occurred and instead each layer was present, but in smaller amounts. Therefore, although sometimes referred to as core compression, it is actually sediment that has been lost by being pushed laterally outwards ahead of the tube (i.e., it is lost before it enters the tube; Fig. 4).

To understand how this takes place, one has to further examine the conditions that are outlined in Figure 1. The kinetic energy of the corer is expended in two ways as the core tube penetrates the sediment (Hvorslev, 1949). First, frictional wall forces are present on the outside and inside of the tube (X_i and X_o, for frictional forces on the inside and outside of the tube, respectively); secondly, deformational forces (Y) occur at the leading edge of the tube as material is displaced to allow the core tube to penetrate. Both forces increase as penetration progresses but for different reasons (Fig. 1). The frictional forces increase simply as a function of tube length (i.e., the surface area of the tube in contact with the sediment). The deformational forces (Y), on the other hand, increase because the sediment at depth is more compact (having less water content) and offers greater resistance to deformation. This increase in compaction with depth also increases the frictional forces (X) to some extent, but this increase is considered negligible for purposes of this explanation.

In the initial coring stages, while the penetration of the core tube is only a few centimeters into the sediment, the frictional forces are low, and little resistance is offered to additional material entering the tube as it descends. The deformation forces (Y) at this point are confined to the immediate area adjacent to the leading edge of the tube.

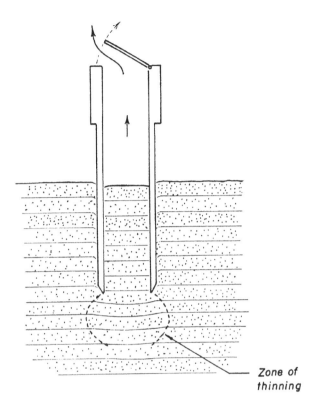

Zone of
thinning

Figure 4. Zone of thinning created by increasing friction on the inside of the core tube, laterally displacing sediment ahead of the advancing core tube. The steady increase in wall friction will produce greater thinning with depth. As the frictional forces approach that of the shear strength of the material in front of the core tube, the core sample will act as a plug and deflect the material ahead of it.

As penetration of the tube into the sediment increases, the frictional forces (X) increase proportionally. The deformational forces at the tube edge (Y) increase as denser sediment is encountered with depth, but are not much greater than in the initial stages. The combined forces, X and Y, increase the entry resistance of the material into the tube, creating a bulb-shaped pressure wave below the core tube (Fig. 4) that radially moves material away from the core axis. This combined force is responsible for the progressive thinning of the material by plastic deformation ahead of the descending tube (Piggot, 1941). If the resisting forces (X) are high enough on the inside of the tube (X_i), then the shear strength of the material about to enter will be overcome, and it will be deflected completely by the pressure wave that will now act as a solid surface with respect to the sediment it encounters. In the majority of cases where short cores in soft sediment are concerned, the plasticity of the material results in partial displacement of the sediment. Under these conditions, material passing into the core tube is deformed (thinned) conforming to the radius of the pressure wave existing in front of the core tube, thus resulting in the situation illustrated in Figure 3. In this way the overall shortening of the core is accomplished by plastic deformation of the sediment before it enters the tube (Hongve & Erlandsen, 1979; Evans et al. 1986; Blomqvist, 1985, 1991).

In reality, sedimentary deposits that consist of layered material may exhibit differing shear strengths at microscopic scales and express directional variability (anisotropy) in the way they deform. The presence of overlying lake water penetrating with the core tube into sediments that have less water content introduces further dynamic changes (Pratje, 1935; Piggot, 1941; Emory & Dietz, 1941; Hvorslev, 1949). Dynamic measurements carried out on sediment behaviour as the core tube is being driven indicate that, even under constant drive rates, the entry of material into the tube is not smooth but follows a stick-slip (i.e., intermittent) motion. This is likely the result of the small-scale variability in shear strength of the layered material being pressurized below the core tube. Although a detailed analysis of these forces is beyond the scope of this chapter, their effects should be considered when one examines the operating features of the coring devices covered in the next section.

Major types of sampling equipment

The equipment described in this chapter is arranged in general order of complexity, with the simplest types first and noting the key features of each. Although some categories overlap, the equipment is classified in the following order:

(1) Open-barrel and gravity corers

(2) Box cores and dredges

(3) Piston corers driven by rods

(4) Cable-operated piston corers

(5) Chamber-type samplers

(6) Freeze samplers

(7) Vibracorers (vibra-drilling)

(8) Percussion or hammer corers

(9) Modified gravity corers

(10) Pneumatic corers

(11) Core catching devices

(12) Extruding equipment

Open-barrel and gravity corers

Some examples of open-barrel and gravity corers are the Kajak-Brinkhurst (K.B.) types (Kajak et al., 1965; Brinkhurst et al., 1969), HON-Kajak type (Renberg, 1991), Glew (1989, 1991), Axelsson-Håkanson (1978), Hongve (Wright, 1990), Limnos (Kansanen et al., 1991), and many others.

Theory of operation

Open-barrel corers, sometimes referred to as open-drive samplers, are the simplest type of coring devices and under many conditions the easiest to use. In its simplest form of operation, a tube (open at both top and bottom) is driven vertically into the sediment. At the end of the drive, the top of the tube is closed and the tube recovered. Sediment is retained in the tube by the sealing of the top before withdrawal (thereby sealing the hydrostatic head), by the cohesion of the sediment in the core tube, or by closing the bottom of the core tube. Driving the tube into the sediment may be accomplished in a variety of ways. If the sediment is relatively soft, the tube can be pushed in and withdrawn by hand using drive rods connected to the top of the core tube. More compact sediment may require that the tube be mechanically driven in (and out) with either a hammer or a weight or advanced by hydraulic pressure (Acker, 1974; Brink et al., 1982; Barendregt & Vreeken, 1983; Missen, 1986). Percussion or hammer corers are described more fully in a subsequent section. In most sediments, the mass of the device itself can be used to drive the core tube into the sediment (e.g., gravity corer).

Other than directly driven corers, which are operated from the surface by connecting rods and therefore limited to relatively shallow water conditions (e.g., <40–50 m), most open-drive samplers are gravity corers. Of these, there are two basic types of devices: close-on-contact types and messenger-operated types. As their name implies, close-on-contact types close automatically once the core is driven into the sediment. An automatic device of this type may seem to be the most efficient, particularly when samples are required to be taken from deep water; however, the close-on-contact types have some deficiencies as outlined below.

The seal at the top of the tube must remain open while the tube descends into the sediment. Once the sample has been taken, the sediment must provide an efficient barrier to any flow of water into the top of the tube. The action of the seal is usually flow- or load-actuated. In the case of the flow-actuated seal, the sealing mechanism operates essentially as a one-way valve, which senses the flow direction of water through the core tube. In this case, two major problems affect sediment recovery. First, the flow-through mechanism operates to some degree as a proportional flow device (i.e., low flow will only open the seal a small amount and greater flow volume will be required to open it fully). At some low velocity (threshold velocity) the seal will close completely, restricting further entry of material into the core tube. Furthermore, the energy required to keep the seal open will produce a pressure differential and create a pressure wave below the lower end of the core tube as it descends. This pressure wave, often referred to as a bow wave, may be large enough to deflect low-density sediment radially away from the core axis, and potentially lose ("blow away") the surface sediments (Crusius & Anderson, 1991; Cumming et al., 1993; Stephenson et al., 1996) or mix the sample material (Fig. 5). This leads to difficult

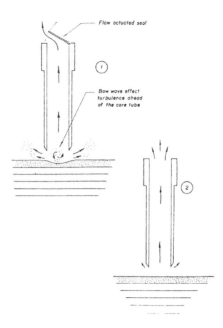

Figure 5. The importance of free flow through the core tube. 1 = back pressure resulting from the operation of the flow actuated seal, creating a turbulent zone ahead of the descending core tube (bow wave), resulting in the loss of surficial sediment from the surface. 2 = near laminar flow conditions of an open tube approaching the sediment surface.

design requirements of the seal, which must minimize flow restriction within the core tube at very low flow-through rates, but must seal completely when flow stops.

Load-actuated closing mechanisms are ones that hold the seal open on the device as the corer is lowered. Once the device begins to enter the sediment, the relaxation of the line tension causes the mechanism to close. Problems associated with this type of mechanism relate to its sensitivity and the tendency of such an arrangement to close prematurely. This results from the fact that tension on the recovery line can be zero when the core tube is still penetrating the sediment. In some designs, the seal can be allowed to float (not seal tightly) during the penetration, and will only be secure once the tension is reapplied during recovery, as in the original Kajak design (Fig. 6), or with a Hongve (Wright, 1990) or Boyle (1995) sampler. However, such an arrangement still reintroduces the problems encountered in flow-through devices, as noted above (Fig. 5).

Messenger-operated gravity corers

The most common solution to the problems described above regarding close-on-contact and flow-through-type devices is to hold the core tube open until it is closed by the operator from the surface by means of a messenger. In its simplest form, the messenger is a weight

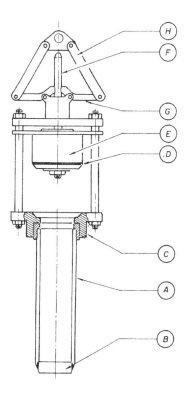

Figure 6. General design of the original Kajak coring design (modified from Kajak et al., 1965). *A* = core tube, 36 mm or 80 mm internal diameter (ID); *B* = cutting shoe; *C* = adapter plate for alternate core tube sizes; *D* = seal; *E* = seal weight; *F* = seal guide stem; *G* = seal closing and release link; *H* = release link.

captive on the recovery line that is released at the surface and is free to fall down the line and contact the coring device on the lake bottom. Messenger-operated coring equipment incorporates some kind of release that uses the energy of the descending messenger to close the top seal on the core tube. Probably the most significant disadvantage of most messenger-operated devices is that the operator must be able to determine when the corer is on the bottom and ceased penetration in order to close the device. This can be a problem in highly organic and loose sediments, especially in deep waters, although even these types of sediments can often be effectively cored using these devices, with some practice. In addition, the recovery line must be kept under a certain amount of tension to enable the free transit of the messenger. Neither of these conditions is easily maintained as operating depths increase (e.g., greater than about 50 m). However, adaptations made to gravity corers can alleviate this problem to some extent. The coring device may be equipped with a support frame or a buoyant upper section to maintain its stability, as well as a mechanism to sense when the core tube has been driven to its limit and to trigger an actuating device to then close the core (Glew, 1995).

Disadvantages with the messenger-triggered devices relate to the operator's ability to determine with certainty when the equipment is at the sediment surface. The ability to determine the position of the device, freely suspended in the water column or penetrating the sediment, and whether the device is open or closed, is often referred to as "feel". In water depths usually between 5 and 50 meters, feel is not usually a problem unless excessive drifting and wave motion are present. To some extent, the line length at these intermediate depths is effective in decoupling such minor motions of waves and drifting. In very deep water, the dynamic aspects of the recovery line in the water tend to mask the motion of the device itself, resulting in the loss of feel. In very shallow water (e.g., <3 m), the moderating effects of the line are insignificant in damping out surface motion such as lateral drift and wave action, often impairing control of the corer.

The main advantage of gravity coring is that it is a simple and rapid method of recovering relatively short cores. The core tubes themselves can be easily removed and replaced, and the core sample can be extruded and subsampled in the field.

Recovery of relatively small volumes of sample and limited penetration (especially in stiff sediments, such as clays or silts) are likely the most significant shortcomings of the open-barrel gravity corer, as size and weight are compromising factors in the design of any such device. Increasing the core tube diameter and increasing the weight of the corer go some way in overcoming these problems, but some limitations should be considered. Core aspect ratios (i.e., core length to core diameter) become problematic when they are about 6:1 or less. This is because the typical shear strength of lacustrine sediment will allow it to flow out of the core tube during lifting if the column of sediment approaches such low aspect ratios. For this reason, smaller diameter core tubes retain sediment more securely than larger diameter tubes. On the other hand, if the tube is too narrow, its recovery of sediment may be incomplete because of excessive internal friction, as previously described. For larger diameter cores, greater penetrating depths require greater weight.

One solution to the sample volume, weight, and core length problem is to use corers that carry multiple core tubes or barrels (Brinkhurst et al., 1969; Tratt & Burne, 1980). The main disadvantage of this approach is that the penetrating resistance is significantly higher for multiple tubes than for single tubes, and that multiple-tube corers are mechanically more complex. Typically gravity corers with weights between five and ten kilograms and core tube diameters of between 40 and 100 mm can efficiently recover cores in the order of 30 to 70 cm long, and sometimes up to 1 m. If the corer is designed with a low centre of mass and is symmetrical about the vertical axis, these devices probably represent the most efficient and reliable type of sampling equipment under a variety of conditions. Increasing the weight and core size can extend the penetration range of gravity corers, but their mechanical operation and ease of handling is compromised.

Box corers and dredges

Box corers operate on the same principle as gravity corers in that they can be driven into the sediment by means of extension rods or by gravity. Some of the earliest types of quantitative sediment samples were of this type (such as the Ekman or Birge-Ekman grab sampler, Murdoch & Azcue, 1995; box corers, Bouma, 1969; etc.). The main differences among these types of corers and those described in the previous section are in their size range; box dimensions for samplers used in lakes are often in the size range of 15 cm × 15 cm and 20 cm deep or larger (for example, limnologists sometimes use samplers primarily developed for marine systems, with dimensions in the range of 1 m^2 of sediment surface), and the sample retention is accomplished by bottom closure of the box rather than sealing of the top of the chamber. In order to do this, sliding or rotating jaws close off the bottom of the box after it enters the sediment (Jenkin & Mortimer, 1938; Jenkin et al. 1941). The hinged or sliding mechanism associated with the jaws and the physical size of the box in most types of box corers limit the depth of penetration of such devices (10–20 cm being typical, although deeper penetration is possible with some types of heavy equipment). In addition, the force required to close the jaws once the device is embedded in the sediment is considerable, and the reactionary force generated on closing can seriously disturb the surficial sediments inside the box (Blomqvist & Boström, 1987). In spite of such disadvantages, these devices can potentially recover large volumes of surface and shallow subsurface sediment with little disturbance under favorable conditions (Flower et al., 1995). In recent decades, some of the most successful marine samplers have been large-scale box corers, such as the USNEL (United States Naval Electronics Laboratory) box corer (Hessler & Jumars, 1974).

Piston coring devices driven by rods

Piston corers were initially developed in the 1920s and were used to sample unconsolidated sediment ahead of drilled holes (Olsson, 1925). In this type, the piston corer was attached to and operated by the same drill rods assembled for advancing the drill.

Piston corers developed specifically for limnological work were introduced in the 1950s (Livingstone, 1955; Vallentyne, 1955; Brown, 1956). These devices operated in a similar way to the original Olsson (1925) corer, except that this corer was designed to obtain only short sections of sediment (<1 m). They were of lighter and simpler construction, utilizing a cable to control the piston instead of rods and core casing. Such an arrangement saves considerable weight and therefore enables the equipment to be used from small boats or improvised platforms. Because a stable platform is required for piston coring, an ice-covered lake surface, when available, is often the preferred method. Wright (1991) offers many suggestions to help paleolimnologists obtain good piston cores.

Theory of operation

The typical piston corer consists of three major components: the piston and cable assembly, the core tube, and the drive head and drive rods (Fig. 7). Core tubes contain a close-fitting adjustable sliding plug or piston. In the driving or closed position (Fig. 7A), the piston is secured at the lower end of the tube by the rods used to drive the device to the sample depth. In this position, the piston keeps the core tube closed and deflects material as it penetrates the sediment. At the selected sampling depth, the drive rods are retracted through the core tube and are locked in the drive head (Fig. 7B), leaving the piston at the bottom of the core tube. A number of mechanisms have been devised to do this, such as the cone clamp (described in Hvorslev, 1949), a latch (Vallentyne, 1955) and a square rod with pegs (Wright, 1967). The piston is independently restrained by a cable passing through the core tube and the drive head to the surface, and secured at its upper end by a winch, clamp, or other means. With the rod string locked to the drive head, the rods and the core tube are driven down past

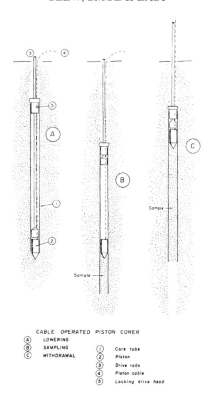

CABLE OPERATED PISTON CORER
(A)   LOWERING
(B)   SAMPLING   (1)   Core tube
(C)   WITHDRAWAL   (2)   Piston
                        (3)   Drive rods
                        (4)   Piston cable
                        (5)   Locking drive head

Figure 7. General operation of a Livingstone-type piston corer, showing lowering (A), sampling (B), and withdrawal (C).

the piston that is prevented from descending by its restraining cable (Fig. 7B). With the drive completed, the piston is at the top of the core tube against the drive head (Fig. 7C), and the tube is filled. The cable is then locked to the drive rods to prevent any downward motion of the piston and contained sample when the device is lifted to the surface.

The original Livingstone (1955) corer, inspired by Kullenberg (1947), had no internal rod to hold the piston in place; rather the piston was retained by its tight fit and was released by a sharp push. Vallentyne (1955) introduced an internal rod with a latch to lock it after retraction, but the latch sometimes failed. Wright (1967) replaced the round rod with a square rod, with pegs and slots in the head to permit a firmer lock.

Advantages of rod-driven piston corers

The chief advantage of the device is related to the operation of the piston and the seal it maintains between the internal surface of the core tube. As the core tube is advanced into the sediment, the frictional forces that, in an open-barrel-type configuration would cause the core to be drawn down with the tube, are countered by the seal of that stationary piston. In this type of arrangement, soft sediment cores can be recovered with no displacement in the core tube. The only disadvantage is the increased mechanical complexity of maintaining an efficient seal between the piston and the core tube wall, and providing the precise control required when driving and recovering the device.

Most piston coring devices have simple piston and drive-head mechanisms that permit small core tube diameters (30–50 mm), enabling them to penetrate compact sediment with relatively small applied driving forces. Smaller tubes, although easier to push, encounter excessive internal frictions in the core, and recovery may be incomplete. As much as 40 m of sediment have been recovered with a piston corer of 5-cm diameter, in 1 m sections, in the same cased hole (H.E. Wright, Jr., personal communication).

Disadvantages of rod-driven piston corers

Like all rod-operated sampling devices, the practical limitations imposed by the length of the connecting rods make these types of "Livingstone" piston corers applicable mainly in relatively shallow water conditions. The mechanical limitations of the strength and stiffness of the material from which the rods are made and the type of couplings used to assemble them determine the ultimate operating depth. Although the use of rigid support casing can partially alleviate these operational limitations, the application of these devices becomes increasingly more difficult in water depths greater than about 30 m, but coring with appropriate lifting equipment and rigid casing is possible in water depths of ~50 m.

Probably more than any other rod-operated sampling device, the Livingstone-style piston corer is susceptible to shortcomings associated with the rod-string and the piston cable. This is because close mechanical coordination must be maintained between the rod-string and the piston control cable, and any misalignment of these two components can result in mechanical failure. The cable can be damaged or broken and the piston lost or misaligned when the compressional forces on the rod string are excessive and the position of the piston with respect to rods is not accurately determined.

Problems associated with the drive head, and the method used to transfer the driving force through it, are also common and not easy to solve. In some mechanisms (Vallentyne, 1955) a latching arrangement requires the rod string to be retracted past a spring-loaded element that locks the rod string to the drive head. In other arrangements, such as the square rod type piston corer (Wright et al., 1965), the rod string is retracted and then rotated into a locked position. In both cases, the precise depth (determined by the length of the rod string) must be known and the subsequent manipulation of the rod string is made based on it. Such operations, when carried out under less than ideal field conditions, are difficult.

Cable-operated piston corers

Some of the problems encountered with deep-water piston coring operations can be overcome by use of a Kullenberg-type corer. Originally designed for marine work, where the operational water depths preclude any remote manipulation or direct control from the surface, the Kullenberg (1947) corer represents a type of device that relies on a driving weight that is allowed to free-fall through a fixed distance to drive the core tube into the sediment in a single stroke. The core tube is equipped with a piston that is restrained by a cable fixed to the core head. The driving weight that is initially restrained by a latch mechanism on the head is held in position by an outrigger cable and weight assembly that is freely suspended alongside the core tube. On reaching the sediment surface, the tension on the outrigger line is removed and the latch opened, allowing the weight to fall and drive the core tube past the restrained piston into the sediment. Kelts et al. (1986) describe a modified Kullenberg-type corer which is more appropriate for lake work.

One of the disadvantages of Kullenberg-type corers is that the uppermost sediments are typically lost. Moreover, the corer is heavy and often requires significant infrastructure for its operation.

Chamber-type samplers

Chamber-type samplers (e.g., Hiller corers, Russian peat corers) were developed as an alternative to vertically driven core samplers. They are often used to recover undisturbed subsurface material, such as peat and similar fibrous sediments. In these devices, the sample is cut and enclosed in a rotating chamber of half-cylindrical form. This chamber takes material adjacent to the drive axis in a rotational motion against the fixed plate. The operation of the chamber sampler avoids some of the systemic problems associated with other subsurface samplers, most notably those employing vertically driven core tubes and their potential for incomplete recovery.

Theory of operation

The chamber corer (Fig. 8) is driven vertically into the sediment in the closed position with drive rods. In the closed position, the half-cylinder that comprises the rotating member is locked against a faceplate that forms one side of the chamber. To take the core sample, the half-cylinder is rotated in a direction away from the plate by applying torque to the drive rods; the asymmetric projection of the faceplate providing the reactionary element against which the cylinder can be rotated. After the half-cylinder has pivoted approximately 180°, it is locked with the sample now enclosed in the chamber. Once locked in this position, the device is then withdrawn to the surface. At the surface, the chamber can be rotated backwards to expose the half-cylinder of sediment supported by the faceplate (Fig. 8).

Advantages of chamber corers

An advantage of this type of sampler is that it can recover discrete subsurface samples from selected depths, bypassing the overlaying material in the same way as the piston corer described earlier. Unlike the vertically driven core samplers, however, the action of the chamber is essentially that of a sidewall sampler acting horizontally against the sediment, thus eliminating any distortions to the sample associated with vertically driven coring devices. An additional advantage is that the recovered sample is easily removed from the chamber (it is not extruded) as the chamber is opened by rotating the half-cylinder in the reverse direction. This allows researchers to inspect the sediment material directly on site.

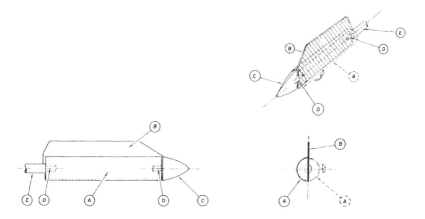

Figure 8. Chamber corer, also known in various versions as the Russian peat corer, showing operation and major elements. The corer is driven to sample depth by control rods, and the rods are rotated to turn the chamber with respect to the plate, thus enclosing the sediment adjacent to the plate surface (shown hatched on the oblique view). *A* = chamber, hemi-cylindrical; *B* = reaction plate, face plate; *C* = nose; *D* = chamber pivots; *E* = control rod.

Disadvantages of chamber corers

Disadvantages of the chamber sampler lie chiefly with its large cross-sectional area, both in the driving and recovery configuration, and in the large torque forces transmitted by the rods when taking a sample. This becomes particularly significant as the length of the rod-string is increased and the degree of rotational control is reduced due to wind-up induced torsion of the rods. In addition, the asymmetric design of the chamber sampler tends to promote uneven forces on the stationary components, particularly the chamber pivots. Nonetheless, many researchers report good success with this type of sampler, even when working with long rod-strings.

The inability to efficiently seal the sample chamber tightly due to its longitudinal closure against the plate and the necessity of opening the device in a horizontal position generally makes it unsuitable for the sampling of the uppermost portions of the sediment profile or in soft sediments with high water contents.

Freeze samplers

Theory of operation

In sampling materials that are very susceptible to disturbance and mixing, such as those that exist at or close to the sediment-water interface, or that contain gases that might disrupt the core profile during the coring process, *in situ* freezing of the sediment profile has proven to be a successful sampling method. The technique of solidification by freezing was first used to recover cohesionless sands and gravels with high and irregular void ratios (Hvorslev, 1949). In this type of sampling, the material was frozen in place, and then subsequently

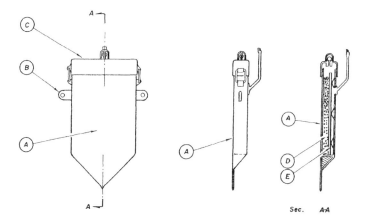

Figure 9. Rectangular freeze corer (modified from Huttunen & Meriläinen (1978). *A* = freezing face; *B* = lifting lugs; *C* = cap with pressure valve; *D* = sectional view showing dry ice and alcohol mixture; *E* = insulating support for mixing chamber.

cored and recovered in the frozen state. For subaqueous sampling of soft sediment with high-water content, a different technique is used. Limnological freeze coring utilizes a low-temperature metallic surface that is lowered into the sediment and held in position long enough to freeze the surrounding sediment on to it (e.g., Shapiro, 1958; Huttunen & Meriläinen, 1978; Renberg, 1981). The device is then recovered, and the frozen sample is removed.

The equipment used for freeze-coring consists of a weighted chamber with an outer freezing surface that may be cylindrical or flat (Fig. 9). A wedge-shaped design is often desirable, as it tends to produce the least amount of disturbance to the sediment profile (Renberg, 1981). The chamber is filled with a freezing medium that reduces the outer surface temperature well below that of the freezing point of water. The medium most commonly used is dry ice (solid carbon dioxide) mixed with an accelerating liquid (alcohol). Other cryogenic liquids have been used, such as liquid nitrogen, but these must be used cautiously. The reaction pressurizes the chamber that is vented to the atmosphere or into the water. After being lowered through the water column and into the sediment, it is held in position long enough to allow sediment to freeze to it (typically about 10 minutes, but this depends on the physical size of the device). The chamber is then recovered with the frozen sample (commonly 5 to 10 cm thick) adhering to the outside surface.

Advantages of freeze corers

The main advantage of the freeze sampler is its ability to spatially preserve material that is suspended in the water column at or close to the sediment-water interface, providing the chamber is fixed in a stationary position. It is a simple device that under certain conditions is easy to use, especially where the sampler can be held in a stationary position (for example, when operating from winter ice cover). Provided the adjacent sediment is not disturbed as the device is lowered, sediment can be recovered easily. If the sampler is designed to expose a large freezing surface, an additional advantage of this method is that a potentially large volume of sediment can be collected. Several high-resolution studies have been accomplished using this technique.

Disadvantages of freeze corers

The problems associated with freeze coring are primarily those associated with the conditions required to hold the device in a fixed position in the sediment for the required period of time (e.g., 10 minutes), and the logistics of transporting the quantities of dry ice and alcohol to the sampling site and transporting the frozen sample from the field site. Adaptations have been developed that address some of these problems. For example, Renberg & Hansson (1993) developed a freeze corer that separated the dry ice/alcohol container from the freezing wedge. Once the corer penetrates the sediment, an electric pump circulates alcohol through the chamber. This device overcomes the problem of ice-crust formation (water freezing to the corer during lowering through the water column) and allows for the collection of a thin freeze wedge. Lotter et al. (1997) further modified this design by including a tripod to stabilize the device at the sediment surface, and a hydraulic system for controlled lowering of the corer into the sediment. Verschuren (2000) developed a freeze corer especially designed for warm-water systems (e.g., tropics) by dropping dry ice pellets steadily into the corer to replace those that have sublimated.

Freeze cores may not be totally suitable when paleolimnological analyses require precise estimates of water content. Stephenson et al. (1996) identified this potential problem when they compared contaminant inventories using different types of sediment samplers. Freeze-cored sediments had higher water contents than tube-cored sediments, possibly due to ice crystal formation, and thus resulting in lower contaminant inventories.

Subsampling the frozen prism of sediment may also present some problems in the field, as a uniform detachment from the device can be difficult due to the uneven melting or mechanical force required to remove it from the freezing face. Transportation of frozen sample material from the field to the laboratory may also present logistical problems.

Vibracorers

Theory of operation

Vibracoring, a technique originally developed in the 1960s for use in marine coastal and shelf oceanographic investigations (Bouma, 1969; Sanders & Imbrie, 1963), has enjoyed considerable popularity as a paleolimnological tool during the past several decades. The ease of operation and portability have made it the equipment of choice for many paleolimnologists (e.g., Smith, 1998; 1992; Thompson et al., 1991) who are attempting to collect moderate length (1 to ~15 m), undisturbed cores of unconsolidated sediments in material that may be difficult to sample (e.g., profiles containing sand lenses). The basic principle of operation of today's modern vibracorer, first described by Lanesky et al. (1979), is straightforward. As the name implies, a vibrating device (most commonly a concrete vibrator powered by a small gasoline engine, but earlier versions of the technique used a variety of other devices; Dokken et al., 1979) is mounted on a core tube (usually aluminum irrigation pipe) which imparts high-frequency, low amplitude standing waves along the length of the tube. This vibration liquefies a thin (1–2 mm) layer of water-saturated sediment at the base of the core tube. The liquefaction and resulting loss of strength allows the core tube to penetrate downward through the sediment column with little or no vertical application of force. Various modifications of the basic design of Lanesky et al. (1979) have been suggested to allow sampling of moderately deep lakes (40–50 m; Smith, 1998; 1984, and recently even deeper (B. Beierle, pers. comm.), and to provide better portability (Smith, 1987), penetration ability (Thompson et al., 1991), and sediment recovery (Reddering & Pinter, 1985).

Advantages of vibracorers

Vibracoring technology has solved many of the problems inherent in attempting to acquire moderate length (>1 m) cores of unconsolidated sediment that is difficult to core using piston or other corers (e.g., due to difficulties in penetrating material, such as sand) within the confines of a limited budget. The system is relatively inexpensive to construct and employs components that are readily available in most urban centers (e.g., cement vibrator and cable, gasoline engine, cable winches, tripod, ladder, etc.) or fabricated in a machine shop. Likewise, the core barrels are inexpensive, commonly available (5 cm or 7.5 cm diameter aluminum irrigation pipe), and can be cut in whatever predetermined length is required for ease of transport. The technique can be used in a wide variety of environments (e.g., on land, on a raft, on an ice cover; through shallow water or through relatively deep water), and the equipment is light enough to be relatively mobile and easily handled by two persons. Because vibracoring works on the basis of inducing fluidization in the material, it can be used to sample a wide variety of sediment types, including sands, silts, diamicts, and clays, provided they are water saturated. We have even had success acquiring cores of greater than 3 m from partly consolidated, coarsely crystalline salt. Fluidization of the sediment and loss of shear strength means that the corer usually advances under its own weight without the necessity of applying significant downward force. Under optimum conditions, sediment penetration and core acquisition can be quite rapid; Smith (1987,1984) reports penetration rates of up to 4 m per minute. Similarly, because little or no vertical force is required, the diameter of the core barrel need not be small. In fact, better penetration rates and better recovery ratios are reported with wider core tubes (i.e., 7.5 cm versus 5 cm; Thompson et al., 1991; Smith, 1987). Finally, the sample is retrieved as a *continuous* sediment sequence, uninterrupted by overlapping or missing sections at core tube breaks.

Disadvantages of vibracorers

The most commonly reported limitations of the vibracorer technique are poor penetration and recovery in some types of sediment and compaction. Vibracoring depends on the core barrel oscillation briefly causing a loss of sediment bearing strength due to temporary fluidization at the base of the tube becoming thixotrophic. This liquefaction requires that the sediment be water saturated. Although coring of non-saturated but moist sediment is possible (Smith, 1984), it is slow and difficult, with poor recovery ratios. Likewise, some types of sediment, even if they are water saturated, do not exhibit thixotropy. The optimum sediment type for vibracoring is sandy silt and muddy sand (Burge & Smith, 1999; Esker et al., 1996; Imperato, 1987; Sly, 1981; Wright et al., 1965). Both coarser and finer grained sediment types are more difficult to core. Gravelly sediment, poorly sorted clayey diamict, and compact clay are extremely difficult to core with this technique (Thompson et al., 1991; Smith, 1984), as they are with other corers. Interbeds of organic litter (peat, twigs, needles, logs) dampen the core tube vibration and make penetration difficult. Desiccation horizons, such as are commonly found in the stratigraphic sequences of playas and lake basins that have periodically dried, can also interrupt coring. Alternatively, in cases where the non-thixotropic and/or organic debris layers are thin enough to allow the vibracorer to be pushed or pounded through, the "difficult" sediment can form a plug in the nose of the tube which prevents further sediment from entering the barrel even though the stratigraphic section is being penetrated by the device. Lanesky et al. (1979) suggested the use of a piston within the core tube to help alleviate this problem and improve recovery, however this requires an additional connecting rod or wire which must be fixed at the surface. Imperato (1987) outlines a method of applying a high-pressure water jet in advance of the core barrel to improve core penetration in these difficult sediments.

As pointed out above, core shortening and compaction of the recovered core relative to the thickness of the stratigraphic interval are problems common to many sampling devices. However, compaction rates with vibracoring can be severe. For example, Thompson et al. (1991) report compaction of nearly 60% (Fig. 10). Molnar & Smith (1993) maintain that compaction values are normally about 10 to 40%. Dott & Mickelson (1995) identify compaction of up to 50%. The average core compaction reported by Smith (1987) using 5 cm diameter core tubes is over 40%.

Other corers and modifications

In the previous sections of this chapter, the coring devices have been grouped by mode of operation. Many of these basic types, however, have been modified and a comprehensive classification of these would be too large to include here. Some basic adaptations are included in this section that are in common use. Most adaptations are designed to increase the operating range or capacity of the device. For example, equipment originally designed for shallow water sampling may be modified to operate in unrestricted depths by use of automated closing mechanisms, or small-capacity samplers may be combined to increase the volume of recovered material. The coring devices discussed below represent some examples of such modifications.

Percussion or hammer corers

One of the most common modifications to open barrel gravity and some piston corers is the addition of some form of driving weight on top of the device to increase penetration of the core tube (e.g., Gilbert & Glew, 1985; Reasoner, 1986, 1993; Nesje, 1992). In this mode of operation, the corer is lowered into the sediment where it penetrates initially

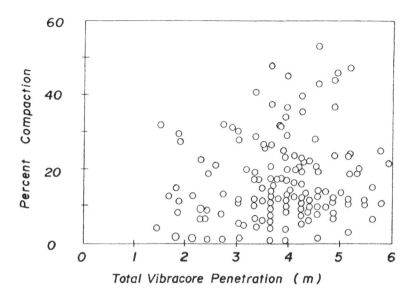

Figure 10. Relationship between penetrating depth and compaction for sediment profiles collected with a vibracorer. Modified from Thompson et al. (1971).

by its own weight. At this point the driving weight (sometimes referred to as a "hammer") can be lifted from the top of the corer and then released to drive the core tube into the sediment. In some arrangements, the recovery line is used as a guide, passing through the weight, and a second line is attached to the driving weight. To operate this efficiently, tension must be maintained on the recovery line and precise control of the weight lifting line must be exercised in order not to interfere with it. In deep water, the operation of two closely parallel lines in such an arrangement is not practical due to their tendency to twist together. An alternative arrangement utilizing a single line requires that the driving weight be mounted on a rigid guide (or jar staff) or similar mechanical support. In order that the force of the weight be controlled effectively, the recovery line itself is operated to lift and release the weight. A number of factors serve to distinguish the percussion (hammer) driven cores from a single drive-type-gravity corer, particularly when considering the potential for disturbance of soft sediments (see Chapter 4 in Hvorslev, 1949, for a detailed examination). The mechanical arrangements of percussion corers are such that they usually preclude the use of messenger closing, so a one-way flow through valve is usually fitted at the top of the core tube. These may perform poorly where large instantaneous pressure changes occur, typical when driving or recovering a core using these techniques.

In addition to the open-barrel percussion types, some piston corers have been modified to extend the penetration depth of the core tube. An additional complication in this type of adaptation is in maintaining the fixed position of the piston during the drive. If a solid ice cover is available, anchoring to the ice is an option.

Modifications to gravity coring equipment

Parachute-controlled devices

In deep water and in situations where control of the coring device may be difficult, one method of providing the device with artificial stabilization is to use a parachute or drogue at the lower end of the recovery line (Cushing et al., 1997). This provides the descending sampling device with a controlled rate of fall through the water column and a degree of vertical stability as it enters the sediment, conditions that would be provided for by the operator in shallow water. The advantage with this type of modification is that it can be added to existing equipment with little or no change in the basic operating mode. However, some provision must be made to keep the stabilizing parachute canopy away from the operating device once on the sediment to prevent it from interfering with the subsequent recovery. This can de done by the inclusion of a float to keep the parachute above the corer when stationary, or a rigid frame that folds to provide minimum resistance when the device is lifted to the surface.

Deep-water automatic gravity corers

A similar modification for operating shallow-water messenger-equipped samplers in deep water is an arrangement that employs a float to stabilize the lower end of the recovery line (in the same way as the parachute attachment noted above) and a messenger release mechanism that senses the tensional force on the line (Glew, 1995). In this way the float assembly acts like an operator on the surface of a shallow lake; once the line tension is reduced, the messenger is released to close the corer. Small devices designed to be operated in shallow water can be effectively deployed in water depths of hundreds of meters.

Pneumatic corers

One of the limiting factors of the piston corers described in the previous section is the use of extension rods. A number of piston-type corers have been devised that do not require rods, the driving force being supplied by inertial weight in the case of the Kullenberg (1947) corer or pneumatic cylinders in the case of the Mackereth-type sampler (Mackereth, 1958, 1969).

The Mackereth sampler

The Mackereth (1958, 1969) corer is a pneumatically driven piston corer designed to recover long cores (e.g., 6 m) in an uninterrupted manner. Because of the large forces required to drive the core tube in this way, the sampler is anchored to the sediment surface by a bell chamber that forms the base of the device. Operation of the corer takes place in two stages. First, the corer is lowered to the sediment surface and water is pumped out of the bell chamber. This action allows the overlaying water pressure (hydrostatic pressure) to force the chamber into the sediment ensuring a rigid platform from which the core can be driven. Second, compressed air is injected from the surface by a compressor into the driving

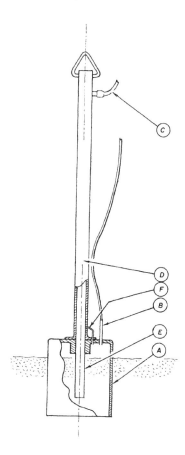

Figure 11. Mackereth piston-type sampler showing main elements (modified from Mackereth, 1969). A = anchor (bell chamber); B = chamber pump-out line; C = compressed air line; D = air driving cylinder; E = core tube; F = air bleed, driving cylinder to chamber. The piston is not shown.

cylinder forcing the core tube past the piston into the sediment in a way similar to that of the rod-operated piston corer (Fig. 11). Air is injected until the core tube is fully driven, at which point the air is automatically bled into the bell chamber. As the bell chamber is pressurized, its buoyancy retracts it and lifts the core tube from the sediment. This is accomplished because the volume of the chamber is of sufficient capacity that when it is air-filled it makes the entire device buoyant and therefore self-recovering (Fig. 11).

   The main disadvantages of the Mackereth corer is that it requires an air compressor and other infrastructure, and so is not very portable. Corers have also been known to become trapped in lake bottoms due to high frictional forces from the sediments. Care must also be taken with this sampler, as when it is released from the sediment, it can reach the water surface at high velocity, break the water surface, and potentially cause injury.

Core catching devices

Core catchers or core retainers are devices that mechanically prevent loss of sediment sample during core retrieval. These are often seen as optional additions to the various corers described in this chapter.

The simplest type of core retainers used for general work are the spring-type devices that consist of a plastic or metal insert that is mounted inside the lower end of the core tube. A variety of shapes exist that have been named after their designers, such as Denison, Emery-Dietz, and Joukowsky (see Hvorslev, 1949). All operate in generally the same way, as they are shaped to conform to the inside diameter of the tube as the core is being driven, but fold or collapse to block sample material from sliding back out of the tube when it is withdrawn. For sampling soft sediment, the mounting of this type of core retainer is problematic; first, because it increases the cross-sectional area of the tube and results in greater driving resistance, and second, because the potential for the retainer to disturb the sediment as it enters the tube is significant due to the pressure required to keep the spring retainer open. A number of designs incorporate mechanical arrangements that can seal the retainer and actuate it only when the core is pulled back on recovery. Such mechanisms greatly increase the cross-sectional area of the core barrel as they require an external shoe and core liner.

For the above reasons, core retainers are rarely used for short core sampling of soft sediment, although sampling of such material would benefit from equipment that could close the end of the core tube (e.g., the Limnos corer; Kansanen et al., 1991). A number of such core-retaining devices have been developed. These serve to close the core tube and support the core sample as it is lifted though the water column. They operate by sliding down the outside of the core tube as it is pulled from the sediment, and close the lower end as it clears the sediment surface. Such devices do not interfere with the inside diameter of the core tube and allow free entry of the tube into the sediment. They do, however, require careful adjustment with respect to tube length, as effective closing is dependent on the final position of the retainer at the end of the tube.

Extruding and subsampling equipment

Extruding equipment

We have thus far emphasized the recovery of relatively undisturbed cores. In many respects, core extruding and subsampling equipment is as important as the coring devices themselves in as much as this equipment must be capable of taking full advantage of the intact nature of the cored material. Once the core is brought to the surface, some reliable method must be available to either preserve it and transport it back to the laboratory for analysis, or subsampling the core on site. For most work involving high-water-content surface cores that are recovered with a portion of the overlaying water column, there is little chance of transporting the core to the laboratory without some significant disturbance. For this reason, these cores are routinely subsampled on site. Even for long cores that are destined to be subsampled in the laboratory and core samples that are robust enough to be transported intact, the extrusion process on site is considered important, as it provides an important check on the performance of the corer and an opportunity for a "first-look" at the core and for some preliminary logging.

Cores of the highly fluid near-surface sediments provide the greatest challenges to the design of extruding and subsampling equipment because of the low strength of the material. Any stratification of the deposit can easily be destroyed by mixing of the sediment during post-collection handling. One method of subsampling short sediment cores is to equip the corer with a segmented tube, such as the one on the Limnos corer (Kansanen et al., 1991). In this arrangement, the segments can be rotated horizontally to subsample the core. Vertical extrusion of the sediment using equipment such as the Glew (1988) extruder (see below) is also a very effective way of sampling even the most watery surface sediments. Another promising technique is to use a gel-forming compound, such as Zorbitrol Plus TM (a polyamide commonly used in clean up of biohazardous laboratory spills; available from Ulster Scientific Inc, New Paltz, NY), to stabilize the water in the core barrel immediately above the sediment-water interface, thereby facilitating transport of the undisturbed sample from the field to the laboratory (D. Livingstone, pers. commun.).

For most other types of equipment, the sample is extruded in a vertical position. The main requirement for an extruder for short near-surface cores is that it can: (1) provide support and containment for the sediment being sampled; (2) repetitively and efficiently subsample accurate volumes of material; and (3) be set-up and operated easily in field situations.

Types of extrusion mechanisms used for core subsampling

Two types of extruders are described here: a common vertical-type extruder used primarily for high-water content sediments (e.g., Glew, 1988), and horizontal-type extruders used for the extrusion of long cores.

Vertical-type extruders

Vertical-type extruders have three key components: 1) a mechanism for securely holding the core tube in a vertical position; 2) a ram or mandrel that is used to push through the core tube to expel the core (in some arrangements the ram is fixed and the core tube is drawn down over it); and 3) a plate or tray onto which the extruded sample can be drawn off, separated, and placed in bags or other containers. Of these three components, the ram poses the greatest design problems as, in most applications, it must be adjusted to the core length (which cannot be predicted *a priori*) and subsequently incremented accurately. In addition, it must function well in field environments in the presence of water, mud, and sometimes ice that can compromise the mechanical action. One of the simplest types of vertical extruders is illustrated in Figure 12.

Incremental extrusion (Fig. 12) is accomplished by pushing the core tube assembly over the ram. With element A locked to the ram and B unlocked, downward motion of the core tube assembly is limited to distance X (set by set-screw C). Locking of B and resetting A against set-screw C repeats the extruding setting X. Locking A and releasing B allows the following increment to be extruded (Fig. 12).

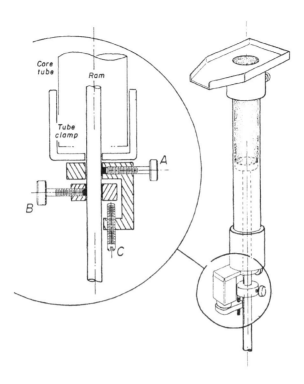

Figure 12. Simple vertical "push rod type" extruder for short cores. Core movements can be preset for repetitive extrusion. With the screw (*A*) in the locked position, the collar (*B*) is set against the adjusting setscrew (*C*), and the screw (*A*) unlocked to extrude the set amount of sediment. Sequential locking of *A* and *B* allows for the core tube to be advanced over the ram.

Long core (horizontal) type extruders

This type of extruder differs from the vertical device used with short cores of highly fluid near-surface sediments in as much as it is not required to incrementally move the core sample for subsampling, because longer and deeper subsurface cores tend to be more compact. Instead, the devices are typically designed to extrude the entire core in a single motion, transferring it from the core tube to a split core tube used as a tray or transporting box for inspection and shipment from the field (Fig. 13).

The actuating mechanisms for long-core extruders range from straight mechanical jacking to pneumatic pumps and hydraulic devices, some generating large forces and high pressures within the core tube. Only two types of basic motion exist however: those operating to draw the tube over the ram and those employing a traveling ram (Fig. 13). Of these types, the fixed ram type presents the least sample disturbance as the emerging core is stationary with respect to its supporting surface. Traveling ram mechanisms can provide the same type of support to the core if the supporting surface is moved with the ram.

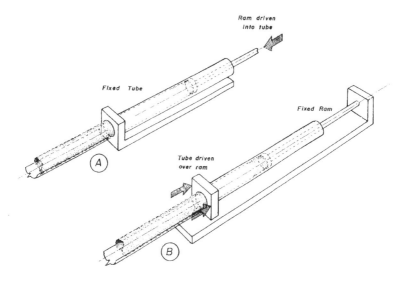

Figure 13. Long core (horizontal) extruders. *A* = fixed tube type; *B* = fixed ram type.

With all types of extrusion, there is the potential for smearing or distortion of the sediment along the core tube wall. This can be minimized by using a close-fitting extruding plug, and is considered to be small when compared to distortion that may occur during the initial drive of the core.

Alternatively, the core may be transported from the field in the core tube and split in the laboratory by cutting the tube longitudinally. This can be carried out on all types of core tubes, as well as with frozen sediment. Because this procedure is carried out in the laboratory, it has the disadvantage of not providing a check or "first look" at the sample material while at the sampling site. The advantage of this approach is that the core is being handled under more controlled conditions.

Sediment handling

Subsequent handling of the core material after acquisition depends upon the nature of the study, type of logging to be done, type and "urgency" of analyses anticipated, distance to the laboratory and/or storage facilities, etc. For large, multiple-investigator, multiple-proxy, multiple-core research projects, considerable preliminary planning must be done and a suitable protocol established in order to properly and efficiently preserve and archive the sediment, as outlined, for example, in Todd et al. (2000), Colman (1996), Tiercelin et al. (1987), and Hsü & Kelts (1986). Although post-collection handling schemes are highly individualized, in general, if the core is not extruded in the field, care should be taken to properly label the tubes with respect to site location, top and bottom, and other necessary orientation information, and to plug and seal the ends of the core barrel to prevent physical loss and chemical alteration. Clearly, such factors as maintaining the core material at relatively constant and cool (but not freezing) temperatures (e.g., 4 °C), avoidance of strong magnetic fields, and minimum agitation must be considered during temporary storage in the field and transport to the laboratory. Once the core is extruded, it is normally split longitudinally and visually logged, photographed, and subsampled immediately. Usually about one half of the core is re-sealed in plastic and/or foil and stored at low temperature for future reference and use, with the other half being used for description and subsampling. Long-term storage schemes must take into account the avoidance of warm temperatures, light, and loss of moisture from the core, factors that significantly alter many of the important paleoenvironmental components of the core.

In the case of multiple-proxy and multiple-investigator projects, it is important to have already established a well-devised subsampling strategy before the core material is acquired. In addition to following an appropriate subsampling protocol, it is also desirable to anticipate and undertake as soon as possible any mechanical and other non-visual logging techniques, such as magnetic susceptibility, GRAPE (gamma ray attenuation porosity estimator), sonic velocity logging, resistivity measurements, etc. Colman (1996) and subsequent chapters in this book summarize many of these post-collection logging techniques.

Summary

Collection of lake sediment cores that are of sufficient quality (e.g., length, resolution) to adequately answer research questions may pose some challenges to paleolimnologists. Nonetheless, a wide spectrum of reliable and well-tested sampling and extruding devices have been developed and modified that are applicable to almost all paleolimnological requirements.

This chapter focused on collecting relatively short sediment cores. Probably the simplest and least expensive coring device available to the paleolimnologist is the gravity corer, which uses the weight of the corer to enter the sediment. If employed correctly in appropriate settings, such devices can provide excellent profiles of recent lake sediments, even if the water content of the sediment is very high. However, careful attention must be paid to the design of these corers to insure complete recovery and to avoid, for example, turbulence ahead of the core tube creating a bow wave. Box corers and dredges can effectively collect large amounts of surface and shallow subsurface sediments. Freeze coring has been employed in a large number of high-resolution studies of recent lake sediments, as this approach freezes a crust of sediment to the core surface *in situ*. Rod-operated piston corers are very commonly used devices, and have become standard pieces of equipment for many paleolimnologists wishing to sample several meters of relatively soft sediment. Open-chamber corers have the advantage of being relatively easy to use, and allow the investigator to easily inspect the sediment profile at the coring site. Vibracoring is especially effective in sampling sandy and silty sediments, whereas other types (e.g., pneumatic corers, percussion corers) can be employed for specific tasks and other problem sampling jobs. In many cases it is advantageous to extrude the core and subsample at the sampling site; considerable progress has been made in the development of extruding systems that can be used to effectively subsample sediment profiles at a resolution determined by the investigator.

Acknowledgements

The authors would like to acknowledge the enormous contribution made to the field of sediment sampling by Juul Hvorslev (1895–1989) who, in cooperation with the Harvard School of Engineering, the American Association of Civil Engineers, and the U.S. Corps of Engineers, produced the seminal work "Subsurface Exploration and Sampling of Soils for Civil Engineering Purposes", over 50 years ago. While advances in technology have revolutionized much of our science, the body of information incorporated in this work still remains the standard reference.

Research in our labs, as well as the development of some of the coring equipment designed at PEARL, was funded by the Natural Sciences and Engineering Research Council of Canada. We are very grateful for many helpful comments and discussion from H. Wright, Jr., D. Livingstone, A. Lotter, V.P. Salonen, S. Leroy, P. Teasdale, B. Beierle, S. Lamoureux, T. Karst, B. Cumming, N. Michelutti, S. Dixit, and other members of our labs. Opinions vary widely on some aspects of the pros and cons of various corers, and the opinions expressed in this chapter are our own.

References

Aaby, B. & G. Digerfeldt, 1986. Sampling techniques for lakes and bogs. In Berglund, B. (ed.) Handbook of Holocene Palaeoecology and Palaeohydrology. John Wiley & Sons, Ltd., Chichester: 181–194.

Acker, W. L., 1974. Basic Procedures for Soil Sampling and Core Drilling. Acker Drill Co. Inc., Scranton, Pennsylvania, 246 pp.

ASTM (American Society for Testing and Materials), 1971. Underwater Soil Sampling, Testing, and Construction Control. ASTM Special Technical Publication 501, ASTM, Philadelphia, Pennsylvania, 241 pp.

Axelsson, V. & L. Häkanson, 1978. A gravity corer with a simple valve system. J. Sed. Petrol. 48: 630–633.

Barendregt, R. W. & W. J. Vreeken, 1983. A coring device suitable for paleomagnetic sampling of unconsolidated subsurface deposits. Can. Geotech. J. 20: 843–848.

Blomqvist, S., 1985. Reliability of core sampling of soft bottom sediment — an *in situ* study. Sedimentology 32: 605–612.

Blomqvist, S., 1991. Quantitative sampling of soft-bottom sediments: problems and solutions. Mar. Ecol. Prog. Ser. 72: 295–304.

Blomqvist, S. & K. Boström, 1987. Improved sampling of soft bottom sediments by combined box and piston coring. Sedimentology 34: 715–719.

Bouma, A. H., 1969. Methods for the Study of Sedimentary Structures. Wiley-Interscience, New York, 458 pp.

Boyle, J. F., 1995. A simple closure mechanism for a compact, large-diameter, gravity corer. J. Paleolim. 13: 85–87.

Brink, A. B. A., T. C. Partridge & A. A. B. Williams, 1982. Soil Survey for Engineering. Clarendon Press, Oxford, UK., 378 pp.

Brinkhurst, R. O., K. E. Chua & E. Batoosingh, 1969. Modifications in sampling procedures as applied to studies on the bacteria and tubificid oligochaetes inhabiting aquatic sediments. J. Fish. Res. Bd. Can. 26: 2581–2593.

Brown, S., 1956. A piston sampler for surface sediments of lake deposits. Ecology 37: 611–613.

Burge, L. M. & D. G. Smith, 1999. Confined meandering river eddy accretions: sedimentology, channel geometry and depositional processes. In Smith, N. D. & J. Rogers (eds.) Fluvial Sedimentology VI. International Assoc. Sed. Spec. Pub. 28, Blackwell Science, London: 113–130.

Colman, S. M. (ed.), 1996. Continental Drilling for Paleoclimatic Records: Recommendations from an International Workshop. PAGES Workshop Report, Series 96–4, 104 pp.

Crusius, J. & R. F. Anderson, 1991. Core compression and surficial sediment loss of lake sediments of high porosity caused by gravity coring. Limnol. Oceanogr. 36: 1021–1031.

Cumming, B. F., J. R. Glew, J. P. Smol, R. Davis & S. A. Norton, 1993. Comment on "Core compression and surficial sediment loss of lake sediments of high porosity caused by "gravity coring" (Crusius and Anderson). Limnol. Oceanogr. 38: 695–699.

Cushing, E. T. & H. E. Wright, 1965. Hand operated piston corers for lake sediments. Ecology 46: 380–384.

Cushing, S. J. Desjardins & J.-M. Fillion, 1997. Parachute-assisted gravity sediment corer (Algonquin Corer). J. Paleolim. 18: 307–311.

Dokken, Q. R., R. C. Circé & C. W. Holmes, 1979. A portable, self supporting, hydraulic vibracorer for coring submerged, unconsolidated sediments. J. Sed. Petrol. 49: 658–659.

Dott, E. R. & D. Mickelson, 1995. Lake Michigan water levels and the development of Holocene beach-ridge complexes at Two Rivers, Wisconsin: stratigraphic, geomorphic, and radiocarbon evidence. Geol. Soc. Amer. Bull. 107: 286–296.

Emery, K. O. & R. S. Dietz, 1941. Gravity coring instrument and mechanics of sediment coring. Bull. Geol. Soc. Am. 52: 1685–1714.

Emery, K. O. & J. Hülsemann, 1964. Shortening of sediment cores collected in open barrel gravity corers. Sedimentology 3: 144–154.

Esker, D., R. E. Sheridan, G. M. Ashley, J. S. Waldner & D. W. Hall, 1996. Synthetic seismograms from vibracorers: a case study in correlating the late Quaternary seismic stratigraphy of the New Jersey inner continental shelf. J. Sed. Petrol. 66: 1156–1168.

Evans, H. E., D. C. Lasenby & P. J. Dillon, 1986. The effect of core compression on the measurement of zinc concentrations and anthropogenic burdens in lake sediments. Hydrobiologia 132: 185–192.

Flower, R., D. T. Monteith, A. W. Mackay, J. M. Chambers & P. G. Appleby, 1995. The design and performance of a new box corer for collecting undisturbed sample of soft subaquatic sediments. J. Paleolim. 14: 101–111.

Gilbert, R. & J. Glew, 1985. A portable percussion coring device for lacustrine and marine sediments. J. Sed. Petrol. 55: 607–608.

Glew, J., 1988. A portable extruding device for close interval sectioning of unconsolidated core samples. J. Paleolim. 1: 235–239.

Glew, J., 1989. A new trigger mechanism for sediment samplers. J. Paleolim. 2: 241–243.

Glew, J., 1991. Miniature gravity corer for recovering short sediment cores. J. Paleolim. 5: 285–287.

Glew, J., 1995. Conversion of shallow water gravity coring equipment for deep water operation. J. Paleolim. 14: 83–88.

Håkanson, L. & M. Jansson, 1983. Principles of Lake Sedimentology. Springer-Verlag, New York, 316 pp.

Hessler, R. R. & P. Jumars, 1974. Abyssal community analysis from replicate box cores in the central North Pacific. Deep Sea Res. 21: 185–209.

Hongve, D. & A. H. Erlandsen, 1979. Shortening of surface sediment cores during sampling. Hydrobiologia 653: 283–287.

Hsü, K. J. & K. R. Kelts (eds.), 1984. Quaternary Geology of Lake Zurich: An Interdisciplinary Investigation by Deep-Lake Drilling. E. Schweizerbart'sche Verlagsbuchhandlung, Contributions to Sedimentology 13, 210 pp.

Huttunen, P. & J. Meriläinen, 1978. New freezing device providing large unmixed sediment samples from lakes. Ann. Bot. Fennici 15: 128–130.

Hvorslev, M. J., 1949. Subsurface Exploration and Sampling of Soils for Civil Engineering Purposes. American Society of Civil Engineers, Waterways Experiment Station, Corps of Engineers, U.S. Army, Vicksburg, Mississippi, 521 pp.

Imperato, D. P., 1987. A modification of the vibracoring technique for sandy sediment. J. Sed. Petrol. 57: 788–789.

Jenkin, B. M. & C. H. Mortimer, 1938. Sampling lake deposits. Nature 142: 834–835.

Jenkin, B. M., C. H. Mortimer & W. Pennington, 1941. The study of lake deposits. Nature 147: 496–500.

Kajak, Z., K. Kacprzak & R. Polkowski, 1965. Chwytacz rurowy do pobierania prób dna. Ekologia Polska Seria B 11: 159–165.

Kansanen, P. H., T. Jaakkola, S. Kulmala & R. Suutarinen, 1991. Sedimentation and distribution of gamma-emitting radionuclides in bottom sediments of southern Lake Päijänne, Finland, after the Chernobyl accident. Hydrobiologia 222: 121–140.

Kelts, K., U. Briegel, K. Ghilardi & K. Hsu, 1986. The limnogeology-ETH coring system. Schweiz. Z. Hydrol. 48: 104–115.

Kézdi, Á., 1980. Handbook of Soil Mechanics, Volume 2: Soil Testing. Elsevier Scientific Publishing Co., New York, 258 pp.

Kullenberg, B., 1947. The piston core sampler. Svenska hydrog.-biol.Komm.Skr. Tredge Seien Hydrografi, Bd. 1: 1–46.

Lanesky, D. E., B. W. Logan, R. G. Brown & A. C. Hine, 1979. A new approach to portable vibracoring underwater and on land. J. Sed. Petrol. 654–657.

Livingstone, D. A., 1955. A lightweight piston sampler for lake deposits. Ecology 36: 137–139.

Lotter, A. F., I. Renberg, H. Hannsson, R. Stöckli & M. Sturm, 1997. A remote controlled freeze corer for sampling unconsolidated surface sediments. Aquat. Sci. 59: 295–303.

Mackereth, F. J. H., 1958. A portable core sampler for lake deposits. Limnol. Oceanogr. 3: 181–191.

Mackereth, F. J. H., 1969. A short core sampler for sub-aqueous deposits. Limnol. Oceanogr. 14: 145–151.

Missen, J. E., 1986. A portable device for continuous sampling in soft sediment. Monash University, Department of Geography, Working Paper No. 22, 18 pp.

Molnar, T. M. & D. G. Smith, 1993. Gamma-ray logging of vibracore holes. J. Sed. Petrol. 63: 758–760.

Murdoch, A. & J. M. Azcue, 1995. Manual of Aquatic Sediment Sampling. Lewis Publishers, Boca Raton, 219 pp.

Nesje, A., 1992. A piston corer for lacustrine and marine sediments. Arct. Alp. Res. 24: 257–259.

Olsson, J., 1925. Kolvborr, ny borrtyp für upptagning av lerpov (Piston sampler, new sampler for obtaining samples of clay). Teknisk Tidskrift 55: 17–20.

Olsson, J., 1936. Methods for taking earth samples with the most undisturbed natural consistency. 2nd Congress on Large Dams (Washington, D.C.) 4: 157–161.

Pierce, J. W. & J. D. Howard, 1969. An inexpensive portable vibracorer for sampling unconsolidated sands. J. Sed. Petrol. 39: 385–390.

Piggot, C. S., 1941. Factors involved in submarine core sampling. Geol. Soc. Am. Bull. 52: 1513–1524.

Pratje, O. 1936. Bohrungen auf den ostpressischen Hafen (Borings in the East Prussian bays). Natur und Volk 66: 587.

Reasoner, M. A., 1986. An inexpensive, lightweight percussion core sampling system. Geogr. Phys. Quat. 40: 217–219.

Reasoner, M. A., 1993. Equipment and procedure improvements for a lightweight, inexpensive, percussion core sampling system. J. Paleolim. 8: 273–281.

Reddering, J. S. V. & R. Pinter, 1985. A simple, disposable core catcher for vibracoring tubes. J. Sed. Petrol. 55: 605–606.

Renberg, I., 1981. Improved methods for sampling, photographing and varve-counting of varved lake sediments. Boreas 10: 255–258.

Renberg, I., 1991. The HON-Kajak sediment corer. J. Paleolim. 6: 167–170.

Renberg, I. & H. Hansson, 1993. A pump freeze corer for recent sediments. Limnol. Oceanogr. 38: 1317–1321.

Sanders, J. E. & J. Imbrie, 1963. Continuous cores of Bahamian calcareous sands made by vibrodrilling. Geol. Soc. Amer. Bull. 74: 1287–1292.

Shapiro, J., 1958. The core-freezer — a new sampler for lake sediments. Ecology 39: 758.

Sly, P. G., 1981. Equipment and techniques for offshore survey and site investigations. Can. Geotech. J. 18: 230–249.

Smith, D. G., 1984. Vibracoring fluvial and deltaic sediments: tips on improving penetration and recovery. J. Sed. Petrol. 54: 660–663.

Smith, D. G., 1992. Vibracoring: recent innovations. J. Paleolim. 7: 137–143.

Smith, D. G., 1998. Vibracoring: a new method for coring deep lakes. Palaeogeogr., Palaeoclim., Palaeoecol. 140: 433–440.

Stephenson, M., J. Klaverkamp, M. Motycka, C. Baron & W. Schwartz, 1996. Coring artefacts and contaminant inventories in lake sediment. J. Paleolim. 15: 99–106.

Thompson, T. A., C. S. Miller, P. K. Doss, L. D. P. Thompson & S. J. Baedke, 1991. Land-based vibracoring and vibracore analysis: tips, tricks, and traps. Indiana Geol. Survey Occasional Paper 58, 13 pp.

Tiercelin, J.-J., A. Vincens, C. E. Barton, P. Carbonel, J. Casanova, G. Belibrias, F. Gasse, E. Gros-didier, F. Melières, R. B. Owen, P. Page, C. Palacios, H. Paquet, G. Peniguel, J.-P. Peypouquet, J.-F. Raynaud, R. W. Renaut, C. Seyve, M. Vandenbroucke & G. Vidal, 1987. Le demi-graben de Baringo-Bogoria, Rift Gregoro, Kenya: 30 000 ans d'histoire hydrologique et sédimentaire. Bull. Centres Rech. Explor.-Prod Elf-Aquitane 11: 249–540.

Todd, B. J., C. F. M. Lewis, D. L. Forbes, L. H. Thorleifson & E. Nielsen (eds.), 2000. 1996 Lake Winnipeg Project: Cruise Report and Scientific Results. Geol. Survey Can. Open File 3470, 842 pp.

Tratt, M. H. & R. V. Burne, 1980. An inexpensive and efficient double-tube hand coring device. BMR J. Austr. Geol. Geophys. 5: 156–158.

Vallentyne, J. R., 1955. A modification of the Livingstone piston sampler for lake deposits. Ecology 36: 139–141.

Verschuren, D., 2000. Freeze coring soft sediments in tropical lakes. J. Paleolim. 24: 361–365.

Wright, H. E., Jr., 1967. A square-rod piston sampler for lake sediments. J. Sed. Petrol. 37: 975–976.

Wright, H. E., Jr., 1980. Cores of soft lake sediments. Boreas 9: 107–114.

Wright, H. E., Jr., 1990. An improved Hongve sampler for surface sediments. J. Paleolim. 4: 91–92.

Wright, H. E., Jr., 1991. Coring tips. J. Paleolim. 6: 37–49.

Wright, H. E., E. J. Cushing & D. A. Livingstone, 1965. Coring devices for lake sediments. In Kummel, B. & D. Raup (eds.) Handbook of Paleontological Techniques. W. H. Freeman and Company, San Francisco: 494–520.

6. CORING AND DRILLING EQUIPMENT AND PROCEDURES FOR RECOVERY OF LONG LACUSTRINE SEQUENCES

SUZANNE A. G. LEROY (suzanne.leroy@brunel.ac.uk)
Department of Geography
and Earth Sciences
Brunel University
Uxbridge UB8 3PH
Middlesex, UK

STEVE M. COLMAN (scolman@usgs.gov)
U.S. Geological Survey
384 Woods Hole Rd
Woods Hole, MA 02543
USA

Keywords: drilling rig, lake, continuous sedimentary sequences, drilling platform, equipment, coring

Introduction

Three types of palaeoclimatic records form the primary basis for most long global palaeoclimatic reconstruction: marine deposits, ice cores and lacustrine sediments. Compared to the other two, lacustrine palaeoclimatic records are still underdeveloped. Drilling and coring projects focused on palaeoclimatic records derived from lake sediments would benefit from a more collaborative approach with the development of relevant new technologies more like their marine (ODP, Ocean Drilling Program) and ice-core (GISP, Greenland Ice Sheet Project, and GRIP, Greenland Ice Core Project) analogues.

At scientific workshops, discussions have emerged on existing drilling equipment — and their limitations — for retrieving long continuous sections of soft sediments from lakes and the difficulty of obtaining published information. What information exists is mostly informal and not widely available. In order to address this deficiency, recommendations for various scientific and technical aspects of lake drilling, including site choice policy, archiving of cores, dissemination of results and analytical methods have recently been published in a Past Global Changes (PAGES, a core project of the International Geosphere-Biosphere Programme, IGBP) report (Colman, 1996; also freely available from http://www.pages.unibe.ch/publications/reports/Contidrill). In addition, the chapter by Glew et al. (this volume) discusses the retrieval of relatively short sediment cores.

In order to drill deep lakes (typically more than 30 or 50 m water depth but there is no fixed definition) such as Lake Titicaca (maximum depth: 284 m), Lake Baikal (max. depth:

W. M. Last & J. P. Smol (eds.), 2001. *Tracking Environmental Change Using Lake Sediments. Volume 1: Basin Analysis, Coring, and Chronological Techniques.* Kluwer Academic Publishers, Dordrecht, The Netherlands.

1640 m) and Lake Malawi (max. depth: 700 m), and to recover long continuous undisturbed sediment cores, international collaboration and the development of new technologies are required.

PAGES sent a questionnaire world-wide in order to inventory existing drilling systems capable of obtaining long lake records (deep lakes and/or long sedimentary records) (Leroy, 1996). In addition, a new technology, the GLAD800 drilling rig, is currently being tested under the auspices of the International Continental Drilling Program (ICDP). This program is a multinational program to further and fund geosciences in the field of Continental Scientific Drilling (http://icdp.gfz-potsdam.de). One of its interests is deep lake drilling to obtain sedimentary records relevant to all possible geological timescales.

Coring and drilling technology and operation from lake surfaces

For the purpose of this paper we consider mostly operational issues in coring and drilling from lake surfaces. Surface sediments in lakes (sediment currently being deposited) offer significant advantages for calibrating palaeoclimatic proxy signals from known modern depositional environments.

The coring or drilling method used must be tailored to individual situations owing to of the wide variety of types of lakes and character of lacustrine sediments. In the following discussion "coring" refers to single-entry sampling and/or the retrieval of relatively short sections (<15–20 m; Glew et al., this volume), whereas "drilling" refers to multiple-entry systems that recover a series of core sections to relatively great depth. These definitions, however, remain flexible. In sampling lacustrine sediments there are three primary variables that determine the type of coring or drilling system that is appropriate: water depth, penetration of sediment required, and type of sediment.

Whereas the second variable depends on the aims of the project, the first is limiting by the additional weight of the extension rods, or drill string required to span the water column. It is also limited by the additional time to lower and retrieve the coring. Re-entering the drill hole necessitates the use of a casing acting as a guide from the water surface to the sediment surface.

Sediment type affects the type of coring or drilling system used and the depth of sediment penetration that is possible with any given system. Lacustrine sediments more than 15–25 m below the sediment surface are generally at a depth and degree of compaction that requires drilling, although hydraulic piston cores can commonly go to depths of 50–200 m before percussion, vibration, or rotation are required. Percussion or vibratory coring or drilling devices are also commonly required for thick sand or gravel layers, chemically precipitated (indurated) sediments, buried soil horizons and volcanic ashes.

Factors that affect the quality of cores (an ideal core preserves all the sediment along with its structures) obtained by drilling on lakes include: 1) stability (heave and horizontal position) of the drilling platform, which is a compromise between three competing factors: stability, portability and cost; 2) presence of sand and gravel, tephras, or indurated layers such as palaeosols or evaporite layers; 3) presence of gas in the sediments — degassing commonly destroys sedimentary structures and in extreme situations, may constitute a safety issue; 4) overlying materials, such as outwash or till, that are difficult to sample; 5) remoteness of location, which affects many of the drilling logistics; and 6) experience and competence of the drillers.

Commercial drillers tend to be oriented towards making holes rather than obtaining cores. The need for high-quality cores distinguishes scientific drilling from commercial drilling. A head drilling engineer who fully understands both the scientific aim of the project, as well as the drilling techniques being used, is extremely valuable. Scientific drilling operators or contract drillers with experience in scientific drilling are desirable.

In large-scale drilling projects, such as those addressed here, the primary technological and operational consideration must be retrieving the drill core in a manner as undisturbed as possible and preserving it in as close to its original condition as possible. It makes no sense to invest large amount of effort to only obtain cores of compromised quality. These considerations apply particularly to soft, near-surface sediments and to the methods used to recover deeper drill core sections.

Adaptation from ODP technology and the development of a new drilling rig

Although a wide variety of drilling systems are applicable to lacustrine sediments, new coring/drilling systems need to be investigated for use in lakes. The most successful model used to date has been the one used by ODP (see below). Some commercial drilling systems could also be adapted and re-designed specifically for lake sediments. It is even possible that lessons could be learned from the ice core drilling operation.

It is apparent that drilling and coring systems for continental drilling, especially in lakes, need to be much more versatile than ODP systems. Variations in lithology of continental sediments require multiple approaches, potentially including hammering, percussion, vibration and rotation.

The recently developed GLAD800 system is a joint venture between the International Continental Scientific Drilling Program (ICDP) and Drilling, Observation and Sampling of the Earth's Continental Crust (DOSECC) , Inc. DOSECC, operator of the GLAD800 system, is a non-profit corporation controlled by a consortium of American universities, which exists to enhance continental scientific drilling. In general, the designs conform to the Ocean Drilling Program (ODP) standards for core quality, recovery and archiving (Kelts, 1998). A series of modifications have been developed from the ODP type high-quality coring tools to recover undisturbed sediment, notably a version of the advanced hydraulic piston corer. Hydraulic piston corers modelled on the ODP version have been included in both the NEDRA lake drilling system and the GLAD800 system, both of which are described below.

Drilling systems

Commercial drilling systems may be useful for some continental drilling projects for palaeo-climatic research, if proper care and arrangements are made to maximise core recovery and quality. Core loss or disturbance from commercial drilling operations most commonly result from a failure to use a core liner (using instead split-spoon or extrusion recovery of the cored material) or from vibrational or rotational disturbance from the coring tool itself. The latter type of disturbance is most severe for soft sediments near the top of the drill hole; other coring methods can often be used in combination with drilling to obtain a composite, minimally disturbed drill core. The use of core liners (thin, transparent plastic pipe inserted inside the core barrel is most useful) is strongly recommended; a variety of wireline, hydraulic and hollow-stem-auger drilling systems can accommodate liners.

The primary technological constraints on drilling operations in lakes are related to the weight of the drill string and the stability of the drilling platform (positioning and heaving). The weight of the drill string is determined by the type of material (e. g., steel, aluminium) used, the water depth and the amount of sediment penetration. This weight affects a variety of technical and logistical factors, including the size of the drill rig, the size of the drilling platform, the type of winch required and others. For systems that use casing or a drill string, the drill rig must be kept in place over the drill hole. The positioning and stability considerations for such systems are discussed in the next section.

Drilling platforms and positioning

A variety of drilling platforms, from large barges to canoes, are available for drilling on lakes. The larger platforms, such as barges or sophisticated rafts, necessary for large-scale operations, are expensive. They can hold large drilling rigs as well as support facilities, such as cooling containers.

For many lakes, portable, lightweight platforms are desirable because of access and logistical considerations. Catamaran-style construction is the simplest and provides a central working aperture (moon hole), which minimises stability problems. Rafts need to be repairable; modular cell systems are useful. Aluminium is excellent for strength, weight and durability, but not as easily repaired (by welding) in remote locations. Successful raft designs include: (1) the GeoForschungsZentrum system, which uses modular heavy-duty plastic cells that are bulky but light and strong, easily assembled and very stable and buoyant (Fig. 1); they are of Austrian manufacture and solve many of the problems associated with raft construction but are relatively expensive; (2) an aluminium frame system with inflatable buoyancy used by a South African group on Pretoria Salt Pan (Partridge et al., 1993) ; this system is light and small enough to be transported by air but is subject to problems of loss of buoyancy if holed and requires calm weather conditions; (3) the system of aluminium/polystyrene sandwich modules used by the Limnological Research Center at the University of Minnesota, which provides both constructional strength and buoyancy.

In addition to the platforms listed above, the Jet-Float system has been used successfully form a platform of about 4×4 m up to 6×6 m. The stability of the platform varies with its size. It can only be used when there is minimal waves activity.

In relatively small or shallow lakes, sufficient positional stability can be obtained by anchoring a raft or other platform, or by tying it to shore. However, the need to avoid significant heave may require calm weather. Heave is minimised by large platform size and by site locations with small wind fetch. In cold regions, thick winter ice can provide sufficient stability and platform capacity, either directly, or by trapping a barge in position.

A major technological barrier exists for large, deep lakes that do not freeze solidly. The necessary platform stability for such lakes can be achieved by dynamic positioning, which typically keeps the raft in place over a transponder on the bottom. Robust, automatic systems, such as that used by the ODP drill ship, are very expensive and require deep water. To our knowledge, they have not been used in major drilling projects on lakes. However, simpler, manual, fair-weather systems are possible and can be very effective in the hands of a skilled operator.

Figure 1. The GeoForschungsZentrum platform system, which uses modular heavy-duty plastic cells, at Meerfeldermaar, Germany (photo A. Brauer).

Accurate site location can be established by using Global Positioning System (GPS) units. Simple, inexpensive receivers can achieve accuracy of less than 10 m and differential GPS systems (using a second receiver fixed at a known location) can achieve accuracy of less than one meter.

Procedure recommendations

Virtually no drilling method can recover undisturbed cores of soft, near surface lake sediments. Each lake-drilling project should thus include box cores, multi-cores and (or) freeze coring of the uppermost meter or so of sediment (see Glew et al., this volume) to provide material for calibration studies. Surface or core-top sediments are typically in high demand for such studies and box cores provide the necessary large volumes of sediment. Depending on the drilling method used and the degree of consolidation of the sediments, the upper 3–10 m of sediments should be sampled by multiple gravity or piston cores. As noted previously, efforts to recover undisturbed cores of the upper part of the section are relatively inexpensive and are critical to the overall success of a drilling project. Skimping on this aspect would make little sense.

Duplicate drill holes (offset and parallel) are recommended to cover gaps between core segments (segment depths should be staggered between two drill holes) and to maximise the amount of material for analytical and dating studies. In some situations, more than two holes may be justified.

Several general procedures can be applied to most types of drilling operations and serve to maximise core recovery and minimise core disturbance. They are oriented toward cased-hole, wireline methods, but can be adapted to many other methods, such as hollow-stem augering. They are: 1) the use of casing wherever the stability of the drill-hole walls is in question (wherever sediments are less than fully consolidated); 2) the coring drive should always precede lowering of the casing in order to minimise disturbance; 3) if the coring stroke is incomplete, casing drive should be adjusted accordingly; 4) the drilling mud pressure should be carefully monitored to minimise disturbance while allowing material generated by the casing lowering to be flushed; 5) liners, preferably transparent plastic pipes, in the coring tool are highly recommended; and 6) multiple cores (multiple drill holes) are relatively inexpensive (because much of the drilling cost is associated with transportation and mobilisation) and can be used to construct composite sections that solve problems such as coring gaps and gas expansion.

Complete and accurate drilling records should be kept for all aspects of the core recovery operation. Experience has shown that human memory is no substitute for thorough hard-copy records. These records should be a combined effort on the part of the on-duty drilling engineer and the scientist, and exact procedures should be agreed upon in advance. A variety of processes, including hole re-entry, incomplete recovery, compaction and gas expansion commonly lead to uncertainties about the exact depths of recovered core sections. Uncertainty about which end of the core is the top are not unprecedented. Such depth uncertainties are critical to later correlation and time series analyses and need to be minimised. Any other information related to the weather, the change of shift, the end of day's work, the condition, treatment, or changes to core sections should be carefully recorded.

Inventory of drilling equipment

A list of drilling systems focused on lake sediments was compiled as part of a PAGES effort to enhance lake drilling activities (Leroy, 1996). This list includes information on 1) drilling and coring systems and techniques; 2) platform design and construction; and 3) most rigs owned by scientific groups and some commercial companies willing to work with scientists in developing specific techniques or having already worked with a scientific purpose.

Some systems that were used to drill well-known sites are not included here. Lake Biwa was drilled by a unique system described by Horie (1987, 1991): a cylindrical drilling pipe engineering method based on the principle of a bottle floating in water. Recent drilling in a Cainozoic lacustrine basin near Fort Yukon, Alaska, by the U. S. Geological Survey, used a truck-mounted, self-contained Portadrill 524 rotary core rig (Ager, 1994).

Several institutions maintain world-wide web pages with information about their coring systems and operations. These include:

- Limnological Research Center, University of Minnesota:
  http://www.geo.umn.edu/orgs/lrc/lrc.html
- Drilling, Observation and Sampling of the Earth's Continental Crust (DOSECC)
  http://www.dosecc.org/

Multiple-entry coring devices designed to retrieve less than about 20 m of sediment are widely available and are not described here (see Glew et al., this volume). Examples include the hydraulic push corer at the Limnological Research Center of the University of Minnesota (as much as 20 m); the Merkt-Streif system (Merkt & Streif, 1970); and the modified Livingstone corer (Livingstone, 1965; Wright, 1967, 1980, 1991; Wright et al., 1965). Under favourable circumstances, these systems can recover substantially more than 20 m of sediment.

The questionnaire

In 1995, PAGES sent an electronic questionnaire world-wide in order to inventory existing drilling systems capable of obtaining long lake records (Leroy, 1996). Six drilling and one coring systems are described here in addition to the new GLAD800 rig. For four of them (NEDRA, Sedidrill, Usinger and GLAD800), at least one scientist or scientific laboratory with PAGES-related research are part of the management structure. Soiltech, Eurodrill, BIP and Tree (R) company are commercial companies, all have already drilled for scientific purposes. A compilation and an update of the answers is presented below. We appreciate that this information may change with time, but it is the best and most up-to-date information currently available. The questions are listed below.

A.- List of equipment available for coring-drilling continental sections for PAGES' purposes.

    1. What is the name of the coring-drilling system?
    2. Where is it located?
    3. What is the name of the contact person (his/her address including e-mail)?
    4. Description of the system:

        a) system type: rotation, piston, gravity;
        b) inner, outer diameter of the core, of the casing;
        c) diameter of the liner, if any, type of liner, transparent or not;
        d) method of extrusion if necessary;
        e) length of the core barrels;
        f) extension rods;
        g) tripod;
        h) power supply/ies, push force of the engine;
        i) winch(es);
        j) platform size, raft resistance necessary to extract the core out of the sediment;
        k) need of a crane;
        l) other.

B.- Transport of the equipment

    1. What is the total weight of the equipment? Total volume?
    2. Is the system portable? How many people? What is the weight of the heaviest piece?
    3. How do you usually carry the equipment from the laboratory to the working site? By truck (yours or rented)? By container? How big?

4. How are the cores transported to the laboratory (cooled, container, etc.)?

C.- Past experience

1. Where was the system used? Location name, type of lake, publication.
2. How deep can your system go? Detail the water depth and the sediment penetration.
3. Can your system work on land as well from a lake surface and from ice?

D.- Potential of the system

1. What, in your opinion, can be achieved with the system in addition to the experience you have until now (water depth, penetration, etc.)?
2. Can you work at high and low temperatures and altitudes?

E.- General information

1. How many people are needed in total to operate the system? How many people with experience in using the system are necessary?
2. Does the sediment rotate inside the core barrel? Is it possible to orient the cores for palaeomagnetic studies?
3. What would be the cost you would ask to rent the equipment as a service for scientific purposes? What is the real cost of operation?
4. What are the other laboratories equipped in the same way?
5. Any other information you think is important.
6. Any other contributor that should be contacted, any other type of equipment to be included.
7. What are according to you the major advantages and disadvantages of your equipment for the drilling-coring of PAGES-type palaeorecords? Technical or other?

The answer to question A. 3 has been placed in Appendix 1 as this information is liable to change. For question E. 3 about costs, only qualifying comments are given.

NEDRA drilling enterprise (Figs. 2 and 3)

A. 1. There are 2 systems with different characteristics: NEDRA-Baikal-600 (first dimension given below) NEDRA-Baikal-2000 (second dimension given, in parentheses)

The Federal State Unitary Enterprise "Scientific-Industrial Center for Superdeep Drilling and Comprehensive Studies of the Earth's Interior "NEDRA" (FGUP NPC NEDRA).

A. 2. Yaroslavl, Russia; operated in Lake Baikal.

A. 4. a) Single rig, riser, special tools with six coring systems available: hydraulic piston coring tool adapted from ODP advanced hydraulic piston core, conventional rotary coring/drilling (rotary with a pilot hole), hydropercussion-vibratory system, hydropercussion-rotary system, and motor-driven percussion system.

Figure 2. NEDRA-Baikal-600 drilling system at Buguldeika site, Lake Baikal, March, 1993 (photo S. Colman).

A. 4. b) Multi-tool, cased, wireline system with aluminium casing, 20 cm iØ (inner diameter).

A. 4. c) Core Ø: 56, 80 mm. Transparent liner: 63, 58 (70, 62) mm. Coring tool Ø: 178 (185) mm, core bit Ø, 212.7 mm.

A. 4. d) No extrusion in normal operations; mechanical extrusion possible.

A. 4. e) The core barrels are 2 (6.0) m-long.

A. 4. f) No extension rods.

A. 4. g) Drilling tower is 15 (28) m in height.

A. 4. h) 75 (200) kW, 220 V, 66 (140) kW drive motor; load capacity 150 (500) kN.

A. 4. i) Winch of 5.0 (30) kW for wireline retrieval of coring tool.

A. 4. j) Platform total dimensions: 6 (12) × 18 (50) × 4 (h) m; mounts on a sledge (4 pieces) 3 × 6 m or a 12 × 36 (67) m barge. Core storage: 3 × 6 × 4 m.

A. 4. k) Drilling tower used instead of crane.

A. 4. l) System capable of drilling a wide range of materials, from unconsolidated sediment to crystalline rock; coring and drilling tools available for many conditions.

B. 1. 300 (3000) kN.

B. 2. The heaviest piece is 150 (135) kN.

B. 3. Air, rail, or truck.

B. 4. Cores brought back refrigerated (not frozen) to core storage in Irkutsk.

Figure 3. View up the drilling tower of NEDRA-BDP 600 drilling rig (photo S. Colman).

C. 1. NEDRA-Baikal 600 system: in 1993, drilling off Buguldeika Saddle, southern Baikal, 365 m water depth, 2 100-m holes, 97% recovery (BDP-93 Baikal Drilling Project Members, 1996). In 1996, two holes drilled and cored on Academic Ridge in 333 m water depth, one 200 m and one 100 m, with about 89% recovery (BDP-96 leg II Baikal Drilling Project Members, 1997). In 1997, two holes drilled and cored on Southern Basin in 1436 m water depth, one 161 m and one 225 m (Kuzmin et al., 1998).

NEDRA-Baikal 2000 system: in 1998, three holes drilled on Academic Ridge in water depth 337 m, one 201 m, one 674 m (600 m of continuous coring), one 53 m with a recovery of 95% (BDP Project members, 2000). In 1999, two holes on Posolskaia Bank in 202 m water depth, one 113 m and one 350 m (continuous coring on 250 m) with 96% recovery.

C. 2. NEDRA-Baikal 600 system coring depth up to 100–300 m and water depth up to 500 m. NEDRA-Baikal 2000 system coring depth up to 1000 m and water depth up to 1000 m.

C. 3. NEDRA-Baikal 600 system is adaptable to land, water (on a barge) and ice. NEDRA-Baikal 2000 is also adaptable to land, water (on a barge) and ice (on a barge).

D. 1. See C. 2.

D. 2. Adaptable to most conditions (+40 °C to −60 °C). Drilling crew experienced from north of Arctic Circle to central Asia.

E. 1. For 24-hour (3 shifts) operation: 12 drillers, 3 engineers and 2–3 geologists for logging and preliminary description of cores. Including 3 specialists with system work experience.

E. 2. The sediment does not rotate inside the core barrel; orientation is possible.

E. 3. System is available; cost dependent on local conditions, fuel costs and other factors.

E. 6. Dr L.A. Pevzner, Deputy Director General, Scientific Research and Dr V.V. Kochukov, Section Leader, Deep water drilling.

E. 7. Continuous, relatively undisturbed coring in deep water with high rates of recovery. Adaptable to a broad range of conditions.

SOILTECH, a division of Franki Africa Ltd

A. 1. Piston coring, Shelby tube and wireline rotary core drilling.

A. 2. The company has equipment in East and South Africa (Burundi, Tanzania and South Africa) available to drill elsewhere such as in the large lakes of Eastern Africa.

A. 4. Several types of equipment, including: drill rigs: Longyear type 44, Bonne Espérance FBE2 and other smaller rigs for shallower holes.

Drilling tools: piston coring: a wide range of Ø available. Shelby tube: this system is employed to obtain undisturbed material from soft and very soft cohesive sediments. Wireline rotary core drilling: a range of systems and Ø is available.

A triple tube core barrel is now available in wireline system. It has three nested tubes for maximum core recovery and minimum disturbance. These systems are available from well-known manufacturers (e. g., Acker, Craelius, Longyear). For example, the Acker triple tube features a 1.17 mm-thick plastic inner liner and an adjustable inner tube to minimize water contact with the core. The transparent liner contains the undisturbed core allowing for immediate examination. Plastic caps are used to seal both ends of the liner which is 1.5 m-long. A built-in check piston is incorporated in the barrel for "blowing out" the liner. Water or air under pressure enters the side ports and pushes on the piston head of the liner. A new liner is then installed through the reamer shell against the piston head and pressed back to the operating position.

A. 4. j) Raft resistance (200 m deep hole) is approximately 38.4 T, including a safety factor of 1.5. Raft positioning and stability offers some problems at present.

B. 1. Longyear 44: 2.6 T and mast and pumps: 2 T; volume: rig itself: 3 × 2 × 1.5 m, mast: 8 m. Bonne Espérance: 5 T, volume: 3 × 2 × 7 m, mast not removable. There is also a lighter drill rig for shallow boreholes.

B. 3. by trucks, train and air freight.

C.- Soiltech has extensive experience in geotechnical sampling and coring for mining purpose in a large range of ore bodies of any nature. In addition, in 1988, they drilled a 200 m deep hole in the Pretoria Salt Pan for palaeoclimatological purpose (Kerr et al., 1993; Partridge et al., 1993). In the top part of the hole piston and Shelby sampling techniques were used. The piston/Shelby samples were 0.5 m-long, requiring advancement of the casing by 0.5 m intervals as each new sample was extracted to prevent the borehole from collapsing. This process was carried out until a depth of 150 m at which stage a layer of boulders was encountered. From this level onwards, rotary wireline core drilling techniques were employed to extract continuous, undisturbed core samples from 150 to 200 m. High quality samples were obtained throughout. Soiltech have also conducted various archaeological drilling projects (including Taung & Sterkfontein Caves; Tobias et al., 1993; Partridge & Watt, 1991) where they extracted continuous cored samples, including oriented cores.

D. 1. The equipment is able to penetrate several hundred meters and should be able to drill in water depths of 50 to 100 m, and possibly more depending on the drilling platform.

D. 2. The equipment is fully adapted to work in hot African conditions.

E. 1. One experienced driller is necessary to operate the system assisted by three to four other persons. The whole project is under the supervision of a drilling manager.

E. 2. With piston coring and the Shelby tube, the core does not rotate inside the core barrel and it is possible, with a lot of care, to orient the core. With rotary drilling, it is impossible to know if the core rotates while drilling.

Depending on the nature of the geology and with new technology and equipment, they are now able to perform core orientations in vertical (or inclined) holes to any depth.

E. 3. The cost of renting the equipment and of labour depends on the type of drilling, the size of holes (Ø and depth) and the site. In addition to the price per day, there are shipping costs, raft renting and related costs. Full cost estimates for specific projects in the large lakes of East Africa can be generated. For large projects, supplementary equipment can be mobilised.

EURODRILL S. A. (Fig. 4)

A. 1. Eurodrill S. A.

A. 2. Belgium.

A. 4. a) Rotation, with rotation of the casing, not of the corer.

A. 4. b–c) Øs of core, casing and liner: version 1: casing Ø: 140 mm-lower tool Ø: 146 mm-core Ø: 100 mm-liner Ø: 106/102 mm. Version 2: casing Ø: 96 mm-lower tool Ø: 100 mm-core Ø: 63 mm-liner Ø: 64/68 mm. In a single drill hole, it is possible to use both corers: 120 m in 100 mm Ø and 100 m in 63 mm Ø.

A. 4. d) Extrusion is not necessary.

A. 4. e) The length of the core barrels in both versions (1 & 2): 1.5 m.

A. 4. f) The length of the extension rods in version: 120 cm, in version 2: 220 cm.

A. 4. g) A tripod is not necessary.

A. 4. h) High power supply: diesel engine 150 hp.

A. 4. i) 2.5 T-cable 250 m.

A. 4. j) The platform size is 9 × 7 m-raft resistance: 12.5 T.

A. 4. k) Crane: not necessary.

Figure 4. EURODRILL system being set up on land (photo G. Seret).

A. 4. 1) Tools and wireline core barrels are Craelius and Atlas-Copco equipments modified for soft or even liquid sediments.

B. 1. Total weight (truck included): 24 T. Drilling-coring equipment: 9 T (6 T for the drilling rig and 3 T for the tools and other small equipment). Estimated weight of 150 m of sediment in 100 mm Ø varies from 2 to 3 T (soft or hard rocks).

B. 2. Not portable but autonomous on caterpillar tracks.

B. 3. Equipment carried by own truck and trailer.

B. 4. Container on the truck for short cores.

C. 1. Land coring: new modified system successful in Spain (69 m — in a karstic lake; Wansard, 1996) and Jura, France (moraine-dammed lake).

C. 3. On land as well as from lake surface. Not from ice.

D. 1. Sediment: 220 m. Water depth: a maximum of c. 100 m. It may be possible to combine the GFZ Usinger corer with the EURODRILL system.

D. 2. Not possible at very cold temperatures.

E. 1. Two persons, including one with experience.

E. 2. The sediment does not rotate. Palaeomagnetic analyses possible.

E. 3. Prices depend on the version. Transport is extra.

E. 4. Are other laboratories equipped in the same way? None as far as is known.

E. 5. Equipment easily available. The system has been built by FRASTE (Italy). They can provide a detailed brochure of the technical characteristics of the system.

E. 7. High mobility: truck, trailer, broad caterpillar tracks (experienced on peat). High capacity of about 20 m/day. Equipment also available for hard rock drilling. Possible exploratory pre-coring test in destructive mode: about 100 m/day.

SEDIDRILL SPRL (Figs. 5, 6 and 7)

A. 1. Sedidrill 500.

A. 2. France.

A. 4. a) Rotation, piston.

A. 4. b) Ø from 67.5 to 100 mm for the corer, 120 mm for the casing 50 m in 119 mm iØ and 123 mm oØ. Left thread.

A. 4. c) Transparent liner available in different Ø.

A. 4. d) Extrusion not necessary with liner.

A. 4. e) Trepans with tricones. Core barrel with wireline. Mazier core barrel with rods. Piston core barrel. Up to 3 m and 4.5 m long rods when lowering and raising set of rods

A. 4. f). Drilling capacities: 200 m in destructive drilling; 110 m with a conventional Mazier or piston corer; 50 m with a wireline corer (up to 110 m in an uncased hole). 50 m in Craelius 60 mm, 100 m possible. Type N.

A. 4. g) Tripod: yes.

A. 4. h) 50 hp.

A. 4. i) With a winch.

A. 4. j) Two platforms are available. They are both composed of up to 13 elements. They can be fitted together by hand by 2 persons. For transport it occupies a quarter of a 20 ft container. The small version fits into of van.

- Platform belonging to LGQ-LBHP, size 10 × 7 m. Buoyancy capacity 72 T. Weight of the platform and the equipment on it: 12.5 T. Net buoyancy: 59.5 T. Transported by lorry and trailer. Assembling time: one day. Disassembling time: one day.

- LGQ-CEREGE modular barge. Weight: 3 T including moorings. For 30 T power. 9 × 6 m (large version) or 5 × 3 m (small version).

A. 4. k) Crane required to put platform on water.

A. 4. l) Other: Mud piston pump: maximum pressure up to 23 bars; hydraulic engine with a driving head of a power of 400 kg/m; hydraulic percussion.

B. 1. 3.5 T (including the truck), 3 m³.

B. 2. System portable by three persons.

B. 3. The drilling rig is attached to a truck.

B. 4. Refrigerated transport of cores.

C. 1, 2. Numerous deep drillings in lakes, mostly in France, since 1987. In 1990, Lac du Bouchet (France) was drilled under 27 m of water and cores up to 65 m long were recovered (Williams et al., 1996). In 1993: Lac d'Annecy, France a 50 m core was taken in 60 m of water. From the dry surface of former Bòbila Ordis Early Pleistocene lake (Spain), a 52 m-long core was taken (Løvlie & Leroy, 1995).

Figure 5. Sedidrill moved by a crane onto a platform on Lake St-Front, France (photo J.-L. de Beaulieu).

C. 3. Land, water and ice.

D. 1. The power of the engine may be increased.

D. 2. Yes.

E. 1. Two or three persons, in addition to the driller, are needed to operate the system. One person with experience is needed.

Figure 6. Sedidrill system at the site of one of the former Bòbila Ordis Early Pleistocene lakes, Spain (photo S. Leroy).

Figure 7. Sedidrill system, Mazier corer, at the site of one of the former Bòbila Ordis Early Pleistocene lakes, Spain (photo S. Leroy).

E. 2. Cores rotate but can be oriented if required.

E. 4. No other laboratories equipped in the same way.

E. 7. Extensive experience with the equipment. The driller has a good knowledge of his equipment. Sampling speed: wireline core barrel: 12 m/day in hard sediment, 15 m/day in soft sediment. Conventional Mazier core barrel: 8 m/day for the 20 first m, then 4 or 5 m/day down to 50 m, lower: 3 to 4 m/day. Piston core barrel: 20 m/day down to 30 m, then 5 m/day down to 100 m.

Tree (R) company

A. 1. Tree (R) model P8-120, made in 1999 by Electromecánica Ambiental in A. G. Chaves

A. 2. in the city of A. G. Chaves, in the Buenos Aires province, Rep. of Argentina

A. 4. The system has a vertical drill capacity of 120 m.

A. 4. a) rotation, tools from 4 to 15 inches ø.

A. 4. b) 200 mm of core Ø. In soft sediment with a casing: 100 mm.

A. 4. c) it is possible to use liners with a ø adapted to the needs

A. 4. d) extrusion by screw (it is also possible to make an hydraulic system for extrusion).

A. 4. e) length of core barrels: 1 m, but adaptable.

A. 4. f) extension rods in 4.5 m

A. 4. g) no tripod but a 7 m-long mast mounted on a trailer with a 6 m-long steel frame with 4 steering wheels.

A. 4. h) two electrical motors, 900 rpm (10 hp) and 1500 rpm (5 hp) and a gear box with 3 forward speeds and one reverse speed.

A. 4. i) two electrical winches: a heavy and low speed one and a fast one.

A. 4. j) not available.

A. 4. k) not available.

A. 4. l) three mud pumps with electrical motors of 3, 5.5 and 12.5 hp mounted on the same frame as g). Several pumps of different capacities for auxiliary work.

Screw air compressor Sullair
Diesel engine, with 5 m^3/minute at 7 kg/cm^2 of pressure, to perform diverse drilling operations, such as hole cleaning, and down-hole hammer driving. The down-hole hammer, type Cop-4 of Cop-6 (models from the Atlas Copco Company) or similar, when adequately adapted to a drilling system, allows to take cores of large diameter ($\geq$ 4 inches).

Other auxiliary equipment
Generator of electricity for the drilling system, with diesel engine and 30 kW of capacity in three phases of 380 volts. Diverse hand tools; electrical and oxy-acetylene welding systems; water tank; jeep; truck, etc. In addition, there is a small system to take piston cores, with 16 m of steel bars, mounted on a small steel frame with 4 wheels.

B.   The weight of the drilling rig is c. 2,500 kg, including the mast. The approximate measurements are 2.10 m of width, 6 m (8 m with the mast) of length and 2.60 m of height. It is transportable on its own wheels. The weight of the compressor is c. 1.200 kg and it is also transportable on its own wheels. The Tree (R) company can transport and operate the system in any part of Argentina and in most of South America.

B. 4.   Transport of cores as required.

C.   Currently, the drilling rig is used to make drilling holes for irrigation in Argentina. The company personnel has extensive experience in coring and drilling operations in soft rocks and lake sampling for palaeoclimatic research.

C. 3.   It is possible to adapt the drilling system to operate from a platform on a lake.

D.   The system could be useful for PAGES purposes in Argentina. Specially, with the incorporation of a free-fall hammer and a down-hole hammer.

D. 1.   Currently, it is planned to take core-samples of 10 cm Ø at Salinas del Bebedero in summer 2000 in the context of the Patagonian Lake Drilling Project of PAGES. It is hoped to reach, at least, 100 meters deep, using telescopic casing.

E. 1.   The system operates with a geologist (Miguel A. González), an experienced, chief drill operator and one assistant.

E. 2.   It is possible to take oriented cores, with an adequate core barrel.

E. 3.   The cost of operation of this system changes largely according to the kind of work: rotary, sampling with a down-hole or with a free-fall hammer. Also it varies according the need (or not) to use a casing in the hole.

The diesel engines of the electrical generation and the air generation are less than the costs of similar equipments using gasoline. Usually, the largest costs are the salary of the chief and the assistant personnel and the cost of casing and other material.

E. 4.   In Argentina there are several similar machines in different private, and official institutions and different provinces. The main inconvenient is to find them.

E. 5.   The equipment is directly useful for PAGES purposes. All the personnel is well-trained to work precisely and fast in any kind of drilling, sampling and casing. The company is willing to expand and to buy a good and bigger drilling machine, to reach at least 300 m of vertical depth in large diameter. In addition, a small workshop has been set up to make any mechanical reparation and improvement.

Usinger piston corer (Figs. 8 and 9)

A. 1.   Usinger Piston Corer.

A. 2.   The most complete systems are located in: Botanisches Institut der Universität, D-24098 Kiel (BIK) and GeoForschungsZentrum (GFZ) Potsdam, both in Germany.

A. 4. a)   Piston corer.

A. 4. b)   Core Ø BIK: 38/40 mm, 53/55 mm, 78/80 mm, 96/98 mm and GFZ: 38/40 mm, 53/55 mm, 78/80 mm; casing (for raft-based coring, from platform to lake bottom), BIK: 94/100 mm, 114/120 mm, GFZ: 150/156 mm.

Figure 8. Usinger system being set up on land near one of the Eifel maar lakes in Germany (photo S. Leroy).

A. 4. c) No liner.

A. 4. d) Extrusion by pushing with a piston in a specially designed "extruder".

A. 4. e) Length of the core barrels: 1 m, 2 m.

A. 4. f) Extension rods: 1 m, 2 m.

A. 4. g) Tripod: for raft-based coring, aluminium. For land-based coring down to approximately 20 m depth, no tripod is necessary.

A. 4. h) Unconsolidated sediments down to a depth of 10 to 20 m can be cored by human power. For deeper drilling, the core barrel is hammered using a standard 2-stroke gas engine.

A. 4. i) For raft-based coring: GFZ: 2 cyl., 4-stroke, gasoline engine, hydraulic coupling, 1.6 T max. force on line; BIK: hand-operated winch, 1.2 T. For land-based coring down to approximately 20 m depth no winch is necessary.

A. 4. j) GFZ: two different rafts, the first a 5 × 5 m steel pontoon platform with wooden planks, buoyancy about 5–10 T; the second, a 6.5 × 6.5 m platform of plastic pontoons with wooden planks (Jet Float System, exactly like the one described earlier); buoyancy about 3–5 T. BIK: a 3 × 4 m aluminium platform supported by inflatable tubes (manufactured by UWITEC, Mondsee, Austria), buoyancy 3 T.

Figure 9. Raft for Usinger system at one of the glacial lakes of Mecklenberg-Vorpommern, Germany (photo S. Leroy).

A. 4. k) Crane for placing the steel platform of GFZ on the lake.

B. 1. These and the following answers depend largely on the kind of coring: raft-based coring and ice-based coring require a casing, land-based coring can be usually without a tripod and a winch. The raft-based coring equipment of GFZ weighs about 5–6 T; it fits into one standard 20 ft container. The raft-based coring equipment of BIK weighs less than 1 T. The basic land-based coring equipment (without tripod and winch; i.e., for c. 20 m of coring) weighs about 25 kg; motor hammer (if necessary): 27 kg. The weight of extension rods is 3 kg/m.

B. 2. Raft-based coring system of GFZ is portable, but transport container and steel platform require a crane. Four to six persons are needed to unload the equipment from trucks and to assemble the platform; the heaviest piece is the winch, about 130 kg. Raft-based coring system of BIK is portable by a small truck. Land-based coring system for c. 20 m of sediment, i.e., without tripod and winch, is easily portable (at least for persons). The heaviest piece is the motor hammer (27 kg), if necessary.

B. 3. Raft-based coring system of GFZ is transported by haulage contractor using for standard 20 ft container (one is a cooling container for the cores). Land-based coring system can be transported by station wagon, van, or small truck.

B. 4. Cores from the raft-based system are stored in a refrigerated container; those from the land-based system of GFZ are stored in core barrels or between PVC half-tubes after extrusion.

C. 1. Raft-based coring: Eifel Maar Lakes (Germany; Hajdas et al., 1995; Leroy et al., 2000), Lac du Bouchet (France), Lago Grande di Monticchio, Italy (Zolitschka & Negendank, 1996; Brauer et al., 2000), dry lake in Mexico, Lago di Vico (Italy), Stechlinsee (maar lakes, caldera lake, glacial lake, Germany); Lake Belau (Schleswig-Holstein, Germany), Lakes Ledniza and Swientokrzyskie (Poland), several lakes in Ireland (O'Conell). Land-based coring: many sites in Germany, Traeth Mawr, Hawes Water, Little Lopham Fen, Rosgoch Common (UK), Boelling-Soe (Denmark), Marais de Limagne (France), Inner Mongolia (Naumann, 1999).

C. 2. Maximum water depth: 55 m; maximum sediment depth: 65 m; maximum total of 78 m.

C. 3. All kinds of coring (land, raft and ice) are possible.

D. 1. Water depths up to 70 m should be possible; sediment penetration up to 75 m should be possible; for a maximum combination of the two, 90 to 100 m should be possible.

D. 2. Can possibly work at high and low temperatures and in altitude.

E. 1. For raft-based coring, two experienced persons and a total of five to six persons. For land-based coring, two to three persons, one with experience.

E. 2. Definitely no rotation; until now no oriented cores taken, but in principle this should be possible.

E. 3. For GFZ operational costs in m of core depend on water depth and sediment penetration. The transport of equipment is usually in addition. The BIK rents its equipment free of charge, transport excluded, with the condition that correct use by experienced operators is guaranteed and that all components lost or damaged will be replaced by components manufactured at Kiel.

E. 4. Geological Survey of Denmark and Greenland, Copenhagen, Denmark; Institut für Geographische Wissenschaften, Berlin, Germany; Geologishes Institut der Universität Köln, Germany; Institut für Ur- und Frühgeschichte, Kiel, Germany; Instituto de Geofisica, UNAM, Mexico.

E. 7. Precise, high quality cores without disturbance; when 2 or 3 parallel cores are taken, long continuous sediment profiles are obtained. Operation limited by water depth.

Bureau for investigation and prospection (BIP)

A. 1. BIP/Geoprobe (Bureau for Investigation and Prospection).

A. 2. Belgium.

A. 4. a) Hydraulic percussion and rotation.

A. 4. b) iØ, 50 mm; oø, 78 mm; other Ø possible.

A. 4. c) Liner Ø, 50 mm; transparent or opaque; other Ø possible.

A. 4. d) Extrusion not necessary.

A. 4. e) length of the core barrels: 0.5 to 2 m.

A. 4. f) The extension rods are 90 to 200 cm-long.

A. 4. g) No tripod; hydraulic mast with maximum height of 3.9 m.

A. 4. h) 25 hp; pull-out strength 12.5 T.

A. 4. i) No winch.

A. 4. j) 5 × 4 m; raft resistance: 20 T.

A. 4. k) No crane.

A. 4. l) Waterproof piston core barrel allows good recovery of very soft sediment.

B. 1. 1.5 T; 2.1 × 1.2 × 1.0 m (folded and without vehicle, which is 4 T).

B. 2. No; heaviest piece: 844 kg.

B. 3. On a four-wheel drive vehicle.

B. 4. Cooled.

C. 1. Never used from lake surface, but numerous drillings from land and salt pans.

C. 2. 25 m water depth; 20 m sediment penetration.

C. 3. Land, lake surface, ice, salt pan.

D. 1. Sediment penetration depends mainly on the nature of the sediment.

D. 2. −10 °C to 45 °C; up to 4,500 m altitude.

E. 1. One experienced driller is necessary to operate the system; one additional person for logging the core.

E. 2. No rotation of the sediment inside the core barrel; orientation of the core is possible.

E. 3. In addition to the price per day, there also is shipping costs and raft rental.

E. 4. There are no other laboratories similarly equipped.

E. 7. High mobility: installation and full deployment of the probe in two minutes. Rapidity: usual performance is 30–40 m/day.

GLAD800 (Global Lake Drilling 800) drilling system

The system has the following characteristics (drill rig and barge photos are posted on the following website: http://www.dosecc.org/GLAD800):

- The drill rig can operate from a variety of types of barge, or can be used on land as a typical diamond core rig for hard rock.
- The barge that will be part of the basic system is intended to be anchored, allowing it to be deployed in waters up to c. 200 m deep.
- Initially, there will be no heave compensation built into the drilling rig.
- The rig will collect cores of 6.2 cm diameter up to a depth of 800 m (water + sediment). Modified to use a smaller diameter drill string, the system could drill to 1200 m or more.

Figure 10. GLAD800 drilling system, including the barge R/V Kerry Kelts, on Great Salt Lake in August, 2000. Photograph by M. Pardey

Surprisingly, it was found that weight and cost do not differ greatly between a rig with 800 m drill string capability and one with 400 m or 200 m, thus it was decided to aim for the 800 m design. This system would not require active heave compensation nor 24 hour shifts nor maintaining operations in bad weather. Pipe and crew would ferry from shore daily. Because an 800 m hole may require only a few days to core, waiting out weather is a cost-efficient option.

A. 1.   GLAD800.

A. 2.   Salt Lake City, Utah, USA.

A. 4.   The general configuration of the system consists of a modified Christensen CS1500 diamond coring rig supported by a barge. The GLAD800 Rig Specifications are posted on the following website: http://www.dosecc.org/GLAD800.

Depth Capacity Coring (Wireline or Conventional)

      HMQ Wireline 1350 m
      DLS Wireline 800 m

Hoisting Capacity

      1- Main

Capacity: single Line-Bare Drum 7,955 kg
Double Line-Bare Drum 15,900 kg
Line Speeds: Bare Drum 40 m/min
Cable Size: 33.6 m × 15.9 mm
2- Wireline
Capacity: Single Line-Bare Drum 1,136 kg
Single Line-Full Drum 382 kg
Line Speeds: Bare Drum 119 m/min
Full Drum 984 m/min
Cable Size: 975 m ×9.5 mm

Feed System

Feed Travel: 3.5 m
Feed Speeds: Fast and Slow with Variable control
Thrust: 6800 kg
Pull: 13,600 kg

Power Unit

Mfg: 1 each Cummins
Power: 175 hp (196 kW)
rpm: 1,800
Engine Type: 6 cyl. Diesel Turbocharged/after cooled complete with clutch
Cooling: Water

Drillhead and Spindle Speeds

Power: Hydraulic Motor — Variable speed/reversible

Mud Pump Hydraulic Driven — Standard Equipment

Core barrel length: 3 m

Iø of liner: 62 mm (ODP standard)

A crane is needed to load barge containers into the water.

The barge: the drill rig will be capable of being mounted on a variety of barges. The one that will be included with the basic GLAD800 system consists of eight 20 × 8 ×8- ft standard shipping containers, with secondary buoyancy (foam or air bladders) inside. The containers will be arranged in a 3 × 3 pattern (60 × 24 ft) with the middle one missing to form a moon pool. In addition to the drilling rig, the barge will hold a rod rack and containers for science, drilling, and shop operations. This platform will provide 77,000 kg of total lift is thought to be adequate for 85% of the world's lakes — those with drill sites in less than about 200 m of water and with conditions that would not generally require heave compensation.

B. Rig Weight: 6,363 kg; Recommended Truck GVW: 14,545 kg.

The drilling system is highly modularised and fit into four or five 20- ft containers in addition to the eight 20- ft containers that comprise the barge.

C. Projects on which the GLAD800 system will be used include the following:

- Koolau Scientific Drilling Project (Hawai) — Drill rig only, May, 2000.
- Great Salt Lake Drilling Project (USA) and Bear Lake Drilling Project (Idaho/Utah, USA). — Test of complete drilling system, August, 2000.
- Tentative future operations: West Hawk Lake, Canada; Lake Titicaca, Bolivia; Lake Malawi, Malawi.

E. 1. two-three scientists and two-three drillers should form an adequate site team, depending on conditions.

E. 2. The sediment does rotate inside the core barrel. It is not possible to orient the cores for palaeomagnetic studies.

E. 3. Access to the drilling rig is via application to DOSECC. Operation of the system is managed by a coordinating committee consisting of one representative each from DOSECC, ICDP and PAGES. Partial funding for drilling operations is available through competitive proposals to ICDP, and ICDP-funded proposals have first priority for use of the system.

E. 4. There are no similarly equipped laboratories.

E. 7. The GLAD800 was developed in response to a need to collect long cores from modern lakes. This system was designed in cooperation with the palaeoclimatic community and PAGES to satisfy their scientific requirements with an easily transported system that includes a drill rig and modular barge.

A brief comparison of the techniques

This coring and drilling equipment survey describes systems located in all continents with the exception of Asia. Many more commercial rigs that could be adapted for scientific drilling are certainly available. The rigs described here can drill through tens to hundreds of meters of water and through similar thicknesses of sediment. The number of persons necessary for drilling and logging varies from 2 to 6, in every case with a minimum of one experienced drilling person. The large operation represented by the NEDRA system requires a team up to 18 person strong and is in a different league to the others. All the systems described use multiple-entry coring and most use core barrels with liners. The inner diameter of the liner is usually comprised between 50 and 106 mm. The core barrel length varies between 0.5 and 3 m, with the exception of the NEDRA system which is up to 6 m. It must be noted that for practical purpose the liners are usually cut in 1.0 to 1.5 m-long sections when returned to the laboratory.

Environmental protection and safety issues

Recently, it has also become the role of the chief driller or chief scientist to advise participants in drilling programs regarding environmental protection (mostly noise and oil pollution) and safety issues. Both can become major issues when using such technologies. Beside preliminary training of all personnel and extreme care during each operation, it is required that everyone wear gloves, hard hats, steel-tipped boots, and if an engine is involved: ear protectors.

Conclusion

Palaeoclimatic records derived from continental sediments are clearly a high priority for climate and global change research. Drilling and coring projects focused on palaeoclimatic records derived from lake sediments would benefit from an approach more like their marine and ice-core analogues than has been their tradition. They need to be larger, broader and better organised if the full potential of lacustrine records of palaeoclimate is to be realised.

New technological developments are required to facilitate lake drilling especially by decreasing its price, time and humanpower requirements. The volume and weight of the drilling systems make them difficult to transport other than by trucks. In consequence road access to many lakes especially in countries with wild areas remains difficult. For palaeomagnetic studies, a secure orientation of the sediment within the core barrel should be found. Light-weight platforms with self-positioning and anti-heaving systems still have to be developed.

Summary

Owing to the increasing interest in continental drilling in retrieving palaeoclimatic archives from extant lakes, a compilation and description of drilling equipment available for continuous and undisturbed sampling (>50 m) of soft sediment from a platform floating on a lake surface is provided. For those systems operated by scientific institutions, at least one scientist with PAGES-related research experience is part of the management structure for the drilling equipment. In the case of commercial companies, an interest has been indicated for cooperative scientific studies. These systems use multiple-entry coring. Most use core barrels with liners. They can drill through tens to hundreds of meters of water and through similar thicknesses of sediment. The recovered sedimentary sequence is of a quality such that palaeoclimatic multiproxy and high-resolution (less than a mm) analyses should be possible.

Acknowledgements

The following persons are gratefully acknowledged for their help in compiling this inventory: J.-L. de Beaulieu, A. Brauer, M. González, A. Govaerts, P. Guenet, B. N. Khakhaev, A. Lannoye, N. Macintosh, J. Negendank, T. Partridge, G. Seret, M. Taieb, H. Usinger and D. Williams. Thanks are also due to the PAGES office. The manuscript has been tested on J. Pilcher, T. Dewez and M. Baillie with no drilling experience. They provided constructive comments to the manuscript.

Appendix 1: contact addresses

The answers to question A. 3 (What is the name of the contact person (his/her address including e-mail)?) is in Appendix 1 as this information is liable to change. All information has been checked and is correct for year 2000.

NEDRA drilling enterprise
Dr B. N. Khakhaev, Dir. General, FGUP NPC "NEDRA", Svoboda 8/38, Yaroslavl -150000, Russia, E-mail: postmaster@nedra.ru; http://nedra.ru; Tel.: +7-852-728101; Fax: +7-852-32-84-71. Prof. D. Williams, Geological Sciences, Univ. of Columbia, Columbia SC 29208, USA. Tel.: +1-803-777-7668, Fax: +1-803-777-6610, E-mail: baikal@geol.scarolina.edu; doug.williams@schc.sc.edu.

SOILTECH, a division of Franki Africa Ltd
N. Macintosh, SOILTECH, 688 Main Pretoria Road, P.O. Box 39075, Bramley, Wynberg 2018, Republic of South Africa. Tel.: +27-11-887-2700, Fax: +27-11-887-5716; franki@iafrica.com. There is also a branch office in Dar Es Salaam.

EURODRILL S. A.
Eurodrill S. A., Rue de Glatigny 35, B-1360 Perwez, Belgium. Tel.: +32-81-656112, Fax: +32-81-657052. Prof. G. Seret, e-mail: seret@geo.ucl.ac.be; home contact: Tel.: +32-10-658077.

SEDIDRILL SPRL
Dr P. Guenet, CPIE, Rue de l'Eglise, F-19160 Neuvic, France Tel.: +33-5- 55 95 93 79; Fax: +33- 5-55 95 96 50. The machine belongs to two laboratories: LGQ and LBHP, Aix-Marseille, France. Scientific committee: Dr J.-L. de Beaulieu (Tel.: + 33-4-91288012; e-mail: beaulieu@dialup.francenet.fr), Dr N. Thouveny, Dr M. Taieb (e-mail: taieb@arbois.cerege.fr), on Technical committee: P. Guenet, D. Arnaud, J.-F. Malaterre. See also: http://www.u-3mrs.fr/cerege/

Tree (R) Company
Lic. M. A. González, Carlzon Caldenius Foundation, C.C. 289 — Sucursal 13 B, 1413 Buenos Aires, R. Argentina. Tel.: +54-11-4581 7038 and +54-2983 15 648233; Fax: +54-11-4682 4315; e-mail: mag@dlgred.com.ar.

Usinger Piston corer
Dr H. Usinger, Botanisches Institut der Universität, D-24098 Kiel. Tel.: +49-431-880 4295, Fax: +49-431-880 1527, e-mail: husinger@bot.uni-kiel.de. Prof. J. F. W. Negendank and Dr A. Brauer, GeoForschungsZentrum, Telegrafenberg A 26, D-14473 Potsdam (Negendank: Tel.: +49-331-288 1300, Fax: +49-331-288 1302, e-mail: neg@gfz-potsdam.de; Brauer, Tel.: +49-331-288 1334, Fax: +49-331-288 1302, e-mail: brau@gfz-potsdam.de).

Bureau for Investigation and Prospection (BIP)
A. Lannoye, Boulevard Neuf 3, B-1495 Villers-la-Ville, Belgium. Tel.: +32-10-656-239, lannoye@pedo.ucl.ac.be; P. Piessens: Fax: +32-71-875-313

GLAD800 (Global Lake Drilling 800) Drilling System
Dr D. Nielson (dnielson@egi.utah.edu), Dr U. Harms (ulrich@gfz-potsdam.de), or Dr
S. Colman (scolman@usgs.gov).

References

Ager, T., 1994. The U. S. Geological Survey Global Change Drilling Project at Fort Yukon, Alaska, 1994. USGS Report, 36 pp.

BDP-93 Baikal Drilling Project members, 1996. Preliminary results of the first drilling on lake Baikal, Buguldeika site, southeastern Siberia. Quat. Internat. 37: 3–17.

BDP-96 Baikal Drilling Project members, 1997. Leg II EOS v. 78, no 51, 597, 601, 604.

BDP-00 Baikal Drilling Project members, 2000. The Late Cenozoic record from Lake Baikal (results investigations of 600 m deep drilling core). Russian Geology and Geophysics 41, 1: 3–32 (in Russian).

Brauer, A., J. Mingram, U. Frank, C. Günter, G. Schettler, S. Wulf, B. Zolitschka & J. F. W. Negendank, 2000. Abrupt environmental oscillations during the Early Weichselian recorded at Lago Grande di Monticchio, Southern Italy. Quat. Intern. 73: 79–90.

Colman, S. M., 1996. Continental Drilling for Palaeoclimate Records. PAGES workshop report, series 96–4, 104 pp.

Hajdas, I., B. Zolitschka, S. Ivy-Ochs, J. Beer, G. Bonani, S. Leroy, J. Negendank, M. Ramrath & M. Suter, 1995. AMS radiocarbon dating of annually laminated sediments from Lake Holzmaar, Germany. Quat. Sc. Rev. 14: 137–143.

Horie, S., 1987. History of Lake Biwa. Institute of Paleolimnology and Paleoenvironment on Lake Biwa, Kyoto University, contribution 553, 242 pp.

Horie, S., 1991. Die Geschichte des Biwa-Sees in Japan. Universitätsverlag Wagner, Innsbruck, 346 pp.

Kelts, K., 1998. Global Lake Drilling: GLAD800. IDEAL Bulletin. Winter 1998/1999. http://lrc.geo.umn.edu/ideal/bulletin/wi98.

Kerr, S. J., I. G. Stanistreet & T. C. Partridge, 1993. The sedimentary facies record from the Pretoria Saltpan crater. S. Afr. J. Sc. 89: 372–374.

Kuzmin, M. I., G. V. Kalmychkov, A. D. Duchkov, V. F. Gelety, A. Y. Golmchtok, E. Karabanov, B. N. Khakhaev., L. A. Pevzner, N. Iochida, N. M. Baginov, Y. A. Diadin, E. G., Larionov, A. Y. Manakov, M. M. Mandelbaum & M. M. I. F. Vashenko, 2000. Gas hydrates from Lake Baikal bottom sediments. Geology of Ore Deposits 42 (1): 25–37. (in Russian).

Leroy, S., 1996. Inventory of drilling equipment for recovery of long lacustrine sequences. Annex A-1. In S. M. Colman, 1996. Continental drilling for palaeoclimate records. PAGES workshop report, series 96–4, 79–101.

Leroy, S., B. Zolitschka, J. Negendank & G. Seret, 2000. Palynological analyses in the laminated sediment of Lake Holzmaar (Eifel, Germany): duration of Lateglacial and Preboreal biozones. Boreas 29, 1: 52–71.

Livingstone, D., 1965. The use of filament tape in raising long cores from soft sediment. Limnol. Oceanogr. 12, 2: 346–348.

Løvlie, R. & S. Leroy, 1995. Magnetostratigraphy of Lower Pleistocene Banyoles paleolake carbonate sediments from Catalonia, NE-Spain: evidence for relocation of the Cobb Mountain sub-chron. Quat. Sc. Rev. 14, 5: 473–486.

Merkt, J. & H. Streif, 1970. Stechrohr-Bohrgeräte für limnische und marine Lockersedimente. Geol. Jb. 88: 137–148.

Naumann, S., 1999. Späät- und postglaziale Landschaftsentwicklung im Bajan Nuur Seebecken (Nordwestmongolei). Erde, Band 130, Heft Nr. 2, Seite 117–130.

Partridge, T. C., S. J. Kerr, S. E. Metcalfe, L. Scott, A. S. Talma & J. C. Vogel, 1993. The Pretoria Salt-pan: a 200,000 year Southern African lacustrine sequence. Palaeogeogr., Palaeoclim., Palaeoecol. 101: 317–337.

Partridge, T. C. & I. B. Watt, 1991. The stratigraphy of the Sterkfontein hominid deposit and its relationship to the underground cave system. Palaeont. afr. 28: 35–40.

Tobias, P., J. Vogel, H. D. Oschadleus, T. C. Partridge & J. K. McKee, 1993. New isotopic and sedimentological measurements of the Thabaseek deposits (South Africa) and the dating of the Taug Hominid. Quat. Res. 40: 360–367.

Wansard, G., 1996. Quantification of paleotemperature changes during isotopic stage 2 in the La Draga continental sequence (NE Spain) based on the Mg/Ca ratio of freshwater ostracods. Quat. Sc. Rev. 15: 237–245.

Williams, T., N. Thouveny & K. M. Creer, 1996. Palaeoclimatic significance of the 300 ka mineral magnetic record from the sediments of Lac du Bouchet, France. Quat. Sc. Rev. 15, 2–3: 223–236.

Wright, H., 1967. A square-rod piston sampler for lake sediments. J. Sedim. Petrol.: 975–976.

Wright, H., 1980. Cores of soft lake sediments. Boreas 9: 107–114.

Wright, H., 1991. Coring tips. J. Paleolim. 6: 37–49.

Wright, H., E. Cushing & D. Livingstone, 1965. Coring devices for lake sediments. In Kummel, B. & D. Raup (eds.) Handbook of Palaeontological Techniques. Freeman, San Francisco: 494–520.

Zolitschka, B. & J. F. W. Negendank, 1996. Sedimentology, dating and palaeoclimatic interpretation of a 76.3 ka record from Lago Grande di Monticchio, southern Italy. Quat. Sc. Rev. 15, 2–3: 101–112.

7. SEDIMENT LOGGING TECHNIQUES

BERND ZOLITSCHKA (zoli@uni-bremen.de)
Geomorphologie und Polarforschung (GEOPOLAR)
Institut für Geographie
Universität Bremen
Celsiusstr. FVG-M
D-28359 Bremen
Germany

JENS MINGRAM (ojemi@gfz-potsdam.de)
GeoForschungsZentrum Potsdam
PB 3.3-Sedimentation and Basin Analysis
Telegrafenberg
D-14473 Potsdam
Germany

SJERRY VAN DER GAAST (gaast@nioz.nl)
& J. H. FRED JANSEN (jansen@nioz.nl)
Netherlands Institute for Sea Research
P.O. Box 59
NL-1790 AB Den Burg (Texel)
The Netherlands

RUDOLF NAUMANN (rudolf@gfz-potsdam.de)
GeoForschungsZentrum Potsdam
PB 4.2-Material Properties and Transport Processes
Telegrafenberg
D-14473 Potsdam
Germany

Keywords: sediment logging, physical sediment properties, magnetic susceptibility, XRF-scanning, laminated sediments

Introduction

Climatic variability, as well as anthropogenic influences on lakes and their catchments, are reflected in the lacustrine sediment record. As a result lake sediments show a considerable downcore variability reflected in physical parameters such as water content, sediment

W. M. Last & J. P. Smol (eds.), 2001. *Tracking Environmental Change Using Lake Sediments. Volume 1: Basin Analysis, Coring, and Chronological Techniques.* Kluwer Academic Publishers, Dordrecht, The Netherlands.

bulk density, grain size, concentration of magnetic minerals and acoustic properties. Such variations are also visible in bulk geochemistry and, of course, in palaeobiological remains. Traditionally, these sediment parameters have been measured on discrete samples obtained from sediment cores with a spatial subsampling resolution of several millimetres. Moreover, analyses were mostly restricted to one profile from each site because of the time commitment needed for measurements and the sediment-"destructive" character of subsampling for these methods. However, for interdisciplinary studies investigating a multitude of parameters, several parallel sediment cores from the same site have to be investigated. To combine various results from different parallel cores, a standard profile needs to be combined from available overlapping core sections after macroscopic core correlation.

During recent years the demands on lacustrine sediments for the reconstruction of continuous and high-resolution records of the terrestrial palaeoenvironment has increased. There is not only the claim to provide precisely dated records of environmental changes with the highest possible time resolution, e.g., at least on a decadal but more often also on an annual or even seasonal time scale, but also the demand of climate modellers for such data to improve their model runs, and finally to improve predictions of future climate conditions based on a better understanding of past climatic variabilities. Additionally, with the advent of new analytical techniques requiring only small amounts of sample material and having a rather rapid sample throughput, more and more analyses are directed towards high or even ultra-high resolution subsampling. However, these methods cannot be applied to the entire recovered sediment profile, but need a focus, e.g., to a time window of special interest.

To fulfil these requirements and to be able to select the appropriate sediment section or time window for more detailed analyses, knowledge of the recovered sediment sequence is necessary prior to subsampling. Additionally, these analyses can be used to compile a continuous composite sediment record based on the best possible correlation between several overlapping core sections yielding a composite depth scale and to avoid disturbed parts of the record.

Recently new logging instruments became available to cope with these scientific demands and to provide a guideline for a focussed subsampling strategy. These instruments are capable to carry out rapid, quasi-continuous, non-destructive and high-resolution analyses of physical sediment properties. They include a core imaging system for archiving purposes and they approximate bulk sediment geochemistry or can even provide qualitative data of major element composition. The potential value for some sediment core logs to palaeolimnological research is already clear, e.g., for magnetic susceptibility (Nowaczyk, this volume). Other logging techniques might produce data of only limited applicability or of presently unclear value for lacustrine sedimentary records. In any case, the combination of available logs provides a valuable diagnostic data set. This logging chapter is intended to provide an overview about available techniques useful for rapid and non-destructive core logging of lacustrine sediments.

Brief history of use and development of logging techniques

The development of soft sediment core logging techniques was initiated by the marine sciences. As the recovery of sediment records from a research vessel is rather fast and provides many long cores to be described and analysed, it was rather obvious from the beginning that any useful parameter that can be measured continuously and rapidly and does not require extra personnel should be measured on a routine basis. Whole core logging thus can improve or even replace laborious core description work (Weaver & Schultheiss, 1990). The first non-destructive core logging began in the 1960s with bulk density using gamma ray attenuation (Evans & Cotterell, 1970; Preiss, 1968). In the 1980s, P-wave velocity logging was introduced (Schultheiss & Mienert, 1987). An example for first automation is the P-Wave-Logger developed to automatically measure and record sediments in plastic liners with fine resolution (Schultheiss & McPhail, 1989).

Gamma ray attenuation, measured with a Gamma Ray Attenuation Porosity Evaluator (GRAPE), basically is a function of sediment wet bulk density. P-wave velocity measures the compressional wave velocity in sediments and is necessary to construct synthetic seismograms for comparison with seismic reflection profiles and/or borehole velocity logs. In addition to providing data for an easier interpretation of seismic records, P-wave as well as bulk density logs enable an accurate correlation of sediment cores.

The next step was to build a multi-sensor whole core logging device that allows to combine all established techniques (e.g., P-wave velocity, gamma ray attenuation and magnetic susceptibility) on a computer-controlled stand with a conveyor or track system that allows a quasi-continuous and fully automated transport of the sediment core along the sensors (Schultheiss & Weaver, 1992; Weber et al., 1997). More recently, the whole core logger has been adapted to be applicable to split cores as well (Gunn & Best, 1998). Other new non-destructive sensors can be attached to available whole core loggers or need their own system. Already available are high-resolution X-ray scanners (Algeo et al., 1994), digital X-ray imaging systems (Migeon et al., 1999), natural gamma ray and spectral gamma ray (Hoppie & Blum, 1994) and thermal conductivity detectors (Blum, 1997), spectrophotometry in the near-UV, visible and near-IR spectrum (Balsam & Deaton, 1996; Malley et al., 1999) and optically stimulated luminescence (OSL) scanning systems (Duller et al., 1997; Poolton et al., 1996). Other systems are emerging, such as computerised tomographic imaging or high frequency acoustic imaging methods, electrical resistivity, radar scattering or neutron activation.

This development towards an improved resolution of logging provides a key element for improving fine-scale correlation between recovered cores and thus a precise determination of sediment depth (see Nowaczyk, this volume). It is also of special importance for detailed studies of climate history allowing recognition of climatically-induced high-frequency cycles where this was not previously possible due to the lack of high-resolution data. In general, these techniques are well established for the use with marine sediment records (Blum, 1997). However, their application to lacustrine sediment sequences is rather rare, mainly because logging devices are expensive systems that pay off only if a large number of cores is processed.

While the marine community is developing new systems like digital X-ray video, microwave image or computer tomographic image systems with at least millimetre resolution (Rack, 1998), the lacustrine community would very much benefit from applying systems already developed and commercially available, such as e.g., the multi-sensor core logger (Fig. 1) (Gunn & Best, 1998; Weber et al., 1997) or the CORTEX XRF-scanner (Fig. 2) (Jansen et al., 1998).

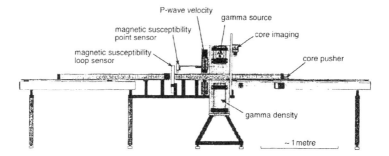

Figure 1. Multi-sensor core logging unit for gamma density, sound velocity and magnetic susceptibility. For image acquisition a linescan camera is mounted. The core, either a split or a whole core, is transported from right to left and passes the sensors one after the other (source of picture: http://www.geotek.dircon.co.uk/mscl.html). Selected results obtained with this core logger are illustrated in Figures 3–4.

Core logging devices

Core logging instruments are computer-controlled and usually come with a conveyor system driven by a track motor that carries the core along in equal steps and pass it by the different stationary sensors which scan the core as it passes (Figs. 1 and 2). The increments at which measurements can be taken are user-defined and vary for most physical data from 1 mm to any larger number. Typical sensors used with such a multi-sensor core logger are for wet bulk density (using gamma ray attenuation), sound velocity (using compressional P-wave velocity) and magnetic susceptibility (Gunn & Best, 1998; Weber et al., 1997). Commonly available are also digital RGB (red-green-blue) cameras providing colour images of the core and data sets of the three colour channels. Used increasingly is also the CORTEX scanner, an X-ray fluorescence (XRF)-scanning system developed at the Netherlands Institute for Sea Research (NIOZ). It allows qualitative determination of the bulk geochemical composition (Jansen et al., 1998). With the exception of the XRF-scanner and image acquisition cameras, which operate only with split core halves, all other mentioned parameters can be determined on split cores and on sediments within their plastic liners. In general, measurements with split cores are preferable, as they provide higher resolution readings and a better control of the scanning process (e.g., sediment disturbances like slumps, gaps or coring artefacts are recognisable). For the use with all sensors on split core surfaces, except cameras and gamma density, the sediment surface needs to be covered with a thin plastic film to prevent sensors from contamination while they touch the sediment. This close contact with the sediment surface is necessary for a proper transmission of the signal.

Ideally, sediment cores should be in the state of best preservation at the time of scanning. Gaps and cracks due to coring artefacts or due to dewatering of aged cores may cause problems, as well as holes in the sediment surface related to degassing or destructive subsampling. Core shrinkage caused by drying may also disrupt the contact of the liner with the sediment necessary to transmit the acoustic signal from the sediment surface through sediment and liner to the receiver and thus making P-wave measurements im-

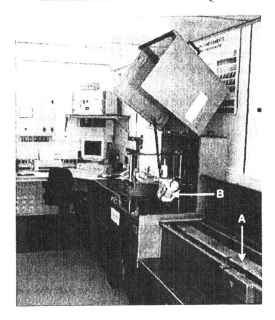

Figure 2. General view of the CORTEX XRF-scanner (top). The split core (A) is transported underneath the X-ray tube and the X-ray detector (B). The whole system is lead-shielded (see open hood) for protection of the operator. The bottom photograph shows a detail of the system with the X-ray tube (C), the helium-flushed prism (D) and the XRF-detector (B). The sensor in the back (E) controls the lowering of the X-ray unit onto the sediment surface (Jansen et al., 1998). Selected results obtained with the CORTEX XRF-scanner are illustrated in Figures 3–6.

possible. Accordingly, the best time to carry out core scans is immediately after the core has been obtained. However, if cores are kept at best possible storing conditions, they can provide good results even after several years of storage. About 30 minutes are needed for a complete run of a 1 m core segment with 1 cm spatial resolution and count rates of 10 seconds.

In addition to their application for stratigraphic core correlation and description, core logging data can also be transformed into or interpreted in terms of palaeoclimatic and palaeoenvironmental proxies.

Digital colour images

Digital images can be acquired by a succession of single shots with a digital camera (Godsey et al., 1999) or by scanning with linescan cameras providing either grey value scans or 24-bit colour images (Saarinen & Petterson, volume 2). The most recent developments with high resolution digital cameras make available core images with a resolution of up to 500 scans cm^{-1} with a linescan camera, a quality not previously attainable (Fig. 3). Commercially available spectrophotometer systems, as widely used and standardised e.g., in the Ocean Drilling Program (Blum, 1997), are also able to measure spectral data downcore. These systems, however, have the disadvantage of low resolution (several mm) and of providing no image but only data for the different spectral channels.

Continuous scanning provides a picture of the entire core that can be manipulated, e.g., it is possible to plot the individual colour channels (Figs. 3 and 4) or the whole picture can be transferred into a greyscale image. As all these data are obtained by computer-based systems, they easily are manipulable by statistical or image analysis tools or can be plotted next to other sedimentological parameters to demonstrate their relation to macroscopically visible sediment colours and structures. The time needed to scan a 1 m core section with a colour linescan camera and a resolution of 100 scans cm^{-1} is 3.5 minutes. Similar to other logging data, such colour scans are used for stratigraphic correlation. They are also a valuable addition for core archiving systems. The digital sediment colour analysis furthermore provides the only possibility to analyse and measure sediment structures like annual laminations down to a sub-millimetre scale. Thus, even high-frequency cycles can be detected in long time series within a reasonable amount of time (Schaaf & Thurow, 1997). However, for all these data, interactive data processing is necessary to remove errors related to, for example, inclined laminations, cracks or voids, and to set thresholds for outlying data to be rejected. In comparison to grey scale scans, colour scans have the advantage of additional information but need a careful colour correction, if data are to be used quantitatively.

One problem arising from core scanning techniques in general, but from colour imaging systems in particular, is the storing and archiving of the obtained data. A colour scan with a resolution of 100 scans cm^{-1} (image width: 10 cm) taken from a 1 m sediment section, for instance, needs 30 MB of disc space on the computer. Therefore, before such a system is set up for routine measurements, it is essential to think carefully about data storage and the use of databases. However, before doing so, an appropriate standard for image formats, compression algorithms, image analyses and publication should be achieved.

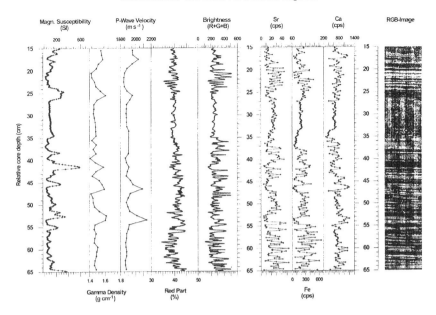

Figure 3. XRF, linescan camera, P-wave, gamma density and separate Magnetic Susceptibility (MS) logs of Holocene sediments from the western shoreline of the Dead Sea (Israel). — Distances between measurements are 1 mm for MS, 2 mm for the XRF-scan, 1 cm for Gamma density and P-wave logs and 0.1 mm for RGB colour scans. All graphs show original data except for the colour scan. These data are averaged with an 11-point running mean. Sediments consist of alternating light-coloured aragonitic and darker mixed clastic layers with some intercalations of greyish gypsum layers (Heim et al., 1997). Peaks in MS are correlated with high gamma densities, high P-wave velocities and low brightness values. These dark layers are clearly distinguishable on the RGB-image and represent increased deposition of allochthonous clastic material. Distinct peaks in brightness and in the red part of the spectrum are mostly related to pale aragonitic layers. These layers also show peaks in strontium (Sr) because Sr is preferably incorporated into aragonite. Additionally, there is one "mixed layer" at 26.5–35.5 cm depth which consists of reworked material and thus shows mean values for most signals.

Gamma ray density

Gamma ray attenuation is applied for non-destructive wet bulk density measurements as one basic parameter for physical characterisation of sediment cores (Fig. 3). In addition to density, fractional porosity or water (liquid) content of the sediment can be estimated. The method was first used as gamma ray attenuation porosity evaluator (GRAPE) within the Deep Sea and Ocean Drilling Programs (Evans & Cotterell, 1970). Today this tool is further developed towards a highly automated instrument with a spatial resolution approaching 5 mm (Gerland & Villinger, 1995).

The principle of this method is attenuation of gamma rays as they pass through the sediment. The source for gamma radiation is ^{137}Cs located in a lead container which requires special safety and health guidelines for the operator. ^{137}Cs has a half-life of 30.2 years and emits gamma rays with an energy of 662 keV. A narrow beam of gamma rays with a given intensity (I_0) is emitted and passes through the core. The intensity of incident gamma rays is altered by three major attenuation processes: photoelectric absorption, pair production

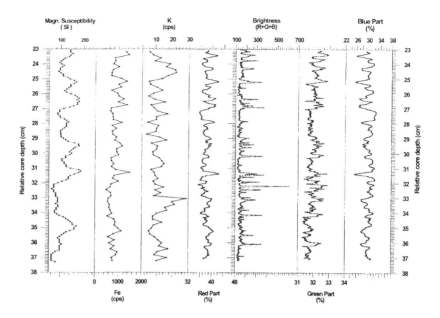

Figure 4. XRF, linescan camera and separate Magnetic Susceptibility (MS) logs of oilshale from the 45 Ma old Eocene Eckfelder Maar (Eifel, Germany). The probably annually laminated sediments consist of submillimetre-scaled organo-minerogenic (typical oilshale) laminae with intercalated and graded minerogenic event-layers and periodical enrichment of siderite (Mingram, 1998). Distances between measurements are 1 mm for MS, 2 mm for the XRF-scan and 0.1 mm for RGB colour scans. All graphs show original data except for the red, green and blue colour channels — these data are smoothed with a 21-point running mean. The most striking feature of this 15 cm long core section is the correlation between MS, iron (Fe) and the red part of the colour scan caused by siderite-rich laminae. There is also good agreement between two 6 mm thick clastic event layers (24.3–24.9 cm and 33.1–33.7 cm relative core depth) and potassium (K). These events have a mm-thick clay top-layer with related peaks in brightness around 24.3 cm and 33.1 cm but with no corresponding peaks in MS. High amounts of potassium are the result of deposition of the clay mineral illite.

and Compton scattering with the latter being the predominant factor. Compton scattering is related to the scattering of incident photons by electrons which results in a partial energy loss (Davidson et al., 1963; Evans & Cotterell, 1970). The intensity of gamma rays (I) detected after passing the core depends on the source intensity of gamma rays (I_0), the thickness (d in cm), bulk density (ρ_B in g cm^{-3}) and the mass absorption or Compton attenuation coefficient (μ in cm^2 g^{-1}) of the sediment core. The equation for calculating bulk density from gamma ray attenuation measurements is:

$$\rho_B = \left(1/-\mu d\right)\ln\left(I/I_0\right). \tag{1}$$

Intensities I and I_0 and the thickness of the core are readily available, whereas the Compton attenuation coefficient (μ) is a material constant. For most minerals in sediments, μ equals 0.0774 cm^2 g^{-1} (Ellis, 1987). However, to reduce the maximum error for bulk density measurements to $\pm 1\%$, for most marine sediments a hypothetical sediment with a porosity of 50% and a resulting μ of 0.0795 cm^2 g^{-1} was suggested (Gerland & Villinger, 1995).

Thus gamma ray attenuation mainly depends on bulk density (ρ_B). Fractional porosity or the liquid content in completely saturated samples (ϕ) can be calculated from bulk density if the sediment is fully saturated and if the density of the mineral component (ρ_g) and the fluid density (ρ_f) are known with equation (2):

$$\phi = (\rho_g - \rho_B)/(\rho_g - \rho_f). \tag{2}$$

The most important application of wet bulk density is to measure density and porosity variations of sediments to develop guidelines for sampling, to characterise the core lithology, to derive a porosity profile, but also for correlation with other physical sediment properties, and as one parameter to calculate accumulation rates. It is possible to derive equations for the calculation of water content and dry bulk density from gamma ray attenuation data (Weber et al., 1997). In general, all these data are useful parameters to examine downcore lithological changes and to correlate multiple cores from the same site. Measurement increments should not be below 0.2 cm as there is no gain of additional information.

P-wave velocity

Measurements of ultrasonic compressional wave or primary wave (P-wave) velocities (Fig. 3) are necessary to understand the physical nature of seismic profiles with their different reflectors. Changes in the impedance-depth function are the reason to reflect energy in the sediments. They are caused either by variations in P-wave velocity or in bulk density or both.

To analyse P-wave velocities from sediment cores, a pair of compressional wave transducers with an active face of 2 cm is vertically mounted on opposite sides of the core. For whole-core measurements, the transducers are spring-loaded against the liner while the core passes by. With split cores the upper transducer moves in and out of the sediment surface between measurements with only the lower one being spring-loaded and in constant contact with the liner. To ensure good contacts of the upper transducer, some of its adjustable weight is taken by the sediment in a way that the upper transducer sits on the sediment. To suit the stiffness of the sediment, it is important to adjust this loading such that the transducer leaves a track behind on the sediment surface not exceeding 0.5 mm in depth. One of the most critical aspects of measurements is the acoustic coupling between transducer faces and the liner. The liner and, if split cores are analysed, the plastic film on top of the sediment surface needs to be sprayed with water to improve the acoustic coupling.

The transducer usually is working with a transmitted pulse frequency of 500 kHz and with a repetition rate of 1 kHz. The system actually detects and amplifies the travel time of the ultrasonic compressional waves across the sediment-filled liner. The P-wave velocity (V_p in m s^{-1}) is calculated as:

$$V_p = (d_R - 2d_L)/(t_R - t_L), \tag{3}$$

where t_R is the recorded pulse travel time (in μs), t_L is the pulse delay time (in μs) related to the travel time through transducer heads and liner walls, d_R is the outer diameter of the liner and d_L is the liner thickness (both in mm). The pulse delay time needs to be recalculated at every time the type of liner is changed. This is done by measuring the P-wave travel time through a liner of the same sort as those containing the sediment but filled with distilled water.

As the thickness of the core is not only necessary for the correct calculation of P-waves, but also for the calculation of bulk density from gamma ray attenuation, core thickness is routinely measured by two displacement transducers attached to the P-wave transducers. After calibration with a solid block of known thickness (e.g., a block of aluminium), thickness determination is accurate to a precision of up to 0.1 mm. To obtain high quality and comparable results, it is important that the temperature of the cores is in equilibrium with the laboratory temperature. As even slight variations in laboratory temperature cause variations in P-wave velocity exceeding those between different sediment facies (Weber et al., 1997), it is necessary to correct all measurements to a temperature of 20 °C (V_{p20}) using an empirical approximation:

$$V_{p20} = V_{pT} + 3(20 - T), \tag{4}$$

where T is the temperature of the core when logged (in °C) and V_{pT} is the P-wave velocity (in m s^{-1}) at temperature T (Schultheiss & McPhail, 1989). The best spatial resolution for P-wave velocity measurements is about 2 cm. Little additional information can be obtained if measurements are carried out at distances less than 2 cm because the transducer face is 2 cm in size.

In addition to improving the understanding of seismic profiles and for purposes of inter-site correlations as carried out for marine sedimentary records, P-waves are applied to characterise sediment properties, such as wet bulk density, grain size or its composition (Bassinot, 1993; Breitzke et al., 1996).

Magnetic susceptibility

Magnetic susceptibility characterises the ease with which sediments can be magnetised. It varies with the content of iron-bearing minerals, especially with magnetite and other members of the iron-titanium solid solution series. As these are common in nearly all rocks and their corresponding weathering products, magnetic susceptibility can be regarded as a proxy for minerogenic contribution to the sediment (Figs. 3 and 4).

In use with the multi-sensor core-logger is a point sensor for split cores or a loop sensor for whole core measurements (Dearing, 1994). The loop sensor is stationary, while the sediment core moves through it. For different sizes of liners it is necessary to use different sizes of loop sensors as well. The point sensor is mounted on an arm extending from the P-wave transducer and they move up and down to the sediment surface together (Fig. 1). Similar to P-wave velocity, it is important that the temperature of the cores is in equilibrium with the laboratory temperature. Otherwise temperature fluctuations, such as when the core is adjusting from cool-room temperatures of +4 °C to much higher laboratory temperatures, may cause instrument drift. The spatial resolution of the magnetic susceptibility point sensor (5 mm) is the best of all sensors mounted on the multi-sensor core-logger. More details about theory and logging of magnetic susceptibility are discussed in Sandgren & Snowball (volume 2) and in Nowaczyk (this volume).

XRF scanner

The CORTEX (core-scanner Texel) is an X-ray fluorescence (XRF) system (Jansen et al., 1998) constructed for the non-destructive analysis of max. 1.5 m long split core sections

(Fig. 2a). The X-ray tube and the energy dispersive detector of this system are oriented at an angle of 45 ° (Fig. 2b). Between the tube and the sample surface, a slit is oriented such that the irradiated sample area can be adjusted to the step size of the track motor. In order to minimise the absorption of the emitted long-wave radiation from lighter elements by air, a helium-flushed hollow prism is mounted between the sample surface and the detector (Fig. 2b). Between measurements, the XRF-unit (X-ray tube, detector and prism) is lifted, the core is moved by the stepping motor and then the XRF-unit is lowered again onto the sediment surface for the next measurement. The lowering is controlled by a sensor (Fig. 2b). The CORTEX system is capable of producing qualitative counts of major elements from potassium (K) to strontium (Sr) including elements important for palaeolimnological studies such as calcium (Ca), titanium (Ti), manganese (Mn) and iron (Fe). The lighter elements aluminium (Al) and silicon (Si), which would need a much longer count rate compared to the heavier elements, are often omitted. The CORTEX produces data about the actual composition of the sediment, in contrast to gamma ray attenuation, magnetic susceptibility and colour images, which are often a multi-interpretative data set depending on a combination of sediment properties. The relative counts produced from split core surfaces correlate highly with concentrations measured with standard atomic absorption spectrometry (AAS) and thus can be calibrated (Jansen et al., 1998). They also correlate with standard XRF-data used for calibration in Figure 5.

Generally, measurements are carried out with high-resolution and within a short period of time and thus, like other data from core scans, can be used for inter-core correlation. They also generate a detailed time-series of major elements (Figs. 3 and 4), which can be applied for palaeoclimate and palaeoenvironment interpretations.

For the element Ca the XRF-scan gives information from a depth of several tens of a μm, and for Fe from a few hundredths of a μm. Therefore, thin films of ferric or carbonaceous precipitates forming on the sediment surface of split cores upon contact with air need to be removed, such that the surface is very clean. The surface also needs to be as smooth as possible. Sand grains are already large enough to produce a rough surface which attenuates the fluorescent radiation (Potts, 1987) and demands a more careful interpretation of results. This usually causes no problems with lake sediments, as their grain sizes are mainly of the silt and clay fractions.

The track motor that guides the way of the core passing the sensor is operating down to step sizes of 1 mm. However, the highest feasible resolution of analyses to be obtained is 2 mm because the slit that determines the irradiated sample surface and that has to be adjusted to the step size of the track motor cannot be closed to produce an irradiated area of <2 mm. Even if this restriction could be overcome technically, such a small exposed surface area would give prolonged counting times to obtain a statistically sound number of counts. At this resolution of 2 mm, the XRF-scanner is capable of recognising high frequency signals of the sediment and not only produces noise or an integration (smoothing) of a larger window of the sediment. This is demonstrated by three scans with different step sizes of analyses across a 0.6 mm thick copper wire placed parallel to the plane of sedimentation (Fig. 6). The copper peak related to the copper wire gets sharper with a decrease in step size, indicating that recorded high frequency variations are really related to discrete changes in sediment composition.

The time needed to carry out one measurement depends on the elemental composition of the sediment core and varies between 60 and 120 seconds. In general, standard measurements for a 1 m core section with 2 cm resolution will take about one hour. However, the time needed for measuring a 1 m core section increases to 9–10 hours with 2 mm resolution. This is a task that can be done overnight as the system runs without supervision.

Conclusion and future directions

The available quasi-continuous core logging techniques now permit researchers to obtain high resolution time series of sediment physical data. For marine records, it was demonstrated that logging data can be interpreted in terms of palaeoenvironmental or palaeoclimatic changes. This is not yet the case for lacustrine records where these methods are now just being established. However, the potential of multi-sensor core logging for palaeolimnological investigations becomes evident with some applications from the marine realm.

Applying gamma ray attenuation and P-wave measurements for sediments from the Antarctic Polar Frontal Zone, it was possible to predict the content of biogenic opal with bulk density and P-wave velocity data (Weber et al., 1997). For another set of sediments from the shelf off Bangladesh, the same authors were able to estimate the sand/silt to clay ratio using the acoustic impedance, which is the product of wet bulk density and P-wave velocity. P-wave velocities were also the focus for another study of carbonaceous sediments from the Indian Ocean. Firstly, sediment cores, as distant as 1700 km from each other, have been successfully correlated and, secondly, changes in P-wave velocity parallel changes in the coarse fraction ($>63 \mu$m) of the sediment, which is mainly composed of foraminifers. As changes in grain size are related to carbonate dissolution, P-wave velocity data provide a quick and high resolution proxy of carbonate dissolution for this depositional environment (Bassinot, 1993). For equatorial Pacific sediments, it was possible to establish an algorithm to calculate a carbonate record from gamma ray attenuation data (Mayer, 1991). With such data obtained from multi-sensor core loggers extremely high-resolution proxies of the past ocean-climate system become available for many sites. This enables modellers to look more closely at the long-term history and evolution of the ocean's response to orbital forcing and to explore its impact on insolation (Mayer et al., 1996). High-resolution magnetic susceptibility and gamma ray attenuation data made available a detailed record of peaks in physical sediment properties from the North Atlantic that correlate with the history of iceberg events and thus with Heinrich events (Chi & Mienert, 1996). Finally, a combination of stable isotope data, carbonate content and XRF-scans of Fe, Ca and Ti provided a high-resolution link between marine and terrestrial climatic signals obtained from a sediment record from the continental slope off northeastern Brazil (Arz et al., 1998).

Applied to lake sediments, these techniques open a complete new field of analyses that can be used rapidly and simultaneously along a sediment core. Such a time resolution was never reached before with standard methods of discrete subsampling. Even if only colour scanning, magnetic susceptibility and CORTEX logging are down to a resolution of 1–2 mm and the other techniques approach "only" 0.5–2 cm, such data allow new insights into palaeoecological variabilities and their forcing factors as recorded in lacustrine sediments.

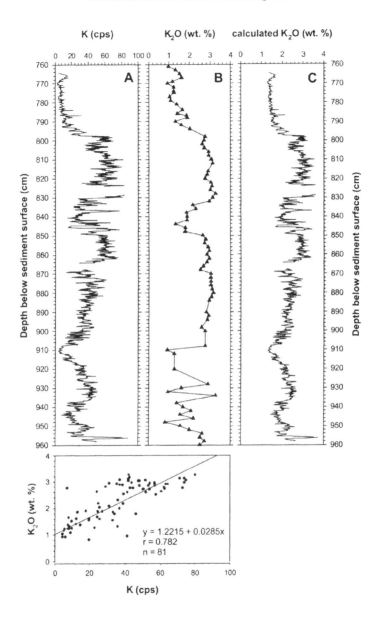

Figure 5. Two independently determined potassium (K) records of a 2 m section from organic Holocene sediments of Holzmaar (Germany) a site with annually laminated sediments (Zolitschka, 1998). A: Measurements with the XRF-scanner (step sizes: 2 mm) produced qualitative data in counts per second (cps); B: Analyses with standard XRF techniques on discrete samples (mean subsampling interval: 20 mm) produced absolute values in weight %. C: Based on a linear regression model developed with 81 data points (lower panel), counts per second can be transferred and quantified into absolute values.

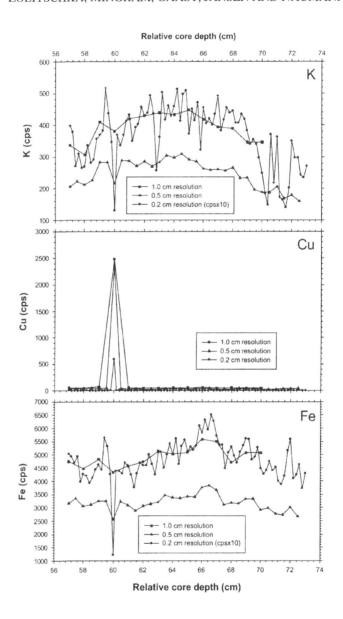

Figure 6. Potassium (K), copper (Cu) and iron (Fe) measurements obtained with the XRF-scanner from a 16 cm long sediment section of Holzmaar. For each element, step sizes of 10, 5 and 2 mm have been selected. The observed decrease in counts per second (cps) with smaller step sizes (higher resolution) is the result of a smaller sediment surface area being irradiated and count rates not being increased. The 0.6 mm thick copper wire placed on the sediment is visible as one point of high Cu values and corresponding dips in Fe and K at 60 cm relative core depth.

Summary

Sediments recovered from lakes have a considerable downcore variability of their physical parameters. In addition to palaeobiological information, such parameters provide a detailed record of palaeoenvironmental changes. New logging instruments now make available continuous, high resolution, and non-destructive physical sediment property data such as sediment colour, gamma ray density, P-wave velocity, magnetic susceptibility or elemental composition. Since many years these techniques have been used for marine sediment core investigations. Here we introduce these methods to palaeolimnological research and provide examples for their applications with laminated Tertiary and Quaternary lake sediments.

Acknowledgements

We would like to thank Rineke Gieles for technical assistance with the CORTEX XRF-scanner and Andreas Hendrich for drawing of figures. Thanks also go to Bill Last and John Smol for the idea of this project and for their patience with the slow progress. To both of them, as well as to an anonymous reviewer, we owe thanks for improving an earlier version of this manuscript. This is a contribution to the European Lake Drilling Programme (ELDP).

References

Algeo, T. J., M. Phillips, J. Jaminski & M. Fenwick, 1994. Research method papers — high-resolution X-radiography of laminated sediment cores. J. Sed. Res. A64: 665–703.

Arz, H. W., J. Pätzold & G. Wefer, 1998. Correlated millennial-scale changes in surface hydrography and terrigenous sediment yield inferred from last-Glacial marine deposits off northeastern Brazil. Quat. Res. 50: 157–166.

Balsam, W. L. & B. C. Deaton, 1996. Determining the composition of late Quaternary marine sediments from NUV, VIS and NIR diffuse reflectance spectra. Mar. Geol. 134: 31–55.

Bassinot, F. C., 1993. Sonostratigraphy of tropical Indian Ocean giant piston cores: toward a rapid and high-resolution tool for tracking dissolution cycles in Pleistocene carbonate sediments. Earth Planet. Sci. Lett. 120: 327–344.

Blum, P., 1997. Physical properties handbook — A guide to the shipboard measurement of physical properties of deep-sea cores by the Ocean Drilling Program. ODP Techn. Note 26: 1–113.

Breitzke, M., H. Grobe, G. Kuhn & P. Müller, 1996. Full waveform ultrasonic transmission seismo-grams: A fast new method for the determination of physical and sedimentological parameters of marine sediment cores. J. Geophys. Res. 101: 22,123–22,141.

Chi, J. & J. Mienert, 1996. Linking physical property records of Quaternary sediments to Heinrich events. Mar. Geol. 131: 57–73.

Davidson, J. M., J. W. Biggar & D. R. Nielsen, 1963. Gamma-radiation attenuation for measuring bulk density and transient water flow in porous materials. J. Geophys. Res. 68: 4777–4783.

Dearing, J., 1994. Environmental Magnetic Susceptibility — Using the Bartington MS2 System. Kenilworth, U.K., Chi Publishing, 104 pp.

Duller, G. A. T., L. Botter-Jensen & B. G. Markey, 1997. A luminescence imaging system based on a CCD camera. Radiation Measurements 27: 91–99.

Ellis, D., 1987. Well Logging for Earth Scientists. Elsevier, New York, 532 pp.

Evans, H. B. & C. H. Cotterell, 1970. Gamma-ray attenuation density scanner. Proc. DSDP, Ini. Rep. 1970/2: 460–471.

Gerland, S. & H. Villinger, 1995. Nondestructive density determination on marine sediment cores from gamma-ray attenuation measurements. Geo-Marine Lett. 15: 111–118.

Godsey, H. S., T. C. Moore, D. K. Rea & L. C. K. Shane, 1999. Post-Younger Dryas seasonality in the North American midcontinent region as recorded in Lake Huron varved sediments. Can. J. Earth Sci. 36: 533–547.

Gunn, D. E. & A. I. Best, 1998. A new automated nondestructive system for high resolution multi-sensor core logging of open sediment cores. Geo-Marine Lett. 18: 70–77.

Heim, C., N. R. Nowaczyk, J. F. W. Negendank, S. Leroy & Z. Ben-Avraham, 1997. Near East desertification: evidence from the Dead Sea. Naturwissenschaften 84: 398–401.

Hoppie, B. W. & P. Blum, 1994. Natural gamma-ray measurements on ODP cores: introduction to procedures with examples from Leg 150. Proc. ODP, Init. Rep. 150: 51–59.

Jansen, J. H. F., S. J. Van der Gaast, B. Koster & A. J. Vaars, 1998. CORTEX, a shipboard XRF-scanner for element analyses in split sediment cores. Mar. Geol. 151: 143–153.

Malley, D. F., H. Rönicke, D. L. Findlay & B. Zippel, 1999. Feasability of using near-infrared reflectance spectroscopy for the analysis of C, N, P, and diatoms in lake sediments. J. Paleolim. 21: 295–306.

Mayer, A., 1991. Extraction of high-resolution carbonate data for palaeoclimate reconstruction. Nature 352: 148–150.

Mayer, L. A., C. Gobrecht & N. G. Pisias, 1996. Three-dimensional visualization of orbital forcing and climate response: interactively exploring the pacemaker of the ice ages. Geol. Runds. 85: 505–512.

Migeon, S., O. Weber, J.-C. Faugeres & J. Saint-Paul, 1999. SCOPIX: A new X-ray imaging system for core analysis. Geo-Marine Lett. 18: 251–255.

Mingram, J., 1998. Laminated Eocene Maar-lake sediments from Eckfeld (Eifel region, Germany) and their short-term periodicities. Palaeogeogr., Palaeoclim., Palaeoecol. 140: 289–305.

Nowaczyk, N. R., This Volume. Logging of magnetic susceptibility. In Last, W. M. & J. P. Smol (eds.) Tracking Environmental Change Using Lake Sediments. Volume 1: Basin Analysis, Coring, and Chronological Techniques. Kluwer Academic Publishers, Dordrecht, The Netherlands.

Poolton, N. R. J., L. Bøtter-Jensen, A. G. Wintle, P. J. Ypma, K. L. Knudsen, V. Mejdahl, B. Mauz, H. E. Christiansen, J. Jakobsen, F. Jorgensen & F. Willumsen, 1996. A scanning optical sensor system for measuring the luminescence of split sediment cores. Boreas 25: 195–208.

Potts, P. J., 1987. A Handbook of Silicate Rock Analysis. Glasgow, Blackie, 622 pp.

Preiss, K., 1968. Non-destructive laboratory measurement of marine sediment density in a core barrel using gamma radiation. Deep Sea Res. 15: 401–407.

Rack, F. R., 1998. Tomorrow's Technology Today — A survey of emerging trends in non-destructive measurments for the geosciences, IMAGES standing committee on "New Technologies in Sediment Imaging", p. 1–31.

Schaaf, M. & J. Thurow, 1997. Tracing short cycles in long records: the study of inter-annual to inter-centennial climate change from long sediment records, examples from the Santa Barbara Basin. J. Geol. Soc., London 154: 613–622.

Schultheiss, P. J. & S. D. McPhail, 1989. An automated P-wave logger for recording fine scale compressional wave velocity structures in sediments. Proc. ODP, Sci. Res. 108: 407–413.

Schultheiss, P. J. & J. Mienert, 1987. Whole core P-wave velocity and gamma ray attenuation logs from ODP Leg 108 (sites 657–668). Proc. DSDP, Ini. Rep. 108: 1015–1017.

Schultheiss, P. J. & P. P. E. Weaver, 1992. Multi-sensor core logging for science and industry. Ocean 92, Mastering the Oceans Through Technology 2: 608–613.

Weaver, P. P. E. & P. J. Schultheiss, 1990. Current methods for obtaining, logging and splitting marine sediment cores. Mar. Geophys. Res. 12: 85–100.

Weber, M. E., F. Niessen, G. Kuhn & M. Wiedicke, 1997. Calibration and application of marine sedimentary physical properties using a multi-sensor core logger. Mar. Geol. 136: 151–172.

Zolitschka, B., 1998. A 14,000 year sediment yield record from Western Germany based on annually laminated sediments. Geomorphology 22: 1–17.

8. LOGGING OF MAGNETIC SUSCEPTIBILITY

NORBERT R. NOWACZYK (nowa@gfz-potsdam.de)
GeoForschungsZentrum Potsdam
Projektbereich 3.3
"Sedimente und Beckenbildung"
Telegrafenberg, 14473 Potsdam
Germany

Keywords: magnetic susceptibility, whole-core logging, split-core logging, sampling techniques, spatial resolution, moving average, frequency analysis, low-pass filtering.

Introduction

Environmental changes due to climatic variability are associated with different weathering and erosional processes, as well as with various transport and deposition conditions, resulting in varying sediment composition with varying content and/or composition of the magnetic minerals as a common constituent of the sediments (e.g., Dekkers, 1997). In a first approximation, the concentration of magnetic minerals can be monitored by measuring the magnetic susceptibility of sediments. During measurement the material to be investigated is subjected to a low magnetic field inducing a small magnetization, that will disappear again when the field is switched off. The proportional factor between magnetic field H and magnetization M is called magnetic volume susceptibility κ:

$$M = \kappa \cdot H. \tag{1}$$

Technically, susceptibilty measurements are performed relative to the ambient magnetic environment. Therefore, first the susceptibility meter is 'zeroed' against the magnetic background. Normal readings without a sample then will yield zero, unless the magnetic active part of the susceptibility meter, generally a coil sensor, is moved away from its position where it was 'zeroed'. Such measurements are non-destructive, fast and easy to perform, yielding characteristic patterns, with values often ranging over several orders of magnitude. Therefore, over the last two decades, logging of magnetic susceptibility has become one of the standard methods applied on stratigraphic sections and sediment cores from marine, lacustrine, and terrestrial environments (e.g., Mead et al., 1986, Kukla et al., 1988, Heller et al., 1991, Bloemendal et al., 1995, Williams et al., 1996, Sun et al., 1998).

Similar to other physical properties, magnetic susceptibility can be determined on un-opened as well as splitted sediment cores or sediments stored in sampling boxes. Since it often reflects changes of the bulk composition of recovered sediments, the obtained logs

W. M. Last & J. P. Smol (eds.), 2001. *Tracking Environmental Change Using Lake Sediments. Volume 1: Basin Analysis, Coring, and Chronological Techniques.* Kluwer Academic Publishers, Dordrecht, The Netherlands.

can be used for correlation of different sediment cores from the same site in order to create composite sections without data gaps, and also for correlation of sediment cores/composites from different sites. This chapter does not give information about different automatic logging systems, which are commercially available. Although such logging devices can help in a fast acquisition of high-quality data, the spectral information that can be obtained is solely determined by the geometry of the applied sensor. Therefore, this chapter mainly deals with the application of different types of susceptibility sensors, their geometric properties and physical limitations, and typical results. A more general overview of analyses of rock magnetic parameters, with magnetic susceptibility being one of them, and their relation to changes in environmental and climatic conditions is discussed in Sandgren & Snowball (volume 2). An introduction to paleomagnetic techniques is given by King & Peck (this volume).

Different methods of logging magnetic susceptibility

In principle, variations in magnetic susceptibility of sediment cores can be monitored by three different techniques. Core processing often starts with the determination of physical properties, such as bulk density, p-wave velocity, etc. (Zolitschka et al., this volume) and of course magnetic susceptibility on un-opened cores (Fig. 1, top). This '*whole-core logging*' can be performed already in the field or onboard of a research vessel or coring raft, and gives first information about relative changes in sedimentation rates and the stratigraphic length of the recovered sediment cores. After opening of the core sections, i.e., splitting them into equal halves, susceptibility can be determined with a special sensor contacting the split surface of one of the core halves (Fig. 1, bottom). This method, therefore, is termed '*split-core logging*'. It can also be performed on sub-cores, such as the so-called 'u-channels' or long boxes taken from an outcrop, provided that direct access to the sediment sequence is possible. Finally, a susceptibility record can be obtained from discrete samples taken from the outcrop, opened core, or, in special cases, from sediments extruded out of the coring tube. Such material is measured independently from the core with sensors especially designed for discrete samples (Fig. 1, right). Sometimes, all three methods are applied on the same core in the order as previously listed, since the results are of different quality and relevance at the actual state of core processing and current analyses.

Figure 2 illustrates the volume affected by the magnetic field during the different types of susceptibility determinations. The response (or weighting) functions of the sensors can be tested by, e.g., moving a thin magnetite-covered round disk through a loop sensor or a half disk passed directly below a spot-reading sensor. Obviously, during measurement with a whole-core logging sensor, at point x_p, a large volume of the sediments is affected by the magnetic field of the loop sensor, as indicated by its broad bell-shaped response function (Fig. 2, top left) reaching from x_a to x_b. The response function of a spot-reading sensor has a similar shape but is much narrower (Fig. 2, middle), so that the affected volume along the core axis is quite small. Therefore, this method yields almost spot-readings of magnetic susceptibility. The weighting function of discrete (cubic) samples, e.g., for paleomagnetic studies, and of core slices (right) are simple rectangles. While logs obtained with the whole-core and split-core methods are generally built up of overlapping readings, the core logs obtained from discrete samples (if not taken from overlapping intervals) are based on independent susceptibility determinations.

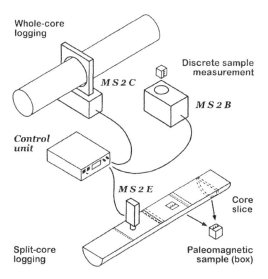

Figure 1. Sketch of the different methods of susceptibility measurements with the Bartington system as an example: whole-core logging is performed with a loop sensor (MS2C) on un-opened core sections (top), split-core logging is performed with a spot-reading sensor (MS2E) directly on the split-surface of a core section (bottom). Paleomagnetic samples and material from core slices subsequently filled into sampling boxes (lower right) can be measured with a discrete sample sensor (MS2B, upper right).

Whole-core logging

After recovery of a sediment core, normally the core is cut into sections that can be handled easily, usually of 1 m in length. For determination of the 'whole-core susceptibility' such a section is moved in equal steps through the loop coil of the susceptibility sensor (Figs. 1 and 2). The inner diameter of the loop sensor should be only a little bit larger than the outer diameter of the core liner in order to have maximum sensitivity and maximum spatial resolution. The measuring interval depends on the maximum spatial resolution that can be obtained from the applied loop sensor. The response function of a loop sensor with respect to a single thin magnetite covered disk is shown in Figure 2 (top left). A sediment core can be looked upon as a succession of many such disks, with the response function acting as a weighting function for the susceptiblity of the individual disks. Therefore, at a certain point x_p, the measuring signal $\kappa_m(x_p)$ is the integral over the real susceptibility distribution $\kappa(x)$ multiplied with the sensor response function $s(x - x_p)$, reaching from x_a to x_b, which can be mathematically described by a convolution integral:

$$\kappa_m(x_p) = \int_{x=x_a}^{x=x_b} \kappa(x) s(x - x_p) dx. \qquad (2)$$

This is equivalent to a weighted running average of the real susceptibility distribution along the core axis, or, in other words, a loop sensor acts as a low-pass filter. Figure 3 illustrates

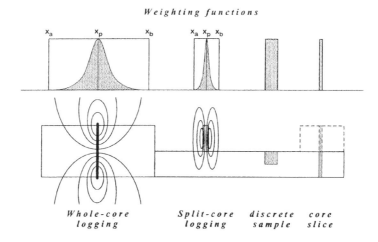

Figure 2. Illustration of the volume affected by the measuring magnetic field of the different possible susceptibility logging methods. From the measuring point x_p, the sensitivity (weighting) function of the loop sensor applied during whole-core logging and of the spot-reading sensor used for split-core logging decay with a bell-shaped characteristic to zero at x_a and x_b, respectively. Susceptibility values obtained either directly from (paleomagnetic) samples or sliced material filled into boxes are rectangular integrals over the sampled depth interval.

the geometric effects that dominate the process of the whole-core logging (note that for reasons of clearity, the response function is enlarged). The dotted rectangle represents air whereas the black rectangle represents a core section. The top (bottom) of the core section is equal to the left (right) end of the rectangle. In Figure 3a the core section is just outside the sensor's sensitivity volume and the reading would be zero, because the sensor is 'zeroed' against air. When the section approaches the sensor (Fig. 3b) with its top end still outside the loop coil, the sensor already 'feels' the sediments, giving a reading different from zero. When the top end of the section is in the centre of the sensor (Fig. 3c), the reading integrates to one half in air and to one half over the top end of the core section. If the section has a constant susceptibility value, the reading would be half of that value. In Figure 3d the section's top end has passed the sensor's centre but it has not yet reached the other boundary of the sensor's sensitivity volume. Therefore, the resulting reading is still influenced by air. Only when the whole response function covers the core section (Fig. 3e) is the reading not biased by the partly measurement of air.

Now, the dotted rectangle shall represent a core section and the black one shall be air (Fig. 3). Then, the succession from Figure 3a to e represents the susceptiblity logging across the bottom end of a core section. As long as the section is covered by the sensor's response function, the readings are not air-biased (Fig. 3a), whereas the values get more and more influenced by the contribution of air (Fig. 3b–d) until the section has left the sensors's sensitivity volume (Fig. 3e).

A third way of looking at Figure 3 is that both the dotted rectangle and the black rectangle are two sections of a core. In this case, the measurements in Figure 3b to d are influenced

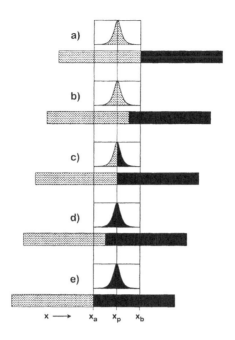

Figure 3. Moving a core section (black) through a loop sensor. The sensitivity volume of the loop sensor reaches from x_a to x_b. Measurements are taken at point x_p. a) Core section just outside sensitivity volume and sensor measuring air only. b) Section top within sensitivity volume of sensor but still outside sensor coil. c) Section top in the centre of the loop sensor measuring half air half core section. d) Section top has passed coil but sensor still partly measuring air. e) sensitivity volume fully covering the core section.

by both sections. The part of the integral that is not covered by the left (dotted) section is 'filled up' by the contribution of the right (black) section and vice versa. So, one possibility to avoid air-biased readings at the ends of the individual core sections is to physically attach the following core section to the one being measured (which is practically done during operation of some commercial logging systems). But this might cause problems, since, in general, the ends of the sections are covered by caps of a certain thickness so that the sediments to be measured cannot be attached completely to each other. A more elegant way is to log the sections separately but starting at a distance that is at least half of the length of the sensor's sensitivity volume in front of the section's top end (Fig. 3a–c for the black section), and stop logging at the same distance behind the section's bottom end (Fig. 3c–e for the dotted section). By superimposing the partly overlapping logs of the individual sections, with respect to a common depth scale, a complete unbiased core log can be obtained. Only the very top and bottom of the whole core is still air-biased. However, if the internal depth scales of the individual core sections are not adjusted correctly, this method might introduce artificial peaks, in case of 'overlapping' core sections (due to erroneous depth scales), or troughs, in case of a positive offset at the core breaks.

Another effect that might occur is drifting of the sensor's level. Due to temperature changes, the working level of the sensor, usually zeroed against air, is drifting, i.e., readings in air do not yield zero anymore after a certain amount of time. In a first approximation, the drift can be looked upon as being linear. In order to detect this possible drift during logging, the individual logs should start even before reaching the sensor's sensitivity volume and should also be extended after leaving it. These two sets of readings should be averaged and fitted with a linear regression. The resulting line then must be subtracted from the whole section log. This processing step must be performed prior to superimposing the section logs to a complete core log. Additional drift readings might be obtained by removing the core section out of the sensor's sensitivity volume at regular intervals.

Whole-core measurements yield informations about relative changes in sedimentation rates and stratigraphic length of different cores even before opening of the cores. This enables the investigators to decide which cores should be processed first. However, since this logging method integrates over quite a large volume, the spatial resolution is the lowest of all methods generally applied on sediment cores. Moreover, the obtained records might be affected by unrecognized disturbances due to the sedimentary enviroment (e.g., slumps) or caused by the coring process itself such as suction phenomena in cases where a piston corer were used. Even clear core liners might not give a sufficient insight to disturbances, since often material is smeared along the inner wall of the tube. This material is also closest to the measuring coil of the loop sensor and, therefore, at least theoretically, decreases the quality of the whole-core logging data.

In principle, the spatial resolution of whole-core measurements can be improved by deconvolution, i.e., inverse filtering of the obtained core logs. However, this requires an exact determination of the loop sensor's response function as well as, preferably, noise-free measurements of the magnetic susceptibility (practically impossible), and an exact adjustment of the depth scales of the individual core sections. However, the results of deconvolution methods are of limited efficiency. Figure 4a shows the response function of a loop sensor with 14.3 cm coil diameter (Bartington MS2C) that can measure cores of up to 12 cm diameter, compared to two different spot-reading sensors (Bartington MS2F and E). The corresponding spectral (or filtering) properties, obtained from fourier analysis of the response functions, are shown in Figure 4b. Obviously, wave lengths shorter than 3 cm, that are still easily detectable with the spot-reading sensors (especially with MS2E), are completely filtered out when using the loop sensor (MS2C). This spectral component cannot be recovered by deconvolution. So, in order to achieve a high-resolution dataset, instead of deconvolving whole-core data, spot-readings of susceptibilitiy should be performed directly, as discussed in the following paragraph.

Split-core logging

Core sections of variable diameter splitted into halves, sub-cores or archive boxes taken from cores or outcrops, can all be subjected to the spot-reading technique of magnetic susceptibility, here referred to as 'split-core logging', since lake sediments are investigated mainly in terms of sediment cores. During logging the sensor is directly lowered onto the split surface of a section half that should be covered by a thin plastic foil in order to prevent the sensor from contamination. Since the sensitivity of the sensor decays exponentially with distance from its sensing plane, the sensor must touch the sediment surface completely,

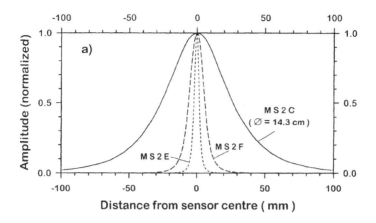

Figure 4. Response functions of a loop sensor (MS2C) and two spot-reading sensors (MS2E and MS2F) with respect to a thin magnetic disk (a) and their low-pass filter properties (b) obtained from fourier analysis of the response functions.

otherwise the obtained susceptibility values would be much too low. In addition, prior to logging and covering with plastic foil, the sediment surface should be cleaned carefully, so that the measurement is not biased by material smeared along the split-surface. Since only a small portion of the core material is measured, the obtained record might not be representative in cases where the recovered sediments show large horizontal inhomogenites, or, if distortions and displacements, such as inclined breaks, which are induced by the stress acting on the sediment during coring and that are typical for the lower part of some gravity corers, are present. On the other hand, this logging method enables the investigator to select undestorted parts that cannot be excluded during whole-core logging. Since the bell-shaped response function of spot-reading sensors are much narrower, only a few readings close to the ends of the split-cores (about 10 mm for MS2E) are influenced by the edge effects as discussed for the loop sensor in Figure 3.

Discrete samples

Discrete samples that are taken for paleomagnetic purposes are also investigated for magnetic susceptibility since it is a basic rock magnetic parameter (see also Sandgren & Snowball, in press). And in fact, the most sensitive susceptibility meters were designed especially for this kind of samples (e.g., the Czech Kappabridge KLY-3). So, if susceptibility is generally very low along an investigated core (or outcrop section), this method still may yield significant variations that can be used for correlation/interpretation. Commonly, cylindric or cubic paleomagnetic samples have a volume in the range of 6 to $10\,cm^3$. Theoretically, the minimum wave length that can be resolved from (non-overlapping) discrete samples is twice the sampling interval, resulting in two readings per cycle, e.g., 4 cm when samples were taken every 2 cm. However, taking a sine wave as an example, no amplitude mudulation will be detected if the boxes are sited exactly at the zero crossing of the sinusoidal variation. So, practically the minimum detectable wave length is rather four times than twice the sampling distance, e.g., 8 cm with 2 cm sample distance. Therefore, the spatial resolution is about the same as with the whole-core logging method.

Another method of logging magnetic susceptibility with discrete samples, and of course many other parameters except paleomagnetism, is extruding the sediment material out of the core tube and cut it into thin slices of e.g., 1 cm in thickness. Such slices then can be filled into small boxes (or cylinders) that fit into susceptibility meters for discrete samples. This is preferably done with sediment cores from the water-sediment boundary, because this material cannot be splitted into halves, which prevents the application of the split-core logging method. Application of the whole-core method might introduce mixing up of the material when the core is measured horizontally and/or if the core tube is not filled completely. Nevertheless, susceptibility can be also determined from slices cut out of a splitted core as indicated in Figures 1 and 2.

Examples

Since susceptibility logging is not restricted to lake sediments, the following examples describe results from marine as well as from lacustrine materials. Figure 5a to c illustrates a typical result from a whole-core logging. The piston core with a diameter of 10 cm and a length of 575 cm was cut into 6 sections and measured with a loop sensor of 14.3 cm coil diameter. Figure 5a shows the result of the susceptibility log of the 5th section. Because of the sensor's drift, measurements with the section outside the sensor's sensitivity volume were different from zero. The drift during measurement of the sections was assumed to be linear, as indicated by the horizontal dashed line in Figure 5a for the 5th section, and subsequently suctracted from all the logs. The drift corrected, overlapping logs of all six sections is shown in Figure 5b. After superposition of these records, the composite curve was multiplied with an empirically determined correction factor of 8.6 in order to yield volume susceptibility values as multiples of 10^{-6} (Fig. 5c).

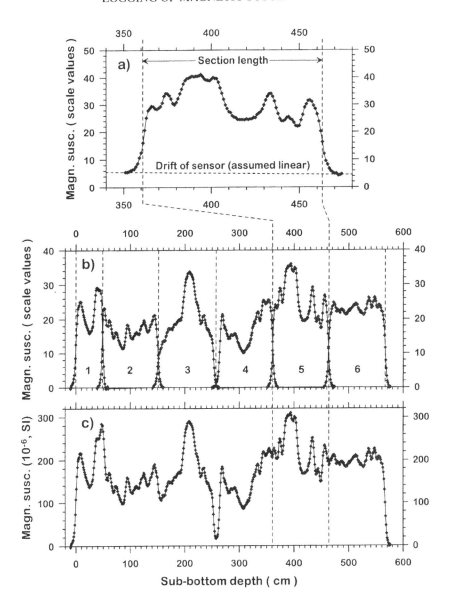

Figure 5. Core PS 2163-1 GPC, Arctic Ocean, typical result from a whole-core logging. Core diameter was 10 cm and coil diameter of susceptibility sensor was 14.3 cm. Measurements were performed each 1 cm. Raw data (a) of the fifth section of the core. Susceptibility values are readings given by the control unit. Because of drifting of the sensor, the log does not start/stop at zero but stays at a certain level with the core section outside the sensor's sentivity volume. The sensor's drift was assumed to be linear, indicated by the horizontal dashed line, and then subtracted, together with the other core sections (b). After stacking the individual logs, the resulting composite was multiplied with a correction factor according to the manual's correction function in order to yield volume susceptibility values expressed as multiples of 10^{-6} (c).

The procedure of drift correction during split-core logging can be performed in a more detailed way. Since the sensitivity volume of such a sensor is quite small, the sensor can be moved away (about 5 cm) from the core section without a change in the reference level, if the magnetic environment is homogenous enough. Thus, the sensor drift can be monitored as often as desired. Figure 6 shows an example of a core section that was measured manually in steps of 2.5 mm (dots in Fig. 6a) with readings against air each 25 mm logging distance (crosses in Fig. 6a). The sensor drift is a function of time rather than a function of logging distance. Therefore, during measurements, the logging time, starting at zero when logging started, was stored together with the susceptibility and the distance readings, and the drift corrections were performed with respect to time. In order to reduce the noise in the air readings, which should be a smooth curve because of its thermal origin, the obtained curve was smoothed with a weighted five point running average (Fig. 6b). After subtraction of the smoothed drift curve, the susceptibility log was also multiplied with an empirically determined correction factor of 8.9 in order to yield volume susceptibility values as multiples of 10^{-6} (Fig. 6c). For comparison, Figure 6d shows the result of the susceptibility log that was obtained from the cubic paleomagnetic samples of 2×2 cm taken each 2 cm from this core section. The symbol size in Figure 6d is proportional to the sample size. The logging results of the complete core is shown in Figure 7. Prior to processing, the 1 m long sections ($\phi = 12$ cm) were scanned with a whole-core sensor (Fig. 7a). After splitting, the sections were scanned with a Bartington MS2F spot-reading sensor, (Fig. 7b) and then sampled for a high resolution magnetostratigraphic investigation. Finally, the sample's susceptibilities were measured with an AGICO Kappabridge KLY-3S discrete sample sensor yielding the log shown in Figure 7c. Obviously, the spatial resolution of the whole-core log and the discrete sample log is about the same, whereas only the high-resolution split-core log reveals the full complexity of the susceptibility pattern within the core (see also Figs. 6c and d). Moreover, it is obvious that the paleomagnetic samples partly intergrate over quite heterogenous sediments.

The whole-core, as well as the split-core susceptibility records of the core shown in Figure 7, were subjected to fast fourier analysis in order to demonstrate the low-pass filtering properties of the applied sensors. The normalized spectra are plotted together with the sensor's frequency response functions (Fig. 8, compare to Fig. 4), clearly showing how the high-frequency components are suppressed by the respective sensors. Spectral lines exceeding the sensor's frequency response functions, such as within the spectrum of the whole-core log (a), are due to noise.

Another example of high-resolution logging is shown in Figure 9. Two parallel cores from a southeast China maar lake were recovered in a way that breaks between adjacent sections of one core lie within the sections of the parallel core, so that a sedimentary record without gaps could be obtained. The core halves were scanned with an automatic system in steps of 1 mm with drift readings each 10 mm and corrections analogous to Figure 6. The drift and volume corrected susceptibility records were then also smoothed with a weighted five point running average in order to remove noise from the logs. Consistently in both cores a high-frequency component within the susceptibility record with a wave length of about 1.2 to 1.5 cm could be revealed. With such high-quality logs it is possible to create site composites with a precision of a few mm. Differences in the amplitudes around 685 cm sub-bottom depth might be due to slight thickness variations of the laminae of higher susceptibility. Such amplitude variations also point to the limits of increased spatial

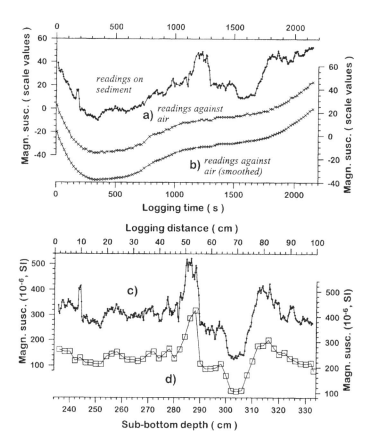

Figure 6. Typical result from split-core logging of a core section (PS 2138-1 SL, Arctic Ocean). The drift of the sensor (crosses) has been monitored each 10 readings of the sediment (dots) in order to get a more exact definition of the sensor's drift. Data are shown versus logging time stored together with logging data. After smoothing the drift readings of the sensor with a weighted five point running average (b), the drift was subtracted from the readings on sediment and multiplied with a calibration factor of 8.9 in order to yield volume susceptibility values expressed as multiples of 10^{-6} (c). For comparison the susceptibility log obtained from paleomagnetic samples taken from the same section is shown in (d). Symbol size is propotional to the sample size.

resolution. Spot-reading logs might be less representative for a sediment sequence than whole-core logs.

Figure 10 gives an example of a short core with sediments that were characterized by a very high water content so that it could not be split into two halves. Therefore, the core was kept vertically and the sediments were extruded from the liner with the help of a piston. The resulting slices of 1 cm thickness were then filled into standard paleomagnetic boxes of $6.2\,cm^3$. Measurements of magnetic volume susceptibility were performed with a discrete sample sensor (Kappabridge KLY-3S). Since the readings often yielded negative

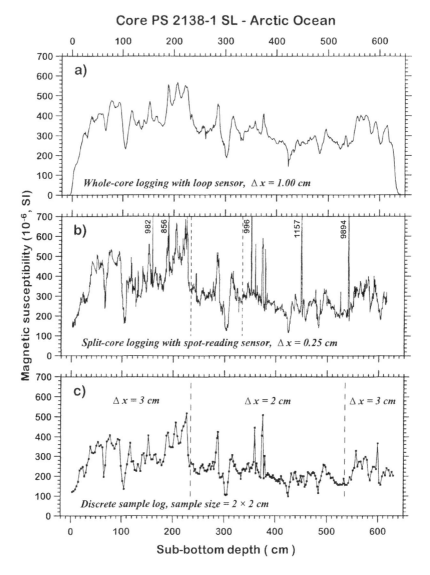

Figure 7. Comparison of three different logging methods performed on the same core. The un-opened core sections (PS 2138-1 SL, Arctic Ocean) with a diameter of 12 cm were scanned with a 14.3 cm coil in steps of $\Delta x = 1$ cm (a). After splitting the sections, a high-resolution spot-reading sensor was applied in steps of $\Delta x = 0.25$ cm (b). Numbers indicate peak values of readings extending beyond the y-axis. The two vertical dashed lines indicate the position of the core section displayed in detail in Figure 6. A third susceptibility record (c) was obtained from paleomagnetic samples that were taken at different spacings of Δx as indicated in the graph. The complex susceptibility features revealed by the spot-reading technique (b) are significantly smoothed in both other records (a and c).

Susceptibility spectrum of Core PS2138-1 SL

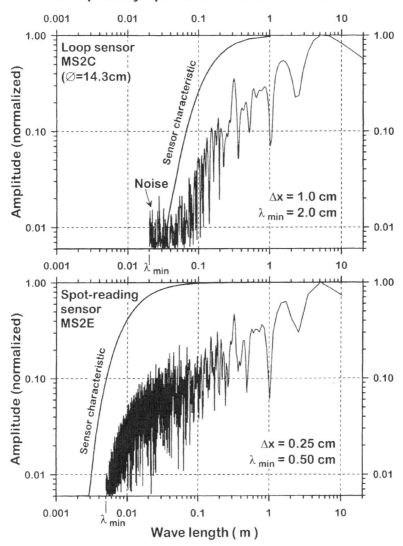

Figure 8. Spectral analysis of the susceptibility logs of a marine core (PS 2138-1 SL, Arctic Ocean) determined with a MS2C loop sensor (a) and with a MS2E spot-reading sensor (b, compare to Fig. 7). After subtracting the average, each susceptibility record was multiplied with a cosine window covering the whole core length and then subjected to fast fourier analysis. The spectra are normalized to their maximum. The minimum wave length λ_{min}, which can be detected during logging, is determined by the sampling rate $\Delta x (= 0.5 \lambda_{min})$, which also influences the spectral resolution. The thick lines represent the sensor's frequency response functions (compare to Fig. 4). Obviously, the whole-core record (a) contains some high-frequency noise, since some peaks in the susceptibility spectrum exceed the sensor's frequency response function.

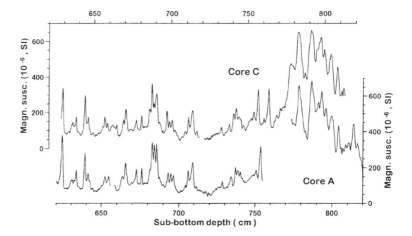

Figure 9. High-resolution susceptibility records obtained from two parallel cores by split-core logging ('H'-Lake, SE-China). Data gaps are due to core breaks but since the cores were recovered with an offset, the gaps can be filled by the data of the parallel core. Logging was performed automatically in steps of 1 mm followed by smoothing with a weighed five point running average. A high-frequency signal of about 1.2 to 1.5 cm wave length is present in the susceptibility logs.

values (Fig. 10), for comparison, an additional paleomagnetic sampling box filled only with distilled water was also measured (dashed line in Fig. 10, top). In order to check the validity of the results, an anhysteretic remanent magnetization (ARM, see also Sandgren & Snowball, volume 2) was imprinted on the samples, yielding a curve roughly parallel to the susceptibility log. Both curves, magnetic susceptibility and ARM intensity, are a measure of the concentration of magnetic minerals. While the ARM is related to the remanent magnetization of the magnetic particles only, magnetic volume susceptibiltiy is based on the measurement of an induced magnetization carried by the bulk sediment including pore water. Therefore, in this case of a sediment with a very high water content, the curve of magnetic volume susceptibility shows an obvious offset towards the susceptibility of water indicated by the dashed line in Figure 10.

Summary

Different methods of susceptibility logging have been established. Depending on the type of recovered material and research interests, various types of sensors are available with different specifications and physical limitations. Since all measurements are based on determinations on a certain volume (or mass) all logging methods are characterized by a more or less low-pass filtering effect. Highest spatial resolution can be achieved by split-core logging method, whereas highest amplitude resolution can be achieved with discrete sample sensors. Whole-core logs with a low spatial resolution are very useful in cases where basic information about relative changes in sedimentation rates is needed from cores that are still closed. Susceptibility measurements are non-destructive, easy to perform, with results often representative for the bulk composition of the investigated sediments.

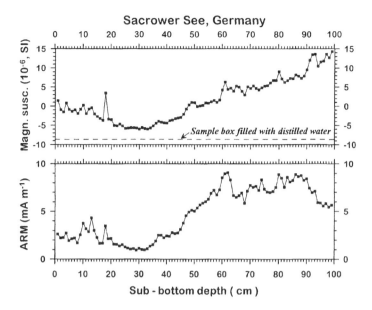

Figure 10. Example of a susceptibility record obtained from discrete samples (Sacrower See, Germany). Due to the high water content, the core had be extruded vertically, and cut into slices. The sediments were cut into 1 cm thick slices and then filled into standard paleomagnetic boxes of 6.2 cm³. The obtained log is obviously strongly influenced by the susceptibility of water, indicated by the dashed line. For comparison, the record of anhysteretic remanent magnetization (ARM, see Sandgren & Snowball, volume 2) is included. ARM is also a parameter related to the concentration of magnetic minerals but based on the remanent and not induced magnetization, as susceptibility is. Since water cannot carry a remanent magnetization both curves have a similar morphology, with the susceptibility record showing an offset caused by the negative susceptibility of water.

Therefore, susceptibility logging became one of the most important methods for monitoring variations in sediment composition in the scope of environmental research.

Acknowledgements

I like to thank Dr. T. W. Frederichs, Dr. U. Frank, and K. Teuber for help during logging and sampling of some of the cores discussed in this paper. Dr. Mingram, M. Köhler, D. Berger, M. Ramrath, R. Scheuss, A. Hofmann and M. Hauff performed coring of sediments in SE China.

References

Bloemendahl, J., X. M. Liu & T. C. Rolph, 1995. Correlation of magnetic susceptibility stratigraphy of Chinese loess and the marine oxygen isotope record: chronological and palaeoclimatic implications. Earth. Planet. Sci. Lett. 131: 371–380.
Dekkers, M. J. 1997. Environmental magnetism: an introduction. Geologie en Mijnbow 76: 163–182.
Heller, F., X. Liu, T. Liu & T. Xu, 1991. Magnetic susceptibility of loess in China. Earth. Planet. Sci. Lett. 103: 301–310.

Kukla, G., F. Heller, X. M. Liu, T. C. Xu, T. S. Liu & Z. S. An, 1988. Pleistocene climates in China dated by magnetic susceptibility. Geology 16: 811–814.

Mead, G. A., L. Tauxe & J. L. LaBrecque, 1986. Oligocene paleoceanography of the South Atlantic: Paleoclimatic implications of sediment accumulation rates and magnetic susceptibility measurements. Paleoceanography 1: 273–284.

Sun, D., J. Shaw, Z. An, M. Cheng & L. Yue, 1998. Magnetostratigraphy and paleoclimatic interpretation of a continuous 7.2 Ma Late Cenozoic eolian sediments from the Chinese Loess Plateau. Geophys. Res. Lett. 25(1): 85–88.

Williams, T., N. Thouveny & K. M. Creer, 1996. Palaeoclimatic significance of the 300 ka mineral magnetic record from the sediments of Lac du Bouchet, France. Quat. Sci. Rev. 15: 223–235.

9. CHRONOSTRATIGRAPHIC TECHNIQUES
IN RECENT SEDIMENTS

P. G. APPLEBY (appleby@liverpool.ac.uk)
Department of Mathematical Sciences
University of Liverpool
Liverpool L69 3BX
UK

Keywords: Dating, lake sediments, radionuclides, ^{210}Pb, ^{137}Cs, ^{241}Am, fallout.

Introduction

Records stored in natural archives, such as lake sediments or peat bog accumulations, are used in a wide range of environmental programs, for example:

- the assessment of changing erosion rates in a catchment arising from disturbances such as afforestation, deforestation, changing agricultural practice;

- determining the history of changes in lake water quality associated with problems such as eutrophication or "acid rain";

- monitoring atmospheric pollution by heavy metals (e.g., Pb, Hg), organic pollutants (e.g., PCBs, PAHs), radioactive emission from nuclear installations, and other contaminants.

Accurate sediment chronologies are of crucial importance in interpreting these archives. One of the most important means for dating recent sediments (0–150 years) is by ^{210}Pb (half-life 22.3 y), a natural radioactive isotope of lead. In a large number of cases the method has proved to be very reliable, particularly in stable environments with uniform sediment accumulation rates where the dating calculations are unambiguous. The method has also been found to give good results at many sites with non-uniform accumulation, though here the problem is more difficult in view of the need to determine an appropriate dating model. There are two simple models, commonly referred to as the CRS and CIC models (Appleby & Oldfield, 1978, Robbins, 1978). Of these, the CRS (constant rate of ^{210}Pb supply) model is perhaps the most widely used. The main principles of this model are exemplified in Appleby et al. (1979) by cores from three Finnish lakes with annually laminated sediments, all of which contained layers recording dilution of the atmospheric ^{210}Pb flux by increased sedimentation. ^{210}Pb dates calculated using the CRS model were in good agreement with those determined by laminae counting. There are, however, circumstances where the CIC

W. M. Last & J. P. Smol (eds.), 2001. *Tracking Environmental Change Using Lake Sediments. Volume 1: Basin Analysis, Coring, and Chronological Techniques.* Kluwer Academic Publishers, Dordrecht, The Netherlands.

| 4.51x10⁹ y | | 1602 y | | 3.82 d | | 22.26 y | | 138.4 d | |

$$^{238}U \xrightarrow{\quad 4.51\times10^9 \text{ y} \quad} {}^{226}Ra \xrightarrow{\quad 1602 \text{ y} \quad} {}^{222}Rn \xrightarrow{\quad 3.82 \text{ d} \quad} {}^{210}Pb \xrightarrow{\quad 22.26 \text{ y} \quad} {}^{210}Po \xrightarrow{\quad 138.4 \text{ d} \quad} {}^{206}Pb$$

Figure 1. ^{238}U decay series, showing the principal radionuclides concerned with the production of ^{210}Pb, and their radioactive half-lives.

(constant initial concentration) model is appropriate, e.g., in a core from Devoke Water (Appleby & Oldfield, 1992) where the CRS model was invalidated by an abrupt discontinuity in the sediment record.

There are many situations in which neither of the simple models is valid. Causes may include mixing of the surficial sediment by physical or biological processes, or variations in the ^{210}Pb supply due to changes in the pattern of sediment focussing. For these reasons it is important to validate ^{210}Pb dates using independent chronological evidence wherever possible. In a very real sense the simple models should, in the first instance, be regarded as tools whose purpose is to determine, as far as is practicable, the processes that have generated the radiometric data contained in the sediment record. Where the simple models are inadequate, reliable dating can still be achieved using various *ad hoc* models appropriate to the particular circumstances. To achieve this goal, it is necessary to understand the transport processes delivering ^{210}Pb to the sediment record, and also to determine one or more independent dates against which to assess the ^{210}Pb chronology.

The most widely used independent dating technique is from sediment records of artificial radionuclides such as ^{137}Cs and ^{241}Am. Fallout on a global scale began in 1954 following the onset of the atmospheric testing of high-yield thermonuclear weapons, and reached a peak in 1963 shortly after the test-ban treaty. Where there is a good qualitative record of atmospheric fallout, sediments recording to these events can be identified and thus dated. More recently, fallout from the Chernobyl reactor accident has been used to identify the 1986 depth.

The objectives of this chapter are to review briefly the origins and transport processes delivering fallout radionuclides to lake sediments, to describe the simple dating models, and to outline procedures for developing alternative methodologies where the simple models prove inadequate. An account of various numerical algorithms used in the calculations will also be given, including the determination of standard errors.

Origins of fallout radionuclides

The ^{210}Pb cycle

^{210}Pb occurs naturally as one of the radionuclides in the ^{238}U decay series (Figure 1). Disequilibrium between ^{210}Pb and its parent isotope in the series, ^{226}Ra, arises through diffusion of the intermediate gaseous isotope ^{222}Rn. A fraction of the ^{222}Rn atoms produced by the decay of ^{226}Ra in soils escape into the atmosphere where they decay through a series of short-lived radionuclides to ^{210}Pb. This is removed from the atmosphere by precipitation or dry deposition, falling onto the land surface or into lakes and oceans. ^{210}Pb falling directly into lakes is scavenged from the water column and deposited on the bed of the lake with the sediments.

Excess ^{210}Pb in sediments over that in equilibrium with the *in situ* ^{226}Ra in sediments decays in accordance with the radioactive decay law,

$$C_{Pb} = C_{Pb}(0)e^{-\lambda t} + C_{Ra}\left(1 - e^{-\lambda t}\right) \qquad (1)$$

(where λ is the ^{210}Pb radioactive decay constant). This equation can be used to date the sediments provided reliable estimates can be made of the *initial* ^{210}Pb *activity* $C_{Pb}(0)$ in each sediment layer *at the time of its formation*.

Modelling and quantifying the processes by which excess ^{210}Pb is produced and redeposited on the earth's surface is an important prerequisite to the development of reliable methods for calculating ^{210}Pb dates.

^{222}Rn *production and diffusion*

^{222}Rn formed by radioactive decay of ^{226}Ra escapes from soil particles into the interstices both by recoil on ejection of the alpha particle, and by diffusion. Most simple models of these processes assume that the interstitial transport of ^{222}Rn in soils occurs by Fickian diffusion (Crank, 1975). According to this process, variations in the ^{222}Rn activity C in pore spaces with time t and depth x below the soil surface are controlled by the diffusion equation

$$\frac{\partial C}{\partial t} = -\lambda C + F + \frac{\partial}{\partial x}\left(D\frac{\partial C}{\partial x}\right), \qquad (2)$$

where F is a parameter characterising the rate of emanation of ^{222}Rn from soil particles into the interstices and D is the diffusivity of ^{222}Rn through the pore spaces. The flux across the land surface boundary to the atmosphere will vary significantly on short time-scales, depending on factors such as rainfall, wind speed and air pressure. On longer time-scales, however, the flux is likely to be relatively uniform, except where there have been major changes in local conditions. Calculations using steady state solutions of the diffusion equation suggest that in well-aerated soils with ^{226}Ra activities in the typical range 30–50 Bq kg^{-1}, the mean ^{222}Rn flux to the atmosphere is of the order of 1400–2400 Bq m^{-2} day^{-1}. Measurements of ^{222}Rn exhalation rates reported in the literature from land surfaces around the world, summarised in Appleby & Oldfield (1992), range from 600–2800 Bq m^{-2} d^{-1}, with a mean value of 1160 Bq m^{-2} d^{-1}. Estimates of the global average are 1270 Bq m^{-2} d^{-1} (Israel 1951) and 1360 Bq m^{-2} d^{-1} (Wilkening et al. 1975). Although there are still considerable uncertainties in these figures, it is reasonable to suppose that the mean ^{222}Rn exhalation rate from the c. 76% of the earth's land surface free of ice sheets and permafrost is in the range 1100–1400 Bq m^{-2} d^{-1} (6000–7700 atoms m^{-2} s^{-1}). Since the total area contributing to this flux is about 110×10^6 km^2 (c. 22% of the total surface of the earth), the total ^{222}Rn flux to the atmosphere is estimated to be between 1.2–1.5×10^{17} Bq d^{-1}.

Atmospheric production and fallout of ^{210}Pb

^{222}Rn in the atmosphere decays via a number of short-lived radionuclides to ^{210}Pb. Measurements of ^{210}Pb concentrations in air indicate that it has a relatively short residence time

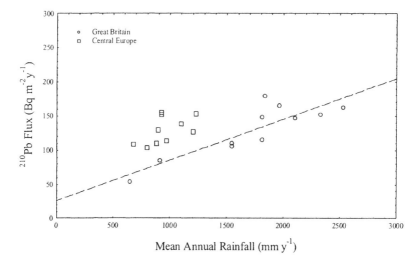

Figure 2. ^{210}Pb flux versus mean annual rainfall at sites in Great Britain and central Europe. The fallout in Great Britain appears to be linearly related to mean annual rainfall. At sites with the same rainfall, fallout in central Europe is around 50% higher than in Great Britain.

in the atmosphere. Individual ^{210}Pb atoms become readily attached to airborne particulate material and are removed both by washout and by dry deposition.

Using simple mass balance arguments, the global mean flux of ^{210}Pb from the atmosphere is:

$$\mathcal{P} = \mathcal{F} \frac{T_{Rn}}{T_{Pb} + T},$$ (3)

where $\mathcal{F}$ denotes the total ^{222}Rn flux to the atmosphere, T_{Rn} and T_{Pb} denote the ^{222}Rn and ^{210}Pb radioactive half-lives and T is a mean residence time for ^{210}Pb in the atmosphere (typically not more than a 1–2 weeks). Using the above value for $\mathcal{F}$, the mean ^{210}Pb flux from the atmosphere is estimated to be c. 2.5×10^{16} Bq y^{-1}, or c. 50 Bq m^{-2} y^{-1} over the whole of the Earth's surface.

The ^{210}Pb flux is normally assumed to be constant at any given site (when averaged over a period of a year or more). Its value may, however, vary spatially by up to an order of magnitude, depending on factors such as rainfall and geographical location. This is illustrated by the measurements of ^{210}Pb fallout in Great Britain and central Europe summarised in Figure 2. These show that, at a regional level, there is a strong correlation with rainfall. Overall levels of fallout in central Europe are however significantly higher than in Great Britain, presumably due to a build of ^{222}Rn concentrations in the atmosphere as the prevailing winds transport air masses over the intervening land surface. The global pattern suggests that there is a consistent west to east increase in ^{210}Pb fallout within the major continents, superimposed on a baseline ^{210}Pb flux of c. 30–40 Bq m^{-2} y^{-1} per metre of rain at sites remote from major land masses. ^{210}Pb fluxes at sites remote from oceanic influences can exceed 200 Bq m^{-2} y^{-1}.

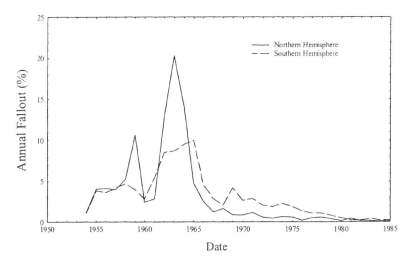

Figure 3. [137]Cs fallout in the Northern (solid line) and Southern (dashed line) Hemispheres from the atmospheric testing of nuclear weapons, shown as percentages of total cumulative fallout to 1985. The two major peaks record the fallout maxima immediately prior to the impact of the 1958 moratorium and 1963 test ban treaty.

Artificial radionuclides

The two main sources of artificial fallout radionuclides have been atmospheric nuclear weapons tests during 1953–63, and the Chernobyl reactor fire in April 1986.

Fallout from the atmospheric testing of nuclear weapons

Widespread global dispersal of artificial radionuclides from the atmospheric testing of nuclear weapons began with the high-yield thermo-nuclear tests in November 1952. These tests injected radioactive debris from the explosions into the stratosphere where it was transported around the world. Re-entry of this debris to the troposphere was followed by fallout onto the Earth's surface. The principal radionuclides in the debris included ^{90}Sr, ^{137}Cs, ^{239,240,241}Pu.

Within each hemisphere, fallout followed a pattern similar to that shown in Figure 3. Throughout the Northern Hemisphere fallout reached significant levels by 1954 and increased rapidly during the succeeding years. There was a brief decline following the 1958 moratorium on testing but renewed testing following the expiry of the moratorium in 1961 resulted in a very sharp increase during 1962–63. After the 1963 Test Ban Treaty, fallout declined steadily from the 1963–64 peak, except for minor increases in the early 1970's due to atmospheric testing by non-Treaty countries. Cambray (1989) reported that, by 1983, fallout in the Northern Hemisphere was below their limits of detection. Fallout in the Southern Hemisphere followed much the same pattern though the peaks were less pronounced and less well resolved. The global record up to 1987 is summarised in Cambray et al. (1989). Although the patterns were similar, the amounts of fallout varied substantially from place to place, depending on factors such as latitude and rainfall.

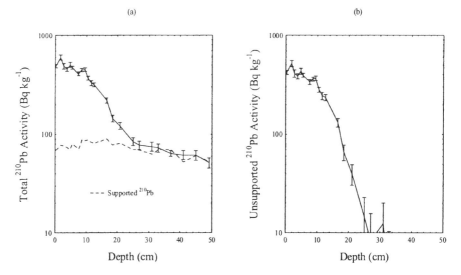

Figure 4. ^{210}Pb activity versus depth in a sediment core collected from Esthwaite Water (U.K.) showing (a) total and supported ^{210}Pb and (b) unsupported ^{210}Pb. Equilibrium between total ^{210}Pb and the supporting ^{226}Ra, corresponding to c. 130–150 years accumulation, occurs at a depth of about 30 cm.

Fallout from the chernobyl reactor accident

The accident at the Chernobyl nuclear power plant in the Ukraine on 26th April 1986 ejected large amounts of burning core debris into the atmosphere. Significant releases of radionuclides, including ^{131}I, ^{137}Cs and ^{134}Cs, continued until 5th May 1986. Fallout occurred at sites thousands of kilometres from Chernobyl (ApSimon et al., 1986, Smith & Clark, 1986). Total deposition of ^{137}Cs throughout the Northern Hemisphere is estimated to have been of the order of 10^{17} Bq (Cambray et al., 1987), compared to c. 4.3×10^{17} Bq (decay corrected to 1986) from weapons tests (Cambray et al., 1989). The distribution of Chernobyl fallout was very uneven and mainly controlled by the path of the plume and its coincidence with rainfall. In parts of Scandinavia, central Europe and Great Britain, deposition of Chernobyl ^{137}Cs during the course of a few days was over an order of magnitude greater than total deposition from weapons fallout.

Sediment records of fallout radionuclides

Fallout radionuclides deposited in a lake directly from the atmosphere or indirectly via its catchment are transported through the water column to the bed of the lake and incorporated in the sediment record.

^{210}Pb *activity in sediment records*

The *total* ^{210}Pb activity in sediments has two components, *supported* ^{210}Pb which derives from *in situ* decay of the parent radionuclide ^{226}Ra, and *unsupported* ^{210}Pb which derives from the atmospheric flux. The supported component will usually be in radioactive equilibrium with the ^{226}Ra. Unsupported ^{210}Pb is determined by subtracting supported activity from the total activity. Figure 4 plots (a) total and supported ^{210}Pb and (b) unsupported ^{210}Pb activity versus depth in a core from Esthwaite Water in Cumbria (UK). High unsupported ^{210}Pb activities in the surficial sediments derive from their exposure to recent inputs from the atmosphere. In deeper layers that have been buried by more recent deposits the total activity C_{tot} decays with time in accordance with the law

$$C_{tot} = C_{tot}(0)e^{-\lambda t} + C_{sup}(1 - e^{-\lambda t}), \qquad (4)$$

where C_{sup} is the supported activity, λ is the ^{210}Pb radioactive decay constant and $C_{tot}(0)$ is the total ^{210}Pb activity of the sediment at the time of burial. In most cases equilibrium between total ^{210}Pb activity and the supporting ^{226}Ra is effectively achieved after a maximum of about 6–7 ^{210}Pb half-lives, that is, 130–150 years. In the Esthwaite core this occurs at a depth of about 30 cm. It is however important to note that in sediments with high ^{226}Ra concentrations, or low initial ^{210}Pb concentrations, limits of detection may be such that disequilibrium cannot be determined in sediments older than 3–4 half-lives (60–90 years). The point at which disequilibrium effectively vanishes can be termed the ^{210}Pb *dating horizon*.

Writing

$$C_{uns} = C_{tot} - C_{sup},$$

the unsupported activity C_{uns} satisfies the simple exponential relation

$$C_{uns} = C_{uns}(0)e^{-\lambda t}. \qquad (5)$$

If the initial activity of the sediment can be estimated, this equation can be used to calculate the time t since burial.

Relationship between sediment records and atmospheric flux

Some of the earliest studies of fallout radionuclides in lake sediments were carried out on cores from lakes in the catchment of Lake Windermere in Cumbria (UK), including Blelham Tarn, Esthwaite Water, Elterwater and Windermere itself (Pennington et al. 1973 & 1976). Further analyses carried out during the 1980s and 1990s at the Liverpool University Environmental Radioactivity Research Centre have resulted in a record of measurements from this region that now spans a period of more than 25 years. The nature and quality of these records is illustrated in Figure 5, which plots ^{137}Cs concentrations versus depth in sediment cores from Blelham Tarn (Cumbria, UK) collected in 1973 (Pennington et al. 1976), 1985 and 1996. In the course of 23 years, the record of the 1963 nuclear weapons fallout peak has moved progressively deeper in the core with relatively little loss of resolution. The 1996 core has an additional peak recording the 1986 Chernobyl accident.

Although the records are qualitatively correct, they do not immediately give accurate quantitative records of the atmospheric flux. The Windermere lakes all have similar rainfall and so may be expected to have similar atmospheric fluxes of ^{210}Pb. Figure 6(a) shows the mean rates of supply of atmospherically derived ^{210}Pb to about 20 cores from all four lakes, collected during 1973–97. In many cases, the ^{210}Pb supply rate is twice as high as the

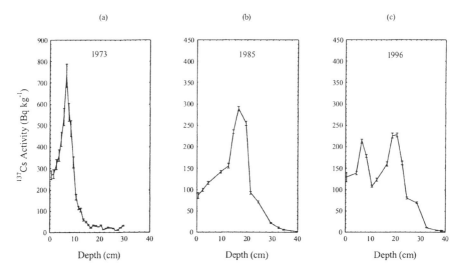

Figure 5. ^{137}Cs activity versus depth in sediment cores collected from Blelham Tarn (U.K.) in (a) 1973, (b) 1985, (c) 1996. The single ^{137}Cs peak in the 1973 and 1985 cores records the 1963 fallout maximum from the atmospheric testing of nuclear weapons. The additional feature in the 1996 core records fallout from the 1986 Chernobyl accident. The 1963 peak has moved progressively deeper in the core with relatively little loss of resolution.

direct atmospheric flux (indicated by the dashed line), and in one case three times as high. These results highlight the fact that transport processes through catchment/lake systems have a strong influence on sediment records of the atmospheric ^{210}Pb flux, and thus on ^{210}Pb dating methodologies.

Factors influencing differences between the atmospheric flux and sediment record may include inputs from the catchment, losses from water column via the outflow, and sediment focussing. The influence of water column processes is shown by comparing inventories of atmospherically derived ^{210}Pb in sediment cores with those of weapons fallout ^{137}Cs. Values of the ^{137}Cs/^{210}Pb inventory ratio in cores from the Windermere lakes (Figure 6(b)) are significantly lower than that expected from direct fallout (dashed line), presumably due to relatively greater losses of ^{137}Cs from the system (via the outflow) arising from its higher solubility.

Transport of fallout radionuclides to the sediment record

Figure 7 shows a simple flow diagram for modelling the mass balance of atmospherically deposited radionuclides in catchment/lake systems. In this diagram, $\Phi(t)$ denotes the atmospheric flux per unit area at time t. If $\mathcal{A}_L$ and $\mathcal{A}_C$ denote the areas of the lake and its catchment, inputs by direct fallout onto these two components are $\mathcal{A}_L\Phi(t)$ and $\mathcal{A}_C\Phi(t)$ respectively. Over a period of time a proportion of the fallout onto the catchment will be released to the lake via streams and overland flow. Writing $\Psi_C(t)$ for the transport rate at

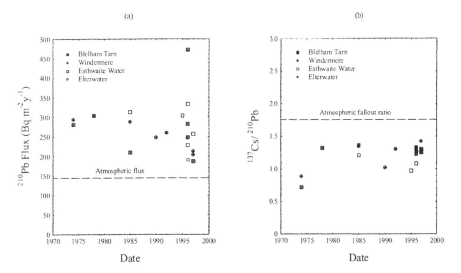

Figure 6. Data from sediment cores from lakes in the Windermere catchment (U.K.) showing (a) the mean ^{210}Pb flux and (b) the ^{137}Cs/^{210}Pb fallout ratio at each core site versus time. Also shown are the atmospheric ^{210}Pb flux and the atmospheric ^{137}Cs/^{210}Pb fallout ratio.

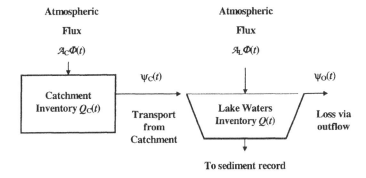

Figure 7. A simple flow diagram showing the transport of atmospherically deposited radionuclides through a lake and its catchment.

time t, total inputs to the lake can be represented as

$$\Psi_C(t) + \mathcal{A}_L \Phi(t) = \mathcal{A}_L (1 + \alpha \eta) \Phi(t), \qquad (6)$$

where $\alpha = \mathcal{A}_C / \mathcal{A}_L$ is the catchment/lake area ratio and η is a catchment/lake transport parameter.

The extent to which radionuclides entering the lake are lost via the outflow or removed to the bed of the lake (mainly on settling particles) depends on the water residence time of the lake T_W, the particle residence time in the water column T_S, and the solubility. Simple box models (cf. Santschi & Honeyman, 1989, Appleby & Smith, 1993) suggest that the fraction entering the sediment record is

$$\mathcal{F} = \frac{f_D T_L}{T_S},\tag{7}$$

where f_D is the fraction of the radionuclide in the water column attached to particulates, and T_L is a residence time of the radionuclide in the water column, given by

$$\frac{1}{T_L} = \frac{f_D}{T_S} + \frac{1}{T_W}.\tag{8}$$

Simple ^{210}Pb dating models

Measurements of ^{210}Pb deposition show that, although fluxes can vary significantly on short (daily or monthly) time-scales, it is relatively uniform on longer time-scales of the order of a year or more. Denoting the mean annual ^{210}Pb flux by $\mathcal{P}$, it follows from equations (6) and (7) that the mean annual rate of supply of fallout ^{210}Pb to the sediments can be written

$$P = \mathcal{F}_{Pb}(1 + \alpha \eta_{Pb})\mathcal{P},\tag{9}$$

where η_{Pb} is a catchment/lake transport parameter for ^{210}Pb and $\mathcal{F}_{Pb}$ is the fraction of ^{210}Pb in the water column transferred to the sediment record.

Lakes with constant sediment accumulation rates

In lakes where erosive processes in the catchment and production rates in the water column are steady, and give rise to constant rates of sediment accumulation, it is reasonable to suppose that each layer of sediment will have the same initial unsupported ^{210}Pb activity. Since sediments at a depth m (measured as cumulative dry mass, g cm^{-2}) below the sediment-water interface will be of age

$$t = \frac{m}{r},\tag{10}$$

where r denotes the dry mass sedimentation rate (g cm^{-2} y^{-1}), the unsupported ^{210}Pb activity will in this case vary with depth in accordance with the formula

$$C(m) = C(0)e^{-\lambda m/r},\tag{11}$$

where λ is the ^{210}Pb radioactive decay constant (0.03114 y^{-1}), and

$$C(0) = P/r\tag{12}$$

is the unsupported activity at the surface of the core. When plotted on a logarithmic scale the resulting ^{210}Pb activity versus depth profile will appear linear. The mean sedimentation rate r can be determined from the slope of the graph using a least-squares fit procedure.

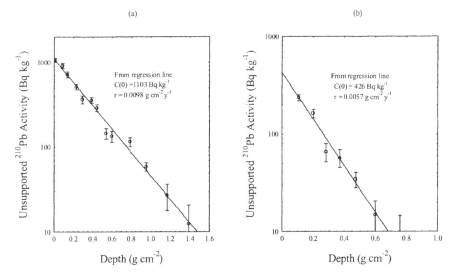

Figure 8. Unsupported ^{210}Pb in lake sediment cores from (a) Øvre Neadalsvatn (Norway) and (b) Braya Sø (Greenland) showing unsupported activity versus depth in g cm^{-2}. Both profiles are closely represented by the exponential equation $C = C(0)e^{-\lambda m/r}$.

This is illustrated by results from Øvre Neadalsvatn (Norway) and Braya Sø (Greenland) shown in Figure 8. In both cores, the unsupported ^{210}Pb activity versus depth relation is closely approximated by an exponential relation, indicating uniform accumulation over a period of more than a century. The mean sedimentation rate for each core is calculated from the slope of the regression line.

Lakes with variable sediment accumulation rates

In view of the dramatic environmental changes that have taken place over the past 150 years, rates of erosion and sediment accumulation are likely to have varied significantly at many sites during this period. Where this has occurred, unsupported ^{210}Pb activity will vary with depth in a more complicated way and ^{210}Pb profiles (plotted logarithmically) will be *non-linear*. Figure 9 shows examples from two sites, Blelham Tarn (Cumbria, UK), and Lac Korba (Tunisia).

Non-linearities in the ^{210}Pb record may be due to a number of different processes:

1) Dilution of the atmospheric fallout by increased sedimentation rates;

2) Varying degrees of sediment focussing;

3) Interruptions to the normal process of sediment accumulation;

4) Mixing by physical, biological or chemical processes.

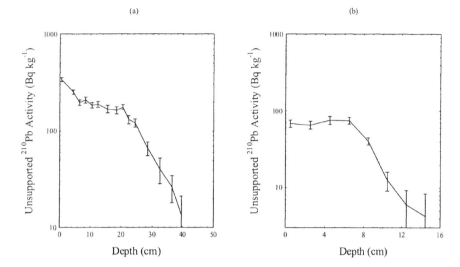

Figure 9. Unsupported ^{210}Pb in lake sediment cores from (a) Blelham Tarn and (b) Lac Korba (Morocco) showing unsupported activity versus depth. Significant deviations from a simple exponential relationship occur in both profiles. The changes occur at depths ranging from 8 cm to 20 cm.

For reliable dating each case must be assessed for the dominant mechanism and dates calculated according to a suitable model.

CRS [210]Pb *dating model*

The methodology developed in the original paper on dating lake sediments by [210]Pb (Krishnaswami et al., 1971) was based on three main assumptions:

a)   the rate of deposition of unsupported ^{210}Pb from the atmosphere is constant,

b)   [210]Pb in fresh waters is quickly removed from solution onto particulate matter so that unsupported ^{210}Pb activity in sediments is essentially that due to overhead fallout from the atmosphere,

c)   the initial unsupported ^{210}Pb activity in sediments laid down on the bed of the lake is not redistributed by post-depositional processes and decays exponentially with time in accordance with the radioactive decay law.

In many of the early published studies, sedimentation rates were relatively uniform, giving rise to the simple exponential ^{210}Pb profiles discussed above. Application of the same assumptions to sites where the sedimentation rate is not uniform led to the development of what has been called the CRS (constant rate of ^{210}Pb supply) model. The methodology for calculating sediment dates lake by this model was developed by Appleby & Oldfield (1978) and Robbins (1978). Its application to particular sites was tested and validated in Oldfield et al. (1978), and Appleby et al. (1979).

Where the CRS model is valid, changes in the sedimentation rate through time will result in changes in the initial unsupported ^{210}Pb concentrations, in accordance with equation (12). In these circumstances, the dates of older sediments are calculated not from their present concentrations but from the distribution of ^{210}Pb in the sediment record. After decay, the present amount of ^{210}Pb remaining in the record from inputs at time τ in the past during the succeeding small time interval dτ is

$$Pe^{-\lambda\tau}\mathrm{d}\tau, \tag{13}$$

where P denotes the ^{210}Pb supply rate. Assuming P constant, the residual ^{210}Pb in sediments of age t or greater is

$$A = \int_t^\infty Pe^{-\lambda\tau}\mathrm{d}\tau = \frac{P}{\lambda}e^{-\lambda t}. \tag{14}$$

The residual ^{210}Pb in the entire record, the ^{210}Pb *inventory*, obtained by putting $t = 0$ in this equation, is

$$A(0) = \frac{P}{\lambda}. \tag{15}$$

It follows that

$$A = A(0)e^{-\lambda t}. \tag{16}$$

The values of A and $A(0)$ can both be calculated by numerical integration of the concentration versus depth profile. If m denotes the depth of the sediment layer of age t,

$$A = \int_m^\infty C(m)\mathrm{d}m, \qquad A(0) = \int_0^\infty C(m)\mathrm{d}m. \tag{17}$$

From their values, the age of sediments of depth m is calculated using the formula

$$t = \frac{1}{\lambda}\ln\left(\frac{A(0)}{A}\right). \tag{18}$$

From equations (5) and (12), for sediments of age t,

$$r = \frac{P}{C}e^{-\lambda t}. \tag{19}$$

Since

$$P = \lambda A(0) = \lambda Ae^{\lambda t}, \tag{20}$$

it follows that the dry mass sedimentation rate at time t in the past can be calculated directly using the formula

$$r = \frac{\lambda A}{C}. \tag{21}$$

It can alternatively be calculated as the mass increment per unit time between adjacent samples.

CIC ^{210}Pb *dating model*

This model assumes that sediments have a constant initial ^{210}Pb concentration regardless of accumulation rates. In consequence the supply of ^{210}Pb to the sediment record must

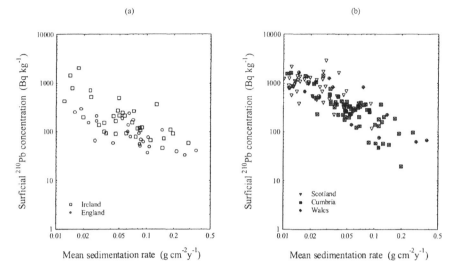

Figure 10. ^{210}Pb concentration in surface sediments from (a) England and Ireland, (b) Scotland, Cumbria and Wales. The higher concentrations in (b) reflect the higher ^{210}Pb flux arising from the higher rainfall in these regions. Broadly speaking, surface concentrations fall as sedimentation rates increase, as indicated by the CRS model.

vary directly in proportion to the sedimentation rate. Under this assumption, the age of a sediment layer of depth m can be calculated using the formula

$$C(m) = C(0)e^{-\lambda t}. \tag{22}$$

This method is implicit in the approach used in Pennington et al. (1976) to date sediments from Blelham Tarn.

Model choice

It is evident from Figure 6 that the ^{210}Pb supply rate to specific sites on the bed of the lake is not simply related to the atmospheric flux. Although this calls into question the basic assumption of the CRS model, there is some evidence that it is generically valid in that the atmospheric flux is the dominant factor controlling the ^{210}Pb supply. At sites with similar atmospheric fluxes, the CRS model suggests that surface concentrations should vary inversely with the sedimentation rate. Figure 10 plots surface concentrations versus mean sedimentation rates for cores from (a) England and Ireland, (b) Wales, Cumbria and Scotland, using logarithmic scales. Sites in each plot have similar annual rainfall totals and hence similar atmospheric ^{210}Pb fluxes. The data generally conform to the CRS model requirement that they should fit a straight line of gradient -1. Deviations from the regression line reflect the impact of processes such as sediment focussing, and the fact that the mean sedimentation rate has been used rather than the contemporary value. To avoid circularity, the mean sedimentation rate has been calculated from the 90% equilibrium depth (Appleby & Oldfield, 1983) and represents the mean accumulation rate during the past 75 years.

Studies by Binford et al. (1993) and Blais et al. (1995) also support the hypothesis that the CRS model is reliable in the majority of cases, though Binford et al. do suggest that it may be more appropriate in seepage lakes than in drainage lakes.

Equation (9) suggests that the rate of supply of fallout ^{210}Pb to the sediments is controlled by two parameters, the catchment/lake transport parameter η_{Pb}, and the water column/bottom sediment transfer fraction $\mathscr{F}_{Pb}$. For relatively insoluble species, such as ^{210}Pb, which are predominantly associated with particulates, the transfer fraction is relatively insensitive to the particular value of the distribution fraction f_D and the bulk of the input is always transported to the bottom sediments (Appleby, 1997).

The influence of catchment inputs is more problematic. Studies of catchment-lake transport (Lewis, 1977; Scott et al., 1985; Dominik et al., 1987) suggest that just 1–2% of the annual ^{210}Pb fallout on the catchment is removed to the lake. This is supported by results from a recently completed study of radionuclides in the sediments of Blelham Tarn (Appleby et al., 1999) which suggests that just over 1% of fallout onto the catchment is delivered to the lake. Because of the large catchment/lake ratio ($\alpha = 42$), inputs via the catchment in this case represent around 30% of total inputs of ^{210}Pb to the lake. The bulk of the catchment derived ^{210}Pb appears however to remain close to the inlet streams. Away from these areas of the lake, the supply appears to be dominated by the direct fallout from the atmosphere with differences between individual sites reflecting the influence of sediment focussing. A reasonable conjecture is that the CRS model is likely to be valid at such sites provided there been have no major hydrological changes in the lake that might affect the normal pattern of sediment focussing, or hiatuses in the sediment record due e.g., to sediment slumps, slides or turbidity currents. Where such events have occurred, the CIC model may provide a valid alternative if primary sedimentation rates from the water column have been constant. At sites where primary sediment accumulation and ^{210}Pb supply rates have both changed, neither model will be appropriate. Since this cannot be determined in advance, the best procedure is to begin by calculating dates using both simple models. Where there are no major differences, it can be assumed that accumulation rates have been relatively constant. Otherwise it is essential to identify as accurately as possible the dominant processes controlling the delivery of fallout ^{210}Pb to the core site. Contra-indications to the use of the CRS model will be ^{210}Pb supply rates that are excessively high or excessively low compared to the atmospheric flux. Contra-indications to the use of the CIC model will be non-monotonic variations in the ^{210}Pb concentration versus depth profile. Such features necessarily indicate changes in initial concentrations.

This approach has been used successfully over many years to date sediment records from a wide range of lakes. In many cases, it has been found that one or the other of the simple models is valid. In more complex situations it may be necessary to apply them in a piecewise way to different sections of the core. The key to success is careful testing and validation on an individual case by case basis. An important outcome of these assessments is that the radionuclide record will often be helpful in determining the quality of the sediment record, as well as its chronology.

An important exception to the normal assumption that catchment inputs are relatively small has been observed by Appleby et al. (1995) in lakes from Signy Island (maritime

Antarctic) where greatly elevated inventories of unsupported ^{210}Pb in the sediments (relative to the atmospheric flux) were explained by the bulk of fallout onto the catchment being delivered to the lakes during the annual thaw. In terms of the above equation, this corresponds to the case where $\eta_{Pb} \approx 1$.

Model testing and validation

Although parameters such as the ^{210}Pb inventory of a core and the surficial concentration, together with features in the unsupported ^{210}Pb activity versus depth profile (changes in slope, non-monotonic "kinks"), all play a role in the assessment of ^{210}Pb data from sites with varying sediment accumulation (Appleby & Oldfield, 1983), independent validation of the chronology is essential to a high level of confidence in the results. Where simple models prove inadequate, deficiencies revealed by the validation process can be used to develop *ad hoc* procedures for dealing with special cases.

The most important means for validating dates for the last 30–40 years is via records of artificial radionuclides. Since fallout from the atmospheric testing of nuclear weapons occurred on a global scale, the presence of a subsurface peak in the ^{137}Cs concentration will in most cases identify the 1963 depth. At sites where there was fallout from the Chernobyl accident, the presence of a second more recent peak will identify the 1986 depth.

Although many cores contain high quality records of both events, there are sites where the value of ^{137}Cs dating has been significantly reduced by post-depositional diffusion within pore-waters. In regions of high fallout, downward diffusion of Chernobyl ^{137}Cs has in some cases obliterated the weapons ^{137}Cs profile. Where this has occurred, measurement of ^{241}Am, another product of fallout from atmospheric nuclear weapons tests, may provide a useful alternative indicator of the 1963 depth (Appleby et al., 1991). The amount of ^{241}Am is very small, less than 0.5% of the ^{137}Cs inventory. In spite of this, ^{241}Am has been detected in many cores and evidence from a growing data set suggests that ^{241}Am is significantly less mobile than ^{137}Cs (Appleby et al., 1991). Up to the early 1990s, identification of the 1986 Chernobyl peak could be confirmed by detection of the short-lived radionuclide ^{134}Cs (half-life 2 y).

This is illustrated by results from a recent core collected from the north basin of Windermere (Cumbria, UK), shown in Figure 11. The ^{137}Cs activity versus depth profile (Fig. 11c) has two well resolved peaks. Identification of the earlier peak as a record of the 1963 fallout maximum from the atmospheric testing of nuclear weapons is confirmed by the presence of a similar (but much smaller) ^{241}Am peak at the same depth. Identification of the second more recent ^{137}Cs peak as a record of the 1986 Chernobyl accident is confirmed by the detection of ^{134}Cs at this depth. Discharges from the Chernobyl reactor fire included both caesium radionuclides in a characteristic ratio of c. 0.6 (Cambray et al., 1987). Figure 12 shows that the dated levels determined from the ^{137}Cs record in the Windermere core are in excellent agreement with the CRS model ^{210}Pb dates.

Validation of older ^{210}Pb dates near the base of a core remains a major problem, particularly where there have been significant late 19th or early 20th century changes in accumulation rates. The most usual method is by independent chronostratigraphic features in the pollen, diatom or trace metal records.

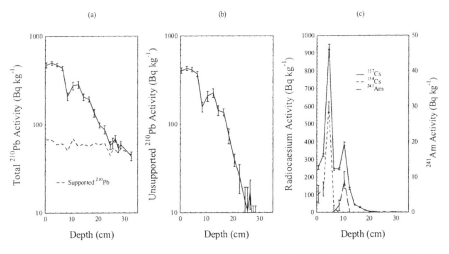

Figure 11. Fallout radionuclides in a recent sediment core from Windermere (U.K.) showing (a) total and supported ^{210}Pb, (b) unsupported ^{210}Pb and (c) ^{137}Cs, ^{134}Cs and ^{241}Am concentrations versus depth.

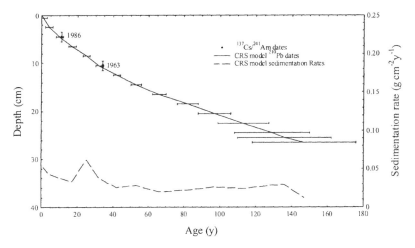

Figure 12. Dates and sediment accumulation rates in the Windermere core calculated using the CRS model. Also shown are the depths of 1986 and 1963 determined from the ^{137}Cs, ^{134}Cs and ^{241}Am stratigraphy.

Composite models

Discrepancies between CRS model dates and those determined from secure chronostratigraphic features necessarily indicate variations in the ^{210}Pb supply rate to the core site. Discrepancies with CIC model dates indicate variations in the initial ^{210}Pb concentration. These can be due to a variety of causes such as flood events, sediment slumps, turbidity currents, major land-use changes. Even where the profile contains non-monotonic variations indicating significant dilution of the atmospheric ^{210}Pb flux, the events recorded by these features may import significant quantities of additional fallout ^{210}Pb. For a brief period the ^{210}Pb supply rates may be significantly higher than at times of normal accumulation, and where this occurs the simple CRS and CIC models will both give erroneous dates.

Where there are discrepancies, changes in the ^{210}Pb supply rate can be calculated using independent dates defined by chronostratigraphic markers. If x_1 and x_2 are the depths of two features in the core with known dates t_1 and t_2, the mean ^{210}Pb flux during the period spanned by this section of the core is

$$P = \frac{\lambda \Delta A}{e^{-\lambda t_1} - e^{-\lambda t_2}},$$
(23)

where ΔA is the ^{210}Pb inventory between x_1 and x_2. Comparisons with the atmospheric flux can help identify the processes causing the change. Further, assuming the flux to be uniform within the section, corrected ^{210}Pb dates and sedimentation rates for intermediate depths can be calculated by applying the principles of the CRS model with the relevant ^{210}Pb flux. From the CRS model equations, having calculated P the date of the sediment layer at a depth x intermediate between x_1 and x_2 is determined by solving the equation

$$\frac{P}{\lambda}e^{-\lambda t} = \frac{P}{\lambda}e^{-\lambda t_1} + \Delta A(x_1, x),$$
(24)

where $\Delta A(x_1, x)$ is the ^{210}Pb inventory between x_1 and x. Since this equation necessarily assigns the dates t_1 and t_2 to the depths x_1 and x_2, dates of intermediate points are unlikely to be greatly in error even if the ^{210}Pb supply rate has not been exactly constant. The sedimentation rate at time t can be calculated using the formula

$$r = \frac{Pe^{-\lambda t}}{C}.$$
(25)

This method is illustrated by results from Knud Sø, Denmark. Figure 13 shows plots of measured radionuclide activities in a sediment core from this lake versus depth. The ^{210}Pb profile has a number of non-monotonic features suggesting irregularities in the process of sediment accumulation. Calculations using the simple CRS model reveal substantial discrepancies between the ^{210}Pb dates and the 1986 and 1963 depths in the core identified by two well resolved peaks in the ^{137}Cs profile (Figure 14). An analysis of the results suggests that the record can be divided into two zones (Table I), below 6.5 cm (pre-1979), and above 6.5 cm (post-1979). The mean ^{210}Pb supply rate in the upper zone is calculated to be 58 Bq m^{-2} y^{-1}, half that in the deeper zone (117 Bq m^{-2} y^{-1}). The decline is due to a significant reduction in the initial ^{210}Pb concentrations in sediments accumulating during the past two decades. The reason for this change is not known. Figure 14 shows the corrected ^{210}Pb chronology calculated by applying the CRS model to each zone separately and using the ^{210}Pb supply rates shown in Table I. The results indicate a relatively uniform sedimentation rate since c. 1900 punctuated by two brief episodes of rapid sedimentation dated c. 1940 and c. 1974.

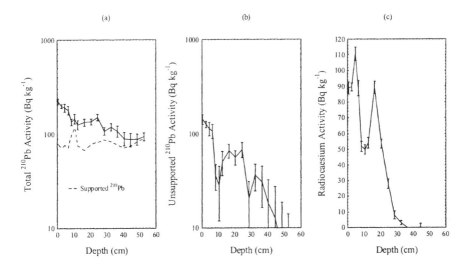

Figure 13. Fallout radionuclides in a lake sediment core from Knud Sø (Denmark) showing (a) total and supported ^{210}Pb, (b) unsupported ^{210}Pb, (c) ^{137}Cs concentrations versus depth.

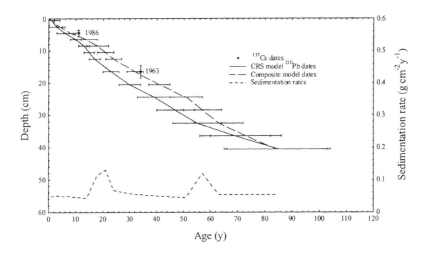

Figure 14. Radiometric chronology of Knud Sø core showing CRS model ^{210}Pb dates together with dates determined from the ^{137}Cs stratigraphy. Also shown are corrected dates and sedimentation rates calculated using the composite model.

Table I. ^{210}Pb inventories and supply rates in Knud Sø sediment core.

Depth	Date	Incremental ^{210}Pb inventory	Mean ^{210}Pb supply rate
cm	AD	Bq m^{-2}	Bq m^{-2} y^{-1}
0–6.5	1979–1997	800	58
> 16.5	< 1979	2145	117

Vertical mixing

Mixing or acceleration?

Radionuclides delivered to the bed of the lake on settling particles may be redistributed within the sediment column by two main processes,

- physical or biological mixing at or near the sediment-water interface,

- chemical diffusion or advection within the porewaters.

Sediment mixing typically results in a flattening of the ^{210}Pb activity versus depth profile in the surficial sediment layers, and degradation of features such as ^{137}Cs or ^{241}Am peaks. The effects of these processes have been documented in many studies and a number of models have been developed to take account of them (Robbins et al., 1977; Oldfield & Appleby, 1984). Since other processes can have similar effects on sediment records, an essential problem in using mixing models is the difficulty in making objective determination of the mixing parameters. Estimates of such parameters should be based on at least two independent records. The importance of this is illustrated by the results from White Lough (Northern Ireland) shown in Figure 15. ^{210}Pb activity in a sediment core was almost uniform throughout the top 20 cm of the core, but below this depth declined exponentially with depth in a manner indicating uniform sedimentation. Although this appears at first sight to be a classical example of rapid intensive mixing, an examination of the ^{137}Cs record showed that this is not the case. The ^{137}Cs profile has a well resolved peak at 8.5 cm depth recording fallout from the 1986 Chernobyl accident, and a further such peak at 20.5 cm recording the 1963 weapons test fallout maximum. These features show that sediment mixing at this site was negligible and that the flattening of the ^{210}Pb profile was almost certainly due to dilution of the atmospheric flux by accelerating sedimentation.

In those cases where mixing has occurred, Appleby & Oldfield (1992) have pointed out that corrections will often only be important in those cases where the CIC model is being used. Where the CRS model is applicable, calculations show that the use of uncorrected ^{210}Pb dates in a sediment core with a mixing zone spanning 10 years accumulation gives a maximum error of less than 2 years.

Where physical mixing is demonstrably negligible, the possibility that flattened ^{210}Pb profiles might be due to chemical remobilisation can be tested by comparing ^{210}Pb dates with those determined from the ^{137}Cs stratigraphy. The excellent agreement between ^{210}Pb

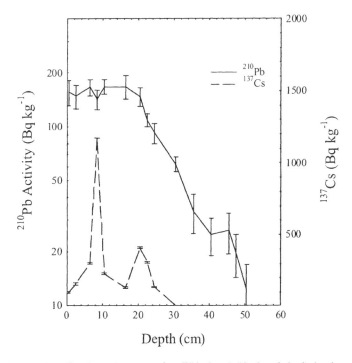

Figure 15. Fallout radionuclides in a sediment core from White Lough (Northern Ireland) showing unsupported ^{210}Pb (solid line) and ^{137}Cs (dashed line) concentrations versus depth. Although ^{210}Pb concentrations are virtually constant in the top 20 cm of the core, the well-resolved ^{137}Cs peaks show that this is not due to sediment mixing.

and ^{137}Cs (or ^{241}Am) in a large number of cases suggests that in most circumstances this is not a significant problem.

Numerical techniques

Calculation of inventories

In order to calculate ^{210}Pb dates using the CRS model, it is necessary to calculate the ^{210}Pb inventory in the whole core, and in that part of the core that is beneath the sediment layer being dated. These are the quantities denoted by $A(0)$ and A in equation (16). Since errors in the inventory calculations can result in substantial errors in the calculated dates, it is essential to adopt sound numerical procedures.

Cumulative dry mass

The first step is to calculate the cumulative dry mass in the core. There are two main methods, depending on whether the dry bulk density in the core (dry mass per unit *in situ* volume) has been measured on all sections, or just on some of them. If all sections have been measured, x_1, x_2, x_3 etc. denote the lower surfaces of consecutive sections, and s_1, s_2, s_3 etc. denote the dry bulk densities in consecutive sections, the cumulative dry mass above sediments of depth x_n is

$$m_n = m_{n-1} + s_n(x_n - x_{n-1}). \tag{26}$$

This equation is used successively to calculate m_1, m_2, m_3 etc. The calculations are easily managed using a spreadsheet. If only some sections have been measured, a mid-point procedure is required. If x_1, x_2, x_3 etc. now denote the mid-points of sections with dry bulk densities s_1, s_2, s_3 etc., the cumulative mass above the sediments of depth x_n can be estimated using the trapezium rule

$$m_n = m_{n-1} + 1/2(s_n + s_{n-1})(x_n - x_{n-1}). \tag{27}$$

Again the calculations are easily managed on a spreadsheet. To calculate the initial value m_1, it is necessary to include an estimate of the surficial density s_0.

Cumulative unsupported ^{210}Pb

The cumulative unsupported ^{210}Pb inventory is calculated by a very similar procedure, though since measurements are rarely made on consecutive samples the mid-point method is used in most cases. A common strategy is to measure samples from about every decade, with closer attention being given to resolving significant stratigraphic details such as ^{137}Cs peaks or non-exponential features in the ^{210}Pb profile. If x_1, x_2, x_3 etc. now denote the mid-points of just those sections that have been analysed for ^{210}Pb, the first step is to tabulate the cumulative dry masses m_1, m_2, m_3 etc. for each of these depths. This will normally be a subset of the values determined by the above calculation, particularly if the mid-point procedure has been used. If C_1, C_2, C_3 etc. denote the unsupported mass specific concentrations in the measured sections, the cumulative unsupported ^{210}Pb above sediments of depth x_n (or m_n) can be estimated using the trapezium rule

$$\hat{A}_n = \hat{A}_{n-1} + 1/2(C_n + C_{n-1})(m_n - m_{n-1}). \tag{28}$$

Some care does however need to be taken over units. If the concentrations are measured in Bq kg^{-1} (or equivalently mBq g^{-1}) and the cumulative dry mass is measured in g cm^{-2}, the above formula will give the ^{210}Pb inventory in mBq cm^{-2}. To obtain values in Bq m^{-2} the result must be multiplied by 10.

The trapezium rule assumes that the concentration varies linearly between measured values. Since variations in ^{210}Pb activity are usually dominated by exponential decay, a possibly more accurate approach is to use the formula

$$\hat{A}_n = \hat{A}_{n-1} + \frac{C_{n-1} - C_n}{\ln(C_{n-1}/C_n)}(m_n - m_{n-1}). \tag{29}$$

This assumes that the concentration varies exponentially with depth between data points.

Total unsupported ^{210}Pb inventory

In principle, the total unsupported ^{210}Pb inventory of the core, denoted by $A(0)$ in the CRS model dating equations, is the value of the cumulative inventory $\hat{A}$ at the point where the total ^{210}Pb activity reaches radioactive equilibrium with the supporting ^{226}Ra. The unsupported ^{210}Pb inventory below the sample being dated, A in the dating equation, is then calculated by subtraction:

$$A = A(0) - \hat{A}. \tag{30}$$

In practice, the equilibrium depth is often not well-defined and this final step is not always straightforward. Since erroneous values of $A(0)$ can lead to substantial inaccuracies in the dating, particularly in the first few samples above the equilibrium depth, it is important to determine this parameter with some care. If m_{equ} denotes the estimated depth at which ^{210}Pb is effectively in equilibrium with ^{226}Ra, and m_N is the last data point prior to reaching this depth, a first estimate for the total inventory is

$$A(0) = \hat{A}_N + A_b, \tag{31}$$

where

$$A_b = 1/2 C_N (m_{equ} - m_N) \tag{32}$$

is the unsupported ^{210}Pb inventory below the depth m_N (or x_N). Since there are usually large uncertainties in unsupported concentrations near the equilibrium depth this method can still result in substantial errors. There are however a number of ways for improving the calculation.

Reference accumulation rates

At sites where ^{210}Pb activity in the older sections is evidently declining more or less exponentially with depth, indicating uniform accumulation, the mean (basal) sedimentation rate for that period, $\bar{r}_b$, can be estimated from the gradient of the relevant section of the ^{210}Pb profile (plotted logarithmically) using the formula

$$\frac{d}{dm} \ln C = -\frac{\lambda}{\bar{r}_b}, \tag{33}$$

where m is the depth measured as cumulative dry mass. The ^{210}Pb inventory A_b below the depth x_N can then be estimated using the formula

$$A_b = \frac{1}{\lambda} \bar{r}_b C_N, \tag{34}$$

(cf. Appleby & Oldfield, 1978) where C_N is the unsupported activity at depth x_N.

Reference dates

The above method cannot be applied at sites where there have been non-uniform sediment accumulation rates in the 19th or early 20th century, or a hiatus in the sediment record. Accuracy in these cases can be improved by using an independently dated reference level. If $\hat{A}_{ref}$ denotes the entire unsupported ^{210}Pb inventory *above* the reference level, determined by numerical integration of the activity versus depth profile, the inventory below the reference level be estimated using the formula

$$A_{\text{ref}} = \frac{\hat{A}_{\text{ref}}}{e^{\lambda t_{\text{ref}}} - 1}, \tag{35}$$

(Oldfield & Appleby, 1984), where t_{ref} is the age of the reference level. The total inventory is then

$$A(0) = \hat{A}_{\text{ref}} + A_{\text{ref}}. \tag{36}$$

The dated reference level must of course be at or above the ^{210}Pb dating horizon. When it is below the ^{210}Pb dating horizon (and there appear to have been significant increases in sedimentation rates in the late 19th century), it may instead be used with the older ^{210}Pb dates to make estimates of the basal sedimentation rate $\bar{r}_b$. The value of A_b can then be estimated by the first method. The process can be iterated to give improved accuracy.

Calculation of standard errors

There are two methods for estimating uncertainties in CRS model ^{210}Pb dates, the calculus of the propagation of errors, and Monte Carlo simulation, both of which are exemplified in Binford (1990). Monte Carlo methods are relatively cumbersome for simple problems and not easily adapted to spreadsheet calculations. The basic methodology of the calculus of errors is simple in principle. If a variable y is calculated from a set of independent variables $x_1, x_2, \ldots, x_N$, each of which is subject to a standard error $\sigma_1, \sigma_2, \ldots, \sigma_N$, assuming that uncertainties are relatively small and follow a normal distribution the standard error in y is given by the formula

$$\sigma_y^2 = \sum_{i=1}^{N} \left(\frac{\partial y}{\partial x_i}\right)^2 \sigma_i^2. \tag{37}$$

If the variables $x_1, x_2, \ldots, x_N$ are not independent, further terms must be included to correct for their mutual dependence. Although the procedure may be difficult in practice for calculations involving multiple steps, in the case of the CRS model the problem can be reduced to a manageable size by proceeding in a set of orderly steps and making use of number of simplifying assumptions.

Standard errors in inventories

The ^{210}Pb inventory is calculated making repeated use of the formula

$$\hat{A}_n = \hat{A}_{n-1} + \Delta\hat{A}_n, \tag{38}$$

where if the trapezium rule is used,

$$\Delta\hat{A}_n = 1/2(C_n + C_{n-1})(m_n - m_{n-1}). \tag{39}$$

Since $\hat{A}_{n-1}$ and $\Delta\hat{A}_n$ both depend on C_{n-1}, these variable are not in fact independent. For the purpose of the error calculation we can however assume that $\Delta\hat{A}_n$ is calculated using the rectangle rule formula

$$\Delta\hat{A}_n = C_n(m_n - m_{n-1}). \tag{40}$$

$\hat{A}_{n-1}$ and $\Delta \hat{A}_n$ are now independent and the standard error in $\hat{A}_n$ can be calculated using the formula

$$\sigma_{\hat{A}_n}^2 = \sigma_{\hat{A}_{n-1}}^2 + \sigma_{\Delta \hat{A}_n}^2. \tag{41}$$

Assuming a constant percentage error p in the dry mass increments,

$$\sigma_{\Delta \hat{A}_n}^2 = \left(\sigma_{C_n}^2 + p^2 C_n^2\right)\left(m_n - m_{n-1}\right)^2. \tag{42}$$

A reasonable value for p is about 7%.

Standard errors in CRS model dates

Since the variables A and $A(0)$ used in calculating CRS model dates are not independent, to calculate standard errors in the dates the equation needs to be rewritten in a slightly different form. If (as usual) $\hat{A}$ denotes the inventory in sediments above the depth x and A denotes the inventory in sediments beneath this depth, since

$$A(0) = \hat{A} + A, \tag{43}$$

the dating equation can be written

$$t = \frac{1}{\lambda} \ln \left(1 + \frac{\hat{A}}{A}\right). \tag{44}$$

Assuming (for the purpose of the error calculations), that $\hat{A}$ and A have each been calculated using the rectangle rule, they can be regarded as being independent. Following through the calculus it can then be shown that

$$\sigma_t = \frac{1}{\lambda} \left[\left(\frac{\sigma_{A(0)}}{A(0)}\right)^2 + \left(1 - \frac{2A}{A(0)}\right)\left(\frac{\sigma_A}{A}\right)^2 \right]^{1/2} \tag{45}$$

Calculation of the standard error in the sedimentation rate r is complicated by the dependence of A on the concentration C, particularly near the base of the core. A good approximation to the percentage standard error in r is given by the formula

$$\frac{\sigma_r}{r} = \left[\left(\frac{\sigma_A}{A}\right)^2 + \left(1 - \frac{\Delta A}{A}\right)\left(\frac{\sigma_C}{C}\right)^2 \right]^{1/2} \tag{46}$$

where ΔA is increment in the cumulative inventory to the next deepest data point. Table II illustrates these calculations using data from the Windermere core shown in Figure 11.

Radiometric techniques

^{210}Pb activity in lake sediments is usually determined either by alpha spectrometry, or by gamma spectrometry.

Table II. Spreadsheet calculation of CRS model dates, sedimentation rates and standard errors for a sediment core from Windermere. The trapezium rule has been used in calculating the cumulative ^{210}Pb inventories. The dashed line indicates the ^{210}Pb dating horizon (above which the ^{210}Pb dates become significant).

		Unsupported ^{210}Pb						CRS model chronology			
	Depth	Activity				Cumulative inventories		Age		Sedimentation rate	
x	m	C		$\hat{A}$		A		t		r	
cm	g cm^{-2}	Bq kg^{-1}	±	Bq m^{-2}	±	Bq m^{-2}	±	y	±	g cm^{-2} y^{-1}	±(%)
0.0	0.00			0		6930	247	0	0		
0.5	0.03	404	31	137	1	6793	244	1	1	0.052	8
2.5	0.21	430	24	884	69	6046	244	4	1	0.044	7
4.5	0.42	415	24	1769	105	5161	234	9	2	0.039	7
6.5	0.70	373	25	2875	146	4055	220	17	2	0.034	8
8.5	1.04	157	20	3768	165	3162	196	25	2	0.063	13
10.5	1.38	207	28	4388	197	2542	180	32	3	0.038	14
12.5	1.67	226	25	5022	215	1908	144	41	3	0.026	12
14.5	1.98	144	18	5593	225	1337	115	53	3	0.029	13
16.5	2.30	137	14	6041	231	888	95	66	4	0.020	13
18.5	2.63	74	14	6391	236	539	79	82	5	0.023	21
20.5	2.99	40	6	6595	237	335	62	97	7	0.026	22
22.5	3.40	26	9	6730	241	200	56	114	10	0.024	40
24.5	3.82	13	6	6811	242	119	40	131	12	0.028	54
25.5	4.02	11	9	6834	243	95	31	138	11	0.028	76
26.5	4.23	17	6	6864	243	65	25	150	13	0.012	48
–	–	–	–	–	–	–	–	–	–	–	–
27.5	4.46	6	6	6891	244	39	21	167	18	0.021	99
28.5	4.71	6	6	6905	244	24	15	181	21	0.013	
32.5	5.72	−1	5	6930	250	0	0				

Alpha spectrometry

This method measures ^{210}Pb via alpha radiation emitted by ^{210}Po, the granddaughter product of ^{210}Pb decay (Fig. 1). ^{210}Po is extracted from the sample by chemical digestion and plated onto silver planchets for assay in a low-background alpha spectrometer. The procedures are described in detail in Flynn (1968), Eakins & Morrison (1978).

^{210}Po can be used directly as a measure of ^{210}Pb activity only if the two radionuclides are in radioactive equilibrium. Since ^{210}Po has a half-life of 138 days, samples should in principle be stored for up to two years (6 half-lives). In practice, *in situ* disequilibrium is likely to be relatively small and smaller storage times will usually be adequate. Where there are any doubts, an alternative approach is to plate ^{210}Pb onto a second planchet made after stripping all [210]Po from the original solution. [210]Pb can then be determined by ingrowth of [210]Po after storage for several months (Benoit & Hemond, 1987).

Gamma spectrometry

The advent of modern semi-conductor devices has lead to the development of techniques for determining ^{210}Pb through their gamma ray emissions. Use of this approach was first illustrated by Gägeler et al. (1976), and, more successfully, by Durham & Oliver (1983) using a hyper-pure planar germanium detector. Well-type coaxial low background germanium detectors equipped with Na(I) escape suppression shields have much lower detection limits (c. 0.01 Bq for both ^{210}Pb and [226]Ra) and are now routinely used for dating sediment records using gram-size samples (Appleby et al., 1986; Schelske et al., 1994).

Figure 16 shows the configuration of a typical gamma spectrometer in the University of Liverpool Environmental Radiometric Laboratory. Background radiation is suppressed by means of a 100 mm thick lead castle with a 3 mm copper lining, and a Na(I) anti-coincidence escape suppression shield. The detector is cooled by liquid nitrogen through a cold finger from an attached Dewar. Samples are packed in cylindrical holders and inserted into the well recess in the crystal. Gamma emissions from the sample are recorded using an ADC interfaced to a PC. Figure 17 shows a typical spectrum of the numbers of photons detected at each photon energy. The sharply resolved peaks in the spectrum identify the characteristic photons emitted by particular radionuclides in the sample. [210]Pb is measured via its gamma emissions at 46.5 keV, and [226]Ra by the 295 keV and 352 keV γ-rays emitted by its daughter isotope [214]Pb following 3 weeks storage in sealed containers to allow radioactive equilibration. [137]Cs and [241]Am are determined by their emissions at 662 keV and 59.5 keV. The absolute efficiencies of the detector at each photon energy are determined using standard sources of known activity, with corrections being made for the effect of self absorption of low energy γ-rays within the sample (Appleby et al., 1992). For a given radionuclide emitting photons of energy E with a yield Y, if N denotes the number of counts in the photo-peak, t the count time, and ε the detector efficiency, the activity of the sample is calculated using the formula

$$A = \frac{N}{\varepsilon Y t}. \tag{47}$$

A detailed account of the methodology is given in Debertin & Helmer (1988)

Choice of method

In most cases the choice will be governed by what is available. Until recently, alpha spectrometry was the most widely used method, though increasing numbers of laboratories are now equipped with gamma spectrometers. Alpha spectrometry is more sensitive and most suitable for small samples of very low activity. The detectors are simpler and less expensive, though it is necessary to have access to radiochemical facilities. A significant

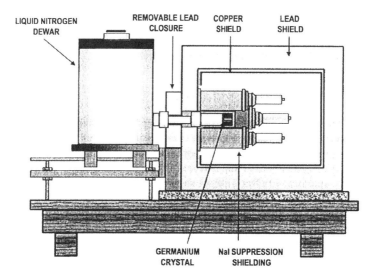

Figure 16. Configuration of a typical well-type gamma spectrometer in the University of Liverpool Environmental Radiometric Laboratory.

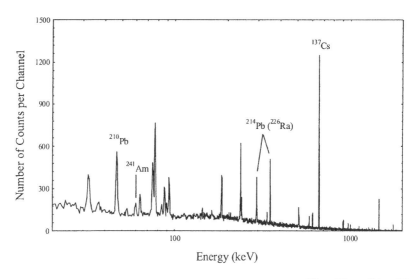

Figure 17. Gamma spectrum from a Blelham Tarn sediment sample showing the ^{210}Pb, ^{241}Am, ^{214}Pb (^{226}Ra) and ^{137}Cs photo peaks.

disadvantage is the time required to establish ^{210}Pb/^{210}Po equilibrium, or to allow for ^{210}Po ingrowth. Further, the method only determines total ^{210}Pb, and ^{226}Ra (supported ^{210}Pb) activity must be determined by a separate procedure, or estimated from the total ^{210}Pb activity in samples below the ^{210}Pb dating horizon. A number of studies have shown that this seriously limits the reliability of sediment chronologies in disturbed catchments where changes in sediment type often give rise to significant changes in ^{226}Ra concentrations (Oldfield et al., 1980).

Advantages of gamma assay include:

- Minimal sample preparation. Since gamma photons can travel significant distances without absorption, measurements can be carried out on dried sediment samples without the need for radiochemical separation.

- Measurements are non-destructive. After gamma assay, samples can be used for further analyses. This is a particular advantage where samples are need for a range of different analyses and the total amount of sediment available is small.

- Simultaneous determination of a range of radionuclides, including ^{210}Pb, ^{226}Ra, ^{137}Cs, ^{241}Am. As pointed out above, measurements of ^{137}Cs and ^{241}Am are invaluable in assessing data from sites with poor ^{210}Pb records, or where there are significant discrepancies between the standard dating models.

Disadvantages of gamma spectrometry include:

♦ higher overheads,

♦ lower sensistivity for some radionuclides,

♦ efficiency calibration is demanding, particularly at low energy.

The importance of ^{226}Ra determinations is illustrated by the two examples shown in Figure 18, from Dalvatn in northern Norway (Norton et al., 1992) and Certova in the Czech Republic (Vesely et al., 1993). In both cases the sediment record contains a discrete layer in which ^{226}Ra activity was significantly higher than in the rest of the core. Where this occurs, ^{222}Rn diffusion, sustained over a number of decades, may result in significant differences between supported (*in situ* generated) ^{210}Pb and ^{226}Ra activity. Calculations using a simple diffusion model show that supported ^{210}Pb activity is reduced in the ^{226}Ra rich sediments and increased in the adjacent layers. Using the corrected values, CRS model calculations indicated that the ^{226}Ra rich sediments were associated with episodes of rapid sedimentation.

For a more detailed account of analytical methods, see e.g. Ivanovich & Murray (1992).

Discussion

It is unlikely that dating by ^{210}Pb will ever be a totally routine procedure. Even in the relatively closed system of a lake and its catchment where the CRS model has an underlying theoretical basis, potential complexities in transport processes are such that neither of

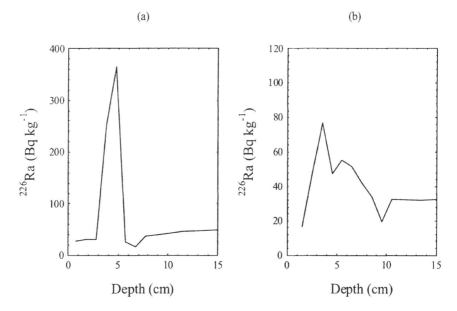

Figure 18. ^{226}Ra activity versus depth in sediment cores from (a) Dalvatn (Norway) and (b) Certova (Czech Republic).

the simple models (CRS and CIC) can be presumed without independent validation. The situation is considerably more difficult in marine environments where there is no *a priori* reason to suppose that ^{210}Pb supply rates are governed by any simple relationship. In these circumstances, the CRS model might just give the best results because of the robustness of the dating parameter. Apart from the fact that integration is a relatively reliable numerical procedure, it also has the effect of smoothing out minor irregularities.

Since validation can, at best, be achieved at a small number of points, good quality results depend on the adoption of good numerical procedures that take account of the strengths and weaknesses of the particular dating model being used. The CIC model will be prone to large errors if there is significant mixing of the surficial sediments. The CRS model may give nonsensical results if there are gaps in the sediment record. Potential dating problems may be highlighted by routine calculation of ^{210}Pb dates using both models. Comparisons between the two sets of dates will help determine appropriate procedures for correcting or improving the initial calculations. These procedures, some of which are outlined above, are best carried out in the light of a good understanding of the potential transport processes controlling the supply of fallout ^{210}Pb to the sediment record.

Summary

The fallout radionuclides ^{210}Pb, ^{137}Cs and ^{241}Am are proving to be of considerable value not just in dating environmental records in lake sediment, but also in assessing their quality, and the nature of the transport processes delivering environmental data to the archive. The simple CRS ^{210}Pb dating model does give reliable chronologies in many instances, but should in all cases be validated by comparison with the CIC model, and with independent chronostratigraphic dates given by, e.g., ^{137}Cs, ^{241}Am, pollen and other markers. Where significant differences occur, possible reasons for the discrepancy should be examined and corrected dates calculated using composite models.

To maximise the possibility of good results the greatest care, and accurate quality control, are needed at all steps in the process, from the coring itself through to the detailed numerical processing of the radiometric results. Of particular importance are reliable sediment density data. Poor quality measurements of water content and dry bulk density seriously compromise the dating calculations, and estimations of fluxes of radionuclides and pollutants.

As the subject develops, important sites are increasingly being resampled and background information from earlier studies is often invaluable in assessing the new data. This process is greatly facilitated by the establishment of data archives that include coring details (date and location) as well as the records they contain. In anticipation of this, it is helpful to encode all relevant information in a simple but robust way at an early stage of the project.

Acknowledgements

I am grateful to colleagues from the Institute of Freshwater Ecology Windermere Laboratory, Environmental Change Research Centre, University College London, Geological Survey of Denmark, the University of Ulster, and the many other institutions who have submitted sediment cores to the University of Liverpool Environmental Radioactivity Research Centre for ^{210}Pb dating. Many of the methodologies and illustrations used in this chapter are based on results from those cores, and on two projects funded by the European Community, the Environment and Climate Programme (MOLAR Project, contract ENV4-CT95-0007), and the Nuclear Fission Safety Programme (Large-scale and long-term environmental behaviour of transuranic elements as modelled through European surface water systems, contract F14P-CT96-0046).

References

Appleby, P. G. & F. Oldfield, 1978. The calculation of ^{210}Pb dates assuming a constant rate of supply of unsupported ^{210}Pb to the sediment. Catena 5: 1–8.

Appleby, P. G., F. Oldfield, R. Thompson, P. Huttenen & K. Tolonen, 1979. ^{210}Pb dating of annually laminated lake sediments from Finland. Nature 280: 53–55.

Appleby, P. G. & J. T. Smith, 1993. The transport of radionuclides in lake-catchment systems. Proc. UNESCO Workshop on Hydrological Impact of Nuclear Power Plant Systems, UNESCO, Paris, 1992: 264–275.

Appleby, P. G. & F. Oldfield, 1983. The assessment of ^{210}Pb data from sites with varying sediment accumulation rates. Hydrobiologia 103: 29–35.

Appleby, P. G., P. J. Nolan, D. W. Gifford, M. J. Godfrey, F. Oldfield, N. J. Anderson & R. W. Battarbee, 1986. ^{210}Pb dating by low background gamma counting. Hydrobiologia 141: 21–27.

Appleby, P. G., N. Richardson & P. J. Nolan, 1992. Self-absorption corrections for well-type germaniun detectors. Nucl. Inst. & Methods B 71: 228–233.

Appleby, P. G., N. Richardson & P. J. Nolan, 1991. ^{241}Am dating of lake sediments. Hydrobiologia 214: 35–42.

Appleby, P. G. & F. Oldfield, 1992. Application of ^{210}Pb to sedimentation studies. In Ivanovich, M. & R. S. Harmon (eds.) Uranium-series Disequilibrium: Applications to Earth, Marine & Environmental Sciences. Oxford University Press, Oxford: 731–778.

Appleby, P. G., V. I. Jones & J. C. Ellis-Evans, 1995. Radiometric dating of lake sediments from Signy Island (maritime Antarctic): evidence of recent climatic change. J. Paleolim. 13: 179–191.

Appleby, P. G., 1997. Sediment records of fallout radionuclides and their application to studies of sediment-water interactions. Water, Air & Soil Pollution 99: 573–586.

Appleby, P. G., E. Y. Haworth, G. Barci, H. Michel, D. B. Short, G. Laptev, J. Merino, B. M. Simon & A. J. Lawler, 1999. Internal project report to the European Commission on 'The large-scale and long-term environmental behaviour of transuranic elements as modelled through European surface water systems', contract F14PCT960046, 45pp.

ApSimon, H. M., H. F. MacDonald & J. J. Wilson, 1986. An initial assessment of the Chernobyl-4 reactor accident release source. J. Soc. Radiol. Prot. 6: 106–119.

Benoit, G. & H. F. Hemond, 1987. A biogeochemical mass balance of ^{210}Po and ^{210}Pb in an oligotrophic lake with seasonally anoxic hypolimnion. Geochim. Cosmo. Acta 51: 1445–1456.

Binford, M. W., 1990. Calculation and uncertainty analysis of ^{210}Pb dates for PIRLA project lake sediment cores. J. Paleolim. 3: 253–67.

Binford, M. W., J. S. Kahl & S. A. Norton, 1993. Interpretation of ^{210}Pb profiles and verification of the CRS dating model in the PIRLA project lake sediment cores. J. Paleolim. 9: 275–96.

Blais, J. M., J. Kalff, R. J. Cornett & R. D. Evans, 1995. Evaluation of ^{210}Pb dating in lake sediments using stable Pb, *Ambrosia* pollen, and ^{137}Cs. J. Paleolim. 13: 169–78.

Cambray, R. S., P. A. Cawse, J. A. Garland, J. A. B. Gibson, P. Johnson, G. N. J. Lewis, D. Newton, L. Salmon & B. O. Wade, 1987. Observations of radioactivity from the Chernobyl accident, Nuclear Energy 26: 77–101.

Cambray, R. S., K. Playford, G. N. J. Lewis & R. C. Carpenter, 1989. Radioactive Fallout in Air and Rain: Results to the End of 1987. AERE-R 13226, Harwell.

Clark, M. J. & F. B. Smith, 1988. Wet and dry deposition of Chernobyl releases. Nature 323: 245–9.

Crank, J., 1975. The Mathematics of Diffusion. Clarendon Press, Oxford, 414 pp.

Debertin, K. & R. G. Helmer, 1988. Gamma and X-ray Spectrometry with Semiconductor Devices. North-Holland, 399 pp.

Dominik, J., D. Burns & J.-P. Vernet, 1987. Transport of environmental radionuclides in an alpine watershed. Earth. Plan. Sci. Lett. 84: 165–180.

Durham, R. & B. G. Oliver, 1983. History of Lake Ontario contamination by sediment radiodating and chlorinated hydrocarbon analysis I. Great Lakes Res. 9: 160–168.

Eakins, J. D. & R. T. Morrison, 1978. A new procedure for the determination of lead-210 in lake and marine sediments. Int. J. Appl. Rad. Isotopes 29: 531–536.

Flynn, W. W., 1968. The determination of low levels of polonium-210 in environmental materials. Anal. Chim. Acta 43: 221–227.

Gägeler, H., H. R. Von Gunten & W. Nyffeler, 1976. Determination of ^{210}Pb in lake sediments and air samples by direct gamma-ray measurements. Earth Planet. Sci. Lett. 33: 119–121.

Israel, H., 1951. Radioactivity of the atmosphere. Compendium of Meteorology, 155–161, Amer. Meteor. Soc., Boston, Mass.

Ivanovich, M. & A. Murray, 1992. Spectroscopic methods. In Ivanovich, M. & R. S. Harmon (eds.) Uranium-series Disequilibrium: Applications to Earth, Marine & Environmental Sciences, Oxford University Press, Oxford: 127–173.

Krishnaswami, S., D. Lal, J. M. Martin & M. Meybeck, 1971. Geochronology of lake sediments. Earth Planet. Sci. Lett. 11: 407–414.

Lewis, D. M., 1977. The use of ^{210}Pb as a heavy metal tracer in Susquehanna River system. Geochim. Cosmochim. Acta 41: 1557–1564.

Norton, S. A., A. Henriksen, P. G. Appleby, L. L. Ludwig, D. V. Verault & T. S. Traaen, 1992. Trace

Metal Pollution in Eastern Finnmark, Norway, as Evidenced by Studies of Lake Sediments. Norsk Institut for Vannforskning Report 487/927, 42pp.

Oldfield, F., P. G. Appleby & R. W. Battarbee, 1978. Alternative ^{210}Pb dating results from the New Guinea Highlands and Lough Erne. Nature 271: 339–342.

Oldfield, F., P. G. Appleby & R. Thompson, 1980. Palaeoecological studies of lakes in the Highlands of Papua New Guinea. I. The chronology of sedimentation. J. Ecol. 68: 457–477.

Oldfield, F. & P. G. Appleby, 1984. Empirical testing of ^{210}Pb dating models. In Haworth, E. Y. & J. G. Lund (eds.) Lake Sediments and Environmental History. Leicester Univ. Press: 93–124.

Pennington, W., R. S. Cambray & E. M. Fisher, 1973. Observations on lake sediments using fallout ^{137}Cs as a tracer. Nature 242: 324–326.

Pennington, W., R. S. Cambray, J. D. Eakins & D. D. Harkness, 1976. Radionuclide dating of the recent sediments of Blelham Tarn. Freshwater Biology 6: 317–331.

Robbins, J. A., J. R. Krezoski & S. C. Mozley, 1977. Radioactivity in sediments of the Great Lakes: post-depositional redistribution by deposit feeding organisms. Earth Planet. Sci. Lett. 36: 325–333.

Robbins, J. A., 1978. Geochemical and geophysical applications of radioactive lead. In Nriagu, J. O. (ed.) Biogeochemistry of Lead in the Environment, Elsevier Scientific, Amsterdam: 285–393.

Santschi, P. H. & B. D. Honeyman, 1989. Radionuclides in aquatic environments. Radiat. Phys. Chem. 34: 213–240.

Schelske, C. L., A. Peplow, M. Brenner & C. N. Spencer, 1994. Low-background gamma counting: applications for ^{210}Pb dating of sediments. J. Palaeolimnol. 10: 115–28.

Scott, M. R., R. J. Rotter & P. F. Salter, 1985. Transport of fallout plutonium to the ocean by the Mississippi River. Earth. Plan. Sci. Lett. 75: 321–326.

Smith, F. B. & M. J. Clark, 1988. Radioactive deposition from the Chernobyl cloud. Nature 322: 690–1.

Vessely, J., H. Almquist-Jacobsen, L. M. Miller, S. A. Norton, P. G. Appleby, A. S. Dixit & J. P. Smol, 1993. The history and impact of air pollution at Certova Lake, Southwestern Czech Republic. J. Palaeolim. 8: 211–31.

Wilkening, M. H., W. E. Clements & D. Stanley, 1975. Radon-222 flux in widely separated regions. In The Natural Radiation Environment 2, chap. 32. U.S. Energy and Research Development Administration, Oak Ridge, Tenn.

10. ^{14}C CHRONOSTRATIGRAPHIC TECHNIQUES IN PALEOLIMNOLOGY

SVANTE BJÖRCK (svante.bjorck@geol.lu.se)
& BARBARA WOHLFARTH (Barbara@geo.su.se)
GeoBiosphere Science Centre, Quaternary Sciences
Lund University, Sölveg. 12
SE-223 62 Lund
Sweden

Keywords: ^{14}C dating, lake sediments, sources of error, chronostratigraphic tool, high resolution dating, ^{14}C calibration, absolute dating techniques, dating strategies.

Introduction

Down-core paleolimnological investigations of lake-sediment sequences require sufficient age control to enable comparisons and correlations on local, regional and global scales. Provided that the sediments are annually laminated, age control may be achieved through counting of annual layers or varves (see Lamoureaux, this volume). However, since only sediments deposited under certain conditions will allow obtaining an annual time resolution, other dating techniques based on, for example, the radioactive decay of certain elements will have to be applied. Among these, radiocarbon dating (^{14}C) is the most widely used and also the earliest radiometric method available. The method was 'invented' in 1951 by Libby (1955), who subsequently introduced and applied it to date the recent geological past. The background and principles of radiocarbon dating have been outlined and discussed in detail in numerous textbooks, journals and articles (see e.g., Lowe & Walker, 1997; Lowe, 1991a; Smart & Frances, 1991; Olsson, 1991; 1986), as well as in specific volumes of the journal *Radiocarbon* and will, therefore, only shortly be summarised here.

^{14}C atoms are continuously produced in the upper atmosphere, where cosmic ray flux leads to the collision of free neutrons with other atoms and molecules. One of the effects of these nuclear reactions is the displacement of protons from nitrogen atoms (^{14}N) to produce carbon atoms (^{14}C). The radioactive ^{14}C isotope survives on average for 8270 years before decaying into the stable element ^{14}N. ^{14}C atoms are rapidly oxidised to carbon dioxide (^{14}CO$_2$), become mixed throughout the atmosphere, and absorbed by oceans and by living organisms during tissue building. During the lifetime of an organism, the carbon used for tissue building will be in isotopic equilibrium with its contemporaneous life-medium (atmosphere, ocean or fresh-water). Upon death, uptake of CO$_2$ stops, while the decay of ^{14}C in the organic tissues continues. Radiocarbon measurements of fossil organic matter are based on this decay process and allow, together with the internationally agreed fixed

W. M. Last & J. P. Smol (eds.), 2001. *Tracking Environmental Change Using Lake Sediments. Volume 1: Basin Analysis, Coring, and Chronological Techniques.* Kluwer Academic Publishers, Dordrecht, The Netherlands.

half-life of the ^{14}C isotope of 5568 years (Mook, 1986), to determine the age of the fossil material. However, because of the relatively short half-life of the isotope, radiocarbon dating can only be applied back to c. 40,000 years. In addition, due to the fairly large measuring uncertainties, varying atmospheric ^{14}C content, combustion of fossil fuels (producing "old" CO_2) and nuclear weapon tests (increased ^{14}C production), the technique can hardly be used for the last few hundred years. For such young sediments, measurements of the short-lived ^{210}Pb and ^{137}Cs radioisotopes (see Appleby, this volume) can complement the ^{14}C method.

Although the three carbon isotopes ^{12}C, ^{13}C, and ^{14}C have natural occurrence ratios, a fractionation of these ratios often occurs in nature. These effects are fairly small, but they can significantly influence radiocarbon ages where the precision is less than 1%. Therefore the ^{13}C/^{12}C ratio is measured and compared with a limestone standard, PDB, which consists of belemnites from the so-called Peedee formation in South Carolina (Craig, 1957), and most terrestrial/lacustrine samples have negative values compared to this standard. During calculation of the ^{14}C age, the ^{14}C activity is normalised in relation to a δ ^{13}C value of -25%, the value for wood. For example, a 5% depletion ($\delta^{13}C = -30\%$) in the ^{13}C/^{12}C ratio implies a 10% depletion in the ^{14}C/^{12}C ratio, which means that the ^{14}C activity should be increased by 10% of the mean lifetime of ^{14}C (8270 years), equivalent to 83 years (Harkness, 1979). Furthermore, all radiocarbon dates are reported with a statistical uncertainty of one-standard deviation, mainly related to uncertainties of measurements and background radiation. If an age is reported as 3560±120 ^{14}C years BP, it means that the 68% confidence interval for the age of this sample ranges at between 3440–3680 ^{14}C BP, while the 95% confidence interval is between 3320–3800 ^{14}C BP.

Lake sediments usually contain a certain amount of organic carbon in the form of terrestrial, telmatic and limnic plant and animal debris and are therefore highly suitable for radiocarbon dating. Since the discovery of the radiocarbon dating method c. 50 years ago, radiocarbon measurements have been performed on numerous lake-sediment sequences in different geographic settings and on all continents, and with a variety of scientific objectives. The majority of these investigations are aimed at obtaining a general age control for the studied sequences by dating selected parts, and only relatively few studies have been directed at recovering very dense sets of ^{14}C dates along the whole sediment column. The scarcity of such high-resolution dated sequences is due to different reasons. The research budget may not have allowed covering the costs for dense ^{14}C dating series or the focus of the study was directed at a specific aim (e.g., dating the onset of sedimentation in a lake basin, its isolation from the sea, or a short, and for some reason more interesting time interval, or a specific event in the lake's history); or, only a rough age estimate was considered necessary, which may be carried out by interpolation between a few dated horizons.

However, the increased demand on studies with good time resolution (e.g., Lowe, 1991b), has led to fewer but more chronologically focussed investigations. Together with the now frequently-used accelerator mass spectrometry (AMS) ^{14}C dating method, this has produced some extremely well-dated limnic sequences, but has also brought attention to many pitfalls connected with the usage and interpretation of both single and large sets of ^{14}C dates performed on various types of materials.

For many years, but no longer, the journal *Radiocarbon* published all radiocarbon dates obtained at different radiocarbon laboratories. It also publishes articles connected to all aspects of radiocarbon dating (e.g., technique developments, dating for archaeology, calibration, ¹⁴C in marine, lacustrine and soil systems), and is thus probably the best source of information for radiocarbon related issues.

Methods and problems

Conventional versus AMS ¹⁴C dating

Two different approaches for radiocarbon dating fossil material are now available. The original method, the so-called 'conventional radiocarbon dating technique' is based on decay counting of the isotope, while the fairly new accelerator mass spectrometry (AMS) technique is based on particle counting. Detailed descriptions of the individual approaches and techniques have been extensively described in the journal *Radiocarbon* and in various textbooks (Lowe & Walker, 1997). The AMS ¹⁴C method, which has been developed and improved over the past 10–20 years, has partly revolutionised radiocarbon dating, because only small amounts of pure carbon are needed to perform a measurement (Linnick et al., 1989).

Before AMS ¹⁴C dating became available, radiocarbon measurements were entirely based on decay counting, i.e., on the 'conventional' technique. For such measurements 0.5–1 g of pure carbon is required (Fig. 1), which means that, in the case of samples with low organic content, fairly large samples are necessary to obtain a ¹⁴C age. Consequently, unless peat, charcoal, large plant/animal fragments or concentrated layers of macrofossils (e.g., moss-rich horizons) could be dated, bulk sediment samples had to be used. Such samples often comprised intervals of 2–10 cm. The resulting chronology was often characterised by a low depth/time resolution and in the case of low organic content, also by fairly large standard errors unless large enough samples were submitted for dating. However, in the case of organic-rich deposits, many attempts were made to overcome this problem by dating very thin sediment slices, and thereby reducing the error within samples. Furthermore, the conventional technique can often be more precise if enough carbon is at hand, which is shown by, for example, the large sets of such high precision dates on tree rings.

A major step forward was achieved with the implementation of the AMS ¹⁴C technique (Fig. 1). The possibility of not only dating very small amounts of sediment material, but also different parts of the sediment (i.e., identified organic material), has reduced many of the errors/uncertainties connected with the conventional dating method. Consequently, AMS measurements can result in a much better time resolution of the individual samples and of the sediment sequence as a whole, and lead to a better understanding of the problems connected with dating bulk sediments. However, its application has also shown that a careful treatment of the individual samples, often consisting of identified macrofossils, is necessary to obtain reliable ages (Fig. 1).

Sources of error

Lake sediments reflect a variety of different deposits ranging on a scale from purely allochthonous to purely autochthonous, minerogenic and/or organic material. They may contain, for example, precipitated and/or in-washed minerogenic matter, terrestrial and

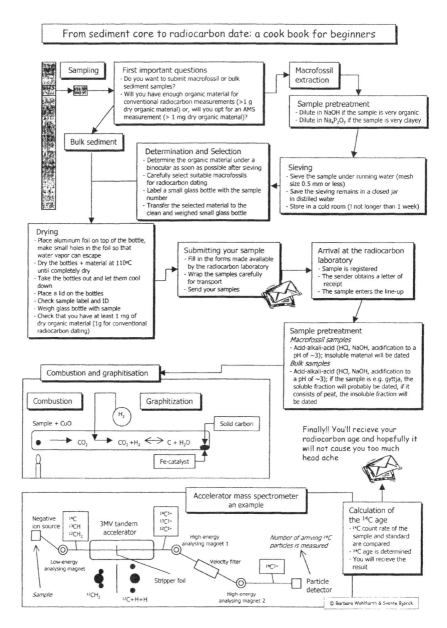

Figure 1. Sketch, illustration the long way from sediment sample to radiocarbon date. The extraction of plant macrofossils from sediment samples and their preparation is according to the technique employed at the Department of Quaternary Geology, Lund University. The shown schematic pictures, which exemplify the combustion/graphitisation process and one type of accelerator mass spectrometer is based upon the techniques used at the Lund University AMS facility.

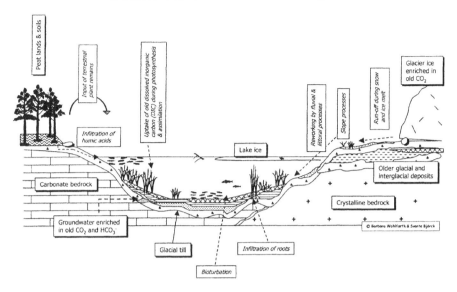

Figure 2. Sketch showing a variety of possible sources of errors, which can influence bulk sediment radiocarbon dates in a hard-water and a soft-water lake, and which are further discussed and exemplified in the text. Although the main difference between the two lake types is related to the carbonate availability from the substratum (carbonate-rich sedimentary rocks and/or sediments vs. non-carbonate bedrock and/or sediments), several other factors may affect the composition of sediments and lake water in both types of lakes. Such are, for example, contamination by old and young organic material (reworking and deposition of older organic deposits, infiltration of humic acids, root penetration, bioturbation), perennial lake ice cover and/or inflow of glacial melt water (enrichment of the lake water by old CO_2). The lake may also be fed by groundwater containing old dissolved inorganic carbon (DIC) or, in volcanic areas the lake water may be enriched in old CO_2 from volcanic emissions. While these latter processes are more obvious in a soft-water lake, they are 'hidden' in hard-water lakes.

aquatic plant and animal remains, including algae, bacteria, fungi, as well as reworked older organic material (Fig. 2). The organic material may also be affected by diagenesis.

Although careful pre-treatments (Fig. 1) are applied to all samples before a radiocarbon measurement (see e.g., Lowe & Walker, 1997), a large number of unknown factors will still influence the resulting radiocarbon age, especially if it has been obtained on the bulk sediment. If only single dates are obtained along a sediment sequence, the likelihood and amount of contamination will, therefore, be difficult to appraise.

Lake water composition is one major limiting factor for obtaining accurate radiocarbon ages on bulk sediment and aquatic plant and animal samples (Olsson, 1991, 1986). A second important factor is contamination by younger carbon through root penetration (Kaland et al., 1984), by percolating younger humic acids, or by downward movement through bioturbation. Bulk sediment samples may also have been contaminated by minerogenic carbon containing "dead" ^{14}C (Fig. 2), such as coal, graphite and chalk (Lowe, 1991b; Olsson, 1968). A further source of error is related to sample storage and sample size (Wohlfarth et al., 1998b; Hedges, 1991; Olsson, 1991; 1979; Geyh et al., 1974). When a sample is submitted for radiocarbon dating (Fig. 1), it is, therefore, very important to know as much as possible about the depositional environment and the post-depositional processes, which may have affected the sample. Furthermore, detailed information on composition, treatment, storage and size of the submitted material usually has to be reported to the radiocarbon laboratory together with the sample (Fig. 1).

A commonly neglected source of error is that ^{14}C dates are often published with 1-sigma (σ) errors. Attempts are then made to fit a best depth-age curve through as many as possible of the reported, or calibrated, ages (with one standard deviation), although statistics tell us that out of a series of, for example 15 dates, only 10 of them should be on the curve. That this is a "forgotten fact" has become more obvious with the increased number of high-resolution studies resulting in highly variable sedimentation rates when the depth-age curve is fitted to as many ^{14}C dates as possible. The other extreme case is the perfectly smoothed curve, often based on some kind of mathematical function fitted to the dates, which leaves out any possibility for sudden sedimentary changes. Such curves may be useful in low-resolution studies of a homogenous sediment sequence, but should be avoided in detailed chronologic studies, since gradual but important changes in sediment focussing may hardly be discernable in the lithologic record but only from detailed dating series. On the other hand, if clear changes in the sediment lithology are observed, a "golden rule" is to try to fit any probable change in sedimentation rate, implied by the (calibrated) datings, to the observed lithologic changes.

Contamination by old and young organic material

An important source of error in lake sediments is related to input of reworked material into the sediments by a variety of natural processes in and around the lake (Fig. 2). This type of contamination thus consists of organic material, which is older than the age of the final sediment deposition. It may, for example, contain older Quaternary organic material (Björck & Håkansson, 1982; O'Sullivan et al., 1973) or pre-Quaternary coal, graphite or lignite particles in the sediments (Wohlfarth et al., 1995b; Björck et al., 1994; Olsson, 1968).

During studies of dating emergence of lake basins (so-called isolations) in the Baltic Ice Lake (Björck, 1979), it was found that datings of organic-poor sediments deposited just prior to the lakes' isolation yielded ^{14}C ages several thousand years older than the datings performed on the more organic-rich sediments from the lowermost post-isolation level. One possible reason for these age anomalies was prescribed to a larger ratio of reworked organic material in the Baltic Ice Lake clays and gyttja clays (Björck, 1979). This suspicion was confirmed by a more systematic study by Björck & Håkansson (1982), who found that the dating error of late-glacial gyttja clays/clay gyttjas is closely related to the amount of re-deposited pollen grains and inversely related to the percentage of organic carbon. Thus radiocarbon dates on sediments not too poor in organic carbon (>3–5%), and with a constant, but small amount of supposedly reworked pollen grains (<1%), yielded negligible dating errors. However sediments, which consisted of a larger portion of reworked grains (i.e., allochthonous organic material), were more subject to dating errors. This was especially true if the sediment contained little organic carbon (<2%), which is often the case with deglacial/late-glacial sediments. To avoid these problems, Björck (1984) extracted large amounts of aquatic mosses from multiple parallel cores by sieving thin clay gyttja horizons rich in mosses. In this way, a high-resolution chronology for the Older Dryas cold period was achieved with the conventional (decay) method. The resulting age turned out to be very similar to the official, and bulk sediment based age of the Older Dryas Chronozone (Mangerud et al., 1974).

However, later studies have been performed to compare radiocarbon measurements on terrestrial macrofossils, i.e., on plants known to incorporate atmospheric CO_2, with those obtained on contemporaneous soft-water bulk sediments. For example, Björck et al. (1998a) could show that late-glacial, soft-water, bulk sediment ages are often at least 200 years, and in some cases up to perhaps 600 years older than the corresponding ages on plant macrofossils (Fig. 3), although organic carbon values are as high as 5–9%. In fact, the results implied that periods of climate change, in spite of increased organic matter, correspond to levels with the largest differences. This was attributed to increased soil erosion and thereby input of reworked, older organic material. Furthermore, the detailed study by Barnekow et al. (1998), who performed AMS ^{14}C dates on more or less carbonate-free bulk sediments (containing c. 20% organic carbon) and on terrestrial plant macrofossils at the same levels along an almost entire Holocene sequence, is a warning example of dating even Holocene, organic-rich, bulk sediments (Fig. 4). The study illustrates a distinct offset and also variability along the sediment core, but the reason is not clear-cut. It may either be due to the presence of reworked old organic material in the sediment, as discussed above, or caused by a lake reservoir effect. In spite of the absence of carbonates in the sediments, the fairly high pH of 7.5–8 in the lake water (Barnekow et al., 1998), shows that it is a hard-water lake and that the dates may suffer from a subtle hard-water reservoir effect (see below).

There can, however, be exceptions to the more or less unspoken rule of preferably dating macro remains from plants and animals in soft-water lakes, or plants utilizing atmospheric CO_2 in hard-water lakes. Some AMS ^{14}C dating series on bulk sediments, complemented by macrofossil dates and well-dated tephras from the Faeroe Islands (Fig. 17) and Iceland, have shown that these often extremely soft-water lakes seem to yield as reliable ages as the macrofossil dates (Björck et al. in prep.). The reason for this anomaly may be that diamicts and glacial sediments in these more northern, less forest covered areas are much poorer in reworked interglacial organic material than what seems to be the case in more southern glaciated regions. For example, Danish glacial tills and clays have been shown to be fairly rich in pollen grains and other reworked microfossils (Iversen, 1936). Similarly, the results by Gulliksen et al. (1998), who performed comparative AMS radiocarbon dates on terrestrial plant macrofossils and on the NaOH-soluble fraction of gyttja on the sediment sequence from Kråkenäs, Norway, show perfect agreement between the two sets. This may also be a good empirical argument for using the NaOH-soluble fraction for sediment dates. In fact, many of the erroneous/questionable sediment dates reported over the years do not seem to have been carried out on the NaOH-soluble fraction, but rather on the whole bulk sediment.

Therefore, to overcome some of the problems mentioned above, it has often been recommended (e.g., Olsson, 1986) to preferably date the NaOH-soluble fraction (mainly humus) in bulk sediments (see more below) to avoid dating the more insoluble old carbon, which is possibly one main reason for too old ages in organic-poor sediments. This procedure is in contrast to the usual procedure of dating the NaOH-insoluble fraction on macrofossils to ensure dating the original organic material.

In arctic oligotrophic lakes from Baffin Island it has been suggested that ^{14}C depleted particulate and dissolved organic carbon (POC and DOC), transported from soils and peat

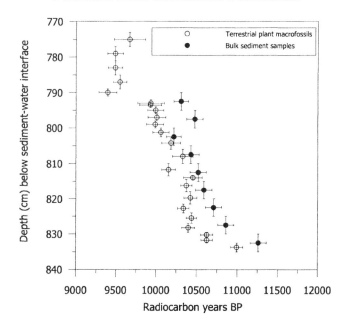

Figure 3. Comparison of results from conventional ¹⁴C measurements on bulk sediment samples (filled circles) and AMS ¹⁴C measurements on terrestrial plant macrofossil material (open circles) at the site Madtjärn in southwestern Sweden (Björck & Digerfeldt, 1991; Björck et al., 1998a). All measurements are displayed with one standard error. For the bulk sediment samples, which were measured earlier with the conventional technique, much more sediment material was needed (Björck & Digerfeldt, 1991), shown by the large vertical bars, compared to the later AMS ¹⁴C dated plant macrofossils (Björck et al., 1998a).

in the watershed of the lakes, may have a large influence on the age of the surface sediments (Abbott & Stafford, 1996). The ¹⁴C age of the sediment-water interface was dated to c. 1000 years BP in three different lakes, while the age of soils and peat varied between 1600–5400 years BP, and the turnover time for organic matter in soil profiles of the watershed was >2000 years. In such extreme environments, with lakes of low aquatic production, the allochthonous organic fraction may make up a large part of the total organic matter, and may thus influence the composition of the sediments. If the allochthonous fraction consists of old organic matter, washed out from peats and soils, it will obviously increase the age of the sediment. Such considerations may be very important for paleolimnologic studies of lakes with low productivity surrounded by a landscape rich in peatlands and thick soils.

Contamination of a sediment sample by younger carbon (Fig. 2) may either arise from roots penetrating into the sediments, from infiltration by younger humic acids or through bioturbation (Olsson, 1991). A study by Kaland et al. (1984) showed that *Isoëtes* roots, which penetrate far down into older sediments, can substantially influence the age of bulk sediment samples. Bulk radiocarbon dates on clayey sediments underlying peat also displayed ages several hundred years younger than expected (Hedenström & Risberg, 1999;

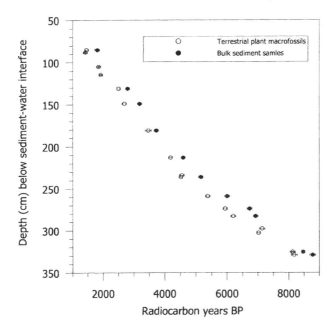

Figure 4. Comparative AMS ^{14}C measurements on bulk sediment and terrestrial plant macrofossils from Lake Vuolep Njakajaure, northern Sweden after Barnekow et al. (1998). The terrestrial plant macrofossil samples comprised fruits, twigs, catskin scales, budscales and leaves of *Betula nana* and *B. pubescens*, leaves of *Salix* sp., fruits of *Alnus incana* and needles of *Pinus sylvestris*. The off set between bulk sediment and terrestrial dates varies considerably (800–300 ^{14}C years), but shows a decreasing tendency upwards in the sequence. According to Barnekow et al. (1998), the bulk organic material in Lake Vuolep Njakajaure, which has a carbonate content of <2–3% and an organic content of 40–50%, consists mainly of algae, the pH of the lake varies between 7.5–8, water depth is 14 m and bottom waters are characterised by oxygen deficiency. The closest outcrop of carbonate bedrock is situated c. 2 km from the lake, but the Quaternary glacial deposits around the lake may contain carbonates.

Åkerlund et al., 1995) and the errors are explained by contamination through either rootlets or humic acids, which percolated down into the sediment. Bacterial action within sediments has generally little effect on radiocarbon measurements, because the carbon atoms are recycled (Hedges, 1991). However, if this recycling involves transport of metabolites in the sediment column, it will result in a type of "molecular bioturbation" (Hedges, 1991).

General lake reservoir effects
If the ^{14}C/^{12}C ratio of the carbon, from which the aquatic plants built up their tissue was lower than the ^{14}C/^{12}C ratio in the CO_2 of the contemporaneous atmosphere, the dating of such plant material is regarded to have been affected by the so-called "lake reservoir effect".

Radiocarbon measurements on bulk sediment samples from soft-water lakes have generally been regarded as giving fairly accurate dates. Less attention has, therefore, been placed on likely errors connected with such samples, although Olsson (1986) and Sutherland (1980) have highlighted these problems already many years ago. An age gradient between soft lake water and the atmosphere can be explained in different ways. It may either be related to the fact that the lake has been efficiently sealed off from the atmosphere by lake ice, that the lake is mainly fed by water from a glacier containing old CO_2, or that "old" groundwater completely "contaminates" the age of the lake water. However, volcanic activity may also contribute to an increased age of the lake water by input of older CO_2 (Hajdas, 1993; Sveinbjörnsdóttir, 1992; Olsson, 1986).

During studies of Antarctic lakes around the Antarctic Peninsula (Björck et al., 1996b; 1993; 1991b; 1991a; 1991c; Zale, 1994; Björck & Zale, pers. comm.), in the Vestfold Hills (Bird et al., 1991), and in the McMurdo Dry Valleys (Doran et al., 1999; 1994; Squyres et al., 1991), major dating problems have been encountered, mainly related to reservoir effects. Some of these studies can be used as examples, although extreme, for these effects. They are mainly caused by the influence of glacial melt-water, containing old CO_2, and the insufficient equilibration between the lake carbon reservoir and atmospheric CO_2. The influence of the former effect is related to how well the "old" glacial melt-water is mixed with atmospheric CO_2. One test by Doran et al. (1999) showed that 7500- year old dissolved inorganic carbon (DIC) of glacier ice was modernized to 600 years in a near-by pool, and the melt-water was completely equalized with modern $^{14}CO_2$ when it reached a lake 3 km downstream.

The second effect is related to the perennial ice cover, which seals off the lake water from the atmosphere, often in combination with influences from glacial melt-water. The age of DIC of the surface water in two of the Dry Valleys lakes (Lake Hoare and Lake Bonney) varies between 1600–2000 years, while DIC of the bottom water has ages of 2700 and 10,000 years BP, respectively, without reducing for the surface reservoir (Doran et al., 1999). Furthermore, ^{14}C dates of microbial mats in a sediment core from Lake Hoare suggest a reservoir age of 2600 years for the surface sediments, which thus fits well with the age of DIC of the bottom waters.

The extreme ^{14}C results from some of the lake studies from Antarctica can thus explain some of the more subtle dating anomalies in less extreme environments. For example, the paleolimnological studies around the Antarctic Peninsula have shown that radiocarbon dates on aquatic mosses from soft-water lakes in this region can generally be regarded as fairly reliable. Their validity could be confirmed by dates obtained on terrestrial mosses and through tephra correlations. In fact, most of the other dated components of the Antarctic Peninsula lake sediments — whole bulk samples, the NaOH-soluble fraction, and the NaOH-insoluble fraction — resulted in highly variable and considerably higher ages than the moss dates (Björck et al., 1991b). The reason for these anomalies is not clear, but may be related to the presence of undetected, fine-grained coal particles. However, not all dates performed on mosses or moss-rich sediments seem reliable. For example, a ^{14}C dating series on bulk sediments rich in aquatic mosses, and pure mosses, from a lake on Horseshoe Island, at the south-western part of the Peninsula, shows an age of c. 2000 years for the surface sediments (Fig. 5) consisting of moss gyttja. With the exception of the lowermost dated level, it also shows that mosses and moss gyttjas attain similar ages. The implied reservoir effect of 2000 years is similar to the one found in the Dry Valleys lakes. The main inflow to this lake comes from a small glacier, which makes up one part of the lakeshore. During the corings, which were performed in the late part of the summer season, most of the lake was ice-free. The carbon reservoir of the lake water, especially the surface water, is thus probably fairly well mixed with the atmosphere in late summer. However, if the lake water mass is vertically separated by a thermocline, with cold glacial water (containing old CO_2)

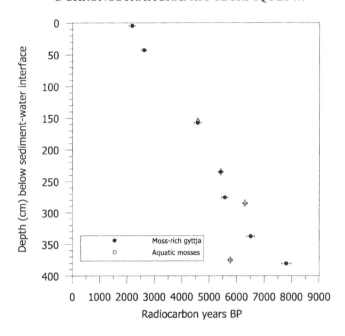

Figure 5. Radiocarbon measurements on bulk sediments abundant in aquatic mosses (open circle) and on aquatic mosses (filled circle) from Lake Zano, Horseshoe Island, Antarctica (Björck et al., 1991b). One sigma standard error is shown for the radiocarbon age as well as depth errors. The moss-rich gyttja at the sediment surface yielded an age of 2170 ± 120 ^{14}C years BP. Note the varying age differences, which may be due to seasonal changes of reservoir effects (see text).

at the bottom, a reservoir effect would be expected for the mosses living on the lake bottom. It is also possible that, if the lake was largely ice-covered during the growing season of the dated mosses, uptake of CO_2 was mainly restricted to the CO_2 of the glacially contaminated lake water, thus causing the aging of the dated material. The anomalous lowermost level (Fig. 5) can be explained by well mixing of the water column with atmospheric CO_2 during the growing season for the mosses at 5700 BP (Fig. 5), while the lake was poorly mixed during the lifespan of most of the other remaining organic sediment-forming organisms.

Another type of lake reservoir effect found by the Antarctic studies is contamination of the lacustrine environment by a marine reservoir effect (e.g., Zale, 1994; Björck et al., 1991b). Because of the high marine reservoir effects of 1000–1700 years (see summary in Berkman et al., 1998) in Antarctic marine mammals and birds, coastal lakes may be severely "polluted" by heavily marine influenced particulate carbon from seals and birds colonising the lakeshore. For example, in Lake Boeckella in Hope Bay, attempts have been made to quantify the effect of penguin guano on the ^{14}C age of the sediments (Zale, 1994), and the apparent age was found to be at least 1600 years. The effect of this type of marine eutrophication and "radiocarbon pollution" is obviously related to lake size/lake volume, size of the animal colonies, the local marine reservoir effect, and the productivity of the lake itself. Although these Antarctic environmental scenarios are fairly extreme, it should not be

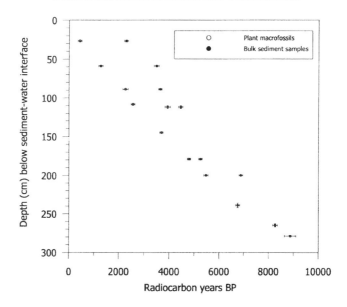

Figure 6. Comparison of AMS ^{14}C measurements on carbonate-rich bulk sediment samples and terrestrial plant macrofossils along the same sediment sequence in Lake Tibetanus, northern Sweden (Barnekow et al., 1998). The terrestrial plant macrofossil samples consisted of leaves of *Dryas octopetala*, leaves, catskin scales and twigs of *Betula* sp. and *Salix* sp. and of *Pinus sylvestris* needles. The lake possibly receives old dissolved inorganic carbon through groundwater and surface runoff from an outcrop of calcite marble situated close by. The smallest age difference between the two dating sets is at 1.78–1.80 m, where the sediment is enriched in terrestrial plant macrofossils. The carbonate content of the sediments is around 90% in the lower part of the sequence and decreases slightly upwards to between 40–80%.

ruled out that many landscapes during, for example, the glacial and deglacial periods might have been subject to similar conditions (Lowe et al., 1988). Consequently, radiocarbon dating of such records has to be carefully scrutinized.

Owing mainly to the presence of old CO_2 in the eruptive gases, the lake water and the surrounding vegetation in volcanic areas may be depleted in [14]C (Olsson, 1986). Radiocarbon ages on lake sediments and plants growing in the vicinity of volcanic emissions have been reported to be several thousand years older than expected (see e.g., references in Olsson, 1986 and Sveinbjörnsdóttir et al., 1992). Comparable AMS radiocarbon dates on terrestrial plant material and aquatic mosses from two lake basins on Iceland by Sveinbjörnsdóttir et al. (1992) varied by several thousand years, and modern mosses yielded radiocarbon ages of 6000–8000 years BP.

Hard-water reservoir effects
In areas with calcareous bedrock and/or soils, the lake water is alkaline and thus rich in bicarbonate ions. Aquatic plants will, during photosynthesis, take up and incorporate "fresh" carbon together with a smaller or larger fraction of this "dead", old carbon (Fig. 2). This so-called 'hard water effect' will result in considerably older radiocarbon dates (Figs. 6 and 7)

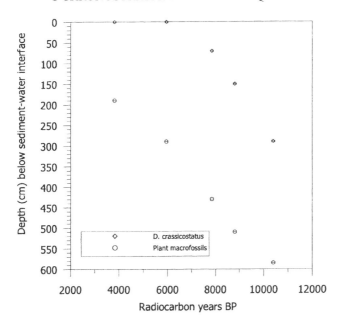

Figure 7. Comparative AMS ^{14}C measurements on aquatic bryophytes (*Drepanocladus crassicostatus*) and terrestrial plant macrofossils from the same sediment sequence in Lake Toboggan, western Canada (MacDonald et al., 1991b). While it has generally been assumed that these bryophytes do not take up ^{14}C-deficient carbon by incorporation of bicarbonates, the radiocarbon measurements on *D. crassicostatus* clearly show the unreliability of these mosses for radiocarbon dating. It is argued that the ^{14}C-deficiency of the mosses could originate from one or several of the following mechanisms: the generation of ^{14}C-deficient CO_2 through isotopic exchange, the formation of CO_2 from bicarbonate by chemical processes, and respiration and decomposition of aquatic organisms that have incorporated ^{14}C-deficient bicarbonate can generate isotopically similar CO_2 (MacDonald et al., 1991b).

compared to the actual time of deposition (Deevey et al., 1954; Shotton, 1972; Olsson, 1986; Andrée et al., 1986; Ammann & Lotter, 1989). This is thus another source of error, specific for hard-water lakes, in addition to the sources mentioned above. Several attempts have been made to quantify the hard-water effect along a sediment sequence. This can either be done by comparing radiocarbon dates on gyttja samples with those obtained on carbonates and plant macrofossils from within the same lake (Andrée et al., 1986; Ammann & Lotter, 1989), by parallel dating of bulk-sediment and terrestrial plant macrofossil samples on the same sediment core (Fig. 6) (see e.g., MacDonald et al. 1991b, 1991a; Barnekow et al. 1998; and Törnqvist et al. 1992), or by parallel radiocarbon measurements of the bulk sediment and the organic fraction (Geyh et al., 1998). These authors showed that the effect can vary considerably through time along a sediment sequence, and it seems thus almost impossible to estimate the 'true' age of carbonate-enriched bulk sediment samples. In some cases the hard water effect may not be obvious, because the effects are subtle and the sediments are more or less devoid of carbonates (Fig. 4). During such circumstances all possible clues have to be taken into consideration: pH of the water, proximity to a carbonate source, water depths, oxygen conditions and sediment type. Oxygen defficient bottom water in Lake Vuolep Njakajaure (Barnekow et al., 1998) may explain the absence of carbonates, and by partly using old DIC for assimilation, algae (which completely dominate the sediments) may account for a possible hard water effect (Fig. 4).

Given the scarcity of terrestrial plant macrofossils in some lake sediments, attempts have been made to use aquatic mosses, which were considered to prefer up-take of atmospheric CO_2 to obtain radiocarbon measurements on hard-water lake sediments. However, MacDonald et al. (1991b) could show (Fig. 7) that, in the case of extreme hard-water lakes, these specific moss samples yielded too old dates, but with a highly varying "error" in relation to the terrestrial dates. The reason for these old moss ages may, however, be complex (see Fig. 7 and figure text).

Sample storage, preparation and size
It is known that bacterial action is common on and in sediment cores stored at room temperatures or generally stored for a too long time. Bacteria may fix substantial quantities of CO_2 from the surrounding atmosphere, which then becomes incorporated in the sediment. Radiocarbon measurements, which are performed on such sediments, very likely result in younger dates than expected (Geyh et al., 1974). Contamination of foraminifera and molluscs by modern carbon during storage has been reported by Olsson (1991), who suggested that such samples should be stored sealed off from the atmosphere. Wohlfarth et al. (1998b) showed that bacterial action and/or contamination by fungi might be a serious problem if terrestrial plant macrofossils are stored too long in water prior to the radiocarbon measurements. Bacteria/fungi are able to use their surrounding medium (water, air) for CO_2 uptake, which will then become incorporated in the radiocarbon sample. Dust particles, which may become mixed with the dated material during the preparation process, may be a further source of error. If a sample contains a fairly large amount of organic carbon and if the contamination is only minor, such an error is hardly detectable. However, samples with a low carbon content and a higher degree of contamination will result in considerably younger ages (Olsson, 1979; Wohlfarth et al., 1998b), as displayed in Figures 8a and b.

In order to avoid contamination of a sample by, for example dust, the preparation, extraction and determination of the macrofossils has to be performed very carefully. Wet storage of the selected macrofossils should be avoided in order to prevent bacterial/fungal activity and the sample should be dried at 100–110 °C as quick as possible (Fig. 1). This can be done either in a clean glass bottle or on aluminium foil. To obtain a reliable measurement, the carbon content of the sample should ideally be >1 mg and >1 g for AMS and conventional datings, respectively.

Calibration of radiocarbon dates and procedures on reporting ages
After a few decades of experiences with radiocarbon dating and an increased number of radiocarbon measurements on tree rings from older tree ring chronologies (Pearson et al., 1986), it became obvious that the obtained ^{14}C ages did not correspond to the tree ring (or dendro) age, i.e. the assumed true calendar year age. We now know that these differences between calendar and ^{14}C ages are related to changes in the production rate of atmospheric ^{14}C, caused by the geomagnetic and solar influence on cosmic ray flux, and changes in global ocean ventilation rates (Stuiver & Brazunias, 1993). The added effects of these processes have led to significant age differences between calendar and ^{14}C years, especially in pre

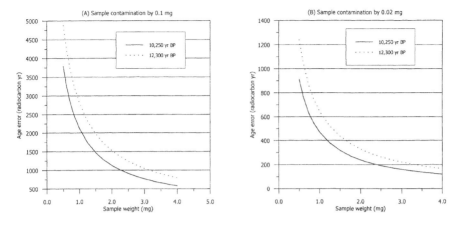

Figure 8. Resulting age errors of samples with expected radiocarbon ages of 10,250 and 12,300 ^{14}C years BP, respectively, and with a varying dry sample weight when contaminated by recent material in the order of (A) 0.1 mg, and (B) 0.02 mg (Wohlfarth et al., 1998b).

mid-Holocene time. Furthermore, these effects are occasionally also expressed as abruptly changing differences between the two time scales.

Apart from the obvious advantage of relating one's records to true ages, it is necessary to use a time scale with a constant length of a year when e.g., rates of change or true sedimentation rates are estimated. This often necessitates the use of calendar years, and therefore different attempts, procedures and programs for converting ^{14}C ages into calendar years have been developed over the years. These so-called calibrations have usually been deployed on ^{14}C dated and (assumed) calendar age based archives, such as annually laminated (varved) sediments and tree rings.

Today's internationally agreed radiocarbon age calibration record, INTCAL98, consists of radiocarbon dated European and American tree ring chronologies (oak and pine), covering the last 12,000 calendar years, in combination with ^{14}C dated laminated marine sediments and ^{14}C/U-Th dated corals (Stuiver et al., 1998) covering the time period between 12,000–24,000 calendar years. This record can be shown as a calibration curve (Fig. 9), displaying the relationship between radiocarbon years and assumed calendar years through time, from which it is possible to obtain a fairly good estimate on the calendar age of a radiocarbon age. However, accurately calibrated ages for radiocarbon dates are obtained by analysing the dates with so-called calibration programs. These programs, which can be downloaded from the web sites of the ^{14}C laboratories in, for example, Oxford, Belfast or Seattle, are designed to give the statistically most likely calendar year time span for a specific radiocarbon age (with its reported confidence interval). The calibration results depend on the extent of the confidence interval of the reported radiocarbon age in combination with the confidence intervals of the radiocarbon and calendar dates of the time in question in the calibration data set, and the detailed structure of the calibration curve. The latter means that periods with rapidly changing ^{14}C ages will be more accurately dated (smaller confidence intervals) than periods with stable ^{14}C ages, so-called ^{14}C plateaux.

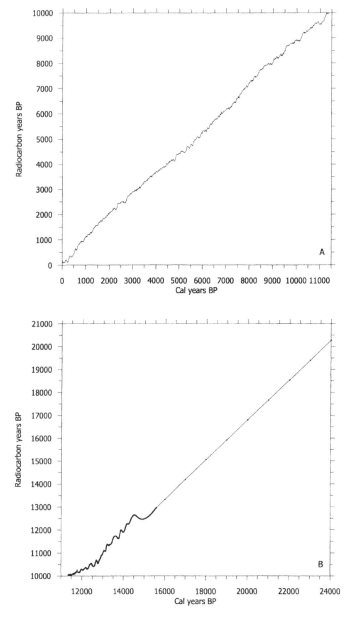

Figure 9. The INTCAL98 radiocarbon calibration curve (Stuiver et al., 1998), divided into (A) a Holocene and (B) a pre-Holocene part. This calibration curve is based on a ^{14}C dated tree-ring chronology (back to 12,000 cal years BP), on a ^{14}C dated varved marine sequence (back to 14,700 cal years BP) and on paired ^{14}C/U-Th measurements on corals (back to 24,000 years BP).

All calendar and radiocarbon ages are related to the year of 1950 (AD 1950). Regarding procedures in reporting calendar and ^{14}C ages, respectively, the former used to be related to BC (before Christ) and AD, while the latter was related to BP (before present=AD 1950). This meant that the calendar ages older than AD had to be added with 1950 years to be comparable with the radiocarbon dates and be related to the present, while the younger AD related ages had to be subtracted from 1950. However, now that many different types of more or less well-dated archives exist, it has lately become more common to clarify which type of years the time scale is based on and relate it to present time, i.e., AD 1950 (BP), such as ^{14}C years BP, varve years BP, ice years BP, calendar years BP, and calibrated (cal.) years BP.

Radiocarbon-dating different fractions of the sediment

The many uncertainties related to radiocarbon-dating bulk sediments and the difficulties that often arise when interpreting and comparing the obtained results have, in general, led to a more careful selection of the material that is submitted for radiocarbon measurements. With the introduction of the AMS ^{14}C technique, it has become possible to date carefully selected fractions of the sediment (Fig. 1).

In the following, we briefly discuss the advantages and disadvantages connected with different types of dateable material. The type of dating technique (decay/particle counting) chosen for the measurement will depend on the amount of material available for dating.

Macrofossils

Macrofossils found in limnic sediments derive either from the lake's catchment, the lakeshore or from the lake itself, and may thus be part of both the autochthonous and the allochthonous fraction of the sediment (Fig. 2). Therefore, the macrofossils in lacustrine sediments are possibly more or less reworked and transported from the terrestrial and telmatic environment into the deeper parts of the lake, where the corings are usually performed. The underlying philosophy behind dating terrestrial macrofossils is, however, that the final reworking and transportation usually takes place fairly soon after the death of the organism in question. For radiocarbon dating, only macrofossil material from plants/animals, which use atmospheric CO_2 for their tissue up-building, should be selected (Törnqvist et al., 1992).

The sediment samples have to be sieved (under running water and through <0.5 mm sieves) and the obtained organic material should preferably be identified to the species level (Fig. 1). This identification is important for a distinct separation between terrestrial, telmatic, and limnic plants. Furthermore, clearly reworked older material, such as highly corroded leaf fragments or wood pieces, may be present among the selected terrestrial macrofossils. It has been shown that especially wood fragments, which can with-stand erosion and thereby could have been part of several redepositional cycles before final deposition, may give considerably older measurements as compared to leaf fragments extracted from the same sediment level (Barnekow et al., 1998; Hajdas, 1993). If suspected old reworked macrofossils have to be submitted, it is advisable to perform a number of additional and reliable dates along the sediment sequence to assess any significant reworking effect on the ages.

In many cases sediments may be poor in organic macro remains. It can, therefore, be difficult to extract enough suitable plant fragments for dating. This has often led to thick columns of sediment being washed for macrofossils and resulted in radiocarbon dates representing perhaps 5–20 cm of sediment. For chronologic details, such dates are often unsatisfactory because of the lack of knowledge about the precise level from where the dated macrofossil(s) originate, but may be used for establishing rougher age estimates. One obvious way of circumventing the problem of detailed age control is to obtain a series of parallel cores at the same coring point. These cores are then precisely correlated with each other, primarily by lining up all the cores against each other, complemented by detailed sediment descriptions and routine sediment analyses such as measurements for organic carbon and magnetic susceptibility (see Sandgren & Snowball, volume 2). If an accurate correlation can be achieved, the same stratigraphic levels may be sampled in all parallel cores with a very dense sampling strategy (0.5–2 cm). In this way, enough plant remains for dating may be extracted for a large set of thin sediment slices (see e.g., Andresen et al., 2000; Björck et al., 1996a), in spite of a restricted abundance of macrofossils.

Although much less common, it is also possible to use faunal remains for radiocarbon dating. For example, terrestrial insects extracted from lake sediments have shown to provide reliable radiocarbon dates (Elias & Toolin, 1990; Elias et al., 1991). A study, that focussed on the possibility to AMS radiocarbon date chironomid remains from soft-water lake sediments showed the potential and limit of this type of material (Jones et al., 1993).

In extremely organic-poor sediments, the search for dateable material may occasionally lead to surprising results. When it was found that a small part of the fine sand fraction in an Antarctic soft water lake on James Ross Island consisted of organic-like spherules of unknown origin (Björck et al., 1996b), they were not originally thought to make up the chronologic basis for the study. However, after several consultations with biologists/limnologists, it was discovered that these features were eggs of a freshwater crustacean, *Branchinecta gainii*, a today extirpated species on James Ross Island. Several hundred eggs were picked out from two sediment levels, which resulted in the possibly two most reliable ^{14}C dates from James Ross Island (Björck et al., 1996b).

Pollen

Pollen can be found in almost all limnic sediments and many local and regional pollen stratigraphies and stratigraphic correlations are based on pollen analytical investigations. It would, therefore, be of great value to directly date the pollen stratigraphic boundaries. Furthermore, the advantage of radiocarbon dating pollen as compared to macrofossils is that pollen are present in many different types of sediments and throughout a whole sediment sequence, while macrofossils may be sparse or irregularly present. Pollen are also predominantly of terrestrial origin and provide radiocarbon dates that are not influenced by lake reservoir effects. It was therefore not surprising that the successive development of the AMS technique and the possibility to radiocarbon date very small samples, resulted in exploring the possibility of obtaining AMS radiocarbon dates on pollen concentrates (Brown et al., 1989; Brown et al., 1992; Long et al., 1992; Regnell, 1992; Richardson & Hall, 1994; Mensing & Southon, 1999). In a pilot study by Brown et al. (1989), the samples were treated following normal pollen analytical preparation procedures, however, omitting the acetolysis step. Pollen concentrates were then obtained through step-wise sieving

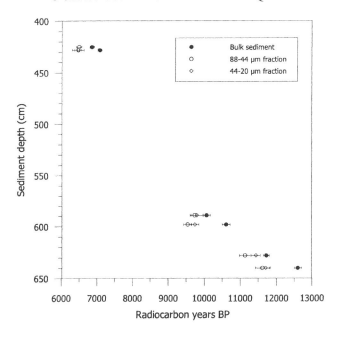

Figure 10. Comparative AMS radiocarbon measurements on bulk sediment samples and pollen concentrates by Brown et al. (1989). The dated fractions of the pollen concentrate (44–88 μm, 20–44 μm) mainly consisted of coniferous pollen grains. The preparation procedure is described in detail by Brown et al. (1989).

(88 μm, 44 μm and 20 μm) and bleaching of the residues. AMS dating was performed on the 88–44 μm and 44–20 μm fractions. The first results by Brown et al. (1989) were very promising and showed the potential of the method, as compared to bulk sediment radiocarbon dates (Fig. 10). Regnéll (1992) slightly modified the method by Brown et al. (1989) and made a comparative study, where the same sediment sample was divided into four sub-samples, which in turn were pre-treated in different ways. Because of the small pollen size in this study, AMS dating was performed on the fraction between 10–20 μm. The results showed a clear age difference between bulk sediment and pollen concentrates, but also between pollen concentrates prepared in different ways (Fig. 11). Although these preparation procedures were able to remove as much non-pollen material as possible to obtain clean pollen concentrates, it was recognised that the concentrates still contained unwanted organic material. Regnéll (1992) therefore suggested that each sample should carefully be examined under the microscope before it is submitted for radiocarbon measurements. Several attempts have since then been made to remove non-pollen components in pollen concentrates. Long et al. (1992) suggested manual separation with a micromanipulator, Regnéll & Everitt (1996) presented a preparative centrifugation method, and Richardson & Hall (1994) a microbiological degradation method. Recently, Mensing & Southon (1999) presented a further development, which is based on a modified version of the pre-treatment procedure described in Brown et al. (1989). However, to clearly

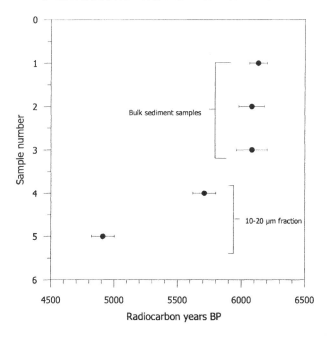

Figure 11. Radiocarbon dated pollen concentrates, based on the 20–10 μm fraction, by Regnéll (1992) with a slightly modified version of Brown et al. (1989).

separate between pollen and organic detritus, which was still present in the concentrates after the different pre-treatment and sieving steps, they picked individual pollen grains under the microscope with the help of a mouth pipette. The picked pollen were stored in a vial, from which they were directly pipetted into quartz combustion tubes and dried in a vacuum centrifuge.

In general, these pollen concentrate studies yielded ages that were usually younger, in some cases significantly younger, than the corresponding bulk sediment ages, but in line with agreed upon ages from terrestrial macrofossils. Clearly, pollen concentrates have the potential to yield good dateable material. It is, however, crucial to know exactly what the samples consist of (i.e., to remove all non-pollen material before radiocarbon dating). In addition, it is a time-consuming method compared to many other ways of separately dating different parts of the sediment.

Alkali soluble (humic) and alkali insoluble (humin) fractions of a sediment

As a consequence of the many problems encountered in radiocarbon dating bulk sediments, Olsson (1979, 1986, 1991) developed a pre-treatment method, which allowed dividing the sediment sample into alkali-soluble (SOL) and alkali-insoluble (INS) fractions (Fig. 12). Her experience showed that the SOL fraction, which was likely to contain organic material from the time of the sediment deposition, yielded reliable ages for sediment samples, while

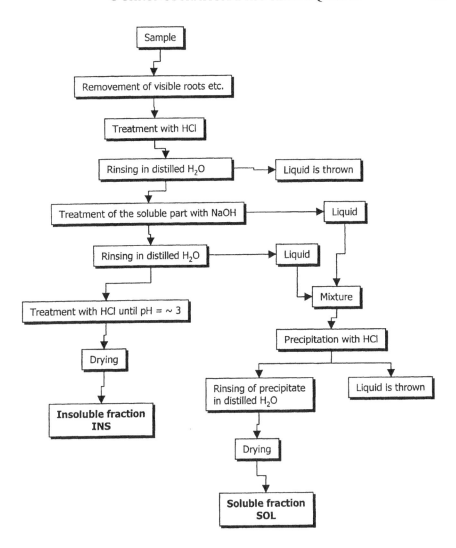

Figure 12. Pre-treatment scheme for bulk and plant macrofossils samples suggested by Olsson (1986, 1991). Although this pre-treatment has since then been slightly modified, the basic steps are performed in most radiocarbon laboratories. SOL = base soluble, humic acid, INS = base insoluble, residual, humin.

the INS fraction, which contained old carbon material (e.g., graphite, coal) resulted in far too old ages. In contrast, for wood, charcoal, peat, and other macrofossils, the INS fraction proved to be more reliable than the SOL fraction.

Lowe et al. (1988) explored the possibility of AMS radiocarbon dating different fractions of sediment samples. They focussed on the humic acid fraction, lipid samples, residues of chlorite treatment (cellulose, mineral component), and residues of HF/HCl

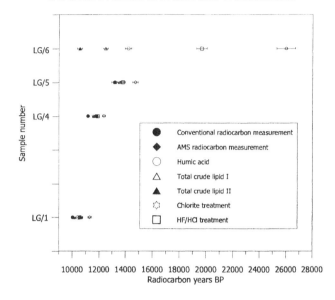

Figure 13. Measured AMS radiocarbon ages of various fractions of four samples from Llyn Gwernan (Lowe et al., 1988). The older than expected ages of the chlorite-treated (cellulose, residual mineral components) and the HF/HCl treated (humin/insoluble fractions, mineral components) samples are due to the mineral carbon error inherent in the sediments.

treatment (humin, mineral component). The resulting ages of the four investigated samples, which are displayed in Figure 13, show a clear tendency towards older ages than expected for the chlorite-treated and the HF/HCl residues. Since these two fractions contain mineral residues, they reflect the mineral carbon error inherent in the sediments. Walker & Harkness (1990) performed a comparative series of radiocarbon dates on the alkali soluble (humic) and alkali insoluble (humin) organic fractions on a sediment sequence from southern Wales. Their results showed, in general, older ages for the humin fraction (Fig. 14). Based on a detailed evaluation of their radiocarbon measurements, Walker & Harkness (1990) regarded the humin fraction, i.e. the alkali-insoluble fraction (INS), as giving more correct ages, which is actually contradicted by most other studies on this topic.

However, as shown by Björck et al.'s (1994) study, radiocarbon dating of different fractions does not always lead to clarifying results (Fig. 15), especially when dealing with sediments in areas with low aquatic productivity. The set of radiocarbon measurements from Lake Boksehandsken, eastern Greenland, which was performed on bulk sediments, plant remains, a marine mollusc, and the INS and SOL fractions, displayed dates for both fractions that were several thousand years older than the assumed deglaciation chronology, and with the INS fraction always being the oldest. It was also found that the age difference between the two fractions was positively correlated to the amount of carbon being burnt at higher temperatures. The organic carbon content of the samples was less than 2.5%, except for the three uppermost bulk samples, which had an organic carbon content of 4–>8.5%.

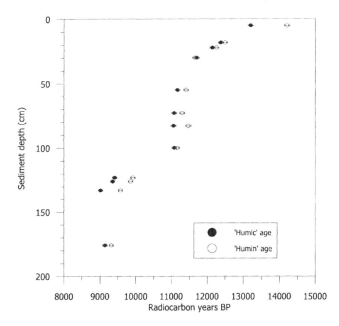

Figure 14. Comparative radiocarbon measurements on the alkali soluble, 'humic' (SOL) and the alkali insoluble, 'humin' (INS) fractions of sediment samples from Llanilid, South Wales according to Walker & Harkness (1990). Except for the lowermost sample, the radiocarbon measurements obtained on the insoluble (INS) fraction are in this study regarded to yield the most reliable dates. This conclusion is based on pollen stratigraphic considerations.

Björck et al. (1994) argued that the samples were contaminated by both old Quaternary reworked organic material, as indicated by an infinite age of plant material (see text in Fig. 15), and local Jurassic coal, of which the latter possibly makes up a significant part of the INS fraction carbon. The effect of these two carbon fractions thus becomes more severe the lower the carbon content of a sample is.

Hedges (1991) and Vogel et al. (1989) discuss the applicability of different sediment compounds for AMS radiocarbon dating and give further references. The study by Vogel et al. (1989) illustrated the large differences in ages obtained from different types of material and the complexity of radiocarbon dating different fractions of a sediment. However, except for dating the alkali soluble (humic) and insoluble (humin) fractions, little research has been performed on dating e.g., carefully extracted lipids.

^{14}C as a chronostratigraphic tool

Long before the radiocarbon method came into general practice in the late 1950's, different ways of using lake sediments as geologic dating tools had already been in use. Changes in, for example, the lithology of the sediments were originally often used for correlations between areas and regions, since these were assumed to be synchronous. These lithostratigraphic correlations were often carried out on late- and postglacial sediments.

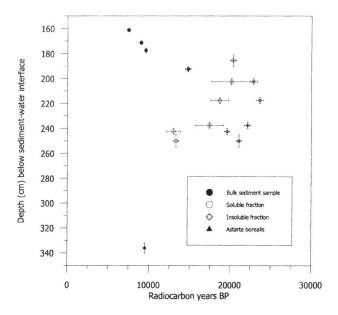

Figure 15. Comparative ^{14}C dates from Lake Boksehandsken, eastern Greenland (Björck et al., 1994). All datings were carried out by the conventional technique, with the exception of the lowermost dated level. Content of total organic carbon is 2–2.5% up to 181 cm, and gradually rises to 8.5% at 160 cm. From regional correlations, the sediments below 200 cm should not be older than 13,000 ^{14}C years BP, but rather be c. 10,000 ^{14}C years old, and are characterised by high sedimentation rates (Björck et al., 1994). Note that two further AMS datings were performed, one on plant remains at 180 cm gave an infinite age, and one dating on wood at 230 cm resulted in a ^{14}C age of 7130 ± 215 BP. While the former is regarded as reworked Quaternary plant material, the latter may be an effect of contamination from higher stratigraphic levels during coring.

Through more or less complicated and dubious correlations to the archaeological time scale, attempts were made to assess the age of the sediments. It was, however, not until the possibility of using pollen grains as a stratigraphic tool was formulated (von Post, 1916) and soon applied, that more secure correlations could tie lake sediments from different areas/regions to each other with the aid of certain distinct pollen assemblage changes. In the light of later knowledge, these pollen stratigraphic correlations, with their implied synchroneity, would, however, cause many problems for detailed chronological assessments and interpretations.

Although the sediment ages were uncertain, it should be noted that by combining an archaeological time scale with the clay-varve based Swedish Time Scale (De Geer, 1912), it was possible to build up a fairly well functioning chronology for the last 13,000 years (a period we now know spans ca 15,000 years (Wohlfarth et al., 1998a)). The general apprehension of the length of the time period after the last glaciation was thus in the right order of magnitude, but the discovery of the ^{14}C method in 1951 (Libby, 1955) meant a revolution in Late Quaternary studies. This was perhaps especially true for scientists working with the often very organic rich lake sediments of the Late Quaternary. When

Libby's (1955) original half-life of 5568 years was found to be erroneous (Godwin, 1962), the accurate half-life is 5730 ± 40 years, it almost resulted in a chaos of reported ages based on the two different half-lives. Since a large number of dates had already been calculated with Libby's half-life, it was fairly soon decided to report all ^{14}C ages, based on a half life of 5568 years (Mook, 1986).

As the number of new ^{14}C laboratories grew during the 1960's, an increasing number of radiocarbon dates became available from Holocene, but also late-glacial, lake sediments. One of the first obvious stratigraphic applications of the increased practice of radiocarbon dating organic-rich, pollen analysed lake sediments was that the previously defined pollen zones or pollen assemblage zones (p.a.z.) were gradually assigned secure ^{14}C ages. Since the Nordic countries had a strong Quaternary research tradition (and the region is abundant in lakes) it was hardly surprising that a ^{14}C-based Nordic chronostratigraphy was soon established for the last 13,000 ^{14}C years (Mangerud et al., 1974). It comprised a detailed chronostratigraphic sub-division of the Holocene time period and the Late Weichselian Substage and was based on a number of radiocarbon dates of significant pollen stratigraphic boundaries in lake sediments, but partly also in peat deposits, mainly from the Scandinavian countries. However, for the oldest dated boundary, the Bølling-Older Dryas chronozone boundary, the scheme was based on dates from Poland and Holland.

The concept of the Nordic chronostratigraphic scheme grew strong and spread rapidly among Late Quaternary stratigraphers, and perhaps especially among Quaternary scientists working with late- and post-glacial lake sediments. In fact, the Nordic focus and basis for the chronostratigraphy was soon neglected or over-looked. Instead, it became a more or less world-wide applied system, being used for subdividing, for example, ice cores, marine sediments or loess deposits, and for correlations between these, and for example, northwest European lake sediments, with their often detailed climatostratigraphy and well-established ^{14}C chronology. The formal construction of this chronostratigraphic system was often challenged. Criticism usually focussed on the poor precision of the ^{14}C method in relation to the strictly defined chronozones (Björck, 1984) or its regional pollen- and climato-stratigraphically defined basis (Gray & Lowe, 1977; Broecker, 1992; Lowe, 1994; Wohlfarth, 1996; Björck et al., 1998b; Walker et al., 1999). It did, however, remain the dominating Late Quaternary chronostratigraphic scheme for more than 20 years. The main reason for this was probably its simply defined and user-friendly chronostratigraphic boundaries. It was obviously an easy system to work with, perhaps especially because many of the boundaries coincided with some main climatic/environmental changes, although this was one of the main points of concern from its critics. However, the gradually more detailed structure of the ^{14}C/dendrochronology curve showed the problems with well-defined ^{14}C dated boundaries; these may correspond to several 100 calendar year long periods of time because of decreasing ^{14}C/^{12}C ratios of atmospheric CO_2, which result in falling Δ^{14}C values. Although these features had been found in densely dated lake (bulk) sediment sequences, they were initially explained as an effect of reworked organic material or extremely rapid sedimentation rates (e.g., Oeschger et al., 1980); we now know that they represent so called ^{14}C plateaux.

With the onset of the 1980's, a new period in radiocarbon dating began (Muller, 1977; Hedges, 1978; Doucas et al., 1978), the era of AMS dating. It came as a result of the realisation that advanced nuclear physic instrumentation most significantly reduces the minimum amount of carbon demanded for obtaining radiocarbon dates.

By a dense series of radiocarbon dated terrestrial macrofossils in Swiss lake sediments (Ammann & Lotter, 1989), the existence of a long radiocarbon plateau in the later part of the Younger Dryas cooling was confirmed, preceded and followed by rapidly declining ages. As stated above, this had already been reported in 1980 from ^{14}C ages on peat (Oeschger et al., 1980), but Ammann & Lotter (1989) also noted a ^{14}C plateau from the Bølling Chronozone. These and other late-, and postglacial ^{14}C plateaux have later been found in several other lake sediment records (Björck et al., 1996a; Goslar et al., 1999; 1995a; Andresen et al., 2000). Periods with such radiocarbon features are obviously unsuitable for defining chronozone boundaries since many of the more distinct ^{14}C anomalies seem to correspond with time periods of climate change (Björck et al., 1996a; Wohlfarth, 1996). Because a large part of the Nordic chronostratigraphy is based on climatostratigraphic correlations, such a chronostratigraphic scheme is clearly outdated with today's high-resolution strategy. Furthermore, the discovery that fairly organic-rich bulk sediments, formed in soft-water lakes, which the chronostratigraphy largely was based upon, often yield too old ^{14}C ages compared to terrestrial macrofossil (see above), made the scheme even more questionable.

Most of the distinct Late Weichselian-Holocene climate events, which are recorded in the oxygen isotope record from the ice (and calendar) year dated Greenland ice cores (Johnsen et al., 1992; Alley et al., 1993), also seem to appear in marine and lacustrine sediment records. It is therefore now proposed to use the Greenland GRIP ice core as a stratotype for an ice year dated event stratigraphy (Björck et al., 1998b; Walker et al., 1999). Climatic events in, for example, lake sediments can thus, through normal geologic correlation principles, be correlated to the calendar (ice) year dated events in the ice cores.

Dating of long (old) stratigraphies

Due to the limits of the radiocarbon method and the availability of suitable sequences, the main focus has generally been placed on establishing good chronologies for the last deglaciation and the Holocene. The maximum age for radiocarbon dating (i.e., the lower limit of counting the activity of a sample) is laboratory dependent but corresponds to eight half-lives or approximately to c. 40,000–45,000 years, after which measurements become infinite (Geyh, 1983; Lowe & Walker, 1997). However, by applying a thermal diffusion isotopic enrichment, where the amount of ^{14}C in a sample is enhanced (Grootes, 1978), radiocarbon ages of up to 70,000 years BP could be obtained (e.g., Woillard & Mook, 1982). This method requires fairly large samples, is time consuming and relatively expensive and has, therefore, not been widely applied.

Good radiocarbon measurements, with uncertainties of $< \pm 200$ years, have been obtained for the time period of the Last Glacial Maximum both with the AMS technique and the conventional decay method (e.g., Kitagawa & van der Plicht, 1998a, 1998b; Woillard & Mook, 1982). For some sequences a fairly low uncertainty has been possible to obtain even for records stretching further back in time (Ramrath et al., 1999; Kashiwaya et al., 1999). However, usually the uncertainty levels become much larger for ages >20,000 years BP, which is due to the decreasing activity of the sample. A nice example, which shows the possibility of radiocarbon dating older sequences, is the study by Woillard & Mook (1982), who performed conventional radiocarbon dating on bulk sediment samples at the La Grande Pile sequence in France. The set of dates, which extends back to 69,500 (+3800/−2600) ¹⁴C years BP, allowed assigning radiocarbon ages to the detailed pollen sequence covering the time period between the St. Germain II stage (Odderade) and the Last Glacial Maximum. The Grande Pile sequence was perhaps the first well-documented evidence for glacial variability, which would later be so well documented from the Greenland ice cores (Johnsen et al., 1992; Alley et al., 1993). While the uncertainty levels for the individual measurements range at around ±300 years between c. 31,000–20,000 ¹⁴C years BP, they amount to up to 1500 at c. 50,000 ¹⁴C years BP (Woillard & Mook, 1982). A number of AMS ¹⁴C measurements on samples from the long Lac du Bouchet sequence, covering the last glacial cycle, are also summarised in Creer (1991).

More recently, a series of AMS ¹⁴C measurements have been obtained on the acid leached total organic fraction of sediment samples, which gave radiocarbon dates as far back as c. 41,000 years BP (Benson et al., 1997). In another study, selected terrestrial plant macrofossils reach ¹⁴C ages of c. 33,000 years BP (Möller et al., 1999). However, the uncertainty levels of the measurements in both examples are fairly high and vary between 130–960 years and 170–2200 years, respectively. That the dating precision increases with the conventional decay counting technique is clearly shown by the ¹⁴C dating series from the 18 m thick and 40,000 year long sediment core from Lake Tulane in Florida (Grimm et al., 1993). For example, five ages between 32,300–39,600 years BP have single sigma errors between 220–650 years. On the other hand, the amount of carbon needed for such precision demanded 10–20 cm thick sediment samples. This low resolution sampling causes a within sample precision of 200–600 years, which is largely inefficient for detailed chronologic assessments. However, by fitting a mathematical function to the 16 finite ages a reasonable age model was obtained (Grimm et al., 1993). Based on this model, a convincing case was made that distinct pollen changes in the Lake Tulane sediments could be correlated to the North Atlantic Heinrich events (e.g., Heinrich, 1988; Bond et al., 1992), and in this respect the Lake Tulane record is a pioneer study.

A large number of more or less contiguous AMS ¹⁴C measurements, covering the time period between 525–41,890 ¹⁴C years BP (Fig. 16), have been performed on plant macrofossils from the long sediment record obtained from Lake Suigetsu in Japan (Kitagawa & van der Plicht, 1998a; 1998b). The high quality of the datings and the relatively low uncertainty levels of the individual measurements (130–820 years for the time period between 20,630–41,890 ¹⁴C years BP) show that it is possible to obtain high quality radiocarbon dates that far back in time (Kitagawa & van der Plicht, 1998a).

High resolution dating and wiggle matching

The increased focus on high-resolution studies can partly be explained as a consequence of the new AMS dating technique, but is also a result of the now fairly good knowledge on the general chronologic framework of the late Quaternary. This has led to an increased focus on chronologic details. However, the discovery and increased apprehension of rapid (climate) change as an important player of the Earth system has probably been the most important factor in this respect.

In spite of very ambitious dating programs, the variable atmospheric ¹⁴C content through time does pose problems in creating a detailed enough ¹⁴C chronology. For the Holocene, this can partly be overcome by the so-called "wiggle matching method", where the varying

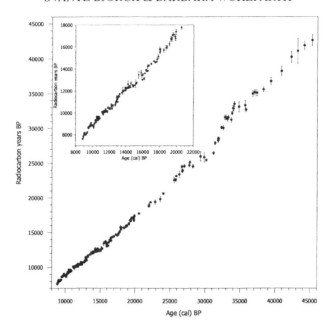

Figure 16. The ^{14}C dating series from the annually laminated sediment record of Lake Suigetsu (Kitagawa & van der Plicht, 1998b). Although the record also spans the whole Holocene, the curve only shows the time period between 7600–42,500 ^{14}C years BP (8800–45,000 varve years BP). The inserted curve shows the details between 7600–17,700 ^{14}C years BP. The laminations are caused by whitish diatom layers (spring growth) in otherwise dark clayey sediments. The annual chronology is mainly established through image analysis of digital pictures, and by different tests the counting error is estimated to <1.5%. The ^{14}C measurements (shown with 1-sigma bars) were performed on leaves, branches and insects of terrestrial origin.

[14]C content is used to tie the ^{14}C dates to the ^{14}C dated dendrochronology (for more theoretical details see Pearson, 1986). It usually involves a dense series of high-precision ^{14}C dates over fairly short time intervals, but is possible (but less secure) also over longer time periods, in attempting to replicate the shape of the ^{14}C/dendro curve. The method demands some extremely reliable ^{14}C dates (i.e., a dated material which was in contact with atmospheric CO_2 during its life-span and was buried in the sediments upon death) and/or one or a few dendrochronologically well-defined sediment horizons, such as well-dated tephra layers (Fig. 17). By, for example, anchoring the most securely ^{14}C dated levels to the calendar chronology via the dendro based calibration curve (Stuiver et al., 1998), it is possible to estimate sedimentation rates (calendar years/mm) for the interjacent sediments. The combination of ^{14}C age and calendar age is then plotted against the official ^{14}C/dendro curve to check the match between the curves. Since this may be closely related to sedimentation rates, the fit often involves minor adjustments of the sedimentation rates, before the most satisfactory solution is found.

To achieve the most secure dendrochronologic anchoring it is, if possible, adviceable to concentrate the ^{14}C dates to periods with large and rapid changes of the atmospheric [14]C

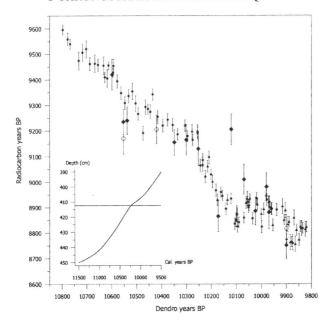

Figure 17. A high-resolution series of 13 ¹⁴C dated bulk sediment samples (filled diamonds) and 3 samples of macrofossils (open circles) from Lake Starvatn on the Faroe Islands (Björck et al., 2001), wiggle-matched to the ¹⁴C/dendro curve (Stuiver et al., 1998), based on ¹⁴C dates on German oaks (filled triangles) and pines (filled circles). The wiggle-matching is based on the inserted sedimentation curve. The sedimentation curve results from the age of the Younger Dryas-Preboreal transition, a few calibrated ¹⁴C ages (where a clear structure appears in the ¹⁴C/dendro curve, see text) and the GRIP ice core age of the Saksunarvatn tephra (Grönvold et al., 1995). The tephra is seen as a horizontal line at 412 cm. All dates are shown with 1-sigma errors. One date falls outside the fit, which may be explained by e.g., reworked organic material, but could also be an effect of statistics.

content. This usually denotes shifts between periods of a radiocarbon plateaux (falling $\Delta^{14}C$ values) and rapidly younger ages (rising $\Delta^{14}C$ values). Since such shifts are often sudden and chronologically well-defined, both in ^{14}C and dendro years, they are thus the perfect anchoring points for the connection between dendrochronology and lake sediments. This means that an optimal wiggle-matching approach can and should be fairly well-planned with respect to ^{14}C dating; by checking the structure of the ^{14}C/dendro curve over the time period of study one can thus chose to date the part of the curve which shows most structure.

With respect to these types of very detailed comparisons between the "perfect" ^{14}C dates of the tree rings and one's own sediment based samples, it is important to present a brief sampling strategy and a simple quality evaluation of the dated material. Firstly, the sampling strategy for detailed ^{14}C work should always include saving a small piece of bulk sample from at least each sieved (for macrofossils) sediment level. This should be done since lake sediments are often barren in macro remains and bulk samples may prove to be reliable. If, for example, bulk sediments are the main basis for the wiggle-matching, which is often necessary to obtain dense enough dates (Gulliksen et al., 1998), they should originate from soft-water lakes. Furthermore, in such cases it is absolutely recommended that at least a few ^{14}C datings are performed on both bulk material and macrofossils from the same levels to test the reliability of the bulk material (Fig. 17). If the ages coincide, the potential for obtaining reliable and detailed chronologies for lake sediments is large. Thus, by combining a wiggle-matching approach with a dense sampling strategy for your lake proxy records, it is possible to achieve a very detailed calendar year based picture of whatever lake story one wants to mediate. It has been suggested that this method may date records within a few years certainty (Pilcher, 1991), but this is questionnable since it should be remembered that it is mathematically impossible to transform ^{14}C years into absolutely certain calendar years. This is due to the combined effect of the uncertainty (i.e. the confidence interval) of the ^{14}C age of several (5–10) tree-rings, and the smaller atmospheric ^{14}C variations.

^{14}C dating versus absolute dating techniques of lacustrine sediments

Annually laminated, varved sediments constitute an important palaeoenvironmental and palaeoclimatic archive, because they allow reconstructing environmental and climatic changes with high time resolution. Although the potential of this type of archive has long been recognised (e.g., De Geer, 1912; 1940; Renberg, 1976; 1981; Renberg & Segerström, 1981; Saarnisto, 1986 and references therein), an increased interest in studying laminated lake sediments can only be observed during the past c. 10–15 years (see references in e.g., Hajdas, 1993; Wohlfarth, 1996; Petterson, 1996; 1999; Kemp, 1996). This owes mainly to the fact that, in addition to being an annual archive for palaeoenvironmental and palaeoclimatic changes, many laminated lake sediments contain sufficient terrestrial plant macrofossils for detailed and more or less contiguous AMS ^{14}C measurements. These new possibilities led to exploring the potential of using AMS ^{14}C dated laminated archives to extend the dendrochronological calibration curve (Kromer & Becker, 1993; Becker, 1993; Stuiver & Reimer, 1993; Stuiver et al., 1998) back in time (Björck et al., 1987; Goslar et al., 1989; 1993; Hajdas et al., 1993; 1995; Kitagawa et al., 1995; Lotter, 1991; Lotter et al., 1992; Wohlfarth et al., 1993; 1995a). At present, well-dated laminated lake sequences are known from Japan (Kitagawa & van der Plicht, 1998a; 1998b), Switzerland (Hajdas et al., 1993; Hajdas, 1993), Germany (Hajdas et al., 1995), Sweden (Wohlfarth et al., 1993; 1995a; 1998a; Goslar et al., 1999), Poland (Goslar et al., 1999; 1995a; 1995b; 1993; 1989) and NW Russia (Wohlfarth et al., 1999). None of these laminated sequences is, however, continuous up to the present and several of them are interrupted by a hiatus or by non-laminated parts, and only a few of them can be regarded as suitable for radiocarbon calibration (e.g., Kitagawa & van der Plicht, 1998b; 1998a; Goslar et al., 1995b; 1995a). However, even these latter sequences are not continuous up to present and have to be tied to the Holocene part of the dendrochronological calibration curve by wiggle matching. Despite these shortcomings, several of the records can be used to reconstruct — for selected time periods — variations in the atmospheric ^{14}C/^{12}C content (Goslar et al., 1995a; 1999; Kitagawa & van der Plicht, 1998b). At present, the laminated sequence from Lake Suigetsu (Fig. 16) (Kitagawa et al., 1995; Kitagawa & van der Plicht, 1998b; 1998a) and the varved marine sequence from the Cariaco Basin (Hughen et al., 1998b; 1998a) are the most detailed ^{14}C dated sediment records, and of these Lake Suigetsu allows extending the dendrochronological calibration curve as far back as 40,000 years BP. Compared to the U/Th calibration curve (Stuiver et al., 1998), it gives a much more detailed picture on radiocarbon variations and radiocarbon plateaux, at least as far back as c. 35,000 ^{14}C years BP.

Radiocarbon dating of laminated lake sediments has also been performed on the most recent part of the Holocene. By a wiggle-matching approach, Oldfield et al. (1997) could simultaneously test the reliability of a lake varve chronology, and explore which of the components of the sediment gave the most reliable ^{14}C ages. A compilation of all available ^{14}C dated varve chronologies were published in a recent volume of *Radiocarbon*.

The increased use of U-Th dating (e.g., corals) from the last glacial cycle by thermal ionisation mass spectrometry (TIMS) has also led to explore its applicability for lacustrine sediments. Such studies have usually been concentrated on more or less pure lake marls, and have encountered problems with high levels of initial Th (Gascoyne & Harmon, 1992; Latham & Schwarcz, 1992; Lin et al., 1996), making age determinations highly uncertain. With a slightly different approach (Israelson et al., 1997), it was found that the algae-rich lake marls from Lake Igelsjön, in south central Sweden, seem to be suitable for U-Th dating (Fig. 18). The sediments contain 55–90% carbonates and 10–50% organic matter. This Holocene lake site is surrounded by Quaternary deposits, which are rich in locally occurring uranium-rich Cambrian alum shale and Ordovician limestone. The sediments show highly variable U-contents (6.8–75.9 ppm), and the peak values occur with the highest organic contents (70–80% of the uranium occurs together with the organic material). Furthermore, algae-rich marl seems to be an almost perfectly closed system; it is very elastic, dense and impermeable, preventing post-depositional movement of uranium. In addition, it is also fairly rich in macrofossils. Bulk samples were dated with U-Th, and macrofossils using atmospheric CO_2 were picked out for ^{14}C dating, followed by calibration. With the exception of a few samples with possibly reworked macrofossils (*Pinus* epidermis) and less dense sediment, the two different dating records coincide almost perfect (Fig. 18). Thus, if suitable sediments are at hand, this comparative ^{14}C/U-Th study shows that the U-Th method can be used in lake sediments as an excellent complementary dating tool to the ^{14}C method. It implies that carbonate-rich bulk sediments may be accurately dated (in calendar years) and the time-range of lake sediment dating could in this way be stretched far back in time, possibly as far back as 300,000 years. The potential of U-Th dating lake sediments older than 40,000 year was also shown in the study of saline lake deposits from Death Valley (Roberts et al., 1997).

Concluding remarks

It is impossible to suggest a universal strategy for obtaining reliable radiocarbon dates, because each approach has to be adapted to the type of lake and sediment material that are to be investigated as well as to the objective of the study. However, before deciding the sampling and dating strategy, the following considerations should be made:

(1) Which type of organic material is available and in which quantity? If, as an example, the sample consists of > 1 g of terrestrial organic carbon, the conventional decay counting technique is probably preferable, because it often gives more precise measurements and is slightly less expensive. In all other cases, the AMS ^{14}C method will be the better alternative (Fig. 1).

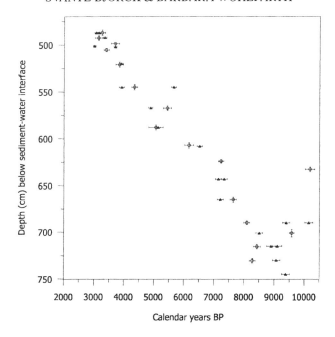

Figure 18. Two different dating sets from the algae-rich marl in Lake Igelsjön, south central Sweden (Israelson et al., 1997). The U-Th ages (filled triangles) were measured on 0.5 cm thin slices of bulk sediments, while the AMS ^{14}C ages (open circles) were measured on washed out terrestrial macrofossils, and calibrated to calendar years BP. In addition, the values for the U-Th ages were subtracted by 46 years; they were measured 1996 and calibrated ^{14}C ages are related to years before 1950. All ages are shown with 1-sigma errors. For more information, see text.

(2) How will the lake type and the sediment composition influence the radiocarbon measurement? Bulk sediment samples are, in general, subject to many uncertainties with respect to the radiocarbon dating result. Sediment samples from soft-water lakes may yield reliable measurements, but may also be exposed to different types of more or less subtle reservoir effects. More serious errors may result from incorporation of old or young carbon in the sediment. Bulk sediment samples from hard-water lakes will be prone to the same contamination effects as soft-water lakes. However, the most significant age error in these types of lakes are due to the hard-water reservoir effect, which usually results in significantly too old ages. Because of these many uncertainties, the radiocarbon measurement should be performed on well-identified material in order to assess any sources of errors related to the dated material.

(3) What are the scientific objectives of the project in terms of dating? In most cases, radiocarbon measurements are applied to obtain a good chronology for the studied section or sediment core. In such a case, samples will have to be taken at certain levels along the entire section or core, but what is the desired time resolution? A detailed chronology demands a large set of dates, and, in our opinion, it is usually much more worthwhile to put the dating efforts into one well-dated sequence, which can also be used as a future reference site, than to split the efforts into several sites with low time resolution. Finally, a good chronology also

means that the time period, which the sample represents, should be smaller than the quoted error obtained from the radiocarbon measurement, i.e., if a radiocarbon date is quoted with a standard error of ±50 years, the sample submitted should ideally comprise <100 years. With the AMS techniques now dominating, this should usually not be a big problem if the right sampling strategy is chosen.

Summary

The application of ^{14}C as a chronostratigraphic technique in paleolimnology is a vast topic and it is impossible to cover all possible aspects of this dating tool in relation to limnic sediments. In this overview, we have tried to focus on the most important and user-orientated aspects of the ^{14}C method. The extensive reference list on each of the addressed issues will, however, make it possible for the reader to explore the subject in more detail.

Lake sediments usually contain a certain amount of organic carbon in form of terrestrial, telmatic and limnic plant and animal debris and are, therefore, highly suitable for radiocarbon (^{14}C) dating. Since its discovery, ^{14}C dating has become the most widely applied chronostratigraphic tool in the study of late Quaternary lake sediment sequences. It allows obtaining an age control for the investigated sequences and enables comparisons and correlations on local, regional and global scales. Radiocarbon measurements are either performed with the conventional decay method, for which c. 1 g of pure carbon is required, or by the accelerator mass spectrometry (AMS) technique, where only a few mg of carbon are necessary. The background to radiocarbon dating, the advantages and disadvantages of both available methods, as well as a short historic overview and a "cook book figure" on the handling of a ^{14}C sample from the field to a dating result, introduce the main text.

Radiocarbon measurements are fairly expensive and often a series of dates will be performed along a sediment sequence. Therefore, each sample has to be carefully scrutinized (type of lake sediments, choice of sample material, sample treatment, size and storage, etc.), in order to obtain good results (Fig. 1). One section has, accordingly, been devoted to different types of errors, such as contamination by old and young organic material, general lake reservoir effects, hard-water reservoir effects, sample storage, preparation and size. Each of these possible errors, which can severely influence the age of a radiocarbon sample, are illustrated by examples from a variety of lake-sediment sequences from different geographic settings and time periods. Based on these examples different sources of error are discussed and special emphasis is placed on a comparison between bulk sediment dates and dates on terrestrial plant macrofossils. We round up this section with discussions on and methods for calibrating radiocarbon dates. The reasons for the calibrations are the highly variable atmospheric ^{14}C/^{12}C content, which leads to large differences between radiocarbon and calendar years. It is therefore necessary to calibrate the radiocarbon dates with a calibration programme, in order to obtain calendar ages for a sequence.

In the following section we addressed and discussed the feasibility of radiocarbon dating different fractions of a sediment sample, which has been made increasingly possible through the development and application of the AMS ^{14}C technique. Terrestrial plant macrofossils, if available in sufficient amounts, are probably most suitable for radiocarbon dating. However, many sediment types may not contain sufficient amounts of plant macrofossils. Therefore, other types of material, such as faunal remains, pollen or other fractions of the sediments (e.g., the humin-insoluble, humic-soluble, lipid fractions), have been tested for radiocarbon measurements.

The final section, which addresses the application of ^{14}C as a chronostratigraphic tool, is introduced by an extended historic overview, including the development of the Holocene and Late Weichselian chronostratigraphic scheme, its applicability, potentials and limits. Based on a number of selected studies, where good quality radiocarbon dates could be obtained as far back as c. 40,000 years BP, the range limit and the accuracy of the radiocarbon method is discussed and illustrated for the time interval between 20,000–40,000 years BP. For the younger part of the ^{14}C time scale we also explore the possibilities for detailed radiocarbon dating. If a contiguous series of high-resolution and high-quality ^{14}C measurements is available, it can be tied to the dendro-based calibration curve through wiggle matching. This approach allows assigning calendar-year ages not only to each ^{14}C measurement, but also to the interjacent sediments and makes it possible to estimate the sedimentation rates in calendar years/mm for a sequence. A short account on ^{14}C dating of annually laminated lake sediments and on comparative ^{14}C and TIMS U-Th measurements on limnic sequences round up the final section. In the future the U-Th method may become the main complementary method to ^{14}C for dating older Quaternary lake sediments.

Useful www addresses

The Radiocarbon Laboratory of the University of Waikato, New Zealand, has a nice home page with information about, among others, the radiocarbon technique, its application, age calculations and radiocarbon calibration. The FAQ section provides understandable information relating to radiocarbon dating for students and lay people who are not requiring detailed information about the radiocarbon dating method. The link section offers extensive links to radiocarbon laboratories on the WWW and to sample preparation labs, however, not completely updated.
http://www.c14dating.com/

The home page of the journal *Radiocarbon* can be found at:
www.radiocarbon.org/

The two radiocarbon calibration programmes which are available can be downloaded from: The Oxford Research Laboratory for Archaeology and the History of Art, calibration programme OxCal v3.4:
http://www.rlaha.ox.ac.uk/

University of Washington or University of Belfast, calibration programme CALIB 4.2:
http://depts.washington.edu/qil/calib/

Acknowledgements

Göran Skog at the ^{14}C Laboratory, Department of Quaternary Geology in Lund, is thanked for reading and commenting the manuscript.

References

Abbott, M. B. & T. W. Stafford Jr., 1996. Radiocarbon geochemistry of modern and ancient Arctic lake systems, Baffin Island, Canada. Quat. Res. 45: 300–311.

Åkerlund, A., J. Risberg, U. Miller & P. Gustafson, 1995. On the applicability of the ^{14}C method to interdisciplinary studies on shore displacement and settlement location. PACT 49: 53–84.

Alley, R. B., D. A. Meese, C. A. Shuman, A. J. Gow, K. C. Taylor, P. M. Grootes, J. W. C. White, M. Ram, E. D. Waddington, P. A. Mayewski & G. A. Zielinski, 1993. Abrupt increase in Greenland snow accumulation at the end of the Younger Dryas event. Nature 362: 527–529.

Ammann, B. & A. F. Lotter, 1989. Late-Glacial radiocarbon- and palynostratigraphy on the Swiss Plateau. Boreas 18: 109–126.

Andrée, M., H. Oeschger, U. Siegenthaler, T. Riesen, M. Moell, B. Ammann & K. Tobolski, 1986. ^{14}C dating of plant macrofossils in lake sediments. Radiocarbon 28: 411–416.

Andresen, C. S., S. Björck, O. Bennike, J. Heinemeier & B. Kromer, 2000. What do ^{14}C changes across the Gerzensee oscillation/GI-1b event imply for deglacial oscillations? J. Quat. Sci. 15: 203–214.

Barnekow, L., G. Possnert & P. Sandgren, 1998. AMS ^{14}C chronologies of Holocene lake sediments in the Abisko area, northern Sweden — a comparison between dated bulk sediment and macrofossil samples. GFF 120: 59–67.

Becker, B., 1993. An 11,000- year German oak and pine dendrochronology for radiocarbon calibration. Radiocarbon 35: 201–213.

Benson, L. V., J. P. Smoot, M. Kashgarian, A. Sarna-Wojcicki & J. W. Burdett, 1997. Radiocarbon ages and environments of deposition of the Wono and Trego Hot Springs tephra layers in the Pyramid Lake Subbasin, Nevada. Quat. Res. 47: 251–260.

Berkman, P. A., J. T. Andrews, S. Björck, E. A. Colhoun, S. D. Emslie, I. D. Goodwin, B. L. Hall, C. P. Hart, K. Hirakawa, A. Igarashi, O. Ingolfsson, J. Lopez-Martinez, W. B. Lyons, M. C. G. Mabin, P. G. Quilty, M. Taviani & Y. Yoshida, 1998. Circum-Antarctic coastal environmental shifts during the Late Quaternary reflected by emerged marine deposits. Antarctic Science 10: 45–362.

Bird, M., A. Chivas, C. Radnell & H. Burton, 1991. Sedimentological and stable-isotope evolution of lakes in the Vestfold Hills, Antarctica. Paleogeog., Paleoclimat., Paleoecol. 84: 109–130.

Björck, S., 1979. Weichselian Stratigraphy of Blekinge, SE Sweden, and Water Level Changes in the Baltic Ice Lake. Thesis 7, Department of Quaternary Geology, Lund University, 248 pp.

Björck, S. & S. Håkansson, 1982. Radiocarbon dates from Late Weichselian lake sediments in South Sweden as a basis for chronostratigraphic subdivisons. Boreas 11: 141–150.

Björck, S., 1984. Bio- and chronostratigraphic significance of the Older Dryas Chronozone — on the basis of new radiocarbon dates. Geologiska Föreningen i Stockholm Förhandlingar 106: 81–91.

Björck, S., P. Sandgren & B. Holmquist, 1987. A magnetostratigraphic comparison between ^{14}C years and varve years during the Late Weichselian, indicating significant differences between the time-scales. J. Quat. Sci. 2: 133–140.

Björck, S. & G. Digerfeldt, 1991. Alleröd-Younger Dryas sea level changes in southwestern Sweden and their relation to the Baltic Ice Lake development. Boreas 20: 115–133.

Björck, S., H. Håkansson, R. Zale, W. Karlén & B. Liedberg-Jönsson, 1991a. A late Holocene lake sediment sequence from Livingston Island, South Shetland Islands, with palaeoclimatic implications. Antarctic Science 3: 61–72.

Björck, S., C. Hjort, O. Ingolfsson & G. Skog, 1991b. Radiocarbon dates from the Antarctic Peninsula — problems and potential. Quaternary Proceedings 1: 55–65.

Björck, S., N. Malmer, C. Hjort, P. Sandgren, O. Ingolfsson, B. Wallén, R. I. Lewis Smith & B. Liedberg-Jönsson, 1991c. Stratgraphic and paleoclimatic studies of a 5500- year-old moss bank on Elephant Island, Antarctica. Arctic and Alpine Research. 23: 361–374.

Björck, S., H. Håkansson, S. Olsson, L. Barnekow & J. Janssens, 1993. Palaeoclimatic studies in South Shetland Islands, Antarctica, based on numerous stratigraphic variables in lake sediments. J. Paleolim. 8: 233–272.

Björck, S., O. Bennike, I. Ingolfsson, L. Barnekow & D. N. Penney, 1994. Lake Boksehandsken's earliest postglacial sediments and their palaeoenvironmental implications, Jamson Land, East Greenland. Boreas 23: 459–472.

Björck, S., B. Kromer, S. Johnsen, O. Bennike, D. Hammarlund, G. Lemdahl, G. Possnert, T. L. Rasmussen, B. Wohlfarth, C. U. Hammer & M. Spurk, 1996a. Synchronised terrestrial-atmospheric deglacial records around the North Atlantic. Science 274: 1155–1160.

Björck, S., S. Olsson, C. Ellis-Evans, H. Håkansson, O. Humlum & J. M. de Lirio, 1996b. Late Holocene palaeoclimatic records from lake sediments on James Ross Island, Antarctica. Palaeogeog., Palaeoclimat., Palaeoecol. 121: 195–220.

Björck, S., O. Bennike, G. Possnert, B. Wohlfarth & G. Digerfeldt, 1998a. A high-resolution ^{14}C dated sediment sequence from southwest Sweden: age comparisons between different components of the sediment. J. Quat. Sci. 13: 85–89.

Björck, S., M. J. C. Walker, L. C. Cwynar, S. Johnsen, K.-L. Knudsen, J. J. Lowe, B. Wohlfarth & INTIMATE members, 1998b. An event stratigraphy for the Last termination in the North Atlantic region based on the Greenland ice-core record: a proposal by the INTIMATE group. J. Quat. Sci. 13: 283–292.

Björck, S., R. Muscheler, B. Kromer, C.S. Andresen, J. Heinemeier, S.J. Johnsen, D. Conley, N. Koç, M. Spurk, & S. Veski 2001. High-resolution analyses of an early Holocene climate event may imply decreased solar forcing as an impotant climate trigger. Geology 29: 1107-1110.

Bond, G., H. Heinrich, W. S. Broecker, L. Labeyrie, J. McManus, J. T. Andrews, S. Huon, R. Jantschik, S. Clasen, C. Simet, K. Tedesco, M. Klas, G. Bonani & S. Ivy, 1992. Evidence for massive discharges of icebergs into the glacial North Atlantic. Nature 360: 245–249.

Broecker, W. S., 1992. Defining the boundaries of the Late-Glacial isotope episodes. Quat. Res. 38: 135–138.

Brown, T. A., D. E. Nelson, R. W. Mathewes, J. S. Vogel & J. R. Southon, 1989. Radiocarbon dating of pollen by accelerator mass spectrometry. Quat. Res. 32: 205–212.

Brown, T. A., G. W. Farwell, P. M. Grootes & F. H. Schmidt, 1992. Radiocarbon AMS dating of pollen extracted from peat samples. Radiocarbon 34: 550–556.

Craig, H., 1957. Isotopic standards for carbon and oxygen correction factors for mass-spectrometric analysis of carbon dioxide. Geochim. et Cosmochim. 12: 133–149.

Creer, K. M., 1991. Dating of a Maar Lake sediment sequence covering the last glacial cycle. Quaternary Proceedings 1: 75–87.

De Geer, G., 1912. A geochronology of the last 12,000 years. Congrès de Geologie International, Comptes Rendues: 241–253.

De Geer, G., 1940. Geochronologia Suecica, Principles. Kungliga Svenska Vetenskapsakademiens Handlingar 18: 1–367.

Deevey, E. S., M. S. Gross, G. E. Huthinson & H. L. Kraybill, 1954. The natural ^{14}C content of materials from hardwater lakes. Proceedings of the National Academy of Sciences of the United States of America 40: 285–288.

Doran, P. T., J. R. A. Wharton & W. B. Lyons, 1994. Paleolimnology of the McMurdo Dry Valleys, Antarctica. J. Paleolim. 10: 85–114.

Doran, P. T., G. W. Berger, W. B. Lyons, J. Wharton, R. A., M. L. Davisson, J. Southon & J. E. Dibb, 1999. Dating Quaternary lacustrine sediments in the McMurdo Dry Valleys, Antarctica. Paleogeog., Paleoclimat., Paleoecol. 147: 223–239.

Doucas, G., E. F. Garman, H. R. M. Hyder, D. Sinclair, R. E. M. Hedges & N. R. White, 1978. Detection of ^{14}C using a small Van de Graaff accelerator. Nature 276: 253–255.

Elias, S. A. & L. J. Toolin, 1990. Accelerator dating of a mixed assemblage of Late Pleistocene insect fossils from the Lamb Spring Site, Colorado. Quat. Res. 33: 122–126.

Elias, S. A., P. E. Carrara, L. J. Toolin & J. T. Jull, 1991. Revised age of deglaciation of Lake Emma based on new radiocarbon and macrofossil analyses. Quat. Res. 36: 307–321.

Gascoyne, M. & R. S. Harmon, 1992. Palaeoclimatology and palaeosea levels. In Ivanovitch, M. & R. S. Harmon (eds.) Uranium-Series Disequilibrium. Oxford Science, Oxford: 553–582.

Geyh, M. A., W. E. Krumbein & H.-R. Kudrass, 1974. Unreliable ¹⁴C dating of long-stored deep-sea sediments due to bacterial activity. Marine Geology 17: M45–M50.

Geyh, M. A., 1983. Physikalische und chemische Datierungsmethoden in der Quartär-Forschung. Clausthaler Tektonische Hefte 19: 1–163.

Geyh, M. A., U. Schotterer & M. Grosjean, 1998. Temporal changes of the ¹⁴C reservoir effect in lakes. Radiocarbon 40: 921–931.

Godwin, H., 1962. Half-life of radiocarbon. Nature 195: 944.

Goslar, T., A. Pazdur, M. F. Pazdur & A. Walanus, 1989. Radiocarbon and varve chronologies of annually laminated lake sediments of Gosciaz lake, Central Poland. Radiocarbon 31: 940–947.

Goslar, T., T. Kuc, M. Ralska-Jasiewiczowa, K. Rózanski, M. Arnold, E. Bard, B. van Geel, M. F. Pazdur, K. Szerocynska, B. Wicik, K. Wieckowski & A. Walanus, 1993. High-resolution lacustrine record of the Late Glacial/Holocene transition in Central Europe. Quat. Sci. Rev. 12: 287–294.

Goslar, T., M. Arnold, E. Bard, T. Kuc, M. F. Pazdur, M. Ralska-Jasiewiczowa, K. Rozanski, N. Tisnerat, A. Walanus, B. Wicik & K. Wieckowski, 1995a. High concentration of atmospheric ¹⁴C during the Younger Dryas cold episode. Nature 377: 414–417.

Goslar, T., M. Arnold & M. F. Pazdur, 1995b. The Younger Dryas cold event — was it synchronous over the North Atlantic region? Radiocarbon 37: 63–70.

Goslar, T., B. Wohlfarth, S. Björck, G. Possnert & J. Björck, 1999. Variations of atmospheric ¹⁴C concentrations over the Alleröd-Younger Dryas transition. Climate Dynamics 15: 29–42.

Gray, J. M. & J. J. Lowe, 1977. The Scottish Lateglacial environments: a synthesis. In Gray, J. M. & J. J. Lowe (eds.) Studies in the Scottish Lateglacial Environment. Oxford: 163–181.

Grimm, C. G., J. G. L. Jacobsen, W. A. Watts, B. C. S. Hansen & K. A. Maasch, 1993. A 50,000- year record of climate oscillations from Florida and its temporal correlation with the Heinrich events. Science 261: 198–200.

Grönvold, K., N. Óskarsson, S. J. Johnsen, H. B. Clausen, C. U. Hammer, G. Bond & E. Bard, 1995. Ash layers from Iceland in the Greenland GRIP ice core correlated with oceanic and land sediments. Earth and Planetary Science Letters 135: 149–155.

Grootes, P. M., 1978. Carbon-14 timescale extended: comparison of chronologies. Science 200: 11–15.

Gulliksen, S., H. H. Birks, G. Possnert & J. Mangerud, 1998. A calendar age estimate of the Younger Dryas-Holocene boundary at Kråkenes, western Norway. The Holocene 8: 249–259.

Hajdas, I., 1993. Extension of the Radiocarbon Calibration Curve by AMS Dating of Laminated Sediments of Lake Soppensee and Lake Holzmaar. ETH Zuerich No. 10157, 1–147.

Hajdas, I., S. Ivy, J. Beer, G. Bonani, D. Imboden, A. F. Lotter, M. Sturm & M. Suter, 1993. AMS radiocarbon dating and varve chronology of Lake Soppensee: 6000 to 12,000 ¹⁴C years BP. Climate Dynamics 9: 107–116.

Hajdas, I., B. Zolitschka, S. Ivy-Ochs, J. Beer, G. Bonani, S. A. G. Leroy, J. Negendank, M. Ramrath & M. Suter, 1995. AMS radiocarbon dating of annually laminated sediments from Lake Holzmaar, Germany. Quat. Sci. Rev. 14: 137–143.

Harkness, D. D., 1979. Radiocarbon dates from Antarctica. British Antarctic Survey Bulletin 47: 43–59.

Hedenström, A. & J. Risberg, 1999. Early Holocene shore-displacement in southern central Sweden as recorded in elevated isolated basins. Boreas 28: 490–504.

Hedges, R. E. M., 1978. New directions of ¹⁴C dating. New Scientist 77: 599.

Hedges, R. E. M., 1991. AMS dating: present status and potential applications. Quaternary Proceedings 1: 5–10.

Heinrich, H., 1988. Origin and consequences of cyclic rafting in the northeast Atlantic Ocean during the past 130,000 years. Quat. Res. 29: 143–152.

Hughen, K. A., J. T. Overpeck, S. J. Lehman, M. Kashgarian, J. Southon & L. C. Peterson, 1998a. A new ^{14}C calibration data set for the Last Deglaciation. Radiocarbon 39: 483–494.

Hughen, K. A., J. T. Overpeck, S. J. Lehman, M. Kashgarian, J. Southon, L. C. Peterson, R. Alley & D. M. Sigman, 1998b. Deglacial changes in ocean circulation from an extended radiocarbon calibration. Nature 391: 65–68.

Israelson, C., S. Björck, C. J. Hawkesworth & G. Possnert, 1997. Direct U-Th dating of organic- and carbonate-rich lake sediments from southern Scandinavia. Earth and Planetary Science Letters 153: 251–263.

Iversen, J., 1936. Sekundäres Pollen als Fehlerquelle. Eine Korrektionsmetode zur Pollenanalyse minerogener Sedimente. Danmarks Geologiske Undersøgelser IV:2: 1–24.

Johnsen, S. J., H. B. Clausen, W. Dansgaard, K. Fuhrer, N. Gundestrup, C. U. Hammer, P. Iversen, J. Jouzel, B. Stauffer & J. P. Steffensen, 1992. Irregular glacial interstadials recorded in a new Greenland ice core. Nature 359: 311–313.

Jones, V. J., R. W. Battarbee & R. E. M. Hedges, 1993. The use of chironomids for AMS ^{14}C dating of lake sediments. The Holocene 3: 161–163.

Kaland, P. E., K. Krzywinski & B. Stabell, 1984. Radiocarbon-dating of transitions between marine and lacustrine sediments and their relation to the development of lakes. Boreas 13: 243–258.

Kashiwaya, K., M. Ryugo, M. Horii, H. Sakai, T. Nakamura & T. Kawai, 1999. Climato-limnological signals during the past 260,000 years in physical properties of bottom sediments from lake Baikal. J. Paleolim. 21: 143–150.

Kemp, A. E. S., 1996. Palaeoclimatology and Palaeoceanography from Laminated Sediments. The Geological Society, London, 1–258.

Kitagawa, H., H. Fukuzawa, T. Nakamura, M. Okamura, K. Takemura, G. Hayashida & Y. Yasuda, 1995. AMS ^{14}C dating of the varved sediments from Lake Suigetsu, Central Japan, and atmospheric ^{14}C change during the late Pleistocene. Radiocarbon 37: 371–378.

Kitagawa, H. & H. van der Plicht, 1998a. A 40,000-year varve chronology from Lake Suigetsu, Japan: Extension of the radiocarbon calibration curve. Radiocarbon 40: 505–515.

Kitagawa, H. & J. van der Plicht, 1998b. Atmospheric radiocarbon calibration to 45,000 yr BP: Late glacial fluctuations and cosmogenic isotope production. Science 279: 1187–1190.

Kromer, B. & B. Becker, 1993. German oak and pine ^{14}C calibration, 7200–9439 BC. In Stuiver, M., A. Long & R. Kra (eds.) Calibration 1993. Radiocarbon: 125–135.

Latham, A. G. & H. P. Schwarcz, 1992. Palaeoclimatology and palaeosea levels. In Ivanovitch, M. & R. S. Harmon (eds.) Uranium-Series Disequilibrium. Oxford Science, Oxford: 423–459.

Libby, W. F., 1955. Radiocarbon Dating. University of Chicago Press, Chicago.

Lin, J. C., W. S. Broecker, R. F. Anderson, J. L. Rubenstone, S. Hemming & G. Bonani, 1996. New ^{230}Th/U and ^{14}C ages from Lake Lahontan (Nevada) carbonates and implications for the origin of their initial Th contents. Geochim. et Cosmochim. Acta 53: 1307–1322.

Linnick, W., P. E. Damon, D. J. Donahue & A. J. T. Jull, 1989. Accelerator Mass Spectrometry: the new revolution in radiocarbon dating. Quat. Internat. 1: 1–6.

Long, A., O. K. Davis & J. de Lanois, 1992. Separation and ^{14}C dating of pure pollen from lake sediments: nanofossil AMS dating. Radiocarbon 34: 557–560.

Lotter, A., 1991. Absolute dating of the Late-Glacial period in Switzerland using annually laminated sediments. Quat. Res. 35: 321–330.

Lotter, A. F., B. Amman, J. Beer, I. Hajdas & M. Sturm, 1992. A step towards an absolute time-scale for the Late-Glacial: annually laminated sediments from Soppensee (Switzerland). In Bard, E. & W. S. Broecker (eds.) The Last Deglaciation: Absolute and Radiocarbon Chronologies. Springer-Verlag, Berlin: 45–68.

Lowe, J. J., S. Lowe, A. J. Fowler, R. E. M. Hedges & T. J. F. Austin, 1988. Comparison of accelerator and radiometric radiocarbon measurements obtained from Late Devensian Lateglacial lake sediments from Llyn Gwernan, North Wales, UK. Boreas 17: 355–369.

Lowe, J. J. (ed.), 1991a. Radiocarbon dating: recent applications and future potential. Quaternary Proceedings 1: 1–89.

Lowe, J. J., 1991b. Stratigraphic resolution and radiocarbon dating of Devensian Lateglacial sediments. Quaternary Proceedings 1: 19–25.

Lowe, J. J., 1994. The objectives of the North Atlantic Seaboard Programme (NASP), a constituent subproject of IGCP-253. J. Quat. Sci. 9: 95–99.

Lowe, J. J. & M. J. C. Walker, 1997. Reconstructing Quaternary Environments. Longman Ltd. 446 pp.

MacDonald, G. M., R. P. Beukens & W. E. Kieser, 1991a. Radiocarbon dating of limnic sediments: a comparative analysis and discussion. Ecology 72: 1150–1155.

MacDonald, G. M., R. P. Beukens, W. E. Kieser & D. H. Vitt, 1991b. Comparative radiocarbon dating of terrestrial plant macrofossils and aquatic moss from the "ice-free corridor" of western Canada. Geology 15: 837–840.

Mangerud, J., S. T. Andersen, B. E. Berglund & J. J. Donner, 1974. Quaternary stratigraphy of Norden, a proposal for terminology and classification. Boreas 3: 109–128.

Mensing, S. A. & J. R. Southon, 1999. A simple method to separate pollen for AMS radiocarbon dating and its application to lacustrine and marine sediments. Radiocarbon 41: 1–8.

Möller, P., D. Y. Bolshyanov & H. Bergsten, 1999. Weichselian geology and palaeoenvironmental history of the central Taymyr Peninsula, Siberia, indicating no glaciation during the last global glacial maximum. Boreas 28: 92–114.

Mook, W. G. 1986, Recommendations/resolutions adopted by the 12th International Radiocarbon Conference. Radiocarbon 28: 799.

Muller, R. A., 1977. Radioisotope dating with a Cyclotron. The sensitivity of radioisotope dating improved by counting atoms rather than decays. Science 196: 489–494.

Oeschger, H., M. Welten, U. Eicher, M. Möll, T. Riesen, U. Siegenthaler & S. Wegmueller, 1980. ^{14}C and other parameters during the Younger Dryas cold phase. Radiocarbon 22: 299–310.

Oldfield, F., P. R. J. Crooks, D. D. Harkness & G. Petterson, 1997. AMS radiocarbon dating of organic fractions from varved lake sediments: an empirical test of reliability. J. Paleolim. 18: 87–91.

Olsson, I., 1968. Radiocarbon analyses of lake sediment samples from Björnöya. Geografiska Annaler 50A: 246–247.

Olsson, I., 1979. A warning against radiocarbon dating of samples containing little carbon. Boreas 8: 203–207.

Olsson, I., 1986. Radiometric dating. In Berglund, B. E. (ed.) Handbook of Holocene Palaeoecology and Palaeohydrology. Wiley, N.Y.: 273–312.

Olsson, I., 1991. Accuracy and precision in sediment chronology. Hydrobiologia 214: 25–34.

O'Sullivan, P. E., F. Oldfield & R. W. Battarbee, 1973. Preliminary studies of Lough Neagh sediments I. Stratigraphy, chronology and pollen analysis. In Birks, H. J. B. & R. G. West (eds.) Quaternary Plant Ecology. Blackwell Scientific Publications, Oxford: 267–278.

Pearson, G. W., 1986. Precise calendrical dating of known growth period samples using a "curve" fitting technique. Radiocarbon 28: 292–299.

Petterson, G., 1996. Varved sediments in Sweden: a brief review. In Kemp, A. E. S. (ed.) Palaeoclimatology and Palaeoceanography from Laminated Sediments. Geological Society Special Publication, London: 73–77.

Petterson, G., 1999. Image Analysis, Varved Lake Sediments and Climate Reconstructions. Thesis Department of Ecology and Environmental Science, Umeå University, 17 pp.

Pilcher, J. R., 1991. Radiocarbon dating for the Quaternary scientist. Quaternary Proceedings 1: 27–33.

Ramrath, A., N. R. Nowaczyk & J. F. W. Negendank, 1999. Sedimentological evidence for environmental changes since 34,000 years BP from Lago di Mezzano, central Italy. J. Paleolim. 21: 423–435.

Regnéll, J., 1992. Preparing pollen concentrates for AMS dating: a methodological study from a hard-water lake in southern Sweden. Boreas 21: 273–277.

Regnéll, J. & E. Everitt, 1996. Preparative centrifugation — a new method for preparing pollen concentrates suitable for radiocarbon dating by AMS. Vegetation History and Archaeobotany 5: 201–205.

Renberg, I., 1976. Annually laminated sediments in Lake Rudetjärn, Medelpad province, northern Sweden. Geologiska Föreningens i Stockholm Förhandlingar 98: 335–360.

Renberg, I., 1981. Formation, structure and visual appearance of iron-rich, varved lake sediments. Verhandlungen des Internationalen Vereins fuer Limnologie 21: 94–101.

Renberg, I. & U. Segerström, 1981. Application of varved lake sediments in palaeoenvironmental studies. Wahlenbergia 7: 125–133.

Richardson, F. & V. A. Hall, 1994. Pollen concentrate from highly organic Holocene peat and lake deposits for AMS dating. Radiocarbon 36: 407–412.

Roberts, S. M., R. J. Spencer, W. Yang & H. R. Krouse, 1997. Deciphering some unique paleotemperature indicators in halite-bearing saline lake deposits from Death Valley, California, USA. J. Paleolim. 17: 101–130.

Saarnisto, M., 1986. Annually laminated lake sediments. In Berglund, B. E. (ed.) Handbook of Holocene Palaeoecology and Palaeohydrology. Wiley, N.Y.: 343–370.

Shotton, F. W., 1972. An example of hard-water error in radiocarbon dating of vegetable organic matter. Nature 240: 460–461.

Smart, P. L. & P. D. Frances, 1991. Quaternary dating methods. Quaternary Research Association 233.

Squyres, S. W., D. W. Andersen, S. S. Nedell & R. A. Wharton Jr., 1991. Lake Hoare, Antarctica: sedimentation through a thick perennial ice cover. Sedimentology 38: 363–379.

Stuiver, M. & T. F. Brazunias, 1993. Sun, ocean, climate and atmospheric $^{14}CO_2$: an evaluation of causal and spectral relationships. The Holocene 3: 189–205.

Stuiver, M. & P. Reimer, 1993. Extended ^{14}C data base and revised CALIB 3.0 ^{14}C age calibration program 1993. Radiocarbon 35: 215–230.

Stuiver, M., P. J. Reimer, E. Bard, J. W. Beck, G. S. Burr, K. A. Hughen, B. Kromer, G. McCormac, J. van der Plicht & M. Spurk, 1998. INTCAL98 Radiocarbon age calibration, 24,000-0 cal BP. Radiocarbon 40: 1041–1083.

Sutherland, D. G., 1980. Problems of radiocarbon dating deposits from newly-deglaciated terrain: examples from the Scottish Lateglacial. In Lowe, J. J., J. M. Gray & J. E. Robinson (eds.) Studies in the Lateglacial of North-West Europe. Pergamon Press, Oxford: 139–149.

Sveinbjörnsdóttir, Á. E., J. Heinemeier, N. Rud & S. Johnsen, 1992. Radiocarbon anomalies observed for plants growing in Iceland geothermal waters. Radiocarbon 34: 696–703.

Törnqvist, P., A. F. M. De Jong, W. A. Oosterbaan & K. Van der Borg, 1992. Accurate dating of organic deposits by AMS ^{14}C measurements of macrofossils. Radiocarbon 34: 566–577.

Vogel, J. S., M. Briskin, D. E. Nelson & J. R. Southon, 1989. Ultra-small carbon samples and the dating of sediments. Radiocarbon 31: 601–609.

von Post, L., 1916. Skogsträdpollen i sydsvenska torvmosselagerföljder. Geologiska Föreningens i Stockholm Förhandlingar 38: 384–390.

Walker, M. J. C. & D. D. Harkness, 1990. Radiocarbon dating the Devensian Lateglacial in Britain: new evidence from Llanilid, South Wales. J. Quat. Sci. 5: 135–144.

Walker, M. J. C., S. Björck, J. J. Lowe, L. C. Cwynar, S. Johnsen, K.-L. Knudsen, B. Wohlfarth & INTIMITE group, 1999. Isotopic 'events' in the GRIP ice core: a stratotype for the Late Pleistocene. Quat. Sci. Rev. 18: 1143–1150.

Wohlfarth, B., S. Björck, G. Possnert, G. Lemdahl, L. Brunnberg, J. Ising, S. Olsson & N.-O. Svensson, 1993. AMS dating Swedish varved clays of the last glacial/interglacial transition and the potential/difficulties of calibrating Late Weichselian 'absolute' chronologies. Boreas 22: 113–128.

Wohlfarth, B., S. Björck & G. Possnert, 1995a. The Swedish Time Scale — a potential calibration tool for the radiocarbon time scale during the Late Weichselian. Radiocarbon 37: 347–360.

Wohlfarth, B., G. Lemdahl, S. Olsson, T. Persson, I. Snowball, J. Ising & V. Jones, 1995b. Early Holocene environment on Björnöya (Svalbard) inferred from multidisciplinary lake sediment studies. Polar Research 14: 253–275.

Wohlfarth, B., 1996. The chronology of the Last Termination: a review of high-resolution terrestrial stratigraphies. Quat. Sci. Rev. 15: 267–284.

Wohlfarth, B., S. Björck, G. Possnert & B. Holmquist, 1998a. A 800-year long, radiocarbon-dated varve chronology from south-eastern Sweden. Boreas 27: 243–257.

Wohlfarth, B., G. Possnert, G. Skog & B. Holmquist, 1998b. Pitfalls in the AMS radiocarbon-dating of terrestrial macrofossils. J. Quat. Sci. 13: 137–145.

Wohlfarth, B., O. Bennike, L. Brunnberg, I. Demidov, G. Possnert & S. Vyahirev, 1999. AMS ^{14}C measurements and macrofossil analysis of a varved sequence near Pudozh, eastern Karelia, NW Russia. Boreas 29: 575–586.

Woillard, G. M. & W. G. Mook, 1982. Carbon-14 dates at Grande Pile: Correlation of land and sea chronologies. Science 215: 159–161.

Zale, R., 1994. [14]C age corrections in Antarctic lake sediments inferred from geochemistry. Radiocarbon 36: 173–185.

11. VARVE CHRONOLOGY TECHNIQUES

SCOTT LAMOUREUX (lamoureux@lake.geog.queensu.ca)
Department of Geography
Queen's University
Kingston, Ontario
Canada K7L 3N6

Keywords: varves, annually-laminated sediments, dating, chronology, errors, cross dating, methods

Introduction

Growing interest in past environmental conditions and the need to obtain a long-term perspective on environmental change is leading to increasing use of high-resolution lake sediment records to obtain the clearest indication of past changes. Varved sediments are particularly important for high-resolution studies because they provide a clear, simple means of identifying one year of deposition within a long sedimentary sequence. The simple, consistent time increment found in varved sequences provides an unparalleled framework for reconstructing and documenting detailed changes in long term environmental conditions, ranging from hundreds to more than ten thousand years into the past. Additionally, varves can be composed of a variety of materials (biological, chemical, mineral) and are found in a wide range of environments. Therefore, varve-based sedimentary records have applications in a diverse range of paleoenvironmental research fields and study locations.

Varves have played a prominent role in numerous geological studies in northern Europe and North America that began in the latter half of the 19th century. Although varved clay deposits were documented in Sweden as early as 1855, it was the linkage between the varves and glacial theories that led the Swedish geologist Gerald De Geer to suggest that varves represented a one-year cycle of deposition. This was largely based on the similarity between the structure of varves and tree rings (Brunnberg, 1995). De Geer suggested that the silt layer represented deposition during the melt season and the clay layer was produced during the cold season under lake ice. By making use of the extensive outcrops of varves found in southern Sweden, De Geer (1912) pioneered the use of varves as a chronological tool to estimate the age of ice retreat. In this and subsequent work, De Geer and other workers measured varve sequences to construct varve thickness chronologies from various locations. These chronologies were correlated to develop a single varve time scale of the postglacial period. This work provided a foundation for future varve work by linking together isolated "floating" chronologies using distinctive marker varves found at multiple sites (De Geer 1912, 1934). Similar to the process routinely used in dendrochronology (Fritts, 1976),

W. M. Last & J. P. Smol (eds.), 2001. *Tracking Environmental Change Using Lake Sediments. Volume 1: Basin Analysis, Coring, and Chronological Techniques.* Kluwer Academic Publishers, Dordrecht, The Netherlands.

the cross-dating technique developed by De Geer continues to be used extensively in varve research.

The growing use of lacustrine varve-derived chronologies in paleoenvironmental research, particularly the construction of multiple varve records from a single region, raises important issues regarding the procedures used to construct and the chronological accuracy for a given record. As research questions continue to resolve finer time scales, direct comparison between lake records will require that chronologies constructed from varves are as accurate and consistent as possible. This chapter will discuss the processes that can be used for varve chronology construction, with emphasis on the methods for assessing the accuracy and consistency of records from any type of varve environment. Because of the variety and complexity of varve formation processes in different lake settings, the reader is directed elsewhere for detailed discussion on varve deposition and preservation (O'Sullivan, 1983, Smith & Ashley, 1985; Saarnisto, 1986).

Methods

Coring

Recovering sediment cores is generally the first step in any paleolimnological study and a variety of methods exist for recovering sediments for specific environments sediment types. Common methods are summarized in Figure 1 and discussed in further detail by Glew et al. (this volume). For the purposes of varve chronology construction, several key issues are worthy of mention. Most coring methods are capable of retrieving largely undisturbed sediment records. Smearing of the sediment along the inside edge of the tube can occur

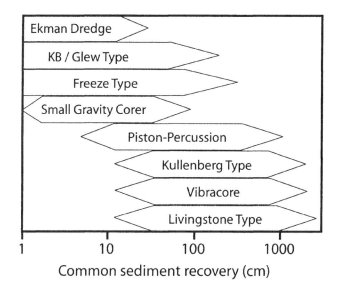

Figure 1. Selected coring methods for obtaining undisturbed sediment samples for varve chronology construction. The probable depth of undisturbed sediment recovery is shown for each method.

with some methods, especially some simple percussion-type systems or when a stiff core catcher is used. However, with the exception of extreme cases of edge disturbance, most coring methods are suitable for obtaining high quality sediment records.

Generally, to obtain the entire varve record, including the recent sediment at the sediment water interface, more than one type of coring method may be required. Methods designed to collect undisturbed surface cores cannot retrieve long (> 1 m) cores, and long-core methods invariably disturb or destroy surface sediments (Fig. 1). Obtaining the surface sediments is critical for several methods used to independently verify the varve chronology (e.g., radioisotopes) and also important for the calibration of sedimentary and fossil parameters to modern processes (Cumming & Smol, 1993; Hardy et al., 1996). Accurately joining multiple cores together by correlating the varve record is readily accomplished by cross dating and is discussed below.

Sample preparation

Obtaining a clear, undisturbed sample of the sedimentary structures is critical for accurate chronology construction. The most frequently used methods for obtaining undisturbed exposures of sediments include: cleaned core faces, photographs and x-radiographs, embedded slabs, thin sections, and, increasingly, advanced imaging techniques such as backscattered scanning electron microscopy (BSE) (Fig. 2). Cleaned core faces are the most common method for working with varved sediments and are usually a requirement for other methods. When the varves are well defined and are thick (ca. >2–5 mm), the exposed core face can be directly used for chronology construction (Saarnisto, 1986). The core is split in half using a sharp knife or wire, followed by a careful removal of the smeared sediment on the core face with a blade or glass edge. An electro-osmotic knife can be used to minimise smearing in clay-rich sediments (Bouma, 1969; Schimmelmann et al., 1990; Pike & Kemp, 1996). Several methods are used to obtain sedimentological information from the core face, depending on the sediment type and clarity of the structures. Renberg (1981) describes a method for cleaning and preparing frozen sediments, by scraping the surface clean using a plane blade or cover slip, with the strokes parallel to the varves. After the surface is clean, thin, clear plastic sheeting is placed over the sediment to minimise reflections and slow the oxidation of the sediments, which obscures the varves. Alternatively, when large quantities of iron or manganese sulphide are present, oxidation of the surface may be required to reveal the sedimentary structures (Renberg, 1981). Other researchers partially dry cores to enhance the contrast between the laminae, particularly in clastic varves. However, drying may not be suitable in many cases, as shrinkage, cracking, and the alteration of physical properties may limit subsequent subsampling.

Photographs of the core face are a simple and effective means for directly measuring varves and archiving the structural information (e.g., Renberg, 1981). When the core is frozen, warming the surface briefly before photographing the sediments can enhance contrast between the laminae (Saarnisto, 1986). If x-ray equipment is available, x-radiographs can also be used for measurements. X-rays from split cores tend to have uneven exposures due to the tube curvature, a problem that can be solved using sediment slabs, embedded if necessary to minimise handling disturbances (Fig. 2f). Finally, direct digital image capture has become a viable and efficient means to obtain core information for storage and analysis (Petterson et al., 1993; Francus, 1998). Although systems vary considerably in features

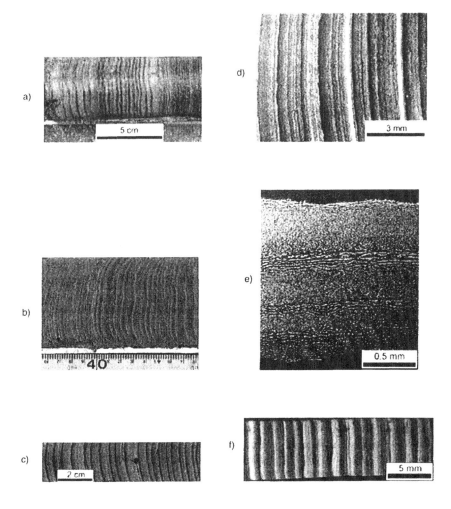

Figure 2. Sample images obtained from common methods for studying varved sediments. The simplest method is to clean the core edge by removing the core tube or liner, Lake Korttajärvi, Finland (a); a split and cleaned core face, Nicolay Lake, Canadian arctic (b); sediment embedded with epoxy resin; cured, cut, and polished, Bow Lake, Alberta, Canada (c); a thin section photomicrograph, Nicolay Lake (d); BSE microscopy image, Sawtooth Lake, Canadian arctic (e, courtesy of P. Francus); and, x-radiograph of an embedded slab, Lake Korttajärvi, Finland (f, courtesy of T. Saarinen).

and cost, the simplest digital camera can be used effectively for obtaining high quality varve images.

Increasingly, researchers are embedding sediments with epoxy resins or similar materials to obtain polished blocks, thin sections, or to archive the samples. A wide range of methods are used for embedding, all of which require replacing the sediment pore water with a liquid media that will cure hard (Table I). Large bodies of literature on

Table 1. Selected methods used for embedding varved lake sediments. Other methods and detailed procedures are described in detail in Bouma (1969) and Murphy (1986).

Method	Advantages	Disadvantages	Source
Freeze drying, epoxy resin under vacuum.	Rapid, minimal shrinkage	May cause cracking in clays, expensive equipment.	Murphy, 1986; Francus, 1998
Liquid solvent dehydration, epoxy resin under vacuum.	Few cracks, high quality samples, vacuum system required.	Sands may collapse without support, sediment blocks may crack or break during handling, time consuming (1 week or more).	Lamoureux, 1999
Liquid solvent dehydration, epoxy without vacuum.	High quality samples, simple equipment.	Similar to above. Requires repeatedly changing resin.	Clark, 1988; Lamoureux, 1994b; Card, 1997
Sample immersed in water-soluble Polyethylene glycol (PEG) for several days.	Simple, minimal equipment, minimal toxicity. Samples can be trimmed with a microtome.	Difficult to prepare thin sections and optical properties not optimal. Shrinkage may occur.	Bouma, 1969; Murphy, 1986

embedding sediments can be found in the soil science, marine geology, and paleolimnology fields, and similar techniques have been developed independently in each discipline. A thorough discussion of the various materials and techniques used to embed different types of sediments can be found in Bouma (1969), Murphy (1986), Lamoureux (1994b), Pike & Kemp (1996), Dean et al. (1999) and Kemp et al. (this volume). Common methods used in paleolimnological studies are summarized in Table I.

Once the embedded sample has cured, it can be cut into blocks and polished for direct measurements, thin sectioned and used for microscopy (Fig. 2). Embedded sediments provide the researcher with high quality surfaces that permit identification of very fine structures through optical and BSE imagery (e.g., Pike & Kemp, 1996; Francus, 1998; Dean et al., 1999). Additionally, slabs can be cut to uniform thickness, which can be used to produce x-radiographs with even exposures and improved clarity (Fig. 2). Often, fine varves and laminae become only apparent through the use of embedded sediments, making this approach particularly valuable for lakes with low accumulation rates or thin structures (e.g., Lamoureux, 1994b).

Logging and assessing the chronological error from one core

The process of identifying varves for counting and measurement is also largely dependent on the specific type of sediment, researcher preference, and the resources available. Once a method to sample and measure the sediments has been determined, the initial step for logging is identifying a process model for varve formation that can be used to objectively and reliably construct an initial chronology. Knowledge about the lake and catchment environments is crucial for the formulation of a process model. Several reviews (Sturm, 1979;

Håkanson & Jansson, 1983; O'Sullivan, 1983; Smith & Ashley, 1985; Saarnisto, 1986; Dean et al., 1999) provide detailed information on the formation of varves with varying composi-tions. Preliminary sedimentological observations and analyses will provide direct evidence for the varve formation model and criteria for varve identification and measurement. Most varve process models used are subjective and are related to visual information contained in the varve couplet. Objective methods for varve formation and recognition have been recently developed and applied, and can be used to improve chronological measurements from sediments with vague or poorly defined laminae (Card, 1997). However, even with objective varve identification criteria, core sections with vague or disturbed laminae should also be assessed carefully for evidence of bioturbation, erosion, or other changes to the depositional environment that could indicate the changing preservation of the varves, and a potentially unreliable record.

Once the varve identification criteria have been formulated, an initial chronology can be generated. For the highest reliability in these measurements, replicate sets of observations should be carried out to identify structures that are vague and to eliminate errors. Where possible, independent measurements by several researchers can further assist in obtaining consistency. In a study of the methods for identifying and measuring the varves from Kassjön Lake, Sweden, Petterson et al. (1993) found minimal differences between the observations by four workers, although they note that this may not be the case for varves where the transitions are not clear and subjective decisions must be made by individuals. Similarly, comparison between two independently derived chronologies from Lake Valkiajärvi, Fin-land showed small differences, except where disturbance obscured structures in one core (Ojala & Saarnisto, 1999).

One typically begins the varve measurement process by identifying prominent marker beds at regular intervals (usually on thin sections). The units interpreted as varves are then counted along four different transects on separate occasions. Notes are taken to describe unusual or vague structures, colour, and the relative quality of each varve is ranked. This process tends to highlight sections of a core that contain potential chronological problems. The varve quality ranking is retained as part of the final measurement set to provide a simple measure of the relative clarity of the sedimentary structures and the potential for chronological errors.

The variance between the replicate varve counts can be used as a measure of the counting imprecision for a given section and cumulatively for the entire core. In cases where the varves are disturbed or indistinct, the deviations between counts for a short segment of varves can be up to 25% (Sprowl, 1993; Lamoureux, 1994a; Zolitschka, 1996; Ojala & Saarnisto, 1999). The variance between replicate segment counts can be used to estimate confidence limits of the counting imprecision for each segment and cumulatively for the core (Fig. 3). The three published examples indicate that imprecision ranges from 3.4–11.7% (Table II), although it is apparent from these studies that core ambiguities and disturbances were responsible for considerable counting variance, particularly in isolated sections of cores (Lamoureux, 1994a; Ojala & Saarnisto, 1999). For example, in the Holocene record from Lake Valkiajärvi, most of the counting discrepancies occurred in a 40 cm section from an older core. For the remainder of the ca. 8500-year chronology, imprecision estimates were less than 3% for individual core segments (Ojala & Saarnisto, 1999). As both extra and missed varves can occur and tend to cancel out, counting imprecision of the total chronology or between independent chronologies can appear to be much smaller than indicated by

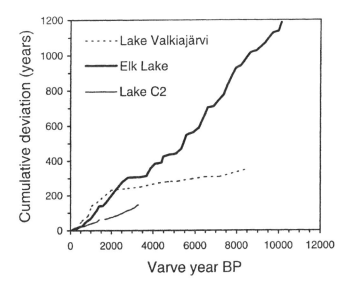

Figure 3. Cumulative deviations estimated for the varve chronologies of Lake Valkiajärvi, Finland (Ojala & Saarnisto, 1999), Elk Lake, Minnesota (Sprowl, 1993) and Lake C2, Canadian arctic (Lamoureux, 1994a). Deviations are calculated using the standard deviation of replicate counts, and represent 95% confidence limits. Cumulative deviations for Lake Valkiajärvi are calculated from the published core segment data using the same method applied to the other two lakes.

short-segment comparisons. These results indicate the importance of evaluating the varve counting precision from within and between individual cores. However, it is important to note that this is not a measure of the chronological imprecision of the varve record. This value cannot be determined from a single core, but must be assessed using other cores (Lamoureux & Bradley, 1996) or other proxy records (Sprowl, 1993).

Cross dating and correlating multiple cores

Given the potential for chronological error introduced by vagaries contained in a single sediment core, a further improvement in the reliability of the chronology can be obtained by correlating and cross dating multiple cores from the same lake. This approach provides an additional means for identifying discrepancies in the varve counts, unconformities and other localized depositional artefacts by matching the chronology from different cores (De Geer, 1934; Lamoureux, 1999). Cross dating is also routinely used in dendrochronology to match records from individual trees (Fritts, 1976). If lake-wide marker beds are available, segments of the record between the markers can be used to correlate and determine errors for shorter sections (Lamoureux & Bradley, 1996). Indeed, the correlation of lake-wide marker beds is a first, but not necessarily required, step in cross dating between cores. Other methods are available for correlating cores (e.g., paleomagnetic stratigraphy, fossils) but do not provide the absolute correlation possible with individual marker beds. Establishing

Table II. Published estimates of varve counting imprecision from two lakes. Lake Valkiajärvi data calculated from core segment differences between two independent chronologies.

Site	Estimated cumulative deviation 95% c.l. (years)	Record length (years)	% Deviation (95% c.l.)	Source
Elk Lake, Minnesota, USA:	1186	10120	11.7	Sprowl, 1993
Lake C2, Ellesmere Island, Canada:				
Core 8	11	323	3.4	Lamoureux, 1994a
Core 84	165	1991	8.2	
Core 85	149	2628	5.7	
Lake Valkiajärvi, Finland:	353	8522	3.4	Ojala & Saarnisto, 1999

a marker bed stratigraphy for the lake permits cross dating between all cores where the markers are visible, and also allows bridging the chronology across core sections that are disturbed (Fig. 4) (Anderson et al., 1993; Lamoureux & Bradley, 1996).

Cross dating can be accomplished using either chart plots of corresponding core sections (referred to as skeleton plots in dendrochronology), with varve thickness plotted on a log-ordinal scale, or by computer programs designed for tree ring analysis that identify missing years in the segments by correlating short (5 or more years) record segments and identifying likely sources of error. The computer programs used for cross dating are relatively simple to use and are available from the University of Arizona Laboratory of Tree-ring Research (http://www.ltrr.arizona.edu).

The plotting method is simple and intuitive to use, but can become time consuming if frequent discrepancies between the cores occur. By studying the patterns of varve thickness variations, it is usually readily apparent where extra or missing years are in a series (Fig. 5). De Geer (1934) recognized the importance of identifying these two types of observational error. He referred to a single varve interpreted to be two or three varves due to deceptive subannual units as a digraph or trigraph (extra varves), respectively, and a multiple varves are interpreted as one year of deposition as a monogram (missing varves).

In the cross dating procedure, prominent thick varves that are offset by one or more years indicate extra or missing varves in one or more cores. In two studies where these errors were considered, missing varves were considered to be more frequent (Sprowl, 1993; Lamoureux & Bradley, 1996). Extra varves (Type A errors, Lamoureux & Bradley, 1996) require observing a varve structure and were generally associated with vague subannual laminae. Varves that were missed were more common and primarily occurred in disturbed core sections (Type B errors, Lamoureux & Bradley, 1996). Using the thickness plots, core discrepancies can be identified, the sediments re-examined, and a decision can be made regarding the presence or absence of a varve. Corrections are made to individual

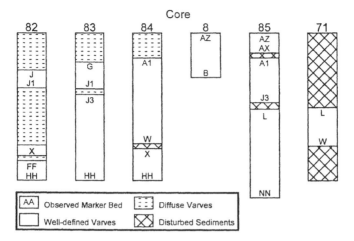

Figure 4. Schematic representation of the marker bed stratigraphy for six varved sediment cores recovered from Lake C2, Canadian arctic. Using marker beds permitted bridging gaps within individual cores and cross dating segments between cores to eliminate counting errors (figure from Lamoureux & Bradley, 1996. Used with permission, ©1996 Kluwer Academic Publishers).

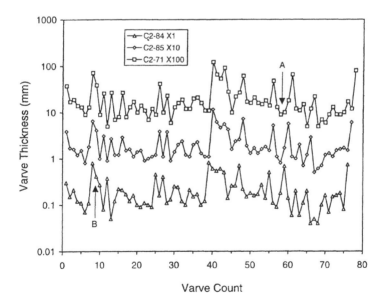

Figure 5. Skeleton plot used to cross date varves between two marker beds. Each core is offset by a factor of ten to separate the plots. Two examples of extra varves (arrows labeled A and B) are shown. In each case, the pattern of varve thickness is offset by a year compared with the other cores. The potential errors identified in the skeleton plot are used to identify individual laminae that may have been misinterpreted in one or more cores. The process is repeated until all discrepancies are resolved (after Lamoureux, 1994a).

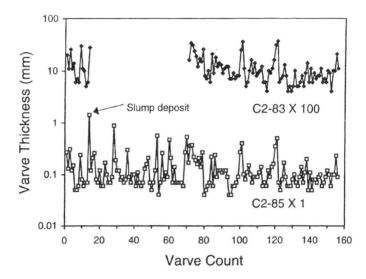

Figure 6. A skeleton plot indicating erosion of 56 varves from core C2-83, Lake C2, Canadian arctic (after Lamoureux, 1994a).

series where appropriate. Where disturbances result in missed varves, spaces are placed in the core series. In some cases, erosional unconformities that are not apparent in the sedimentary structures can be recognized by cross dating (Fig. 6). Where more than two cores are compared, errors are generally apparent only in one core, thereby isolating the likely source of error (Fig. 5). The visual correlation procedure is repeated until no errors remain for the core segments. Depending on the thickness and clarity of the varves, cross-dating cores can reveal chronological errors of 0.5–17.1% (Lamoureux & Bradley, 1996; Zolitschka, 1996). Where the varves are exceptionally clear, or where the sample preparation and measurements are optimal, the level of chronological error revealed by cross dating can be <0.5% (Lamoureux, 1999).

The error that remains after cross dating all of the cores from a lake (Type C errors, Lamoureux & Bradley, 1996) is difficult to assess. Error attributed to extra (Type A) and missed (Type B) varves are relative errors, and do not account for interpretive problems that are applied lake-wide. Potentially, years without deposition or subannual laminae interpreted as varves in all cores could introduce errors into the chronology that would not be apparent in the cross dating process. Therefore, other methods should be used to independently assess the varve chronology for Type C errors and the cumulative impact they have on the chronological error.

Chronology verification

Considerable reduction in chronological error can be accomplished by replicate counts and cross-dating multiple cores. However, without independent verification of the varve chronology, the remaining (Type C, previous section) error in the chronology remains unknown. In some cases, particularly where the laminae have been shown not to be varved (e.g., Lemmen et al., 1988), the Type C error could be substantial so independent verification of the varve chronology is crucial.

Several chronological methods are especially useful for independently verifying varve chronologies and are discussed in other chapters in the chronostratigraphic techniques section of this volume. In particular, ^{137}Cs and ^{210}Pb profiles from the upper sediments are routinely applied to verify a varve chronology for the past ~50 and ~200 years, respectively (see Appleby, this volume). However, application of these isotopes is not without difficulties, particularly in lakes with highly variable accumulation rates (Appleby & Oldfield, 1983). For older sediments, radiocarbon, uranium-thorium, and optical and thermoluminescence dating methods may provide validation for varve-based ages. Bradley (1999) provides a thorough discussion of these methods and their application to paleoenvironmental research.

Other methods can also be used to provide verification of varve ages. In some regions with a long history of human occupation, historical events that generate prominent marker beds in the lake can be dated with some certainty (e.g., logging activity and lake flooding, Anderson et al., 1993). Longer varve records can be compared with the cultural development and succession of the region, providing a good measure of the relative accuracy of the varve chronology (e.g., Saarnisto et al., 1977). Alternatively, other proxy indicators can be used to identify significant historical events, such as major weather events (e.g., Card, 1997) or diagnostic shifts in fossil indicators related to land use changes (e.g., the Ambrosia rise, Anderson et al., 1993; Card, 1997).

An increasingly viable means of comparing varve chronologies is by direct comparison of a proxy common to both records, particularly paleomagnetic variations recorded in the sediment. Recent work using varved sediments has shown the utility of comparing paleomagnetic records between lakes, although the temporal resolution is limited by the sampling interval, and the comparison of paleomagnetic records depends on the accuracy of the master (or regional) paleomagnetic curve chronology (Saarinen, 1994). New paleomagnetic secular variation records from varved lakes will provide a means to assess the consistency of the chronological accuracy in each record and an estimate of the Type C error associated with a given lake. As more accurately dated paleomagnetic records become available and finer resolution magnetic measurements become possible, this approach will provide an important means for assessing chronological errors and developing regional paleoenvironmental records.

Summary and future directions

The methods required to construct an accurate and reliable varve chronology have been presented. Several types of error can occur in a chronology, depending on the number of cores used and the ease and consistency in interpreting the varve structures. A number of steps are recommended for evaluating these errors and improving the reliability of the final chronology. Replicate varve counts from a single core, particularly by several observers, can be used to identify problematic sections of the chronology and assess the repeatability of measurements from a single core. It is important to note that different types of error tend to cancel out, therefore, it is important to evaluate reproducibility of varve counts for short core segments (e.g., Sprowl, 1993; Lamoureux & Bradley, 1996; Ojala & Saarnisto, 1999). Where multiple cores are available, cross dating can be used to further reduce measurement errors and bridge disturbed or vague core segments. Although these methods cannot remove all potential chronological error from a record, they can be used to identify and remove measurement errors. Error related to systematic misinterpretation of sedimentary structures can be identified using a variety of independent measures, including isotope geochronology techniques, correlating with historical events, and comparison with paleomagnetic records from other sedimentary records.

Increasingly, image analysis techniques are being applied to varve identification and measurement (e.g., Petterson et al., 1993; Pike & Kemp, 1996; Francus, 1998; Dean et al., 1999; Saarinen & Petterson, volume 2) and these techniques hold promise for increasing the speed and accuracy of chronology preparation. Moreover, the use of objective tests of varve structure (Card, 1997) is another promising approach that can further clarify difficult sections and to identify seasonal depositional cycles independent of structure.

Application of a systematic approach to identifying and removing error during varve chronology construction, together with advanced studies of varve structure (Dean et al., 1999), offer the potential to generate accurately dated, high-resolution sediment records for assessing rapid and long term environmental change.

Acknowledgements

The author gratefully acknowledges the support of a National Science and Engineering Research Council Postdoctoral Fellowship during the preparation of this manuscript. P. Francus, University of Massachusetts at Amherst, and T. Saarinen, Geological Survey of Finland, generously provided the BSE and x-radiograph images, respectively. Two anonymous reviewers and the editors provided helpful comments on the manuscript.

References

Anderson, R. Y., J. P. Bradbury, W. E. Dean & M. Stuiver, 1993. Chronology of Elk Lake sediments: coring, sampling, and time-series construction. In Bradbury, J. P. & W. E. Dean (eds.) Elk Lake, Minnesota: Evidence for Rapid Climate Change in the North-central United States, Special Paper 276. Geological Society of America, Boulder, CO: 37–45.

Appleby, P. G. & F. Oldfield, 1983. The assessment of ^{210}Pb data from sites with varying sediment accumulation rates. Hydrobiol. 103: 29–35.

Bouma, A. H., 1969. Methods for the Study of Sedimentary Structures. John Wiley & Sons, New York, 458 pp.

Bradley, R. S., 1999. Paleoclimatology, Reconstructing Climates of the Quaternary. International Geophysics Series, Volume 64, Academic Press, San Diego, 613 pp.

Brunnberg, L., 1995. Clay-varve chronology and deglaciation during the Younger Dryas and Preboreal in the easternmost part of the Middle Swedish Ice Marginal Zone. Quaternaria, Series A (2), 94 pp.

Card, V. M., 1997. Varve-counting by the annual pattern of diatoms accumulated in the sediment of Big Watab Lake, Minnesota, AD 1837–1990. Boreas 26: 103–112.

Clark, J. S., 1988. Stratigraphic charcoal analysis on petrographic thin sections: application to fire history in northwestern Minnesota. Quat. Res. 30: 81–91.

Cumming, B. F. & J. P. Smol, 1993. Development of diatom-based salinity models for paleoclimatic research from lakes in British Columbia. Hydrobiol. 269–270: 179–196.

De Geer, G., 1912. A Geochronology of the Last 12 000 Years. Proceedings of the International Geological Congress, Stockholm, 241–253.

De Geer, G., 1934. Geology and geochronology. Geografiska Ann. 1: 1–52.

Dean, J. M., A. E. S. Kemp, D. Bull, J. Pike, G. Patterson & B. Zolitschka, 1999. Taking varves to bits: scanning electron microscopy in the study of laminated sediments and varves. J. Paleolim. 22: 121–136.

Francus, P., 1998. An image-analysis technique to measure grain-size variation in thin sections of soft clastic sediments. Sed. Geol. 121: 289–298.

Fritts, H. C., 1976. Tree Rings and Climate. Academic Press, London, 567 pp.

Håkanson, L. & M. Jansson, 1983. Principles of Lake Sedimentology. Springer-Verlag, Berlin, 316 pp.

Hardy, D. R., R. S. Bradley & B. Zolitschka, 1996. The climatic signal in varved sediments from Lake C2, northern Ellesmere Island, Canada. J. Paleolim. 16: 227–238.

Lamoureux, S. F., 1994a. Paleoclimatic Reconstruction from Varved Lake Sediments. Lake C2, Ellesmere Island, Canada. M.S. thesis, University of Massachusetts at Amherst, 160 pp.

Lamoureux, S. F., 1994b. Embedding unfrozen lake sediments for thin section preparation. J. Paleolim. 10: 141–146.

Lamoureux, S. F. 1999. Spatial and interannual variations in sedimentation patterns recorded in nonglacial varved sediments from the Canadian High Arctic. J. Paleolim. 21: 73–84.

Lamoureux, S. F. & R. S. Bradley, 1996. A late Holocene varved sediment record of environmental change from northern Ellesmere Island, Canada. J. Paleolim. 16: 239–255.

Lemmen, D. S., R. Gilbert, J. P. Smol & R. I. Hall, 1988. Holocene sedimentation in glacial Tasikutaaq Lake, Baffin Island. Can. J. Earth Sci. 25: 810–823.

Murphy, C. P., 1986. Thin Section Preparation of Soils and Sediments. A B Academic, Berkhamsted, UK, 149 pp.

O'Sullivan, P. E., 1983. Annually-laminated lake sediments and the study of Quaternary environmental changes — a review. Quat. Sci. Rev. 1: 245–313.

Ojala, A. E. K. & M. Saarnisto, 1999. Comparative varve counting and magnetic properties of the 8400-yr sequence of an annually laminated sediment in Lake Valkiajärvi, Central Finland. J. Paleolim. 22: 335–348.

Petterson, G., I. Renberg, P. Geladi, A. Lindberg & F. Lindgren, 1993. Spatial uniformity of sediment accumulation in varved lake sediments in northern Sweden. J. Paleolim. 9: 195–208.

Pike, J. & A. E. S. Kemp, 1996. Preparation and analysis techniques for studies of laminated sediments. In Kemp, A. E. S. (ed.) Palaeoclimatology and Palaeoceanography from Laminated Sediments. Geological Society Special Publication No. 116: 37–48.

Saarinen, T., 1994. Palaeomagnetic Study of the Holocene Sediments of Lake Päijänne (Central Finland) and Lake Paanajärvi (North-West Russia). Geological Survey of Finland, Bulletin 376, Espoo, 87 pp.

Saarnisto, M., 1986. Annually laminated lake sediments. In Berglund, B. E. (ed.) Handbook of Holocene Palaeoecology and Palaeohydrology. John Wiley & Sons, London, 343–370.

Saarnisto, M., P. Huttunen & K. Tolonen, 1977. Annual lamination of sediments in Lake Lovojärvi, southern Finland, during the past 600 years. Ann. Bot. Fennici 14: 35–45.

Schimmelmann, A., C. B. Lange & W. H. Berger, 1990. Climatically controlled marker layers in Santa Barbara Basin sediments and fine-scale core-to-core correlation. Limnol. Oceanogr. 35: 165–173.

Smith, N. D. & G. M. Ashley, 1985. Proglacial lacustrine environment. In Ashley, G. M., J. Shaw & N. D. Smith (eds.) Glacial Sedimentary Environments. Society of Paleontologists and Mineralogists, Tulsa, OK: 135–215 pp.

Sprowl, D. R., 1993. On the precision of the Elk Lake varve chronology. In Bradbury, J. P. & W. E. Dean (eds.) Elk Lake, Minnesota: Evidence for Rapid Climate Change in the North-central United States, Special Paper 276. Geological Society of America: 69–74.

Sturm, M., 1979. Origin and composition of clastic varves. In Sturm, M. (ed.) Moraines and Varves, Origin/Genesis/Classification. A.A. Balkema, Rotterdam: 281–285.

Zolitschka, B., 1996. Recent sedimentation in a high arctic lake, northern Ellesmere Island, Canada. J. Paleolim. 16: 169–186.

12. LUMINESCENCE DATING

OLAV B. LIAN (olian@sfu.ca)
D. J. HUNTLEY (huntley@sfu.ca)
Department of Physics
Simon Fraser University
Burnaby, B.C. V5A 1S6
Canada

Keywords: Luminescence dating, optical dating, thermoluminescence dating, OSL, TL, geochronology, lacustrine

Introduction

Luminescence dating is now a well established method for determining the time elapsed since quartz or feldspar grains were last subjected to sufficient heat or light. It is based on the fact that minerals contain structural defects, some of which can trap unbound (free) electrons. If a mineral is heated to a high enough temperature (ca. 400 °C), or exposed to sufficient sunlight, some or all of these traps will be emptied of electrons. Free electrons are produced when radiation emitted during radioactive decay interacts with matter. This radiation consists of alpha and beta particles and gamma rays. The rate at which free electrons are produced, and traps are filled, is proportional to the concentration of radioactive elements in the minerals being dated, and in their surroundings. If one determines the rate at which the radiation energy is absorbed by the mineral(s) of interest, and makes a measure of the number of electron traps that are full, which provides an estimate of the total absorbed radiation energy, the time elapsed since the minerals were last exposed to sufficient heat or sunlight can be calculated. The luminescence age is simply equal to the total absorbed energy divided by the energy absorption rate. Luminescence dating is therefore commonly used to date the last time a sediment was heated, or the time of final burial, which in most cases corresponds to the age of the sedimentary unit in which the sediment occurs. The established age-range of luminescence dating is from about 1 ka to 150 ka. However, in rare cases, ages as young as a few years, or as old as $\sim$ 800 ka, have been calculated.

In this chapter we discuss the basic physical concept behind luminescence dating, review the techniques used for estimating the paleodose, including established multiple-aliquot methods and recently introduced single aliquot/grain techniques, we then explain how the environmental dose rate is evaluated. This is followed by a discussion of the practical aspects, including sample collection, what types of deposits are suitable, with examples from studies of lacustrine environments, and finally we review some of the problems that might be encountered.

W. M. Last & J. P. Smol (eds.), 2001. *Tracking Environmental Change Using Lake Sediments. Volume 1: Basin Analysis, Coring, and Chronological Techniques.* Kluwer Academic Publishers, Dordrecht, The Netherlands.

The mechanism responsible for luminescence

Luminescence refers to the light that is emitted by matter in response to some external stimulation. We shall restrict ourselves here to the cases of the external stimulation being heat or exposure to light. Physical models used to explain this luminescence phenomena are complex, and many of the details are not yet understood. An in-depth review that is relevant to dating is provided by McKeever & Chen (1997), while a more general discussion can be found in Aitken (1998). Below we briefly outline the basic concept.

Natural crystals (e.g., a mineral) are not perfect, and contain structural defects and impurities, some of which can trap unbound (free) electrons. Such electrons may therefore be held (trapped) at these locations for a time that depends on the thermal lifetime of the trap. At environmental temperatures (ca. 20 °C), traps with long thermal lifetimes (sometimes referred to as "deep traps") may hold electrons for millions of years or more, whereas traps with short thermal lifetimes ("shallow traps") may hold electrons for only a few days or less. There are many different kinds of defects, and these all have different thermal lifetimes.

When a mineral is subjected to an external source of energy, for example heat from a campfire or a volcanic flow, or photons of sunlight, a trapped electron may acquire enough energy to escape from its trap. Once evicted, an electron diffuses around the crystal lattice until it eventually becomes re-trapped, or encounters another type of site attractive to electrons referred to as a recombination centre; this process is nearly instantaneous. When an electron encounters a recombination centre, excess energy is given off as either phonons (lattice vibrations or heat), or as photons (luminescence).

Traps also have various sensitivities to light. There are some traps that can be emptied with only a few seconds of sunlight exposure, and there are others for which exposure to sunlight has little or no effect.

External radiation that is found in the natural environment consists of alpha, beta, gamma and cosmic rays. All of these have an energy that is much higher than the energy needed to free electrons from their normal states in atoms, and thus a dose of radiation will free a great many electrons. Some of these will become trapped in the "traps" mentioned above.

In the laboratory luminescence can be invoked under controlled conditions, either by heating the crystal or by exposing it to photons of a specific energy or energy range. When a crystal is heated, light is emitted; it consists of two components, incandescence and thermoluminescence (TL). Incandescence is the light emitted by anything when it is heated, such as that emitted by an incandescent light bulb or a red-hot stove. In contrast TL is the light emitted by a mineral while it is heated and that is a result of a prior dose of radiation which has resulted in electrons in traps. It is a transient phenomenon, and after the mineral has been heated, a repeat heating will not result in further TL emission. During laboratory heating, both incandescence and TL are emitted; in practice, infrared and red absorbing optical filters are used to reduce the amount of incandescence measured to a level that can be tolerated. This works well because usually the TL one wants to measure is blue or ultraviolet.

When a crystal is exposed to photons of light, some of this light is transmitted by the crystal, some of it is scattered, and some is absorbed. In addition, new photons are emitted. These new photons consist of three components: ordinary photoluminescence, Raman-scattered photons[1] and photons that result from a prior radiation dose that has resulted in electrons being present in traps. It is this latter that is of interest here; it is often referred to as optically-stimulated luminescence, although it can also be called photoluminescence. Some of these photons have energies higher than those of the scattered incident photons, ordinary photoluminescence, and Raman-scattered photons, and can therefore be separated using optical filters.

Dating and estimation of the paleodose

The age that is determined is the time elapsed since the traps measured were emptied. This means that the event being dated is that of the last heating to at least $\sim 400\,°C$, or the last exposure to sunlight.

A luminescence age is calculated from:

$$\text{(past radiation dose)} = \text{(radiation dose rate)} \times \text{(age)}. \qquad (1)$$

The radiation dose rate is calculated from a knowledge of the concentrations of the radioactive elements in the sample and its surroundings, and the cosmic-ray intensity. Its evaluation is discussed later. Here we are concerned with the evaluation of the past radiation dose.

Radiation dose is strictly defined as the amount of radiation energy absorbed per unit mass of material; the SI unit is the gray (Gy) where $1\,\text{Gy} = 1\,\text{J·kg}^{-1}$. Equation (1) cannot, however, be used as it stands for the following reason: while the amount of luminescence one measures is the same for a given dose of beta particles or gamma rays, it is not the same for the same dose of alpha particles. A given dose of alpha particles results in about one-tenth of the amount of luminescence as would the same dose of beta or gamma radiation. For this reason, the term "equivalent dose" has been introduced; it is the laboratory beta or gamma dose that would result in the same luminescence as did the environmental dose. The age equation used in practice is therefore:

$$\text{(past equivalent radiation dose)} = \text{(equivalent radiation dose rate)} \times \text{(age)}. \qquad (2)$$

In practice the term "equivalent" is frequently omitted from "equivalent radiation dose rate", and the term "equivalent dose" (commonly abbreviated D_e, D_{eq} or ED) is sometimes incorrectly referred to as the paleodose; "paleodose equivalent" would be appropriate.

The equivalent dose can be determined either by measuring the TL or the optically-stimulated luminescence; the latter is often abbreviated "OSL". These do not measure the same thing and in some cases will yield different equivalent doses. This is because some trapped electrons are readily evicted by the incident photons, whereas others are not. The term "light-sensitive" is often used to refer to those traps for which the incident photons most readily evict the electrons. Thus, when dating the last exposure to sunlight (i.e., for unheated sediments) using optical excitation is in most cases preferred because

---

[1] We refer here to "ordinary" photoluminescence as the luminescence that results when atoms in the crystal, excited by incident photons, return to their ground state. Raman-scattered photons refers to photons of incident light that have undergone a change in energy (wavelength) as the result of inelastic scattering off phonons (lattice vibrations in the crystal lattice).

light-sensitive traps are preferentially sampled, whereas, with thermal excitation the TL arises from electrons originating from both light-sensitive and light-insensitive traps. For this reason optical dating has largely superseded TL dating for sediments. An exception to this is that in some instances quartz grains that have remained buried for a relatively long time can only be dated using TL. Below we describe how an equivalent dose is found in practice. Because there is still significant utilization of TL in dating sediments, we include a discussion of TL dating methods.

Thermoluminescence dating

TL dating was developed in the early 1960's as a means of dating fired pottery. It was later recognised that it could also be used to date the last time sediments were exposed to sunlight (historical accounts can be found in Dreimanis et al., 1978; Wintle & Huntley, 1982; and Aitken, 1985: 3–4). There are currently several procedures available for determining the equivalent dose. Detailed discussions can be found in Aitken (1985), Berger (1988, 1995), or Wagner (1998: 235–262). Of these, the most established are the additive-dose, total bleach, partial bleach (sometimes called the R-gamma or R-beta method), and regeneration methods. The first three methods are discussed below while the regeneration method is explained later in the context of optical dating.

The additive-dose method is used for sediment that has been exposed to heat so that all of the relevant traps have become empty. The procedure involves preparing several aliquots of the separated mineral fraction. Some are left unirradiated (these are referred to as "naturals"), while the remainder are given various laboratory gamma or beta doses. This is done to define how the TL emitted from the minerals of interest responds to increasing radiation doses (the dose response). TL is recorded as a function of temperature, and the resulting data of TL intensity versus laboratory dose are fitted with an appropriate function, commonly a straight line or a saturating exponential, and extrapolated to zero TL intensity — the dose intercept. In practice dose-intercepts are determined at successively increasing temperatures, for example from 200 to 450 °C. When a temperature range associated with thermally stable traps has been reached, the dose-intercepts are expected to be constant, forming a plateau. The equivalent dose is usually taken to be the average value of the dose-intercepts that fall within the plateau. If a plateau is not obtained, then it could be an indication that some of the mineral grains had received insufficient heat while buried, the presence of "spurious TL" (e.g., Aitken, 1985: 6–7), or that the dose-response model used is not appropriate. It may also be an sign of anomalous fading (discussed at the end of this chapter), but there are indications that anomalous fading does not necessarily destroy a plateau. Achieving a plateau in the dose-response versus temperature data is therefore a necessary condition for providing a TL age. An example of the additive-dose method is shown in Figure 1.

For samples for which the event to be dated is the last exposure to extended sunlight, the total bleach method (Singhvi et al., 1982) is often used. The concept here is that the TL intensity, I, can be considered as the sum of two components:

$$I = I_0 + I_d. \tag{3}$$

Here, I_0 is the TL intensity that would have been measured had the sample been collected immediately after burial, and which results only from electrons in light-insensitive traps.

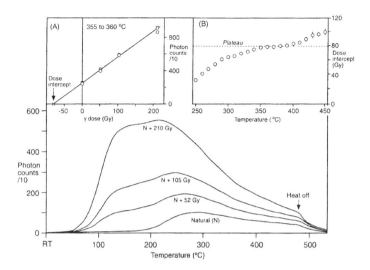

Figure 1. Example of the additive-dose method on a sample of ca. 15 ka fluvial silt that had been heated by fire. The main graph shows thermoluminescence (TL) intensity as a function of temperature for a "natural" aliquot, as well as three others that have been given various laboratory gamma doses. These "glow curves" exhibit peaks that represent different electron traps. RT is room temperature. The inset (A) shows the dose-response constructed from the TL integrated between 355 and 360 °C. A line is fitted to the data and extrapolated to where it intersects the dose axis — the dose-intercept. The inset (B) is a plot of the dose intercepts as a function of temperature. The dose intercepts become constant at about 350 °C, and give an equivalent dose of about 80 Gy. The rise above 400 °C could be due to spurious TL, or due to the use of an inappropriate fit to the additive-dose data.

I_d is the TL intensity resulting from the environmental radiation dose since burial. Such a precise division of the traps into two such types probably does not exist in reality, though in practice it appears to be a good approximation. In this case, determination of the equivalent dose involves extrapolating the dose-response curve to the TL intensity of a natural aliquot that has received an extended exposure to sunlight or a comparable laboratory light; the intention here is to determine I_0, in other words, to produce an aliquot that will give the same TL intensity as would an aliquot prepared from the same material immediately after deposition. To this end a light spectrum similar to that which it would have experienced in its natural environment is used, although it is not clear that this will always produce the desired result.

Equation (3) is clearly an approximation. As well, there are traps with a wide range of light sensitivities and for some samples not all such traps will have been emptied. In addition there is some concern that the extended sunlight exposure used in the total bleach method may have some unwanted side effects. All these concerns are taken care of with the partial bleach method (Wintle & Huntley, 1980). The concept here is to measure the TL reduction caused by a relatively brief and spectrally-restricted light exposure, to plot this reduction against laboratory radiation dose, and extrapolate to zero reduction to find the equivalent dose. In practice the way this is done is to construct the additive dose response as above, and also a second dose-response using aliquots that have received a

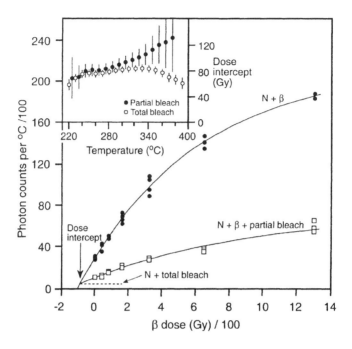

Figure 2. Example of partial bleach and total bleach methods for TL dating. Shown are data for silt-sized grains from loess sample BP-25 ($\sim$ 25 ka) of Shulmeister et al. (1999) (modified from their Fig. 3). The data are from TL recorded at 280 °C. The sample was preheated for 4 days at 130 °C prior to the final measurements. The inset shows the TL plateau plots for both methods.

short exposure to laboratory light following laboratory dosing. Both data sets are fitted with appropriate curves, and these are extrapolated to where they intersect. As with the additive-dose procedure, dose-intercepts are determined at successively increasing temperatures, until a plateau is achieved. An example is shown in Figure 2.

The partial-bleach method is designed to separate only the TL resulting from the most light-sensitive traps. For quartz it has been shown that there exists a component of the TL emitted at about 325 °C that arises from traps that are orders of magnitude more sensitive to yellow light than the traps that give rise to the rest of the TL between 300 and 500 °C (Franklin & Hornyak, 1990; Prescott & Mojarrabi, 1993). It is therefore more prudent to use the TL emitted around 325 °C to estimate the equivalent dose. A thorough discussion can be found in Wintle (1997).

The energy (wavelength, colour) of the photons of the emitted TL depends on the minerals present in the sample and on the types of defects that form the luminescence centres. In practice, optical filters are often used to pass a restricted band of photons that is of interest. There is, however, no consensus at to what band one should use for any particular mineral and for reasons that are not understood it seems that the equivalent dose

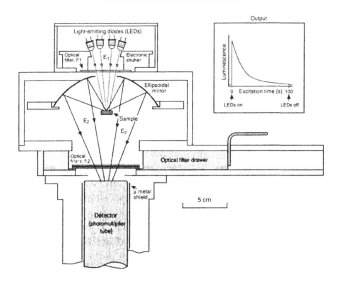

Figure 3. Schematic representation of an apparatus used to measure optically stimulated luminescence (design of Baril, 1997). Photons of energy E_1 are directed from high-power light-emitting diodes onto the sediment sample; these grains respond by emitting photons (luminescence) of energy E_2 which are directed onto the detector (photomultiplier tube) by an ellipsoidal mirror. The longer the time since the sample has been shielded from sunlight, the greater the luminescence. The optical filter F1 is used to absorb photons from the diodes that have an energy higher than E_1. Optical filters F2 are used to block scattered photons of the excitation beam from reaching the photomultiplier tube. These filters must also absorb ordinary photoluminescence and Raman-scattered photons. The sample aliquot is inserted manually using a holder that slides perpendicular to the drawing. The right-hand figure shows the luminescence intensity decreasing with time as the traps are emptied.

one obtains may depend on which band of photons one measures, a situation that can hardly be called satisfactory. A comprehensive discussion of this can be found in Berger (1995).

Optical dating

Introduction

Optical dating was introduced by Huntley et al. (1985). Its fundamental difference from TL dating is that optical excitation is used instead of heat. Photons of a specific energy, or energy range, is used to excite electrons from the traps. This allows the most light-sensitive traps to be preferentially sampled and leads to the ability to date sediments that have only been exposed to a few seconds of direct sunlight prior to burial, and to date samples much younger than possible using TL. Furthermore, because no heating is required, the apparatus used to make the luminescence measurements is much simpler (and can be made cheaper) than that used for TL, and this leads easily to automation. An apparatus designed to collect the maximum number of photons possible from an aliquot is shown in Figure 3. Commercial systems are discussed by Bøtter-Jensen (1997), Aitken (1998), Bøtter-Jensen & Murray (1999), and Bortolot (1999).

Table I. Excitation and luminescence photon energies commonly used to date quartz and potassium feldspar.

Mineral	Energy of excitation photons	Energy of luminescence photons
Quartz	2.2–2.4 eV* (510–560 nm, green)	3.35 eV (370 nm, ultraviolet)
Potassium-feldspar	∼ 1.4 eV (∼ 880 nm, infrared)	3.1 eV (400 nm, violet)

* 1 eV (electron volt) = 1.602 × 10^{-19} joules

In the initial experiments, green light from an argon laser was used to excite both quartz and feldspar samples (Huntley et al., 1985). It was later shown that potassium feldspars could be excited using ∼ 1.4 eV photons which are in the near infrared (Hütt et al., 1988), and that satisfactory ages could be attained this way. There is now no need to use expensive lasers for the excitation as much cheaper blue, green, red and infrared light-emitting diodes are available with sufficient power. A rather nice feature of the use of infrared excitation is that since it does not excite the relevant trapped electrons in pure quartz, it can be used to selectively sample feldspars in polymineral samples, or feldspar inclusions within quartz grains (Huntley et al., 1993a). Table I lists the commonly used excitation energies, and resulting luminescence energies used for dating quartz and potassium feldspar.

A comprehensive review of optical dating can be found in the book by Aitken (1998). Other reviews with varied emphases include those of Berger (1995), Roberts (1997), Prescott & Robertson (1997), Wagner (1998: 262–277), Stokes (1999), and Huntley & Lian (1999).

Multiple aliquot techniques

In practice, as for TL dating, the equivalent dose is found by constructing a dose-response curve and extrapolating it toward the dose axis. However, unlike for TL dating, there is no way of selecting only traps with long thermal lifetimes during the luminescence measurements. To overcome this the sample aliquots are heated ("preheated") before the final measurement. The preheat is done in a laboratory oven, or in a TL apparatus, and is designed to evict electrons that have ended up in shallow traps during laboratory irradiation. The required preheat temperature and duration depends on the nature of the traps and is therefore somewhat dependent on the mineral and (or) sample being dated. For potassium feldspars, 140 °C for 7 days has been found to be appropriate in many cases, whereas for quartz, 225 °C for 5 minutes has sometimes been found to achieve the desired results. Various preheat protocols have been tested with various degrees of success; some these are discussed by Aitken (1998: 190–192) for feldspar, and by Wintle & Murray (1998) and Aitken (1998: 193–194) for quartz. In practice, however, it is prudent to experiment with different preheat protocols, increasing its severity until the calculated equivalent dose becomes constant.

Unfortunately, the preheat has the unwanted side effect of transferring electrons from light-insensitive traps to thermally-stable light-sensitive traps. The quantity of electrons available for this thermal transfer depends on the amount of light exposure prior to burial, and there is an additional contribution from the laboratory dose given to the sample. To allow for this a correction is needed. This is accomplished by employing a second set of aliquots which are given a short exposure to laboratory light after laboratory irradiation. This light exposure should be designed to empty or nearly empty the light-sensitive traps that one wishes to sample, but not empty any of the traps that are the source of the thermal transfer. The additive-dose set of aliquots and the thermal-transfer correction set are preheated and measured together. The dose responses of the two sets are extrapolated to where both curves intersect, which of course must be above the dose axis, and the dose there is taken as the dose intercept. This correction also allows for any residual scattered light from the excitation beam, other photoluminescence, and Raman scattered photons that pass through the optical filters. It also allows for background (dark) counts intrinsic to the photon detector (photomultiplier tube). For old samples exclusion of the thermal-transfer correction may not seriously affect the equivalent dose, but for young samples where the dose-response is linear, or nearly so, omission of this correction can result in the equivalent dose (and optical age) being much too large, in some cases by a factor of two or more (for examples see Huntley & Clague, 1996 and Lian & Huntley, 1999). An example of the additive-dose method, with thermal transfer correction, is shown in Figure 4.

It is good practice to determine dose intercepts for consecutive time intervals after the excitation is switched on. A plot of dose intercepts versus excitation time is sometimes referred to as a "shine plateau" if it is flat (Fig. 4), and it is not to be confused with a TL plateau plot, which is something entirely different. If the dose intercepts are found to rise with excitation time, then something is amiss. One interpretation is that the sample consists of grains that had not received adequate sunlight exposure before burial. Unfortunately, such a test does not always detect the presence of unbleached grains, for if the sample consists of a mixture of grains that have been well bleached, and grains that had not been bleached at all, then dose intercepts are expected to be constant with excitation time, just as if would occur if all the grains had been well bleached.

For samples that are approaching dose saturation, extrapolation of the additive-dose curve will be large, and may therefore not be properly defined by the chosen fit. If this is the case the regeneration method is preferred. The best routine for this is currently that which is discussed by Huntley et al. (1993b) and Prescott et al. (1993). Sometimes referred to as the "Australian slide method", it involves subjecting some of the sample aliquots to a long laboratory bleach, which empties almost all of the relevant traps, and constructing a second growth curve from those. An example is shown in Figure 5. The regenerative set of data ideally includes that sample's response between zero absorbed dose, which is that expected at the time immediately after burial, and the dose absorbed by the sample since it was buried. The two aliquot sets are preheated and measured together. The equivalent dose is then taken as the shift along the dose axis that brings the regeneration data and additive dose data into alignment (Fig. 5), along with a small correction for incomplete laboratory bleaching. For some samples the laboratory bleach can alter the dose response of the regeneration data, and this must be tested for. This can be done by including an intensity scaling parameter in the curve fitting algorithm. If this parameter is statistically consistent with unity, then one deduces that no significant sensitivity change has occurred.

For all multiple-aliquot techniques up to 60 aliquots are typically needed to construct the required dose-response curves. Because these aliquots cannot be made to be identical, and because of intrinsic sensitivity variations in the minerals, all the aliquots need first to be

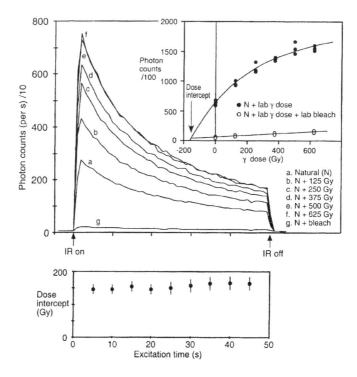

Figure 4. Luminescence decay as measured under 1.4 eV (infrared) excitation for glaciolacustrine silt from the Fraser River valley British Columbia (sample BBCS5 of Lian & Hicock (2001), apparent optical age ~ 85 ka). The inset (upper curve) shows the dose-response of the sample using the luminescence integrated over the first 5 seconds of excitation; the lower curve is the thermal-transfer correction. The lower graph shows how the dose-intercepts vary with excitation time. The preheat was 160 °C for 4 hours.

normalized. This is done by exposing each aliquot to the excitation beam for a short period of time. Usually not more than one second is needed. For each aliquot, the luminescence recorded during the short exposure is divided by the average luminescence of the entire set, thus producing normalization factors that are employed to reduce scatter in the final measurements. For samples that exhibit low luminescence intensity normalization times may need to be relatively long (e.g., a few seconds), and in those cases the resulting decay must be corrected for when the equivalent doses are calculated from the dose-intercepts.

Single aliquot and single grain techniques

When optical dating was introduced, it was suggested that the method could potentially be used to date a single aliquot consisting of only a few milligrams of sediment grains (Huntley et al., 1985). Since then various protocols have been developed that are based on one aliquot (in some cases extra aliquots are needed to monitor sensitivity change) or even a

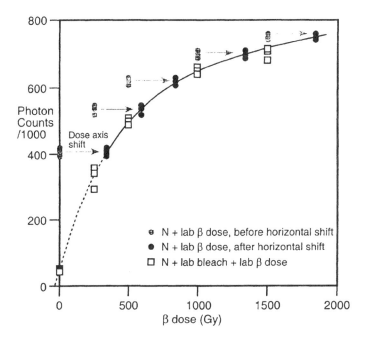

Figure 5. Example of the regeneration method. Shown is the beta dose response for both the additive dose (solid grey points) and regeneration (open square points). The black points are the additive dose data shifted along the dose axis for best fit, the magnitude of the shift being equal to the equivalent dose after a correction for normalization decay and incomplete laboratory bleaching. The dashed part of the curve is that which would have had to have been extrapolated if the additive-dose procedure was used. The sample is ca. 80 ka silt deposited in ancestral Lake Poukawa near Hawkes Bay, New Zealand.

single sand-sized grain. As with multiple aliquot techniques, the paleodose equivalent must be estimated by constructing a dose-response curve. Unlike multiple-aliquot techniques, single aliquot/grain methods involve repeated irradiation, preheating, and excitation of the same aliquot or grain.

The first published studies of dating single aliquots of feldspar was that of Duller (1994, 1995) while the potential for dating single aliquots of quartz was first studied by Liritzis et al. (1994) & Galloway (1996). The potential for dating single grains of feldspar was first investigated by Lamothe et al. (1994) and Lamothe & Auclair (1997), and for quartz by Murray & Roberts (1997). Detailed discussion of some of these techniques can be found in Roberts et al. (1999), Galbraith et al. (1999), Bøtter-Jensen & Murray (1999); a broader review is provided by Aitken (1998: 97–107).

The advantages of these techniques are that very little sample is required, normalization is avoided, and for single grains there is potential for dating only those grains that have received sufficient sunlight exposure (e.g., Roberts et al., 1998; Clarke et al., 1999). This gives promise for dating deposits that might be otherwise deemed inappropriate for multiple

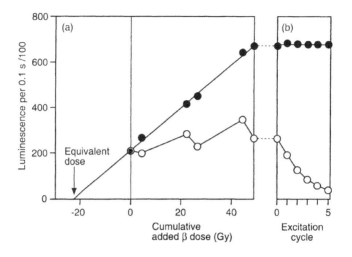

Figure 6. Example of one of the single-aliquot methods, modified from Figure 1 of Murray & Roberts (1997). The sample is quartz. The protocol involves using additive dosing, preheating, and short excitation times. (a) open circles are the raw data, while the solid circles are the same data, but corrected using the average exponential decay of the last six raw data points, open circles in (b). The solid circles in (b) are the open circles in (b) corrected using the same decay used in (a), and are used as a check. The dose intercept is taken from the line extrapolated to zero intensity in (a). In practice, the procedure is repeated using increasingly severe preheats until the measured dose-intercepts become constant. For this sample this occurred at ~ 260 °C. See Murray & Roberts (1997) for a thorough discussion.

aliquot techniques. However, there are also drawbacks. These include low levels of luminescence, undetected, or underestimated, sensitivity changes that may result from repeated heating and measurements on the same aliquot or grain, and the lack of a correction for the presence of thermal transfer. One study has found that such malign affects are present in many quartz samples and that current techniques do not correct for all of them (Stokes et al., 2000). Nevertheless, in several cases, there has been good agreement between equivalent doses found using single aliquots or grains and those found using standard multiple-aliquot techniques (e.g., see Murray et al., 1997), and this suggests good promise. One of the single aliquot methods is illustrated in Figure 6.

Evaluating the environmental dose rate

The dose absorbed by the sediment comes from alpha beta, and gamma radiation from the decay of ^{235}U, ^{238}U, ^{232}Th, ^{40}K, and ^{87}Rb, and their daughter products, both within the mineral grains and their surroundings. There is also a contribution from cosmic rays, which diminishes with depth. The environmental dose rate not only depends on the concentration of these radioisotopes and the intensity of cosmic rays, but also on the amount of water and organic matter in the sediment matrix, as these attenuate or absorb radiation differently from the mineral matter. All of these quantities have to be evaluated and accounted for

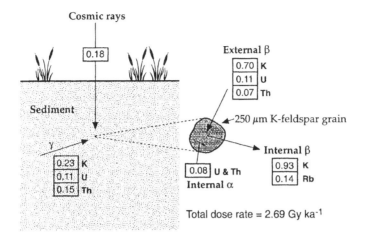

Figure 7. Illustration of the various contributions to the dose rate, based on Figure 4 of Huntley & Lian (1999). Shown is a typical 250 μm diameter potassium feldspar grain at a depth of 1 metre. The sediment contains 1% K, 1 μg·g^{-1} U, 3 μg·g^{-1} Th, and 5% water. The potassium feldspar grain has 13% K, 250 μg·g^{-1} Rb, 0.2 μg·g^{-1} U, and 0.4 μg·g^{-1} Th. "Internal" and "external" refer to the radiation originating inside and outside the grain, respectively.

in the dose rate calculations. A typical example of the different contributions to the dose rate for a potassium feldspar sand grain are shown in Figure 7. Equations used to calculate the dose rate due to U, Th, and K can be found in Aitken (1985, 1998) or Berger (1988); the latest dose-rate conversion factors are those of Adamiec & Aitken (1998). Formulae modified to deal with the presence of organic matter can be found in Divigalpitiya (1982) or Lian et al. (1995), and the formula used to estimated the dose rate due to cosmic rays is found in Prescott & Hutton (1994).

Radioisotope concentrations are commonly determined by taking a subsample of the sediment used for dating and using various laboratory analyses that give the concentration of the parent isotopes, or of various daughter products from which the concentration of the parents can be determined. However, although such analyses may provide good precision, it is usually assumed when the dose rate is calculated that each radioisotope is (and has been) in secular equilibrium. For some sediments, such as dry sands, this is almost always the case (e.g., Prescott & Hutton, 1995) but for others, such as those that occur in wet organic-rich environments, there is a strong possibility that parent isotopes and (or) daughters have been removed from, or introduced into, the sediment matrix. One example is that ^{222}Rn (a daughter product of ^{238}U) can leave by gaseous diffusion, and another is that U, which is soluble in groundwater, can be taken up by organic matter. Although one cannot always determine whether or not there was radioactive disequilibrium sometime during the history of the deposit, there are methods for determining if it existed at the time of sample collection. This involves measuring the activities of various parts of the decay chains and looking for anomalies. If radioactive disequilibrium is detected, it can be allowed for in

(A)

Figure 8. (**A**) Portable gamma-ray spectrometer (Exploranium model GR256). Attached to the spectrometer electronics (**a**) is the probe (**b**) and an optional hand-held computer (**c**) which can be used to control the spectrometer and to upload the data. (**B**) Laminated silt deposited in an ancient glacial lake (apparent optical age ∼ 185 ka) in the Fraser River valley, south-central British Columbia, Canada. A sample block has been removed and the gamma-ray spectrometer probe is shown inserted.

the dose rate calculation. In most cases, however, where disequilibrium is encountered, it is minor and does not significantly affect the dose rate. Discussion and case studies of the affects of radioactive disequilibrium on luminescence dating can be found in Prescott & Hutton (1995) and Olley et al. (1996).

Attention must also by given to the homogeneity of the sediment matrix near the sample location. For example, if there are localized concentrations of heavy minerals or if bounding strata are within about 50 cm of the sample location, then *in-situ* measurements of the gamma ray dose rate is highly desirable. This can be done with TL dosimeters (e.g., Aitken, 1998: 64–65) or, preferably, by using a portable gamma-ray spectrometer, such as that shown in Figure 8.

The largest uncertainty in the dose rate calculation is commonly that associated with water content. This is especially so for sediments occurring in previously glaciated regions and environments where standing or flowing water once existed, such as former lake bottoms, stream beds, or bogs. In such cases often the best that can be done is to use an estimated average value with an uncertainty that will cover both extremes of ignorance (e.g., dry and water-saturated) at two standard deviations. In such extreme cases uncertainty of past water content may contribute $\pm 10\%$ to the uncertainty in the final age.

Sample collection and preparation

Sample collection is ideally done by someone from the dating laboratory together with a geoscientist familiar with the geological history of the site. One method is to gently hammer a metal or hard plastic container (typically a cylinder 5–10 cm in diameter) into the cleaned section face, and then remove it by careful excavation. Alternatively, if the sediment is cohesive, blocks may be removed (e.g., as in Fig. 8) and wrapped in several layer of aluminium foil for protection. If the sediment is loose it can be shovelled into a light-tight container at night, or quickly under an opaque tarp in the daytime. Several things must be kept in mind during sample collection. These include: (i) the portion of the sample to be dated must not be exposed to light nor allowed to lose moisture during collection and transport to the laboratory. (ii) The local stratigraphy of the section must be recorded, and if the sample comes from within 50 cm of strata, or zones, of different material, samples of the bounding material must also be collected for dosimetry. If the sedimentology at, and near, the sample site is complex, *in-situ* gamma-ray spectrometry is highly advisable. (iii) It is best to use a portion of the actual sample used for dating for determining the contents of the radioactive elements, water and organic matter, and sufficient material should be collected for this. It is good practice to collect more than twice as much material as one thinks one is going to need. (iv) If the sample is within a few metres of the surface, knowledge of major unconformities, significant changes in the sedimentation rate, or thickness of former ice cover, are needed to properly evaluate the dose rate arising from cosmic rays. (v) Knowledge of occurrence of significant wet and dry periods throughout the history of the deposit must be ascertained.

Typically about one kilogram of sediment is sufficient for dating. But there have been cases where several kilograms is not enough, and others where one gram has been adequate. How much material is needed depends on the available amount of quartz or potassium feldspar of the desired grain size. Therefore, some prior knowledge of the mineral and grain-size distribution of the deposit can be useful. In general, samples that have not been collected with luminescence dating in mind will be rejected by the laboratory. There are of course rare exceptions, but for those cases reduced accuracy should be expected.

Samples are prepared in a dedicated laboratory under subdued orange or red light. Preparation typically involves removing carbonates and organic matter, deflocculation (if necessary), and chemical and mechanical separation of the desired mineral types, and grain sizes. The most common grain sizes used for dating are 4–11 μm diameter (fine silt), or a range in the sand-sized fraction such as 90–125 μm diameter (fine sand) or 250–350 μm diameter (medium sand). These size fractions are used because the microdosimetry is well understood (Aitken, 1985), but other grain sizes can be used in principle. Detailed descriptions of sample preparation techniques can be found in Wintle (1997). For samples that contain a significant organic component, see also Lian et al. (1995) or Lian & Shane (2000).

What types of depositional environments are suitable for luminescence dating?

Any sedimentary deposit that contains an adequate amount of quartz or potassium feldspar grains that have been exposed to sufficient sunlight or heat prior to final burial has the potential for providing an accurate age. It is therefore no surprise that aeolian deposits, especially loess, have received the most attention. What is "sufficient sunlight" in many ways depends on the event being dated, the age of the deposit, and the desired accuracy. An illustrative example was given by Huntley & Lian (1999) which we repeat here: suppose that the optically-stimulated luminescence from potassium feldspar from of a certain $\sim 70,000$ year-old deposit decreases to 1% of its initial value after 5 seconds of direct sunlight exposure. If this sample is eroded and exposed to 5 seconds of sunlight, and then once again buried, it would yield an optical age of 700 years, not zero as one would want. If this sediment remained buried for 1000 years, it would yield an optical age of 1700 years, which is too old by 70%. However, if the sediment were to remain buried for 50,000 years, it would give an optical age of 50,700 years, which may be considered satisfactory, especially when the analytical uncertainties are taken into account. An optical age of 50,700 years could also be achieved if 99% of the grains had received a long sunlight exposure, and 1% were not exposed at all.

There are factors to consider other than duration of sunlight exposure. These include the possibility of opaque mineral or organic coatings on the sediment grains, the occurrence of grain aggregates (cf. Lian & Huntley, 1999), and the mode of transport which can affect the intensity and spectrum of the light reaching the grains. For example, Rendell et al. (1994) demonstrated that for quartz and feldspar grains there was adequate bleaching after three hours at depths of ca. 10 m of relatively clear water. However, in turbulent turbid waters, thought to be similar to those found in proglacial environments, no significant bleaching was found of potassium feldspar grains even after a 20 hour exposure to light at a depth of only 75 cm (Ditlefsen, 1992).

The importance of the mode of sediment transport in resetting the TL signal has also been illustrated by Berger (1985) and Berger & Easterbrook (1993) who showed that for some known-age ancient glaciolacustrine rythmites, interpreted to be varves, that clay-rich laminae produced TL ages much closer to, or consistent with, those which were expected than did the silt-rich layers. This was thought to occur because the clay-rich layers were probably deposited primarily as slow rainout from overflows, whereas the silt-rich layers were derived mainly from underflow. Successful TL dating of rainout-deposited silt in a proglacial lake was also demonstrated by Berger & Anderson (1994). However this distinction was apparently found to be less obvious when long inflow (fluvial) distances were present (Berger, 1990). Indeed, one study has demonstrated that the infrared-excited luminescence from potassium feldspar silt can be reset in glaciofluvial streams, but only over long transport distances (several kilometres), and only if transport occurs during daylight hours (Gemmell, 1999). These studies, however, only indicate that resetting of the luminescence signal in such environments is possible; there is no known way to ascertain from the facies of the sedimentary deposit whether or not all the constituent mineral grains have received sufficient sunlight exposure prior to burial. One can only deduce from the sedimentology the probability of adequate sunlight exposure. In sediment-laden glaciofluvial or glaciolacustrine systems, often the best that can be done is to collect samples from within one sedimentary unit, and from distinct beds representative of the lowest energy conditions, and take the youngest luminescence age as an upper age limit. This approach was utilized by Lian & Hicock (2001) in dating potassium feldspar in silt deposited in an ancient glacial lake in south-central British Columbia. Apparent optical ages ranged from 183 ± 18 ka to 52 ± 5 ka. These ages, together with independent knowledge of when local and global glaciations occurred, indicated that deposition must have occurred during the last or possibly the penultimate glaciation, but not before. Such information can be valuable, especially if there is no other method available to date the deposits, as was the case here. An alternate approach could have been to locate a coarser facies, and attempt to date single grains of sand from that.

There has been relatively little work on luminescence dating of nonglacial lake deposits. This is surprising because sedimentation rates in nonglacial lakes are expected to be much lower, and the common modes of sediment input are much more amenable to extended sunlight exposure. Emphasis on glaciolacustrine environments has probably been due to the fact that they often account for a significant component of the sediment cover in formerly glaciated areas, and that there is a paucity of methods available to date them. Furthermore, nonglacial lake deposits commonly accumulate organic matter suitable for radiocarbon dating. However, there have been some notable exceptions that illustrate the potential and value of using luminescence to date nonglacial lacustrine sediments. These include presently ongoing work to reconstruct the paleoenvironment of New Zealand from information obtained from long cores: Shulmeister et al. (1999) documented environmental and sea-level changes from lagoonal and lacustrine sediments from a 75 m-long core on Banks Peninsula. The age data was provided mainly by TL dating of potassium feldspar in silt, ranged from 2–200 ka, and was broadly consistent with radiocarbon ages and paleoecological information. Similar work has recently been completed on a 250-metre-long core near Hawkes Bay where optical dating of potassium feldspar silt has yielded ages in the range 25–85 ka, most of which appear to be in good agreement with independent information (Shulmeister et al., 2001; Shane et al., 2001).

What can lead to an inaccurate optical age?

It should always be kept in mind that if sunlight exposure was less than adequate, then the luminescence age will lie somewhere between the true age of the deposit and the apparent age of its parent material. If an age is wrong, and is too old, then this is the most likely cause. In order to test for this possibility, several luminescence ages should be utilised together with some independent age information, such as that from radiocarbon dating, known-aged marker beds, or paleoecological information. If a modern analogue is available, it should be sampled and tested to see if it returns an age consistent with zero. In some cases a detailed study of the luminescence data can reveal if a sample contains some inadequately bleached grains. Since minerals contain electron traps with a wide range of sensitivities to light, it might be expected that for a poorly bleached sample only the most light-sensitive traps have been emptied. If this was the case, then dose-intercepts would be expected to rise with excitation time. The degree of scatter in the dose-response data (e.g., Huntley & Berger, 1995), or analysis of variations in the equivalent dose arising from individual grains (e.g., Lamothe & Auclair, 1997; Roberts et al., 1999) or aliquots (e.g., Clarke et al., 1999), may also be used to tell whether or not a sample consists of a mixture of well-bleached and poorly-bleached sand grains.

In addition to the possibility of failing to recognise such sedimentological problems, one must also evaluate the methods being employed by the luminescence dating laboratory. Luminescence dating is a rapidly evolving field with new techniques constantly being tested. Even the most established methods should not be considered routine, and one must pay special attention to the details. For relatively young samples, for where the dose-response curve is (or is nearly) linear, a thermal transfer correction must be made. If this is not done, the calculated age must be considered a maximum. If the regeneration method is used, a test must be made to see if the regeneration data and the additive-dose data define the same dose-response curve. If they do not, or if such a test has not been made, any calculated age should be viewed with caution. For feldspars for which the dose-response curve is near saturation, as for relatively old samples (> 100 ka) there is evidence that the calculated ages may be underestimates. This may be because the relevant traps are not stable at environmental temperatures over such time ranges. For such samples independent age control for at least some of the samples is highly desired.

Lastly, for feldspars one must be aware of so-called "anomalous fading". First observed in TL by Wintle (1973), anomalous fading refers to the fact that electrons in some traps, expected to be thermally stable over geological time, instead lose their electrons over much shorter times (e.g., see Spooner, 1994). It is likely that all feldspars exhibit anomalous fading to various degrees. For some samples the effect may be insignificant as far as dating is concerned. But for other samples it is clear that anomalous fading leads to a serious underestimation in the age. The effects of this phenomena can be reduced by long time delays (several weeks) between laboratory irradiation/preheating and the final measurements, and for TL dating it is apparent that certain preheat protocols help. Nevertheless, tests must be done to ascertain the degree of anomalous fading as a function of time. If significant fading is observed, the resulting age must be treated as a minimum. Lamothe & Auclair (1999) have recently introduced a method that corrects for anomalous fading, making use of the different fading rates for different feldspar grains. Correction for anomalous fading using measured fading rates has recently been tested by Huntley & Lamothe (2001), and the method shows enough promise to be worth pursuing. Alternatively, a sample can be tested by dating known-aged sediment from the same sedimentary unit. If a satisfactory luminescence age is attained for the known aged sample, then it is likely that the other ages are also acceptable.

Summary

Luminescence (optical and TL) dating can be regarded as an important technique for determining the age of formation of sedimentary deposits consisting of quartz or feldspar grains that have been exposed to adequate sunlight or heat before burial. For unheated sediments optical dating is superior because much less sunlight exposure is required and the measurements are simpler. However, although luminescence dating can be considered an established technique, it should not be considered routine in the same sense as radiocarbon and some other radiometric dating techniques. That is, with luminescence dating, each study should be considered a test of the method for the particular geological situation being studied. This is because minerals are highly heterogeneous, even within one mineral type such as quartz or feldspar, and this can lead to unexpected effects that may be unrecognisable by even accomplished luminescence dating practitioners. In each case appropriate tests must be undertaken. These include checks, or corrections, for incomplete sunlight exposure, thermal transfer, and anomalous fading. If the regeneration method is being used, a test for sensitivity change in the regeneration data must be made. Dating at least one known-age sample from the same lithostratigraphic unit, consisting of the same mineral(s) derived from the same source rocks, is highly advisable. If the appropriate tests are not reported then the luminescence ages must be considered with caution, and in most cases can only be interpreted as upper or lower limits.

Finally, the science of luminescence dating is developing rapidly and new techniques and laboratory protocols are frequently being reported. In light of this one must exercise an appropriate amount of vigilance in accepting ages that are produced using methods that have not been thoroughly tested.

Acknowledgements

We thank W. M. Last, J. P. Smol, and an anonymous referee for comments on an earlier version of the manuscript. Financial support from the Natural Sciences and Engineering Research Council of Canada is gratefully acknowledged.

References

Adamiec, G. & M. J. Aitken., 1998. Dose-rate conversion factors: update. Ancient TL 16: 37–50.

Aitken, M. J., 1998. An Introduction to Optical Dating. Oxford University Press, Oxford, 267 pp.

Aitken, M. J., 1985. Thermoluminescence Dating. Academic Press, London, 359 pp.

Baril, M. R., 1997. Optical Dating of Tsunami Deposits. Unpublished M.Sc. thesis, Simon Fraser University, Burnaby, 122 pp.

Berger, G. W., 1995. Progress in luminescence dating methods for Quaternary sediments. In Rutter, N. W. & N. R. Catto (eds.) Dating Methods for Quaternary Deposits. Geological Association of Canada, GEOtext 2: 81–104.

Berger, G. W., 1990. Effectiveness of natural zeroing of the thermoluminescence in sediments. J. Geophys. Res. 95: 12375–12397.

Berger, G. W., 1988. Dating Quaternary events by luminescence. In Easterbrook, D. J. (ed.) Dating Quaternary Sediments. Geological Society of America Special Paper 227: 13–50.

Berger, G. W., 1985. Thermoluminescence dating applied to a thin winter varve of the late glacial South Thompson silt, south-central British Columbia. Can. J. of Earth Sci. 22: 1736–1739.

Berger, G. W. & P. M. Anderson, 1994. Thermoluminescence dating of an Arctic lake core from Alaska. Quat. Sci. Rev. 13: 497–501.

Berger, G. W. & D. J. Easterbrook, 1993. Thermoluminescence dating tests for lacustrine, glaciomarine, and floodplain sediments from western Washington and British Columbia. Can J. Earth Sci. 30: 1815–1828.

Bortolot, V. J., 2000. A new modular high capacity OSL reader system. Rad. Meas. 32: 751–757.

Bøtter-Jensen, L., 1997. Luminescence techniques: instrumentation and methods. Rad. Meas. 27: 749–768.

Bøtter-Jensen, L. & A. S. Murray, 1999. Developments in optically stimulated luminescence techniques for dating and retrospective dosimetry. Rad. Prot. Dos. 84: 307–315.

Clarke, M. L., H. M. Rendell & A. G. Wintle, 1999. Quality assurance in luminescence dating. Geomorphology 29: 173–185.

Divigalpitiya, W. M. R., 1982. Thermoluminescence Dating of Sediments. Unpublished M.Sc. thesis, Simon Fraser University, Burnaby, 93 pp.

Dreimanis, A., G. Hütt, A. Raukas & P. W. Whippey, 1978. Dating methods of Pleistocene deposits and their problems: 1 thermoluminescence dating. Geoscience Canada 5: 55–60.

Ditlefsen, C., 1992. Bleaching of K-feldspar in turbid water suspensions: a comparison of photo- and thermoluminescence signals. Quat. Sci. Rev. 11: 33–38.

Duller, G. A. T., 1995. Luminescence dating using single aliquots: methods and applications. Rad. Meas. 24: 217–226.

Duller, G. A. T., 1994. Luminescence dating of sediments using single aliquots: new procedures. Quat. Sci. Rev. 13: 149–156.

Franklin, A. D. & W. F. Hornyak, 1990. Isolation of the rapidly bleaching peak in quartz TL glow curves. Ancient TL 8: 29–31.

Galbraith, R. F., R. G. Roberts, G. M. Laslett, H. Yoshida & J. M. Olley, 1999. Optical dating of single and multiple grains of quartz from Jinmium rock shelter, northern Australia. Part I: experimental design and statistical models. Archaeometry 41: 339–364.

Galloway, R. B., 1996. Equivalent dose determination using only one sample: alternative analysis of data obtained from infrared stimulation of feldspars. Rad. Meas. 26: 103–106.

Gemmell, A. M. D., 1999. IRSL from fine-grained glacifluvial sediment. Quat. Geoch. (Quat. Sci. Rev.) 18: 207–215.

Huntley, D. J. & G. W. Berger, 1995. Scatter in luminescence data for optical dating — some models. Ancient TL 13: 5–9.

Huntley, D. J. & J. J. Clague, 1996. Optical dating of tsunami-laid sands. Quat. Res. 46: 127–140.

Huntley, D. J. & M. Lamothe, 2001. Ubiquity of anomalous fading in K-Feldspars and the measurement and correction for it in optical dating. Can. J. Earth Sci. 38: 1093–1106.

Huntley, D. J. & O. B. Lian, 1999. Using optical dating to determine when a sediment was last exposed to sunlight. In Lemmen, D. S. & R. E. Vance (eds.) Holocene Climate and Environmental Changes in the Palliser Triangle: A Geoscientific Context for Evaluating the Impacts of Climate Change on the Southern Canadian Prairies. Geological Survey of Canada Bulletin 534: 211–222.

Huntley, D. J., D. I. Godfrey-Smith & M. L. W. Thewalt, 1985. Optical dating of sediments. Nature 313: 105–107.

Huntley, D. J., J. T. Hutton & J. R. Prescott, 1993a. Optical dating using inclusions within quartz grains. Geology 21: 1087–1090.

Huntley, D. J., J. T. Hutton & J. R. Prescott, 1993b. The stranded beach-dune sequence of south-east South Australia: a test of thermoluminescence dating, 0–800 ka. Quat. Sci. Rev. 12: 1–20.

Hütt, G., I. Jaek & J. Tchonka, 1988. Optical dating: K-feldspars optical response stimulation spectra. Quat. Sci. Rev. 7: 381–385.

Lamothe, M. & M. Auclair, 1997. Assessing the datability of young sediments by IRSL using an intrinsic laboratory protocol. Rad. Meas. 27: 107–117.

Lamothe, M. & M. Auclair, 1999. A solution to anomalous fading and age shortfalls in optical dating of feldspar minerals. Earth Plan. Sci. Let. 171: 319–323.

Lamothe, M., S. Balescu & M. Auclair, 1994. Natural IRSL intensities and apparent luminescence ages of single feldspar grains extracted from partially bleached sediments. Rad. Meas. 23: 555–561.

Lian, O. B. & S. R. Hicock, 2001. Lithostratigraphy and limiting optical ages of the Pleistocene fill in Fraser River valley near Clinton, south-central British Columbia. Can. J. Earth Sci. 38: 839–850.

Lian, O. B. & D. J. Huntley, 1999. Optical dating studies of post-glacial aeolian deposits from the south-central interior of British Columbia, Canada. Quat. Sci. Rev. 18: 1453–1466.

Lian, O. B. & P. A. Shane, 2000. Optical dating of paleosols bracketing the widespread Rotoehu tephra, North Island, New Zealand. Quat. Sci. Rev. 19: 1649–1662.

Lian, O. B., J. Hu, D. J. Huntley & S. R. Hicock, 1995. Optical dating studies of Quaternary organic-rich sediments from southwestern British Columbia and northwestern Washington State. Can. J. Earth Sci. 32: 1194–1207.

Liritzis, I., R. B. Galloway & P. S. Theocaris, 1994. Thermoluminescence dating of ceramics revisited: optically stimulated luminescence of quartz single aliquot with green light-emitting diodes. J. Rad. Nucl. Chem., Articles 188: 189–198.

McKeever, S. W. S. & R. Chen, 1997. Luminescence models. Rad. Meas. 27: 625–661.

Murray, A. S. & R. G. Roberts, 1997. Determining the burial time of single grains of quartz using optically stimulated luminescence. Earth Plan. Sci. Lett. 152: 163–180.

Murray, A. S., R. G. Roberts & A. G. Wintle, 1997. Equivalent dose measurement using a single aliquot of quartz. Rad. Meas. 27: 171–184.

Olley, J. M., A. Murray & R. G. Roberts, 1996. The effects of disequilibria in the uranium and thorium decay chains on burial dose rates in fluvial sediments. Quat. Sci. Rev. 15: 751–760.

Prescott, J. R. & J. T. Hutton, 1995. Environmental dose rates and radioactive disequilibrium from some Australian luminescence dating sites. Quat. Sci. Rev. 14: 439–448.

Prescott, J. R. & J. T. Hutton, 1994. Cosmic ray contributions to dose rates for luminescence and ESR dating: large depths and long-term time variations. Rad. Meas. 23: 497–500.

Prescott, J. R. & B. Mojarrabi, 1993. Selective bleach: an improved partial bleach technique for finding equivalent doses for TL dating of quartz sediments. Ancient TL 11: 27–30.

Prescott, J. R. & G. B. Robertson, 1997. Sediment dating by luminescence: A review. Radiation Measurements 27: 893–922.

Prescott, J. R., D. J. Huntley & J. T. Hutton, 1993. Estimation of equivalent dose in thermoluminescence dating — the *Australian slide* method. Ancient TL 11: 1–5.

Rendell, H. M., S. E. Webster & N. L. Sheffer, 1994. Underwater bleaching of signals from sediment grains: new experimental data. Quat. Sci. Rev. 13: 433–435.

Roberts, R. G., 1997. Luminescence dating in archaeology: from origins to optical. Rad. Meas. 27: 819–892.

Roberts, R. G., M. Bird, J. Olley, R. Galbraith, E. Lawson, G. Laslett, H. Yoshida, R. Jones, R. Fullagar, G. Jacobsen & Q. Hau, 1998. Optical and radiocarbon dating at Jinmium rock shelter in northern Australia. Nature 393: 358–362.

Roberts, R. G., R. F. Galbraith, J. M. Olley, H. Yoshida & G. M. Laslett, 1999. Optical dating of single and multiple grains of quartz from Jinmium rock shelter, northern Australia. Part II: Results and implications. Archaeometry 41: 365–395.

Shane, P. A., O. B. Lian, P. Augustinus, R. Chisari & H. Heijnis, 2001. Tephrostratigraphy and geochronology of a c. 120 kyr terrestrial record at Lake Poukawa, North Island, New Zealand. Glob. Planet. Change, in press.

Shulmeister, J., P. A. Shane, O. B. Lian, M. Okuda, J. A. Carter, M. Harper, W. Dickinson, P. Augustinus & H. Heijnis, 2001. A ca. 60 ka record from Lake Poukawa, Hawkes Bay, New Zealand. Implications for a significant interstadial at the beginning of isotope stage 3. Palaeogeogr. Palaeoclim. Palaeoecol., in press.

Shulmeister, J., J. M. Soons, G. W. Berger, M. Harper, S. Holt, N. Moar. & J. A. Carter, 1999. Environmental and sea-level changes on Banks Peninsula (Canterbury, New Zealand) through three glaciation-interglaciation cycles. Palaeogeogr. Palaeoclim. Palaeoecol. 152: 101–127.

Singhvi, A. K., Y. P. Sharma & D. P. Agrawal, 1982. Thermoluminescence dating of sand dunes in Rajasthan, India. Nature 295: 313–315.

Spooner, N. A., 1994. The anomalous fading of infra red-stimulated luminescence from feldspars. Rad. Meas. 23: 625–632.

Stokes, S., 1999. Luminescence dating applications in geomorphological research. Geomorphology 29: 153–171.

Stokes, S., A. E. Colls, M. Fattahi & J. Rich, 2000. Investigations of the performance of quartz single aliquot D_E determination procedures. Rad. Meas. 32: 585–594.

Wagner, G. A., 1998. Age Determination of Young Rocks and Artifacts. Springer-Verlag, New York, 466 pp.

Wintle, A. G., 1997. Luminescence dating: laboratory procedures and protocols. Rad. Meas. 27: 769–817.

Wintle, A. G., 1973. Anomalous fading of thermoluminescence in mineral samples. Nature 245: 143–144.

Wintle, A. G. & D. J. Huntley, 1982. Thermoluminescence dating of sediments. Quat. Sci. Rev. 1: 31–53.

Wintle, A. G. & D. J. Huntley, 1980. Thermoluminescence dating of ocean sediments. Can. J. Earth Sci. 17: 348–360.

Wintle, A. G. & A. S. Murray, 1998. Towards the development of a preheat procedure for OSL dating of quartz. Rad. Meas. 29: 81–94.

13. ELECTRON SPIN RESONANCE (ESR) DATING IN LACUSTRINE ENVIRONMENTS

BONNIE A. B. BLACKWELL
(bonnie.a.b.blackwell@williams.edu)

Department of Chemistry
Williams College
Williamstown
MA, 01267
USA

Keywords: ESR (electron spin resonance) dating, ESR microscopy, paleolimnology, teeth, mollusc shells, ratite egg shells, travertine, authigenic carbonates, authigenic salts, heated flint.

Introduction

Electron spin resonance (ESR) dating can provide chronometric (absolute) dates over a substantial time range, from as young as 0.5 ka to about 5–10 Ma, currently with 2–10% precision (e.g., Blackwell, 1995a; Huang et al., 1995). ESR, like its sister methods, thermo- (TL), optically stimulated (OSL), and radio-luminescence (RL; e.g., Aitken, 1985, 1992; Shilles & Habermann, 2000), relies on detecting trapped charges induced by radiation in crystals. ESR can be used to date many materials that are commonly encountered at limnological sites, as well as samples curated in museums, and new applications are constantly being added. Although not yet widely used in limnological research, ESR's importance in dating Quaternary and Pliocene sites has now been well demonstrated in archaeological contexts where it has dramatically changed our understanding of human origins and cultures (e.g., references in Table I).

In paleolimnological settings, ESR provides several advantages over rival methods. For example, it can date fossils much older than the ^{14}C dating limit ($\sim$40–50 ka). ESR does not require a nearby volcano to produce datable rocks or ash like $^{39}Ar/^{40}Ar$ does, because ESR can also date fossils and sediment. Unlike the uranium (U) series methods, ESR can date most mollusc species accurately, as well as some authigenic cements, clays, and aeolian sediment. Unlike TL, OSL and RL (e.g., Li, 1999), ESR does not require that signals be completely zeroed for most applications and signals do not suffer anomalous fading. ESR's potential to date a wide variety of sample types will undoubtedly make it an important research tool in late Cenozoic paleolimnological units.

W. M. Last & J. P. Smol (eds.), 2001. *Tracking Environmental Change Using Lake Sediments. Volume 1: Basin Analysis, Coring, and Chronological Techniques.* Kluwer Academic Publishers, Dordrecht, The Netherlands.

Table I. Limnological Applications for ESR Dating.

Sample Type (Age, Stability, Dosimetry Limits) - Application	Location	References
Travertine ($500 \text{ a}^1 - 1 \text{ Ma}^2$; ± 8-15%)		Baietto et al., 1999; Bartoll et al., 2000; Blackwell, 1995a; Blackwell & Blickstein, 2000; Y.J. Chen et al., 1994; Grün, 1985c, 1989a, 1989b, 1989c, 1997; Grün & de Cannière, 1984; Hennig & Grün, 1983; Idrissi et al., 1996; Ikeya, 1985a, 1988, 1993b, 1996; Poupeau & Rossi, 1989; Rambaud et al., 2000; Regulla, 2000; Rink, 1997; Schwarcz, 1994; Skinner, 2000b; Stößer et al., 1996; *Wieser et al., 1985, 1993.
- archaeology	China	P.H. Huang et al., 1985.
	Middle East	Hennig & Hours, 1982.
	Germany	Brunnacker et al., 1983.
- paleoclimatology	China	Li, 1988.
	Germany	Schwarcz et al., 1988a.
	Hungary	Hennig et al., 1983b, 1983c.
	Italy	Grün et al., 1988b; Radtke et al., 1986.
	Jordon	*Wieser et al., 1993.
	Spain	Martinez Tudela et al., 1986.
- geyserites	China	Y.J. Chen et al., 1993, 1994.
- paleoenvironmental reconstruction	Italy	Radtke et al., 1986.
	Japan	Kai & Miki, 1991, 1992; Miki et al., 1993.
Calcrete, caliche ($100 \text{ ka}^1 - 1.5 \text{ Ma}^2$; ± 10-15%)		Blackwell, 1995a; Brückner & Radtke, 1985, 1986; Y.J. Chen et al., 1988b, 1989a, 1989b, 1994; Ikeya, 1985a, 1988, 1993b; Joppard, 1984; Kai et al., 1993; Lloyd & Lumsden, 1987; Özer et al., 1989; Radtke et al., 1988b; Skinner, 2000b.
- archaeology	Spain	Porat et al., 1990.
- paleo sea levels	Mexico	Gerstenhauer et al., 1983.
	Italy	Radtke, 1985a.
	Spain	Radtke et al., 1988b.
- stratigraphy	Australia	Y.J. Chen et al., 1988a.
	Brazil	Y.J. Chen et al., 1989b.
	China	Y.J. Chen et al., 1989a.
- pisolites	Turkey	Özer et al., 1989.
- soils	China	*Y.J. Chen et al., 1988b, 1989a, 1994.
	Israel	Porat et al., 1994a.
- loess	China	*Y.J. Chen et al., 1994.
- paleogeomorphology	Australia	*Jacobson et al., 1988.
	Spain	*Radtke et al., 1988b.
- impact craters	USA	Boslough et al., 1982; Vizgirda et al., 1980.
- K/T boundary	Denmark	Miúra et al., 1985.
- microscopy		Furusawa et al., 1991.
- dosimetry: boiler scale	(> 0.5 Gray)	Wieser et al., 1994b.
- dosimetry: chalk	(> 2 Gray)	Wieser et al., 1994b.
Speleothems ($500 \text{ a}^1 - 1 \text{ Ma}^2$; ± 8-15%)		Bahain et al., 1994a, 1994b; Barabas et al., 1989; Bartoll et al., 2000; Blackwell, 1995a; Blackwell & Blickstein, 2000; de Cannière et al., 1985, 1986, 1988; Debuyst et al., 1984, 1990, 1991; Geyh & Hennig, 1986, 1987; Göksu et al., 1989;

Table I. Limnological Applications for ESR Dating (continued).

Sample Type (Age, Stability, Dosimetry Limits) - Application	Location	References
Speleothems (cont'd)		Goslar & Hercman, 1988; Grün, 1989a, 1989b, 1989c, 1997; Grün & de Cannière, 1984; Grün & Schwarcz, 1987a; Hennig & Grün, 1983; Hennig et al., 1981a, 1983a, 1985; Hirai et al., 1994; Ikeda et al., 1992a, Ikeya, 1975, 1976, 1985a, 1988, 1993b, 1994a, 1996; Jacobs et al., 1989; Kai & Miki, 1992; Kohno et al., 1994a, 1994b; Li, 1988; Liang et al., 1989; *Liritzis & Maniatis, 1989; Lyons, 1988a, 1988b, 1993, 1996; Lyons & Brennan, 1989, 1991; Lyons et al., 1985, 1988, 1989, 1992; Miki et al., 1993; Poupeau & Rossi, 1989; Rambaud et al., 2000; Regulla, 2000; Regulla et al., 1985; Rink, 1997; Rossi & Poupeau, 1989; Schwarcz, 1986, 1994, 1999; Schwarcz & Grün, 1988b; Shopov et al., 1985; Skinner, 1983, 2000b; Smith et al., 1985a, 1985b, 1986, 1989; Stößer et al., 1996; Wieser et al., 1985.
- archaeology	France	Apers et al., 1981; Hennig, 1982; Hennig et al., 1981a; Masaoudi et al., 1997; Valladas et al., 1981; *Yokoyama et al., 1981b, 1982, 1983a, 1983b, 1985b, 1988.
	Greece	Hennig et al., 1981b, 1982; Ikeya, 1977, 1978a, 1980a, Ikeya, 1982c, 1983a; Ikeya & Poulianos, 1979; Karakostanaglou & Schwarcz, 1984; Liritzis, 1982; Papastefanou et al., 1986; Poulianos, 1982; Xirotiris et al., 1982.
	Slovenia	Ikeya et al., 1982a.
- growth rates	Bulgaria	*Shopov et al., 1985.
	Europe	Smart et al., 1988; Smith et al., 1985a, 1985b, 1989.
	Germany	Grün, 1987; Pick & Ikeya, 1980.
	Japan	*Arakawa & Hori, 1989; Ikeya, 1975, 1976; Ikeya & Miki, 1985b; Miki & Ikeya, 1978; Miki & Kai, 1991.
	Slovenia	Ikeya et al., 1982a.
	Spain	Grün, 1985a, 1985c, 1986.
- paleoclimatology	China	Li, 1988.
	Europe	Bluszcz et al., 1988; Hennig et al., 1983c.
	Japan	*Arakawa & Hori, 1989; Ikeya, 1978b; Morinaga et al., 1985.
- microscopy		Furusawa et al., 1991; Ishii et al., 1991.
Vein calcite		Blackwell, 1995a; Ikeya, 1985a, 1988, 1993b; Ikeya et al., 1983.
	China	Y.J. Chen et al., 1994.
Dolomite	Nauru I.	Y.J. Chen et al., 1992.
- metamorphic dolomite		Lloyd & Lumsden, 1987.

continued ....

Notes:

*	not used successfully		
1	depends on U concentration	4	calcite only
2	usually more limited due to diagenesis	5	species dependent
3	older dates need to be corrected for fading or retrapping	6	aragonite only

Table I. Limnological Applications for ESR Dating (continued).

Sample Type (Age, Stability, Dosimetry Limits) - Application	Location	References
Algae, Stromatolites		Schramm & Rossi, 1996.
Bird Egg shells (> 0.5 Gray)		Blackwell, 1995a; Kai et al., 1988; Ikeya, 1988, 1993b, 1996; Grün, 1997; Oduwale et al., 1993; Regulla et al., 1994, 1995; Wieser et al., 1994a, 1994b.
- archaeology: ratites	Middle East	Kai et al., 1988; Wendorf et al., 1987.
- extinctions: *Aepyornis*	Madagascar	Schwenninger & Rhodes, 1999.
- dosimetry: birds		Regulla et al., 1994; Wieser et al., 1994b.
Molluscs (terrestrial & freshwater) (5 ka – 500 ka^{3,4,5}; 2 to 5^5 – 1500 Gray^6)		Blackwell, 1995a; Blackwell & Blickstein, 2000; Kai, 1996; Katzenberger, 1985; Katzenberger et al., 1989; Molodkov, 1993, 1996; Ostrowski et al., 1997; Rambaud et al., 2000; Rink, 1997; Skinner, 2000b.
- lacustrine sediment	Europe	Katzenberger, 1985; Katzenberger et al., 1989;
	Japan	Kai & Ikeya, 1989; Kai & Miki, 1989.
- archaeology	Europe	Molodkov, *1993, *2001.
- lake stratigraphy & paleosalinity	Estonia	Molodkov, 1996.
	USA	Mirecki et al., 1994; Skinner & Mirecki, 1993.
Molluscs (marine) (5 ka – 500 ka^{3,4,5}; 0.5 to 5^5 – 1500 Gray^6; ± 6-10%)		Albarrán et al., 1987; Baffa & Mascarenhas, 1985b; Bahain et al., 1994b, 1995; Barabas, 1991; Barabas et al., 1988a, 1992a, 1992b, 1993; Bartoll et al., 2000; Blackwell, 1995a; Blackwell & Blickstein, 2000; Brumby & Yoshida, 1994a, 1995; de Cannière et al., 1986; Goede & Hitchman, 1987; Grün, 1985b, 1985c, 1989a, 1989b, 1989c, 1991a, 1993, 1997; Grün et al., 1992; Huang et al., 1989; Hütt & Jaek, 1989; Ikeya, 1980b, 1981, 1984a, 1985a, 1988, 1993b, 1994a, Ikeya, 1994b, 1996; Ikeya et al., 1985; Imai & Shimokawa, 1993; Kai & Ikeya, 1989; Katzenberger, 1985; Katzenberger & Willems, 1988; Katzenberger et al., 1989; Liang et al., 1989; Malmberg & Radtke, 2000; Miki et al., 1985a; Molodkov, 1988, 1989, 1993; Molodkov & Hütt, 1985; Murata et al., 1993a, 1993b; Nakajima et al., 1993a; Ninagawa et al., 1985; Ostrowski et al., 1997; Poupeau & Rossi, 1989; Radtke, 1985d, 1988; Radtke et al., 1985; Rambaud et al., 2000; Regulla, 2000; Rink, 1997; Schellmann & Radtke, 1999a; Schramm & Rossi, 1996; Schwarcz, 1994; Skinner, 1984, 1986, 1989, 2000b; Takano & Fukao, 1994.
- archaeology	Brazil	Baffa & Mascarenhas, 1985a, 1985b; *Brunetti et al., 1999.
	Japan	Nakajima et al., 1993b.
	Sri Lanka	Abeyratne et al., 1997.
	South Africa	*Goede & Hitchman, 1987.
- terrace chronology	Argentina	P.H. Huang et al., 1989; Schellmann & Radtke, 1997, 1999, 2000, 2001.
	Australia	*Hewgill et al., 1983, 1985; Murray-Wallace & Goede, 1995; Radtke & Brückner, 1991.

Table I. Limnological Applications for ESR Dating (continued).

Sample Type (Age, Stability, Dosimetry Limits) - Application	Location	References
Molluscs (marine) (cont'd)		
- terrace chronology	Baltic coast	Molodkov & Raukus, 1988; Molodkov et al., *1998.
	Chile	Schellmann & Radtke, 1997; Radtke, 1985b, 1987.
	China	*Y.J. Chen et al., 1989a.
	Italy	Radtke, 1983, 1985a, 1986; Radtke et al., 1981, 1982;
	Japan	Ikeya & Ohmura, 1981, 1984; Ninagawa et al., 1985; *Shimokawa et al., 1992; Tsuji et al., 1985.
	Mexico	Gerstenhauer et al., 1983.
	Middle East	Ivanovich et al., 1983.
	Spain	Brückner & Radtke, 1985, 1986; Radtke, 1985c.
	USSR	Molodkov, 1988.
- site stratigraphy	Arctic	*Grün, 1985b; Katzenberger, 1985; Katzenberger & Grün, 1985.
	Bahamas	Skinner & Shawl, 1994; Skinner & Weicker, 1992.
	Baltic coast	Hütt et al., 1983, 1985; Molodkov, 1986.
	Brazil	Baffa & Mascarenhas, 1985b.
	China	Peng et al., 1989.
	Europe	Barabas et al., 1988b; Linke et al., 1985; Sarnthein et al., 1986; Schwarcz & Grün, 1988a.
	Italy	Radtke et al., 1981.
	Japan	Nakazato et al., 1993; *Takano & Fukao, 1994.
	South Africa	*Goede & Hitchman, 1987.
	USA	Skinner, 1986, 1989.
	USSR	Molodkov & Hütt, 1985.
- cephalopod sepia		Duliu, 2000.
- in till		Molodkov et al., 1998.
- microscopy		Ikeya, 1991, 1994a, 1994b; Ikeya & Furusawa, 1989; Miyamaru & Ikeya, 1993.
- dosimetry		Duliu, 2000; Ikeya et al., 1984; Nakajima et al., 1993a; Raffi et al., 1996; Stachowicz et al., 1995.
- tectonism		Murray-Wallace & Goede, 1995.
Diatoms		Schwarcz, 1994.
Corals ($500\ a^1 - 1.0\ Ma^3$; $\pm 8\text{-}10\%$)		Bahain et al., 1994b; Bartoll et al., 2000; Blackwell, 1995a; Blackwell & Blickstein, 2000; Geode & Hitchman, 1987; Grün, 1989a, 1989b, 1989c; Grün et al., 1992; Ikeda et al., 1992a, 1992b; Ikeya, 1984a, 1985a, 1988, 1993b, 1994a, 1994b, 1996; Jones et al., 1991, 1993; Kai & Miki, 1989; Liang et al., 1989; Malmberg & Radtke, 2000; Miki et al., 1985a; Radtke & Grün, 1988; Rink, 1997; Rambaud et al., 2000; Schramm & Rossi, 1996; Schwarcz, 1994; Skinner, 1985, 1988, 2000b; Walther et al., 1992; Yoshida & Brumby, 1999.

continued ....

Notes:

*	not used successfully		
1	depends on U concentration	4	calcite only
2	usually more limited due to diagenesis	5	species dependent
3	older dates need to be corrected for fading or retrapping	6	aragonite only

Table I. Limnological Applications for ESR Dating (continued).

Sample Type (Age, Stability, Dosimetry Limits) - Application	Location	References
Corals (cont'd)		
- terrace chronology	Argentina	Schellmann & Radtke, 1999b.
	Cook I.	Grey et al., 1992.
	Barbados	Radtke & Grün, 1988; Radtke et al., 1988a; Skinner, 1985, 1986, 1988; Walther et al., 1992.
	China	Peng et al., 1989; Ye et al., 1991.
	Hawaii	Jones et al., 1991, 1993; Sherman et al., 1993.
	Huon Penn.	Grün et al., 1992; Yoshida & Brumby, 1999.
	Indonesia	Hantora et al., 1994; Pirazzoli et al., 1991, 1993.
	Jamaica	Mitchell et al., 2000.
	Japan	Ikeda et al., 1991, 1992b; Ikeya, 1983a; Ikeya & Ohmura, 1983; Koba et al., 1985, 1987; Radtke et al., 1996.
	Nauru I.	Y.J. Chen et al., 1992.
	New Hebrides	Radtke & Grün, 1988.
- paleoenvironments	Japan	Ikeda et al., 1992a.
- tectonic uplift	Huon Penn.	Grün et al., 1992; Yoshida & Brumby, 1999.
- microscopy: growth bands, rates		Furusawa et al., 1991; Ikeda et al., 1992a; Tsukamoto & Heikoop, 1996.
Foraminifera (100 a – 3.5 Ma; ± 10-20%)		Barabas, 1991; Barabas et al., 1988b, 1992a, 1992b; Blackwell, 1995a; Blackwell & Blickstein, 2000; Ikeya, 1985a, 1988, 1993b, 1996; Mangini et al., 1983; Mudelsee et al., 1992a; Rambaud et al., 2000; Rink, 1997; Sato, 1983b; Schwarcz, 1994; Siegele & Mangini, 1985; Skinner, 2000b.
- pelagic sediment		Ikeya et al., 1985; Mangini et al., 1983.
- core stratigraphy	Arctic	Hoffmann et al., 2001.
	Pacific	Mudelsee et al., 1992a; Sato, 1981, 1982, 1983a, 1983b; Takeuchi & Saeki, 1985; Siegele & Mangini, 1985; Wintle & Huntley, 1983
Echinoderms		Yamamoto & Ikeya, 1994
Tooth enamel (1 ka[1] – 2 Ma[2]; 0.05 – 2500 Gray; ± 4-12%)		Blackwell, 1989, 1994, 1995a, 1995b; Blackwell & Blickstein, 2000; Blackwell & Schwarcz, 1993a, 1993b; Blackwell et al., 1998, 2000, 2001a, 2001c, 2001e; Brennan, 2000; Brennan et al., 1997a, 1997b, 2001; Bouchez et al., 1988; Brik et al., 1996, 2000; Callens et al., 1986, 1987, 1995; Çetin et al., 1994; Chong et al., 1985, 1989; Desrosiers et al., 1989; Egersdörfer et al., 1996a, 1996b; Fattibene et al., 2000; Grün, 1985c, 1988, 1989a, 1989b, 1989c, 1989d, 1991a, 1993, Grün, 1995, 1997, 1998a, 1998b, 2000; Grün & McDermott, 1994; Grün & Jonas, 1996; Grün & Taylor, 1996; Grün et al., 1987, 1988a, 1997a, 1997c; Haskell et al., 1996b, 1997a, 1997b; W.P. Huang et al., 1995; Ignatiev et al., 1996; Ikeya, 1985a, 1988, 1993b, 1994a, 1994b; Ikeya et al., 1997a; Jonas, 1995, 1997; Jonas & Grün, 1997;

Table I. Limnological Applications for ESR Dating (continued).

Sample Type (Age, Stability, Dosimetry Limits) - Application	Location	References
Tooth enamel (cont'd)		Jonas & Marseglia, 1997; Jonas et al., 1994; Koshta et al., 2000; Kenner et al., 1998; Komura & Sakanoue, 1985; Lee et al., 1997; Liidja, 1999; Liidja et al., 1996; *Marsh & Rink, 2000; McDermott et al., 1993a; Mellars et al., 1997; Miyake et al., 2000; Murata et al., 1993a, 1993b, 1996; Nakamura & Miyazawa, 1997; *Oduwale & Sales, 1991, 1994; Oliveiri et al., 2000; Onori et al., 2000; Pike & Hedges, 2001; Porat & Schwarcz, 1994; Porat et al., 1990; Poupeau & Rossi, 1989; Rambaud et al., 2000; Regulla, 2000; Rink, 1997; Rink & Hunter, 1998; Rink & Schwarcz, 1994, 1995; Robertson & Grün, 2000; Romanyukha & Regulla, 1996; Rossi & Poupeau, 1990; Scherbina & Brik, 2000; Schramm & Rossi, 2000; Schwarcz, 1985, 1994, 1999; Serezhenkov et al., 1996; Sholom & Chumak, 1999; Sholom et al., 1998a, 1998b, 2000; Skinner, 2000b; Skinner et al., 2000a, 2001a, 2001b; Vanhaelewyn et al., 2000; Vugman et al., 1995; Wieser et al., 2000a, 2000b; Yamamoto & Ikeya, 1994.
- archaeology, human paleontology		Porat & Schwarcz, 1994; Schwarcz, 1999.
	Africa	Grün & Stringer, 1991; *Oduwale & Sales, 1994; Stringer & Grün, 1991.
	China	Blackwell & Skinner, 1999a; Blackwell et al., 2000b; T.M. Chen et al., 1994, 1997, 1999; Grün et al., 1997b. 1998; *P.H. Huang et al., 1993; W.P. Huang et al., 1995; *Oduwale & Sales, 1994.
	Croatia	Rink et al., 1995.
	Czech Rep.	Rink et al., 1996d.
	Europe	Grün & McDermott, 1994.
	France	*Bahain et al., 1993; Blackwell & Schwarcz, 1993a, 1993b; Blackwell et al., 1992, 1994a, 1994c, 2000, 2001d, 2001e; Bouchez et al., 1988; *Falguères et al., 1997; Grün et al., 1991b, 1998; Mellars & Grün, 1991; Schwarcz & Grün, 1988b, 1989.
	Germany	Grün & Brunnacker, 1987; Grün & Invernati, 1985; Schwarcz et al., 1988a.
	Greece	Grün, 1996a.
	Hungary	Blackwell & Skinner, 1999a.
	Italy	Bahain et al., 1992; Schwarcz et al., 1991a, 1991b.
	Japan	Chong et al., 1985, 1989.
	Java	Grün & Thorne, 1997; Swisher et al., 1996, 1997.

continued ....

Notes:

*   not used successfully
[1]   depends on U concentration
[2]   usually more limited due to diagenesis
[3]   older dates need to be corrected for fading or retrapping

[4]   calcite only
[5]   species dependent
[6]   aragonite only

Table I. Limnological Applications for ESR Dating (continued).

Sample Type (Age, Stability, Dosimetry Limits) - Application	Location	References
Tooth enamel (cont'd)		
- archaeology, human paleontology (cont'd)	Middle East	Blackwell & Schwarcz, 1993b; Grün & McDermott, 1994; Grün & Stringer, 1991; Grün et al., 1991a, 1998; Kai et al., 1988; McDermott et al., 1993a, 1993b; Pilbeam & Bar-Yosef, 1993; Schwarcz, 1985; Schwarcz & Grün, 1989, 1992, 1993b; Schwarcz et al., 1988b, 1989; Stringer & Grün, 1991; Stringer et al., 1989; Ziaei et al., 1990.
	Molodova	Borziac et al., 1997.
	Mongolia	Blackwell & Skinner, 1999b; Blackwell et al., 2001b.
	Morocco	Rhodes et al., 1994.
	New Zealand	Oduwale et al., 1993.
	Olduvai	Skinner et al., 2001b.
	Polynesia	Oduwale et al., 1993.
	Slovenia	Lau et al., 1997.
	South Africa	Blackwell, 1994, 1995b; Blackwell et al., 1998, 2001c; Grün et al., 1990a, 1990b, 1996a, 1998; Schwarcz, 1999; Schwarcz & Grün, 1992; Schwarcz et al., 1994.
	Spain	Porat et al., 1990; Rink et al., 1996a, 1997; Schwarcz, 1999; Volterra et al., 1999.
	Sri Lanka	Abeyratne et al., 1997.
	Sudan	McDermott et al., 1996.
	Turkey	Blackwell et al., 1994b; Çetin et al., 1994; Rink et al., 1994, 1996b.
	Ukraine	Marks et al., 1997; Monigal et al., 1997; Rink et al., 1998.
	UK	Jacobi et al., 1998; Schwarcz & Grün, 1993a.
	Vietnam	Ciochón et al., 1996.
	Zaire	Brooks et al., 1995.
	Zambia	Blackwell & Schwarcz, 1993a; Blackwell et al., 1993.
- paleontology	Brazil	Baffa et al., 2000.
	Canada	Grün et al., 1987; Schwarcz & Zymela, 1985; Zymela, 1986; Zymela et al., 1988.
	Caribbean	MacPhee et al., 1989.
	England	Grün, 1991b; Jacobi et al., 1998; Rink et al., 1996c; Zhou et al., 1997.
	Germany	Debuyst et al., 2000b; Wieser et al., 1988.
	Hungary	Berman et al., 2000; G. Chen et al., 1996.
	USA	Blackwell et al., 1994a, 2001d; Johnson et al., 1991.
	Vietnam	Ciochón et al., 1996.
- paleoenvironments		Grün, 1985c; Grün & Invernati, 1985; Grün et al., 1988a.
	France	Blackwell et al., 2000, 2001e.
	USA	Blackwell et al., 1994a, 2001d.
- dosimetry		Aldrich & Pass, 1988; Brennan, 2000; Brennan et al., 1997a, 1997b, 2000; Brik et al., 2000a, 2000b; Brillibit et al., 1990; Chumak et al., 1996; Göksu et al., 1989; Haskell et al., 1996b, 1999, 2000; Hayes & Haskell, 2000; Hayes et al., 1997, 1998, 2000; Hochi et al., 1993; Ignatiev et al., 1996; Ikeya & Ishii, 1989; Ikeya et al., 1984;

Table I. Limnological Applications for ESR Dating (continued).

Sample Type (Age, Stability, Dosimetry Limits) - Application	Location	References
Tooth enamel (cont'd)		
- dosimetry (cont'd)		Ishii & Ikeya, 1990; Ishii et al., 1990; Ivannikov et al., 2000; Kenner et al., 1998; Koshta et al., 2000; Nakamura & Miyazawa, 1997; Nishiwaki & Shimano, 1990; Onori et al., 2000; Pass & Aldrich, 1985; Polyakov et al., 1995; Rodas et al., 1985; Romanyukha & Regulla, 1996; Romanyukha et al., 1994, 1996, 2000; Rossi et al., 2000; Shimano et al., 1989; Sholom & Chumak, 1999; Sholom et al., 1998a, 1998b, 2000; Skvortzov et al., 1995, 2000; Tatsumi-Miyajima, 1987; Wieser et al., 2000a, 2000b; Yamanaka et al., 1991, 1993.
- microscopy		Hochi et al., 1993; Ikeya, 1991, 1994a, 1994b; Ikeya & Furusawa, 1989; Miyamaru & Ikeya, 1993; Oka et al., 1997.
Bones, Antler, Horn (20 ka[1] – 2 Ma[2]; 5 – 1500 Gray; ± 10-20%)		Blackwell, 1995a; Blackwell & Blickstein, 2000; Brik et al., 2000; Caddie et al., 1985; Caracelli et al., 1986; Copeland et al., 1996; Dennison & Peake, 1992; *Dennison et al., 1985, 1997; Desrosiers, 1996; Desrosiers et al., 1989; Grün, 1989a, 1989b, 1989c, 1989d, 1997; Grün & Schwarcz, 1987b; Hayes et al., 1998; Hennig et al., 1981b; Houben, 1971; Ikeya, 1980b, *1982a, 1985a, *1985b, 1988, 1993b, 1994b; *Ikeya & Miki, 1980a, 1980b, 1981b; Kai et al., 1988; Komura & Sakanoue, 1985; *Liritzis & Maniatis, 1989; Michel et al., 1998; *Oduwale & Sales, 1991, 1994; Oduwale et al., 1993; Onori et al., 1996; Ostrowski et al., 1994, 1996; Papastefanou & Charlambous, 1978; Pike & Hedges, 2001; Poupeau & Rossi, 1989; *Raffi et al., 1989; Rambaud et al., 2000; Regulla et al., 1985; Rink, 1997; *Robins et al., 1985a, 1987; Sales et al., 1985, 1989; Schwarcz, 1994; Skinner, 2000b; Stuglik & Sadlo, 1995; Stuglik et al., 1994; Wieser et al., 1994a, 1994b; Zeller et al., 1988.
- archaeology, human paleontology	Brazil	*Baffa & Mascarenhas, 1985b; Mascarenhas et al., 1982.
	China	*P.H. Huang et al., 1985, 1993; *Ikeya, 1985b; *Ikeya & Miki, 1981a, 1981b; *Oduwale & Sales, 1991; *Robins et al., 1985a, 1987; *Sales et al., 1989.
	Europe	Oduwale & Sales, 1991.
	France	Ikeya & Miki, 1981a, 1981b; *Masaoudi et al., 1997; *Yokoyama et al., 1981a.

continued ....

Notes:

*   not used successfully
[1]   depends on U concentration
[2]   usually more limited due to diagenesis
[3]   older dates need to be corrected for fading or retrapping

[4]   calcite only
[5]   species dependent
[6]   aragonite only

Table I. Limnological Applications for ESR Dating (continued).

Sample Type (Age, Stability, Dosimetry Limits) - Application	Location	References
Bones, Antler, Horn (cont'd)		
- archaeology, human	Germany	*Ikeya, 1982a, 1983a, 1985b.
paleontology (cont'd)	Greece	*Hennig et al., 1981b, 1982; Papastefanou et al., 1986.
	Italy	*Falguères et al., 1990.
	Java	*Ikeya, 1985b.
	Japan	*Ikeya, 1985b.
	Middle East	Kai et al., 1988.
	New Zealand	*Dennison et al., 1985; Dennison & Peake, 1992; Oduwale et al., 1993.
	Polynesia	*Dennison et al., 1985, 1997; Oduwale et al., 1993.
- paleontology	Australia	*Goede & Bada, 1985.
	USA	*Robins et al., 1988.
- paleothermometry		Sales et al., 1989.
- crystallinity		Ostrowski et al., 1994, 1996.
- dosimetry		Breen & Batista, 1995; Brik et al., 2000a; Dennison et al., 1997; Desrosiers & Simic, 1988; Dodd et al., 1988; *Duarte et al., 1995; Gray et al., 1990; Houben, 1971; Lea et al., 1988; Raffi et al., 1989; Regulla et al., 1985; Rossi et al., 1992; Schauer et al., 1996; Wieser et al., 1994a.
Dentine, Tusk ivory, Cementum ($20 \text{ ka}^1 - 2 \text{ Ma}^2$; 5 – 1500 Gray; ± 10-20%)		Blackwell & Blickstein, 2000; Chong et al., 1989; Haskell et al., 1995; Kenner et al., 1998; Rambaud et al., 2000; Romanyukha & Regulla, 1996; Whitehead et al., 1986.
- site dating,	Japan	*Chong et al., 1989.
stratigraphy	Polynesia	*Whitehead et al., 1986.
	Yukon	*Porat et al., 1991.
- microscopy: cementum		Skaleric et al., 1998.
- dosimetry		Haskell et al., 1995; Romanyukha & Regulla, 1996.
Fish scales (> 200 ka)	kettle lakes	Blackwell, 1995a; Blackwell et al., 1994a, 2001d; Schwarcz, 1994.
Conodonts		Morency et al., 1969.
Crustacean chitin		Desrosiers, 1996; Stewart & Gray, 1996.
Apatite minerals		Ishii & Ikeya, 1993; Miki & Kai, 1991; Moëns et al., 1993.
	Mexico	Ishii & Ikeya, 1993; Ishii et al., 1991; Meguro & Ikeya, 1992.
Phosphates, phosphorites		Y.J. Chen et al., 1989a, 1992; Miki & Kai, 1991; Nambi, 1982.
	Nauru I.	Y.J. Chen et al., 1992.
Coal		Matsuki et al., 1985; *Übersfeld et al., 1954.
Plants		Ikeya, 1985a, 1988, 1993b, 1994a, 1996; Ikeya & Miki, 1985a.
- wood (5 a - 10 ka)		Ikeya, 1982b; Miki et al., 1985a.
- paper (< 100 a)		Ikeya & Miki, 1980a, 1985c.
- resins		*Robins et al., 1985b; Serzhant et al., 1996.
- phytoliths	New Guinea	Ikeya & Golson, 1985.
- grain		Muñoz et al., 1994; Murietta et al., 1996; Robins et al., 1985b.
- textiles		Ikeya & Miki, 1985a.

Table I. Limnological Applications for ESR Dating (continued).

Sample Type (Age, Stability, Dosimetry Limits) - Application	Location	References
Plants (cont'd)		
- dosimetry: pits, seeds		Bustos et al., 1996; de Jesus et al., 1996, 2000; Desrosiers, 1996; Desrosiers & McLaughlin, 1989; Ghelawi et al., 1996, Ikeya et al., 1989.
- dosimetry: plants		Desrosiers, 1996; Desrosiers & McLaughlin, 1989; Rossi et al., 1992.
- dosimetry: sugar (> 0.5 Gray)		Wieser et al., 1994b.
Quartz (10 ka – 2 Ma, using Al or E' centres; ?1 Ma – ?1 Ga using Shotky-Frenkel centres; ± 4-10%)		Ariyama, 1985; *Bella, 1971; Bershov et al., 1975, 1978; Blackwell, 1995a; Blackwell & Blickstein, 2000; Brumby & Yoshida, 1994; Buhay, 1987, 1991; Buhay et al., 1988; Y.J. Chen et al., 1993, 1997a; *Ehlermann, 1996; Friebele et al., 1979; Falguères et al., 1994; Fukuchi, 1989a, 1989b, 1992a, 1996c; Fukuchi et al., 1985; Gamarra et al., 1998; Garrison, 1989; Göksu et al., 1989; J.H.E. Griffiths et al., 1954; Grün, 1989a, 1989b, 1989c, 1989d, 1993; Grün et al., 1994; Halliburton et al., 1993; Hataya et al., 1997a, 1997b; P.H. Huang et al., 1988; Ikeya, 1983b, 1985a, 1988, 1992, 1993b, 1994a, 1996; Ikeya & Golson, 1985; Ikeya et al., 1986, 1995; Imai et al., 1985, 1986; Jani et al., 1983; Jin et al., 1993; Kislyakov et al., 1975; Lee, 1994; Lee & Schwarcz, 1993, 1994a, 1994b; Liang et al., 1989; Matsuoka et al., 1993; Mackey, 1963; *McMorris, 1969, 1970, 1971; Miallier et al., 1994a; O'Brien, 1955; Moiseyev & Rakov, 1989; Odom & Rink, 1989; Ogoh et al., 1994; Poupeau & Rossi, 1989; Rakov et al., 1985, 1986; Rambaud et al., 2000; Regulla et al., 1985; Rink, 1997; Sato et al., 1985; Sueki et al., 1993, 1996; Schwarcz, 1994, 1999; Shimokawa & Imai, 1987; Skinner, 2000a, 2000b; Smolyanskiy & Masaytis, 1979; Tanaka & Shidahara, 1985; Tanaka et al., 1985, 1995; Toyoda & Hattori, 2000; Toyoda & Ikeya, 1989, 1991b; Toyoda & Schwarcz, 1996, 1997a, 1997b; Toyoda et al., 1996, 1999, 2000, 2001; Walther & Zilles, 1994a, 1994b; Weil, 1984; Wieser & Regulla, 1989; Wieser et al., 1991; Wright et al., 1963; Yamanaka et al., 1996; Ye et al., 1993a, 1996a; *Yokoyama et al., 1985c; Zhao et al., 1991.
- aeolian sediment	Australia	Tanaka et al., 1995; Yugo et al., 1998.

continued ....

Notes:
* not used successfully
[1] depends on U concentration
[2] usually more limited due to diagenesis
[3] older dates need to be corrected for fading or retrapping

[4] calcite only
[5] species dependent
[6] aragonite only

Table I. Limnological Applications for ESR Dating (continued).

Sample Type (Age, Stability, Dosimetry Limits) - Application	Location	References
Quartz (cont'd)		
- beach dunes or		Toyoda et al., 2000; Schwarcz, 1999; Tanaka et al., 1997.
sediment	France	Laurent et al., *1998; *Yokoyama et al., 1985c.
	Germany	Walther & Zilles, 1994a.
	Japan	Tanaka et al., 1985, 1997.
- terrestrial sediment	Brazil	Matsuoka et al., 1993.
- marine sediment		Laurent et al., *1998; Tanaka et al., 1997; Ye et al., 1993a, 1993b, 1996a, 1996b.
- fluvial sediment	France	Laurent et al., *1994, 1998.
- loess	China	P.H. Huang et al., 1988; Jin et al., 1993; Li et al., 1993a, 1993b; Ye et al., 1996a.
- hearth sands	France	*Monnier et al., 1994.
- baked sediment, rocks	France	Falguères et al., 1991, 1994; *Yokoyama et al., 1985a, 1985c.
	Japan	Ikeya & Toyoda, 1991; Shimokawa et al., 1988.
	USA	Schwarcz, 1999; Yamanaka et al., 1996.
	USSR	*Pivovarov et al., 2000.
- tektites		Hennig & Grün, 1983.
- impact craters	Chad	Mialler et al., 1997.
- xenoliths in pumice		*Mialler et al., 1994b.
- volcanic ash, tuff, ash		Berger, 1986.
breccias, tephra	Europe	Falguères et al., 1994.
ignimbrites	Japan	Imai & Shimokawa, 1985, 1988, 1989; Imai et al., 1985, 1986, 1992; Shimokawa & Imai, 1985; Shimokawa et al., 1988; Toyoda & Ikeya, 1989, 1991a, 1991b.
	New Zealand	Buhay et al., 1992.
	Turkey	Ulusoy & Apaydin, 1996.
	USA	Cowan et al., 1993; Toyoda et al., 1995.
- volcanic rocks	France	Mialler et al., 1994a.
(< 500 ka)	Germany	Woda et al., 2001.
	Japan	Ikeya & Toyoda, 1991; Ikeya et al., 1983; Imai & Shimokawa, 1987; Shimokawa & Imai, 1985, 1987; Shimokawa et al., 1984, 1988.
	USA	*Ogoh et al., 1993, 1994.
- fault movement &		Buhay, 1991; Buhay et al., 1988; Schwarcz, 1999; Toyoda & Schwarcz, 1996, 1997a, 1997b.
frequency (fault		
gouge & mylonite)	California	Buhay, 1987; Buhay et al., 1992; Grün, 1992; Lee, 1994; Lee & Schwarcz, 1993, 1994a, 1994b, 1995, 1996; Schwarcz et al., 1987.
	China	Y.J. Chen et al., 1997a; Jin et al., 1993; *Lin et al., 1985; Ye et al., 1996a.
	Japan	Ariyama, 1985; Fukuchi, 1988, 1989a, 1989b, 1991, 1992a, Fukuchi, 1992b, 1993, 1996c, 2001; Fukuchi et al., 1985, 1986; Grün, 1992; Hataya et al., 1997a, 1997b; Ikeya, 1983b; Ikeya et al., 1982b, 1983, 1992, 1995; Ito & Sawada, 1985; Kanaori et al., 1985; *Kosaka & Sawada, 1985; Miki & Ikeya, 1982; Sato et al., 1985; Tanaka & Shidahara, 1985; Tani et al., 1996; Toyoda et al., 1993a; *Toyokura et al., 1985.

Table I. Limnological Applications for ESR Dating (continued).

Sample Type (Age, Stability, Dosimetry Limits) - Application	Location	References
Quartz (cont'd)		
- fault movement & frequency (fault gouge & mylonite, cont'd)	Korea	Lee, 2001.
	New Zealand	Buhay et al., 1992.
- granite, ryolite		Fukuchi, 1988; Toyoda et al., 1999a, 1999b.
	Japan	Shimokawa et al., 1988; Toyoda & Ikeya, 1994a, 1994b; Toyoda et al., 1992, 1993a.
	Libya	Odom & Rink. 1989; Rink & Odom, 1991.
	Russia/USSR	Grün et al., 1999; Rakov et al., 1985.
	USA	Odom & Rink, 1989; Rink & Odom, 1991; Toyoda & Goff, 1996; Toyoda & Ikeya, 1994a, 1994b; Toyoda et al., 1995a.
- vein quartz		Toyoda et al., 1995b.
- uranium ore	USSR	Kislayokov et al., 1975; Korolov et al., 1977; Moiseyev, 1980; Moiseyev & Rakov, 1977; Moiseyev et al., 1982.
- gold ore	Australia	Vanmoort & Russell, 1987.
- ceramics		Maurer et al., 1981.
- geothermometry	China	Y.J. Chen et al., 1997b.
	Japan	Fukuchi, 1989a; Miyakawa & Tanaka, 1985; Toyoda & Ikeya, 1989; Toyoda et al., 1993b.
	USA	Ogoh et al., 1993, 1994.
	USSR	Bershov et al., 1975, 1978; Grün et al., 1999; Smolyanskiy & Masaytis, 1979.
- metamorphic rocks	Turkey	*Ulusoy, 2000.
- provenance studies		Toyoda & Goff, 1996; Yugo et al., 1998.
- denudation rates, uplift	USSR	Grün et al., 1999.
- lunar rocks		*Tsay et al., 1972.
- trace element analysis		Gotze & Plotze, 1997.
- dosimetry		Ikeya & Ishii, 1989; Usatyi & Verein, 1996.
- testing other methods: TL	Brazil	Matsuoka et al., 1993.
Flint, Chert (10 ka – 200 ka?; $\tau = 10^6$-10^9)		Blackwell, 1995a; Blackwell & Blickstein, 2000; Garrison, 1989; D.R. Griffiths et al., 1982, 1983; Grün, 1989a, 1989b, 1997; Imai et al., 1994; Porat & Schwarcz, 1991, 1995; Rambaud et al., 2000; Rink, 1997; Schwarcz, 1994, 1999; Schwarcz & Rink, 2001; Skinner, 1995, 2000a, 2000b; Toyoda et al., 1993b.
- heated archaeological artefacts		D.R. Griffiths et al., 1982, 1983; Grün, 1989a, 1989b, 1989c; Robins et al., 1978, 1982.
	Europe	Walther & Zilles, 1994b.
	Middle East	Porat & Schwarcz, 1991, 1995; Porat et al., 1994b.
	Japan	Tani et al., 1998.

continued ....

Notes:

*	not used successfully			
1	depends on U concentration		4	calcite only
2	usually more limited due to diagenesis		5	species dependent
3	older dates need to be corrected for fading or retrapping		6	aragonite only

Table I. Limnological Applications for ESR Dating (continued).

Sample Type (Age, Stability, Dosimetry Limits) - Application	Location	References
Flint, Chert (cont'd)		
- heated archaeological	Turkey	Özer, 1985.
artefacts (cont'd)	USA	Skinner, 2000a; Skinner & Rudolph, 1996a, 1996b; Toyoda et al., 1993b.
- Devonian		Garrison et al., 1981.
- geothermometry	Japan	Tani et al., 1998.
	Middle East	Porat & Schwarcz, 1991.
Silcretes		Grün et al., 1994; *Radtke & Brückner, 1991.
- stratigraphy		*Y.J. Chen et al., 1993; *Radtke & Brückner, 1991; Schwarcz, 1994.
Feldspar		*Y.J. Chen et al., 1989a; Ikeya et al., 1985; Jaek et al., 1995; Sasaoka et al., 1996.
- ash, tuff, breccias	Japan	Imai & Shimokawa, 1985.
		Toyoda & Ikeya, 1991a.
Laterite (10 ka – 1 Ma)	Goa (India)	Nambi & Sankaran, 1985.
Art		Ikeya, 1988, 1993b, 1994a, 1994b.
- lacquers (< 100 a)		Ikeya & Miki, 1985a.
Zircon		Kasuya et al., 1990; Zeller, 1968.
	Japan	Taguchi et al., 1985.
- microscopy		Furusawa et al., 1991.
Clay minerals ($\tau = 10^3$-10^7 y; 0.1 - 20 kGray; ± 20%)		Angel et al., 1974; Blackwell, 1995a; Blackwell & Blickstein, 2000; *Y.J. Chen et al., 1989a; Fukuchi, 1996a, 1996b; Jones et al., 1974; Rambaud et al., 2000; Stößer et al., 1996.
- montmorillonite: fault movement	Japan	Fukuchi, 1996a, 1996b, 1999.
- kaolinite: provenance & fingerprinting clays	Cameroon	Ildefonse et al., 1990; Muller & Callas, 1989.
- porcelain		Göksu et al., 1996.
- brick		Ginsbourg et al., 1996.
- pottery		Bartoll & Ikeya, 1997.
Gypsum, Gypcrete, Anhydrite		Bershov et al., 1975; Blackwell, 1995a; Blackwell & Blickstein, 2000; Y.J. Chen et al., 1988b, 1989a; Furusawa et al., 1991; Grün, 1985c; Ikeya, 1993b, 1996; Ikeda & Ikeya, 1992, 1993; Kasuya et al., 1991; Kohno et al., 1996, Nambi, 1982, 1985; Rambaud et al., 2000.
- stratigraphy	Australia	Y.J. Chen et al., 1988a; Jacobson et al., 1988; Kasuya et al., 1991.
	France	*Hannß, 1985.
- fault movement	USA	Ikeda & Ikeya, 1992.
- microscopy		Furusawa et al., 1991.
- microscopy: crystal growth		Ikeya, 1991, 1994a, 1994b; Omura & Ikeya, 1995.
- dosimetry		Haskell et al., 1996a.

Table I. Limnological Applications for ESR Dating (continued).

Sample Type (Age, Stability, Dosimetry Limits) - Application	Location	References
Diamond		Baker et al., 1997; Griffiths et al., 1954; Lea-Wilson & Lomer, 1996.
Sapphire		Libin et al., 1996.
Barite - microscopy		Kohno et al., 1996; Rao et al., 1996. Furusawa et al., 1991.
Salts		Alcalá, 1996; Blackwell, 1995a; Blackwell & Blickstein, 2000; Ikeya, 1985a, 1988, 1993b; Rambaud et al., 2000; Stößer et al., 1996; Vainshtein & den Hartog, 1966.
- NaCl, NaHCO₃	Searles L. CA	Ikeya & Kai, 1988.
- monohydrocalcite		Debuyst et al., 1993.
	Australia	Debuyst et al., 2000a.
- dosimetry: baking powder, NaHCO₃ + NaHPO₄ (> 3 Gray)		Wieser et al., 1994.
- dosimetry: NaCl (> 0.5 Gray)		Wieser et al., 1994.
Solid CO₂, SO₂		Blackwell, 1995a; Hirai et al., 1994; Kanosue et al., 1996, 1997; Norizawa et al., 2000; Tsukamoto et al., 1993.
- outer planets		Ikeya, 1993a, 1993b, 1994a, 1996; Ikeya et al., 1997b; Tsukamoto et al., 1993.
Forensics, QI/QC (± 10-15%)		Blackwell, 1995a; Ikeya, 1988, 1993b.
- blood (< 100 days)		Miki et al., 1985b, 1987a, 1987b, 1988; Sakurai et al., 1989.
- skin (< 100 days)		Ikeya, 1984b; Ikeya & Miki, 1980b, 1985a; Miki et al., 1985b.
- food oil (< 100 days)		Ikeya & Miki, 1980a.
- milk: dosimetry		*Wieser et al., 1994b.
- nylon: dosimetry		*Wieser et al., 1994b.
- plastics: dosimetry		*Wieser et al., 1994b.
- engine oil (months)		Ikeya,1985a, 1988, 1993b.
- teflon: microscopy		Ikeya & Furusawa, 1993.

Notes:

*	not used successfully		
1	depends on U concentration	4	calcite only
2	usually more limited due to diagenesis	5	species dependent
3	older dates need to be corrected for fading or retrapping	6	aragonite only

Brief history of ESR dating

In 1936, Gorter and colleagues (Gorter, 1936a, 1936b; Gorter & Kronig, 1936) delineated the basic principles of ESR spectroscopy. Early attempts to date coal, clay, quartz, and lunar basalts (Übersfeld et al., 1954; O'Brien, 1955; Duschene et al., 1961; McMorris 1969; Bella, 1971; Tsay et al., 1972) were unsuccessful, despite having been considered analogous to TL (Zeller, 1968; Zeller et al., 1967). Finally, Ikeya (1975) successfully dated a stalagmite.

A flurry of research quickly followed in which geochronologists tried to date everything from fossils to dried blood, and quartz to engine oil (see Table I), much of it led by Ikeya and other Japanese scientists (Table I). Important early applications included attempts to date fault gouge, burnt flint, teeth, and bones. Unfortunately, some early inaccurate applications to controversial archaeological sites (e.g., Yokoyama et al., 1981a, 1981b) hampered its acceptance by scientists. Currently, some 50 laboratories worldwide, 25 in Japan alone, research ESR dating and dosimetry, but only about 10 routinely perform dating. Its most common and reliable applications today include tooth enamel, molluscs, corals, and quartz from fault gouge (see Table I), but research for food irradiation and retrospective dosimetry (e.g., Desrosiers, 1996; Regulla et al., 1994) is producing numerous theoretical studies that may lead to geological and paleontological uses. Developments in ESR imaging and microscopy promise many new mineralogical and paleontological applications.

Principles of ESR analysis

When minerals experience natural radiation, they gradually accumulate trapped unpaired electrons and positively charged "holes" (Fig. 1a), which each produce characteristic ESR signals detectable with an ESR spectrometer. Several such signals result from defects in the crystalline structure associated with trace contaminants. If the ESR signal height (intensity) for a radiation-sensitive signal can be converted into an accumulated dose (Fig. 1b), and the radiation dose rate experienced by the sample during its deposition is known or can be modelled, a date can be calculated. ESR dates can be obtained using any material, which has a radiation-sensitive ESR signal, provided it satisfies the following criteria:

1. At the time of interest, the mineral's ESR signal was initially, or was reset to, 0.0.

2. The signal lifetime, τ, is long compared to the site age.

3. The accumulated dose, $\mathcal{A}_\Sigma$, is small compared to the saturation level in the material.

In limnological contexts, tooth enamel, clean carbonates (travertine, mollusc shells, calcrete), fish scales, heated or bleached siliceous rock (flint or quartz) have several applications. Many salts may eventually produce valid dates, but the techniques have not been developed yet (Blackwell, 1995a). Sediment dates have been attempted, but problems related to incomplete zeroing must still be resolved. Pilbrow (1996, 1997), Wertz & Bolton (1986), Abragam & Bleaney (1986) detail the underlying ESR spectroscopic theory. This discussion will focus on limnological applications, illustrated where possible by lacustrine examples. It omits other applications, although Table I does list references for readers interested in other uses. Several recent reviews (e.g., Blackwell, 1995a; Rink, 1997; Skinner, 2000b) discuss other applications.

A few technical terms become essential here. An ESR spectrometer uses a microwave signal to create resonance between the unpaired electrons in minerals and an externally applied strong magnetic field. Lande's factor, called the g value, is a dimensionless number that uniquely describes the ESR characteristics for any peak. Pulsed or Q-band ESR may ultimately improve our ability to separate interference signals (e.g., Grün et al., 1997c; Skinner et al., 2000, 2001a). Although other bands, such as Q- or L-band, are occasionally

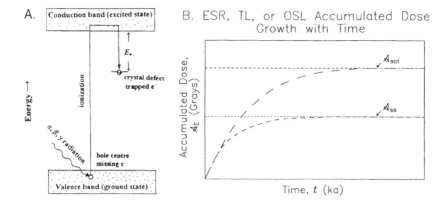

A.

B. ESR, TL, or OSL Accumulated Dose Growth with Time

Figure 1. ESR signal production. With increased irradiation, the ESR signal's intensity grows, eventually reaching saturation: (a) After absorbing energy from incident radiation, excited electrons move through the conduction band. Although most return to the ground state, a few become trapped in charge site defect (traps, often at trace elements substituents in the crystal lattice) that each have specific energies above the ground state. ESR signals result from the magnetic fields generated by such unpaired electrons and the empty holes they have left behind. With irradiation, such trapped electrons and charged holes, which each produce characteristic signals, gradually accumulate in the materials. (b) With natural irradiation, the signal saturates at its maximum (saturated) accumulated dose, $\mathcal{A}_{\Sigma,sat}$, or at a lesser dose, a steady state accumulated dose, $\mathcal{A}_{\Sigma,ss}$, where signal fading loss equals signal production (modified from Blackwell & Schwarcz, 1993b).

used to examine signals in more detail, for most ESR dating, spectra are analyzed in the X-band at 1–10 mW power using microwave frequencies near 8–10 GHz under a 100 kHz field modulation. Under these conditions, most geologically or archaeologically interesting ESR signals fall within $3 < g < 1.9$ (Table 2 in Blackwell, 1995a).

Zeroing reduces an ESR signal's intensity to a level indistinguishable from background levels. Most newly formed minerals have ESR signals with zero intensities. Several physical processes can also zero a signal in a mineral with an accumulated dose (i.e., a measurable signal; $\mathcal{A}_{\Sigma} > 0$). Strong heating to temperatures above 250–500 °C, depending on the mineral, will also zero signals in all minerals (Fig. 2b; e.g., Skinner, 2000a; Skinner et al., 2000). For some signals in a few minerals, exposure to intense sunlight can zero (bleach) the signal (Fig. 2a; e.g., Buhay et al., 1988; Walther & Zilles, 1994a; Tanaka et al., 1995). Luckily, for the radiation-sensitive signals in most minerals, sunlight causes little or no signal loss. High pressure or strain that builds up in faults can partially reset some signals, as can the strain developed in grain comminution during an earthquake or grinding for sample preparation (Fig. 3; Lee & Schwarcz, 1993, 1994a, 1994b; Moëns et al., 1993). Remineralization and diagenesis add new minerals whose radiation-sensitive signals will be zero at formation. Therefore, if the original and new minerals have signals with similar g values, the resultant complex signal may be impossible to resolve, adding inaccuracies to the age determination. If, however, the new signals do not interfere with the original signals, as is true for tooth enamel, only the dating signal's intensity is reduced, thereby reducing the discriminatory range and dating limits for the technique (Skinner et al., 2000).

The reliability of the dating method depends on the signal's thermal stability. Signals which zero easily at typical Earth surface temperatures have little value for dating, but may

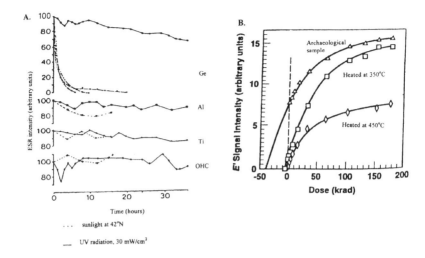

Figure 2. Zeroing in quartz and chert. In quartz, several signals can be zeroed using different techniques: (a) Exposure to intense UV radiation and sunlight can completely bleach the Ge signal and partially bleach the Al, Ti, and OHC (after Buhay et al., 1988). (b) Heating archaeological chert to high temperature can zero the E' signal, reducing its accumulated dose, $\mathscr{A}_\Sigma$, to 0. After zeroing, the signals can regrow if given more irradiation (after Blackwell & Schwarcz, 1993b).

provide other information. The mean signal lifetime, τ, must exceed the desired dating range by at least 2–3 orders of magnitude to ensure reliable ages. In tooth enamel, for example, $\tau \approx 10^{19}$ y (Skinner et al., 2000), sufficiently long, in theory at least, to date anything within the history of the universe. Unlike TL, no datable ESR signal appears to suffer anomalous fading. In practice, however, most signals have a finite saturation limit, beyond which no new, trapped electrons are formed. Many minerals also have a steady state level, somewhat lower than their saturation level, caused by electron loss and retrapping (Fig. 1b). The mean signal lifetime and the steady state limit or saturation limit define the maximum datable age, while the ability of the ESR spectrometer to discriminate between the dating signal and its surrounding background determines the minimum dating limit. Both limits differ depending on the mineral and its habit in the material to be dated. The radiation dose rates experienced by the sample determine how those limits are translated into an actual age. If samples experience high radiation dose rates, the minimum datable age will be relative low, but so will its maximum datable age, and conversely, low radiation dose rates mean higher minimum and maximum limits.

The ESR signal height (Fig. 1) is proportional to the number of trapped charges at that lattice site, and, therefore, to the total radiation dose, $\mathscr{A}_\Sigma$, that the material has experienced. The ESR age, t_1, the time that has elapsed since the mineral formed and began to accumulate charges then is calculated from Equation (1):

$$\mathscr{A}_\Sigma = \mathscr{A}_{\text{int}} + \mathscr{A}_{\text{ext}} = \int_{t_0}^{t_1} D_\Sigma(t)\,dt = \int_{t_0}^{t_1} \left(D_{\text{int}}(t) + D_{\text{ext}}(t) \right) dt \qquad (1)$$

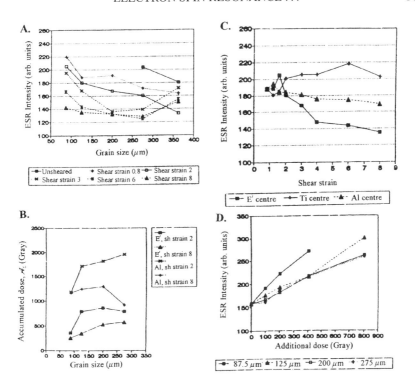

Figure 3. Effects from shear strain on ESR signals in quartz. Shear strain will reset most ESR signals (modified from Lee & Schwarcz, 1993): (a) As strain increases, the differences in ESR intensity between different grain size fractions decreases. (b) At a normal stress of 10 MPa, the measured accumulated (equivalent) dose, $\mathscr{A}_\Sigma$, decreases with decreasing grain size for both the E' and Al signals, until at a small grain size the two signals give equal $\mathscr{A}_\Sigma$ determinations. (c) While the E' signal is the most easily reset, strain also affects the Al signal. The Ti signal appears unaffected. (d) During artificial irradiation for producing a growth curve, the smaller grain sizes show the greatest sensitivity and the most well behaved growth curves.

where $\mathscr{A}_\Sigma$ = the total accumulated dose in the sample,

$\mathscr{A}_{int}$ = the internally derived accumulated dose component,

$\mathscr{A}_{ext}$ = the externally derived accumulated dose component,

$D_\Sigma(t)$ = the total dose rate,

$D_{int}(t)$ = the total dose rate from internal sources: U, its daughters, and any other radioisotopes,

$D_{ext}(t)$ = the total dose rate from the external environment: sedimentary U, Th, and K, and cosmic dose rate,

t_1 = the sample's age,

t_0 = today.

For samples in which the total dose rate, $D_\Sigma(t)$, is constant, this reduces to:

$$t_1 = \frac{\mathscr{A}_\Sigma}{D_\Sigma(t)} \tag{2}$$

Sample collection

An ideal ESR sample should be as pristine as possible. To improve precision and accuracy, both the dating sample and any associated sediment samples should not experience the following treatments during or after excavation:

1. Glues, shellacs, and other preservatives can add contaminant U to the sample that reduces the accuracy of internal dose rate measurements, as well as organic compounds that might cause ESR signal interference.

2. Washing may remove U, datable mineral, and sediment. Sediment attached to the sample may offer the only chance to measure the external β dose rate.

3. If used to remove samples from cemented sediment, acid dissolution can dissolve the sample and leach its U.

4. Removing attached bone from teeth reduces the accuracy of the external dose rate measurements.

5. Removing attached sediment from any sample reduces the accuracy of the external dose rate measurements.

6. Sample numbering uses inks and paints that can add contaminant organic compounds if applied to the sample.

7. Allowing clay samples to dry necessitates extensive grinding during subsequent sample preparation, which can partially bleach some ESR signals.

8. Packing samples for shipment with materials, such as old newspapers, dyed paper, etc., can cause trace element or organic contamination if they contact the sample.

Although preservatives, if available, can be analyzed to correct for contamination effects, the resulting age will still have reduced precision. Fossils can be cast, providing that the casting resin and powder have been tested for contamination potential first (Blackwell & Skinner, 1999a).

For all dating samples except teeth, diagenesis or signal interference may cause some samples to be unsuitable (Table II). Since fossils can be easily reworked into younger depositional units (e.g., Blackwell, 1994), any sampling program should collect at least 8–10 samples from each stratigraphic unit to increase the chance that the samples analyzed provide dates related to the event of interest. Although the required sample weight varies depending on the auxiliary analyses necessary (Table II), the ESR analysis itself, and the associated NAA or geochemical analyses to measure the internal dose rate, require 1–2 g of

Table II. ESR Sample Types.

Sample Type	Minerals[1]	Zeroing req'd?	Isochrons possible?[1]	Species effects?[1]	Best type or species?	Minimum Sample for Standard ESR[2,3] (g/subsample)	Signal Intensity	Inter-ference?	Inaccurate Ages?	Incomplete Zeroing?	Effects from Grinding?
Enamel	HAP	no	yes	no	large teeth; no milk teeth	3-4	decreased	rarely	no	n/a	possibly
Dentine	HAP	no	yes	no	no milk teeth	5-10	decreased	rarely	no	n/a	possibly
Cementum	HAP	no	no	no	no milk teeth	0.1[4]	decreased	rarely	no	n/a	possibly
Bone	HAP	no	yes	no	cortical bone	10-20	decreased	possibly	possibly	n/a	possibly
Tusk, antler	HAP	no	yes	no	densest	10-20	decreased	possibly	possibly	n/a	possibly
Gar fish scales	HAP	no	no	unkn	only gar	2-5[4,5]	decreased	rarely	no	n/a	possibly
Molluscs	cct, argt	no	unkn	yes	large valved[6]	5-10[5]	may increase	likely	possibly	n/a	interference
Ratite egg shells	cct	no	theor	unkn	unkn	5-10[5]	may increase	likely	possibly	n/a	interference
Coral, echinoderms	argt	no	theor	yes	unkn	5-10[5]	may increase	likely	possibly	n/a	interference
Foraminifera, ostracodes	cct, argt	no	unkn	unkn	unkn	10-20[5]	may increase	likely	possibly	n/a	interference
Travertine, speleothem	cct, argt	no	yes	n/a	densest	50-100[7,8]	may increase	likely	possibly	n/a	interference
Calcrete, caliche, stromatolites	cct	no	unkn	n/a	densest	50-100[7,8]	may increase	likely	possibly	yes, if clasts CO3 also	interference
Authigenic cement	cct, argt	no	unkn	n/a	densest	100 mg[4]	may increase	likely	possibly	n/a	interference
Phytoliths, diatoms, radiolarians	qtz	no	unkn	unkn	unkn	5-10[5] or 0.1[4,5]	may increase	possibly	possibly	n/a	reduced intens
Fault gouge, mylonite	qtz, fspar	yes	yes	n/a	qtz or fspar separates	50-100[7,8]	may increase	possibly	possibly	n/a	reduced intens
Ash/tuffs	qtz	yes	theor	n/a	thickest units	50-100[7,8]	may increase	likely	possibly	yes	reduced intens
Flint/chert (burnt)	qtz	yes	yes	n/a	avoid patina	5-10[8]	may increase	possibly	possibly	yes	reduced intens

Table II. ESR Sample Types (continued).

Sample Type	Minerals[1]	Zeroing req'd?	Isochrons possible?[1]	Species effects?[1]	Best type or species?	Minimum Sample for Standard ESR[2,1] (g/subsample)	Effects from diagenesis, secondary mineralization, or cementation				Effects from Grinding?
							Signal Intensity	Inter-ference?	Inaccurate Ages?	Incomplete Zeroing?	
Hearth sand	qtz	yes	unkn	n/a	closest to hearth	50–100[7,8]	may increase	possibly	possibly	yes	reduced intens
Silcrete, laterite	qtz	yes	unkn	n/a	qtz separates	50–100[7,8]	may increase	likely	possibly	yes, if clasts qtz also	reduced intens
Beach, fluvial sediment, loess	qtz, flint, chert	yes	unkn	n/a	qtz separates	50–100[7,8]	may increase	possibly	possibly	yes	reduced intens
Authigenic cement	qtz	no	unkn	n/a	densest	0.1[4]	may increase	likely	possibly	n/a	reduced intens
Dolomite (1°)	dmt	no	unkn	n/a	densest	50–100[7,8]	may increase	likely	possibly	n/a	unkn
Dolomite (2°)	dmt	no	unkn	n/a	dmt separates	0.1[4]	may increase	likely	possibly	n/a	unkn
Gypsum, gyperete	gyp	no	unkn	n/a	gyp separates	50–100[7,8]	may increase	likely	possibly	n/a	unkn
Anhydrite	anhy	no	unkn	n/a	densest	50–100[7,8]	may increase	likely	possibly	n/a	unkn
Halite	hal	no	unkn	n/a	densest	50–100[7,8]	may increase	likely	possibly	n/a	unkn

[1] Abbreviations:

1°	= primary		2°	= secondary
HAP	= hydroxyapatite		qtz	= quartz
cct	= calcite		argt	= aragonite
dmt	= dolomite		gyp	= gypsum
anhy	= anhydrite		hal	= halite
CO₃	= carbonate		unkn	= unknown
intens	= intensity		theor	= theoretically

[1] Sizes assume little or no diagenesis is present. For diagenetically altered samples, larger samples are needed.

[2] For isochron analysis, the sample size must be increased by a factor of 5-8.

[4] This uses a special ramped irradiation technique involves reirradiating some aliquots, but takes up to 2 years to complete.

[5] For species that have not been tested for ESR applicability, another 100–200 g is necessary.

[6] Smaller species may require special techniques or mixing multiple individuals into one subsample.

[7] Large sample sizes ensure sufficient pristine mineral for analysis after mineral separation and for XRD or petrographic analysis to check for recrystallization

[8] For samples from new study sites or sample types not yet tested for ESR applicability, another 100–200 g may be necessary.

pristine datable mineral per standard ESR subsample. For some materials, especially those prone to diagenesis, it is necessary to check for secondary mineralization and remineralization, which affect ESR signal intensities (Table II), requiring larger samples. For samples needing to be separated into discrete mineral phases, such as authigenic cements, caliche, calcrete, and gypcrete, the pristine mineral must be separated from the adjacent sediment, often necessitating much larger samples. For ESR dating sediment, pristine sample blocks of ~0.5 kg cut from thick or extensive units provide the best results, if available. Salt samples need to be stabilized to prevent remineralization or recrystallization during transport, as can occur with some hydrated salts. For very small samples (100–200 mg), the ramping irradiation technique can be used in which several aliquots are reirradiated several times, but this special handling does lengthen the total analysis time significantly.

Most curated museum samples require isochron analysis (see below), because sediment has not usually been preserved. For samples intended for isochron analysis, samples should be photographed before shipping to ensure that broken samples can be reconstructed to maximize the number of viable subsamples. Samples should be packed tightly with minimal air to reduce sample breakage and bag destruction.

Sediment dosimetry and associated sediment samples

Whenever possible, the external dose rates should be assessed using at least two procedures from among isochron analysis, sediment geochemistry, *in situ* γ or TL dosimetry. For TL or γ dosimetry, if dosimetry cannot be completed before collection, sampling locations need to be marked and preserved to permit future dosimetry. Effective TL dosimetry requires that the site be unaffected by further excavation or deflation for 6–12 months. Due to the high probability of equipment tampering or theft in lacustrine sites, either γ dosimetry or sedimentary analysis is preferred over TL dosimetry. Isochron analysis is still experimental for many materials.

With dosimetry by sedimentary geochemistry, the external dose field can be mathematically modelled reasonably accurately. In sediment, β particles can penetrate about 2–3 mm, and γ radiation ~30 cm (Fig. 4). The sediment immediately attached to or surrounding the dating sample usually provides the only direct measurement for calculating the β radiation dose rate. When using TL or γ dosimetry, this sediment must still be analyzed geochemically to provide the external β dose rate. Several sediment samples may be needed to represent the sphere influenced by γ radiation 30 cm in radius around the dating sample.

Sediment sampling protocols vary with the bed or unit thickness, its mineralogy, and its grain size (see Table III; Fig. 5). In many sedimentary contexts, the concentrations of radioactive elements can vary dramatically over short distances if the sediment contains large clasts of several different minerals (Fig. 4d). This makes it necessary to collect several samples from each unit or bed that might be contributing to the dating sample's external dose rate. If the sediment contains a homogeneous grain mixture of fine to medium grained clasts, ~5–10 g are sufficient for each associated sediment sample. For coarser sediment types, sediment samples should include representative portions of cobbles mixed with the matrix, or be submitted as separate matrix and cobble samples. In units with fossils or artefacts, these must be considered as radioactive cobble sources and analyzed also (Blackwell & Blickstein, 2000; Blackwell et al., 2000, 2001a). Generally, the larger the grains, the larger

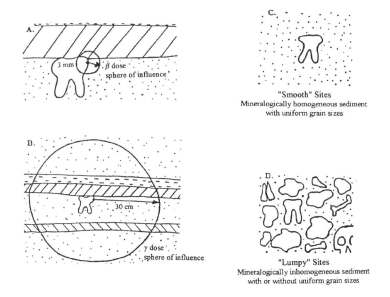

Figure 4. Factors affecting the effective radiation dose field around dating samples. Radiation can reach the dating sample from radioactive decay occurring within the sphere of influence for the particular radiation type: (a) β particles deliver to a sample a significant, but variable, component in the total radiation dose, both externally and internally. Since the penetration range for a β particle averages 1–3 mm, comparable to the sample thicknesses, dose calculations must consider β attenuation within the sample. The sphere of influence for the contributions from β radiation will usually not include more than two or three sedimentary units. (b) Since γ irradiation can penetrate ~30 cm, the sphere of influence for the contributions from γ radiation can include several sedimentary units, which may produce very different dose rates. (c) In "smooth" sites with homogeneous sediment, the dose rate calculation is trivial. (d) In "lumpy" sites, different minerals or clasts within the sediment, which may contain different concentrations of radioactive elements, can contribute dose at very different rates. In all situations, the external dose rate, $D_{ext}(t)$, calculation must volumetrically average the dose rate from each source relative to its importance and location within the sphere of influence each stratigraphic unit or sediment type (after Blackwell & Skinner, 1999a).

the sediment mass that will be needed. In well cemented sedimentary units (e.g., "breccias", etc.), a block of sediment ($\precsim$ 20 cm on a side) showing all representative grains, matrix, and cements on the surfaces often provides the best sample.

If all the sediment samples preferred in the ideal circumstance are not available, sediment from the same or similar beds as close as possible to the dating sample can still be used to assess the radiation dose field's variability and estimate external dose rates. For museum samples, any samples from nearby outcrops may provide valuable clues. Accurately recording and photographing each sediment sample relative to the dating sample ensures accuracy in modelling the external dose field. All *in situ* sediment samples should be placed in tightly sealed jars or doubly bagged in plastic "zip-lock" bags immediately after collection to retain sediment moisture for water concentration analysis. For sections that have been exposed for a long time, or archived sediment, sediment moisture content is not analyzed.

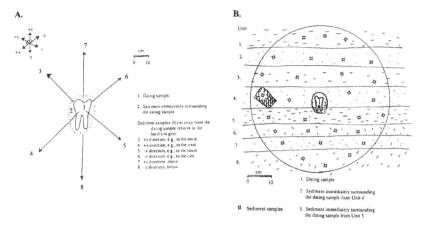

Figure 5. Collecting protocols for associated sediment samples for ESR dating (after Blackwell & Skinner, 1999a). (a) Thickly bedded homogeneous units ("Thick smooth" units, Table III): Assuming that the dating sample lies at least 35 cm from the nearest sedimentary unit boundary, sediment should be collected from four to six of the six orthogonal positions. In pictured example, the associated sediments were collected from the six orthogonal positions that coincide with the site grid plan. (b) Thinly bedded or inhomogeneous units ("Thin smooth", "thin lumpy", or "thick lumpy" units, Table III): The sample for dating (1) sits within Units 4 and 5 (2, 3). In this circumstance, separate samples need to be collected from the two surrounding units. When sampling the surrounding sedimentary units, three to five sediment samples should be collected from each unit, distributing the samples throughout the unit as it falls within the γ sphere of influence. Ideally for each unit, a few should come from along the cut face, one from behind, and one from in front of the cut face in order to sample a somewhat even distribution for each bed.

ESR analysis

Calculating an ESR age requires considering some 30 different parameters, which affect the accumulated dose, the internal and external dose rates. Although improved spectrometers and ancillary equipment (e.g., Dolo et al., 1996; Grün & Clapp, 1997; Haskell et al., 1997a) have sped the process and improved precision somewhat, the basics of ESR analysis were established in the 1980's. Standard analytical protocols for all mineralogies require powdered samples. Although some ESR labs have developed "nondestructive" analyses for tooth enamel (e.g., Grün, 1995; Grün et al., 1996a; Robertson & Grün, 2000; Oka et al., 1996; Miyake et al., 2000; Haskell et al., 2000), even these cause some sample degradation.

Determining the accumulated dose, $\mathcal{A}_\Sigma$

For each sample, the accumulated dose, $\mathcal{A}_\Sigma$, is determined using the additive dose method (Fig. 6a; Grün & MacDonald, 1989). This requires about 0.2–0.5 g of pristine prepared mineral sample (Table II) in order to provide 10–15 aliquots of powdered, homogenized sample. Using fewer than 10 measurements causes significantly lower precision (Grün & Brumby, 1994). Except for one, each aliquot is irradiated to a different precisely known artificial radiation dose, usually from a ^{60}Co γ source (e.g., Blackwell, 1989, 1999). The added doses used usually range from 1–10 Grays for the lowest added dose to 1–40 kGray for highest, depending on the $\mathcal{A}_\Sigma$. Older samples generally get higher doses. The selection

Table III. Sampling for Associated Sediment.

Sedimentary Unit or Site[1]			Sediment Grains (Clasts)		1° Dosing Unit(s)[4]		Sediment Samples		Samples of Clasts >0.5 cm in Diameter[6]	Fig.
Character[1]	e.g. Fig.	Type[1]	Mineral Compositions[2]	Grain Size Range[3]	Thickness (cm)	Mass[5] (g)	from 1° Dosing Unit(s)[4]	from 2° Dosing Units[4]		
"Smooth", thickly bedded sites	4c[7]	Thick smooth	Homogeneous	Uniform	>65	5-10	4-6 orthogonally oriented	none	1-3 for gravel-sized matrix only	5a[7]
"Smooth", thinly bedded sites	4b[7]	Thin smooth	Homogeneous	Uniform	<65	5-10	4-6 orthogonally oriented	3-5 for each unit ≤ 30 cm from dating sample	1-3 for gravel-sized matrix only	5b[7]
"Lumpy", thickly bedded sites	4d[7]	Thick lumpy 1	Homogeneous	Non-uniform	>65	100-1000	4-6 orthogonally oriented	none	1-3 per unit	5a[7]
"Lumpy", thinly bedded sites	4d[7]	Thin lumpy 1	Homogeneous	Non-uniform	<65	100-1000	4-6 orthogonally oriented	3-5 for each unit ≤ 30 cm from dating sample	1-3 per unit	5b[7]
"Lumpy", thickly bedded sites	4c[8]	Thick lumpy 2	Inhomogeneous	Uniform	>65	50-100	4-6 orthogonally oriented	none	1-3 for gravel-sized matrix only	5a[8]
"Lumpy", thinly bedded sites	4b[8]	Thin lumpy 2	Inhomogeneous	Uniform	<65	50-100	4-6 orthogonally oriented	3-5 for each unit ≤ 30 cm from dating sample	1-3 for gravel-sized matrix only	5b[8]
"Lumpy", thickly bedded sites	4d[8]	Thick lumpy 3	Inhomogeneous	Non-uniform	>65	500-1000	4-6 orthogonally oriented	none	1-3 for each lump mineralogy per unit	5a[8]
"Lumpy", thinly bedded sites	4d[8]	Thin lumpy 3	Inhomogeneous	Non-uniform	<65	500-1000	4-6 orthogonally oriented	3-5 for each unit ≤ 30 cm from dating sample	1-3 for each lump mineralogy per unit	5b[8]

[1] Sampling strategy and site character definition is governed by the most inhomogenous unit present. If one "lumpy 3" bed occurs within 35 cm of the sample, the whole sedimentary package is treated as a "lumpy 3" site.

[2] Mineral compositions in the units within 35 cm of the dating sample:
  Homogeneous = all a single mineral, e.g., all calcite or all quartz
  Inhomogeneous = mixed sediment with several mineral or rock fragment types, e.g., mixed limestone and bone, till with quartz sand and gravel-sized granite clasts

[3] Clast (grain) sizes in the units within 35 cm of the dating sample:
  Uniform = all one or two ϕ size classes, e.g., all medium-coarse sand or all silt-fine sand
  Non-uniform = several or a range of ϕ size classes, e.g., diamicton, breccia, most fossiliferous units, till

[4] Dosing units are sedimentary units within the 30 cm γ sphere of influence (Figures 4, 5):
  1° (primary) dosing unit(s) = the one or two unit(s) touching the dating sample that contribute both β and γ dose to the external dose rate affecting the sample.
  2° (secondary) dosing units = all units ≤ 35 cm from the dating sample that contribute only γ dose to the external dose rate affecting the sample.

[5] assuming that sediment matrix is sand-sized or smaller; larger matrix grain size requires larger sample mass.

[6] assuming that the clasts are collected separately from the matrix.

[7] assuming that grains of only one mineral constitute all the components in the sedimentary unit(s).

[8] assuming that grains of several different minerals occur in the sedimentary unit(s).

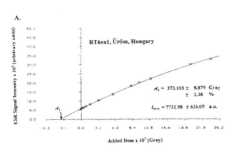
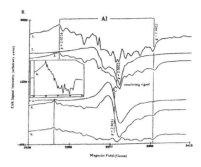

Figure 6. Determining the accumulated dose, $\mathscr{A}_\Sigma$. The additive dose method is used to calculate the accumulated (or γ-equivalent) dose, $\mathscr{A}_\Sigma$: (a) Under artificial irradiation during analysis, the HAP signal saturates at its maximum intensity, I_{max}. Plotting the signal intensity *versus* the added radiation dose produces a growth curve. The x-intercept for this curve gives $\mathscr{A}_\Sigma$. This mammoth tooth from the Üröm pond deposit, Hungary, has a substantial accumulated dose as expected for a Middle Pleistocene site (Berman et al., 2000). (b) For signals suffering interference, signal subtraction is used to remove the interference: Curve 1: A pure Al signal is unaffected by interference signals. Curve 2: An organic radical signal, Ċ, interferes with the Al signal. Curve 3: Unidentified interference signals affect the Al signal. Curve 4: The Al and Ċ signal in a natural archaeological sample. Curve 5: The same sample as Curve 4 heated for 10 minutes at 320 °C to zero the Al signal. Curve 6: When Curve 5 is subtracted from Curve 4, the resulting signal shows the hyperfine splitting typical for the Al signal (see inset; modified from Porat & Schwarcz, 1991).

of added doses does affect the statistics for the curve fitting, and hence, the precision for $\mathscr{A}_\Sigma$ for enamel (e.g., Grün & Rhodes, 1991, 1992; Chumak et al., 1996; Grün, 1996b; Hayes et al., 1997; Lee et al., 1997), and presumably other materials as well. In the ramping technique, only 4–5 aliquots are used, but at least two are used to calibrate the spectrometer with each set of measurements, and two or three are successively irradiated to ever higher added doses.

After measuring the ESR signal heights for both the natural and irradiated aliquots, the added dose is plotted *versus* the signal intensity to produce a growth curve (Fig. 6a). Usually, the points are weighted inversely with intensity (peak height; e.g., Grün, 1998a). In some materials, however, signal subtraction is necessary to isolate the dating signal from the interference in order to measure an accurate peak height (Fig. 6b). Despite a recent controversy over whether peak heights should be measured from integral or derivative spectra with or without deconvolution (e.g., Grün & Jonas, 1996; Jonas, 1995, 1997; Jonas & Grün, 1997), derivative spectra actually provide better resolution (Lyons, 1997; Lyons & Tan, 2000). Most evidence also suggests that deconvolution is not necessary for many dating peaks (e.g., Skinner et al., 2000, 2001a). The dose, $\mathscr{A}_\Sigma$, required to produce the observed natural ESR signal intensity equals the x-intercept for the growth curve. Within some materials, such as caliche, crystals may vary greatly in their $\mathscr{A}_\Sigma$. If some regions are at or near saturation, age underestimation may also occur, because the dose response is nonlinear near saturation (Fig. 1b). This is not a problem for tooth enamel where linear behaviour persists to large doses (Brennan, 1999). Generally, $\mathscr{A}_\Sigma$ can be measured with 0.8 to 5% precision depending on the spectrometer's calibration (Nagy, 2000), the radiation source calibration (Barabas et al., 1993; Regulla et al., 1993), the sample's age and diagenetic state (e.g., Blackwell et al., 2000b, 2000c).

Determining the internal dose rate, $D_{int}(t)$

To calculate the internal dose rate, $D_{int}(t)$, the radiation sources (any U, Th, K, etc.) within the sample are measured, usually using neutron activation analysis (NAA) or any geochemical technique able to measure at the ppb level. Then, $D_{int}(t)$ is derived from theoretical calculations (e.g., Nambi & Aitken, 1986; Adamiec & Aitken, 1998). For samples containing U or Th, those calculations must also consider the increased radioactivity due to ingrowth of the U or Th daughter isotopes (Fig. 7a) over time using an iterative procedure (Grün et al., 1988a). $D_{int}(t)$ calculations also consider radiation attenuation by water within the sample, α and β dose attenuation due to mineral density, and radon (Rn) loss for U- or Th-rich samples (Fig. 7b; e.g., Brennan et al., 1991, 1997b; Faïn et al., 1998).

For samples such as tooth enamel, bone, molluscs, and fish scales, that the sample's internal dose rate derives solely from U absorbed during its burial history requires that the calculated ESR age account for U uptake: Either the sample must be dated by U-series analysis or U/Pb, which allows a unique uptake model to be selected, or the U uptake model must be assumed. Without calibrating dates, four models are commonly used (Figs. 7c, 7d; Blackwell et al., 1992; Brennan et al., 1997b; Grün et al., 1988a):

Early uptake (EU)	assumes that the sample absorbed all its U soon after burial, providing the youngest age given the accumulated dose, $\mathscr{A}_\Sigma$, and external dose rate, $D_{ext}(t)$.
Linear uptake (LU)	assumes that the sample absorbs U at a constant rate throughout its burial history, giving a median age.
Recent uptake (RU)	assumes U uptake very late in the sample's burial history, which reduces its internally generated dose, $\mathscr{A}_{int}$, to a minor contribution compared to $\mathscr{A}_\Sigma$. This gives the maximum possible age.
Coupled uptake (CU)	assumes that the enamel, dentine, cementum, and any attached bone in teeth absorb U by different models: usually LU for the enamel, and EU for the dentine, cementum, and any attached bone (CLEU), producing ages somewhat younger than strict LU, but older then strict EU models.

Other models have also been suggested (e.g., Bahain et al., 1992; Ikeya et al., 1997a), and coupled EU/RU or LU/RU models are also theoretically possible for teeth, along with any intermediate model between EU, LU, or RU, or methods that combine multiple uptake and/or leaching events (Blackwell et al., 2000, 2001e). In teeth, LU or CLEU ages, or models between LU and RU, or those combining two uptake events often agree most closely with ages determined by other means for samples older than 80 ka, but, within a site, the uptake model can vary, since it depends strongly on local paleoenvironmental conditions (e.g., McDermott et al., 1993a; Grün & McDermott, 1994; Blackwell et al., 2001e). Although a coupled ESR-^{230}Th/^{234}U calculation can constrain the U-uptake history in teeth (e.g., McDermott et al., 1993a; Grün & McDermott, 1994), only samples younger than 300–400 ka give finite ^{230}Th/^{234}U ages (Blackwell & Schwarcz, 1995; Schwarcz & Blackwell, 1992). U/Pb can date some uraniferous samples older than 1–2 Ma to delineate

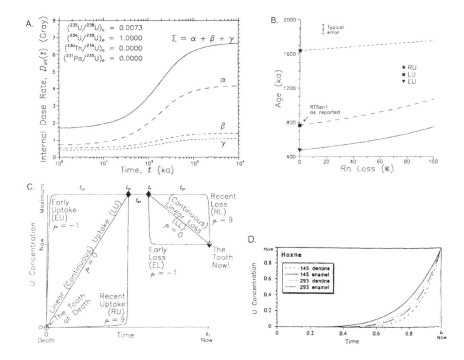

Figure 7. Factors affecting the internal dose rate, $D_{int}(t)$. For bones, teeth, molluscs, and other materials containing or capable of absorbing U, U uptake must be measured or modelled. For minerals or fossils capable of losing U or other U daughter products by leaching or degassing, these must also be modelled or measured: (a) $D_{int}(t)$ increases as the sample ages simply from ingrowth of the U daughter isotopes. This plot assumed an early uptake model U absorption of 10 ppm, with no initial Th or Pa. (b) Radon (Rn) gas, produced when U decays, can escape from samples during diagenesis and fossilization, causing $D_{int}(t)$ to decrease, and therefore, affecting the accuracy in the calculated ages. Assuming 0% Rn loss will not contribute significant errors to age calculation for most samples, except those with very high U concentrations. In this mammoth molar from a pond deposit in Hungary, the uptake model significantly affects the age calculation, because the dentine contains relatively high U concentrations, producing significant differences in the various calculated model ages (Berman et al., 2000). (c) A combined model for U uptake and leaching: Some time, t_{st}, after the end of the initial U uptake event at t_m, and over t_{pr} years, a leaching event can occur. The fossil absorbs all its U immediately after death in the early uptake (EU) model, but it absorbs almost no U until just before attaining its maximum U concentration in the recent uptake (RU) model. Under linear uptake, the fossil absorbs U continuously and constantly throughout the uptake time, and linear leaching assumes an analogous continuous, constant U loss through the leaching period. Under early leaching (EL), the fossil loses U in a geological instant some time before the fossil is discovered, whereas under recent leaching, the loss occurs just before discovery (modified from Blackwell et al., 2001a). Secondary uptake events may also occur in which the U uptake rate differs from that experienced in the first event. More complex models can be devised by combining several uptake and leaching events. (d) U uptake in teeth from Hoxne, England: Recent uptake models are applicable in some situations (modified from Grün & McDermott, 1994).

the uptake model, but coupled ESR-U/Pb ages have not yet been reported. U leaching may also present problems for some samples and may require complex models (Fig. 7c; Blackwell et al., 2001a, 2001e). Precisions for $D_{int}(t)$ depend strongly on the precision for U concentration measurement. Delayed neutron counting (DNC) neutron activation analysis (NAA) can routinely provide precisions and detection limits as low as ± 0.02 ppm, whereas instrumental NAA averages ± 0.2 ppm for precision and ± 1 ppm for detection limits, which makes dating young samples impossible.

Determining the external dose rate, $D_{ext}(t)$

$D_{ext}(t)$ strongly affects the calculated ESR ages (Fig. 8a), especially for samples that have low internal dose rates, $D_{int}(t)$. To derive the external dose rates, $D_{ext}(t)$, four methods can be used:

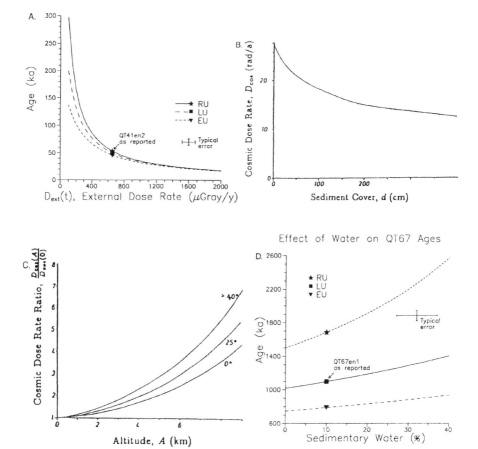

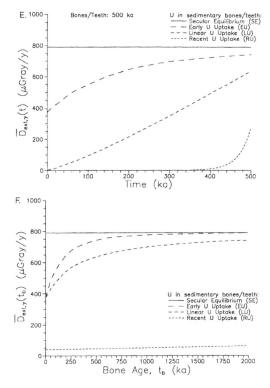

Figure 8. Factors affecting the external dose rate, $D_{ext}(t)$, and ESR ages. $D_{ext}(t)$ is a function of many variables, including the water in the sediment and the cosmic dose impacting the sample: (a) Miscalculated $D_{ext}(t)$'s can dramatically affect the calculated ages, especially for the RU ages. For this tooth from a Mongolian cave, as the external dose rate increases, all the model ages decrease exponentially approaching 20 ka (EU), 25 ka (LU), and 33 ka (RU) at 2000 mGray/y. A 200 mGray/y (100%) change in the measured $D_{ext}(t)$ would introduce changes of approximately 4 ky (9–11%) in the calculated EU age and 10–14 ky (15–20%) in the LU ages. These are insignificant compared to the uncertainties in the age calculation. Even a 100 mGray/y misestimate is significantly greater than the 2σ uncertainty envelop for the RU age (after Blackwell et al., 2001b). (b) As sediment depth increases above a sample, the cosmic dose contributes less to the total external dose rate. For samples covered by 10 m of sediment, the cosmic dose is negligible (modified from Blackwell, 1995a). (c) At higher altitudes and higher latitudes, the cosmic dose increases (modified from Blackwell, 1995a). (d) Sedimentary water attenuates the external dose reaching the sample. As the sedimentary water concentration increases, the calculated standard ESR age increases for this tooth from a South African cave for all model ages. Generally, changing the sedimentary water concentration by <5 wt% does not significantly affect the calculated model ages, especially for samples where the external dose rate represents a small percentage of the total dose rate. If, however, the sedimentary water concentration increases by >10 wt%, the model ages will exceed the reported values by more than the 2σ uncertainty (after Blackwell et al., 2001c). (e) Growth in the time-averaged external dose rate, $\overline{D}_{ext,\gamma}(t)$ as a function of time for sedimentary bones that have been deposited for 500 ka. The external γ dose rate produced by bones in the sediment depends on the U uptake model assumed for the bones. In the case of the linear (LU) and recent (RU) model calculations, they also depend on the age assumed for the bone or tooth in the sediment. For a sedimentary bone component that is 500 ka in age, the dose rates calculated assuming early U uptake (EU) have almost reached those calculated assuming secular equilibrium (no U uptake) in the bone after 500 ky of U daughter ingrowth. The dose rates from LU and RU are much lower throughout the bone's history (after Blackwell & Blickstein, 2000). (f) The (maximum) $\overline{D}_{ext,\gamma}(t_b)$ production for sedimentary bones and teeth as a function of bone age, t_b. Because the LU and RU model U uptake rates depend on the age assumed for the sedimentary bones and/or teeth, their production rates never equal the SE model production rate (after Blackwell & Blickstein, 2000).

1. TL dosimeters placed in the site to measure the current external dose rate, $D_{ext}(t_0)$, from sedimentary β, γ, and cosmic sources over 0.5–2.0 years (e.g., Valladas & Valladas, 1987).

2. γ spectrometers measure the current dose rate, $D_{ext,\gamma}(t_0)$ from sedimentary γ and cosmic sources over 1–2 hours.

3. Bulk geochemical analysis (i.e., by NAA, XRF, α counting, etc.) using powdered sediment collected in conjunction with the sample (Blackwell & Skinner, 1999b) measures the U, Th, K, and other significant radioisotope concentrations in any layers which may have contributed to the sample's $D_{ext}(t)$ (Figs. 4 and 5). The radioisotope concentrations are used to mathematically calculate (e.g., Nambi & Aitken, 1986; Adamiec & Aitken, 1998) the current dose rates, $D_{ext,\gamma}(t_0)$ and $D_{ext,\beta}(t_0)$ which include corrections for β and γ dose attenuation due to mineral density, and backscattering (e.g. Baltakmens, 1975; Kalefezra & Horowitz, 1979). Such $D_{ext}(t)$ calculations also require a measurement for, or assumptions about, $D_{cos}(t)$, the cosmic dose rate (Figs. 8b, 8c; Prescott & Stephen, 1982) for samples buried less than 10 m and also the average sedimentary water concentration to correct for radiation attenuation by water in the sediment (Fig. 8d). In sites with thinly layered deposits or inhomogeneous sediment, $D_{ext}(t)$ calculations ideally should consider each unit or sediment component individually by determining volumetrically averaged dose contributions (e.g., Fig. 4b; Blackwell et al., 1992a, 1994a, 1994b; Blackwell & Blickstein, 2000; Brennan et al., 1997a).

4. An isochron age for a large sample obviates the need for a $D_{ext}(t)$ calculation, because it gives both the sample age, t_1, and $\overline{D}_{ext}(t)$, the time-averaged dose rate, simultaneously (Fig. 9; Blackwell & Schwarcz, 1993a; see below).

For adjacent U-rich or Th-rich layers or sediment components, the measurement or calculation is corrected for possible U uptake, U daughter isotope ingrowth, and potential Rn loss (e.g., Fig. 7; Blackwell & Blickstein, 2000; Blackwell et al., 2000, 2001e).

Assuming that $D_{ext}(t)$ has remained constant throughout the burial history, as many early ESR and TL studies did, may be naïve (Blackwell et al., 1994c, 2001a, 2001c; Rambaud et al., 2000). Changing water or radioactive element concentrations in the sediment (Fig. 8d; e.g., Ottey et al., 1997), increasing burial depth (Fig. 8b), or variable $D_{cos}(t)$ (Prescott & Hutton, 1994), among others, can all affect the $D_{ext}(t)$ experienced by the sample (Blackwell et al., 2001d, 2001e), requiring that $D_{ext,\beta}(t_0)$ and particularly $D_{ext,\gamma}(t_0)$ be corrected for such significant variations. At limnological sites where sedimentary water concentration variations can be significant, or where sediment accumulation or deflation can alter the depth of sediment cover, these considerations become significant, but not insurmountable.

In using geochemical analysis (i.e., NAA) at sites with very inhomogeneous sediment units (lumpy sites, *sensu* Schwarcz, 1994), the inhomogeneity in the dose field (Brennan et al., 1997a) requires volumetric analysis in which the contribution from each component depends on its abundance in order to calculate the actual contribution to $D_{ext}(t)$ from different sedimentary components or layers within the β and γ "spheres of influence". That still, however, does not consider the potential temporal variation in $D_{ext}(t)$ due changes in radioisotope concentrations within the sedimentary components. In lumpy sites, some

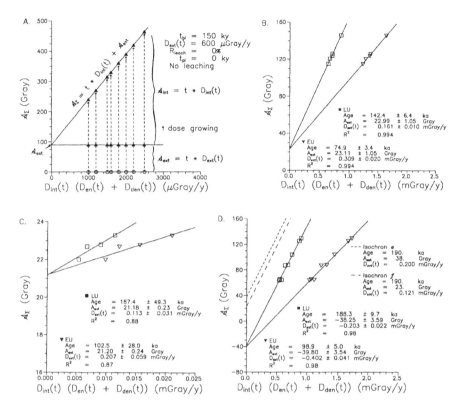

Figure 9. ESR isochrons. (a) A theoretical plot: When the total accumulated dose, $\mathcal{A}_{\Sigma,i}$, for each subsample, i, is plotted *versus* the time-averaged internal dose rate, $\overline{D}_{int,i}(t)$ the slope of the line gives the sample's age, t, while the y-intercept represents the external accumulated dose, $\mathcal{A}_{ext}$. After the first iteration of this technique, $\overline{D}_{int,i}(t)$, t, and $\mathcal{A}_{ext}$ all depend on the U uptake model selected (modified from Blackwell & Schwarcz, 1993a). (b) An isochron plot for a tooth from Bau de l'Aubesier, Provence: In practice, each uranium uptake model produces a line, which all converge on $\mathcal{A}_{ext}$. Isochron analysis can yield ages with uncertainties as low as 4% (after Blackwell et al., 2001d). (c) An isochron for a tooth from tufa deposit associated with a thermal spring and lake at Longola, Zambia: Despite measureable Th in the water, the isochron gives archaeological reasonable ages (modified from Blackwell et al., 1993). (d) An isochron for a tooth from Bau de l'Aubesier, Provence: If a sample, such as this tooth, has experienced U leaching, the isochron's intercept often becomes negative. In this example, the leaching event must have occurred recently, because the isochron age agrees well with ^{230}Th/^{234}U age on adjacent stalagmitic horizons. The accurate isochron age with a negative intercept argues agains proportional or diffusional uptake. Recent secondary uniform uptake following an initial LU event would imply Isochrons h or j as possible isochrons to model initial uptake event (after Blackwell et al., 2001d).

sedimentary components which may be able to absorb U (e.g., peat, teeth, bones, mollusc shells) can constitute a significant sedimentary fraction (Blackwell & Blickstein, 2000). If they can absorb U, $D_{ext}(t)$ will probably have changed, because:

1. Many sedimentary components only absorb U, but not all its daughters, which ingrow later (Fig. 7a).

2. If the uptake occurred early in the sediment's history, its effect will be greater than if it occurred recently. This requires that U uptake into the sediment be modelled analogously to that into teeth (see $D_{int}(t)$ models above; Fig. 7c).

3. U may have been leached from these components, requiring modelling to assess the effect on $D_{ext}(t)$ (e.g., Pike & Hedges, 1999; Fig. 7c) or Rn may have been lost (Fig. 7b).

4. More than one discrete uptake or leaching event may have affected these components (Fig. 7c).

These sedimentary processes can affect ESR, TL, OSL, and RL age calculations alike, by producing significant differences in the calculated $D_{ext}(t)$ (Blackwell & Blickstein, 2000; Blackwell et al., 2000, 2001a, 2001e). Therefore, the isochron method is preferred whenever possible, because the sample acts as its own dosimeter (Blackwell & Schwarcz, 1993a), theoretically compensating for inaccuracies due any change in $D_{ext}(t)$.

When isochron analysis cannot be used for dosimetry in sites with >2 vol% uraniferous sedimentary components, then the U uptake in those components much be modelled in order to correct $D_{ext}(t)$ measured in the modern sediment for the uptake. Blackwell & Blickstein (2000) proposed four limiting models to account for U uptake in sedimentary components. The EU, LU, and RU models are analogous to those used in calculating $D_{int}(t)$ for teeth, while the secular equilibrium (SE) assumes that all sedimentary components have secularly equilibrated U series isotopes, providing a maximum volumetrically averaged external dose rate, $\overline{D}_{ext,\gamma}^{VG,SE}(t_1)$. The presence of bones, teeth, mollusc shells, and peat in the sediment usually preclude this model, which is the one usually used for most $D_{ext}(t)$ calculations in sites without U-absorbing components. EU provides the maximum volumetrically averaged external dose rate, $\overline{D}_{ext,\gamma}^{VG,EU}(t_1)$, possible assuming some sedimentary U uptake, while RU yields the minimum volumetrically averaged external dose rate, $\overline{D}_{ext,\gamma}^{VG,RU}(t_1)$.

Precision depends on the method used to measure $D_{ext}(t)$ and relative concentrations of the radioactive elements. For γ and TL dosimetry, precision tends to average 3–10%, whereas for geochemical or volumetric sedimentary analysis, uncertainties normally range from 5 to 15%. Unlike for standard ESR analysis, precision for $D_{ext}(t)$ in isochron analysis will exceed that associated with the isochron age, because $D_{ext}(t)$ is derived from the age, rather than vice versa. The different measurement protocols do often yield different estimates for $D_{ext}(t)$ (Lyons et al., 1993; Blackwell et al., 2000, 2001a).

The isochron method

Isochrons have been applied to teeth (e.g., Blackwell et al., 1993, 1994a, 2000a), fault gouge minerals (e.g., Ikeya et al., 1995; Tani et al., 1996), and stalagmites (Karakostanoglou & Schwarcz, 1983). With the isochron method, a sample that can yield at least five subsamples is analyzed by standard ESR analysis. If the accumulated doses, $\mathcal{A}_{\Sigma,i}$, plotted against the time-averaged internal dose rate, $\overline{D}_{int,i}(t)$ for each subsample, i, give a straight line (Fig. 9a; Blackwell & Schwarcz, 1993a), its slope equals the sample's age, t_1, while the y-intercept yields the accumulated dose due to external sources, $\mathcal{A}_{ext}$, from which can be derived the time-averaged external dose rate, $\overline{D}_{ext}(t)$.

In teeth, the method gives a family of lines which converge on $\mathcal{A}_{ext}$, but whose ages and $\overline{D}_{ext}(t)$. each depend upon the U uptake model used to calculate $\overline{D}_{int,i}(t)$. (Figs. 9b and 9c). Tests have shown that, if the points fit a line well using a least-squares determination, the slope gives an age consistent with other dating methods (Blackwell et al., 1994b). The isochron method is limited to samples whose internal dose rate, $D_{int}(t)$, constitutes a significant fraction of $D_{\Sigma}(t)$, effectively requiring the sample to contain $\gtrsim$ 2 ppm U. Precisions for isochron analysis ages and $\mathcal{A}_{ext}$ can range as low as 3–4%, but normally tend to be less precise than standard ESR analyses, while minimum uncertainties for $\overline{D}_{ext}(t)$ tend to be ~5–6%.

If samples have lost U or experienced two uptake events, however, isochron analyses may give erroneous ages and/or $\overline{D}_{ext}(t)$ values (Fig. 9d). Millard & Hedges (1996), Millard & Pike (1999), Pike & Hedges (2000) all demonstrated that diffusional U leaching occurs in some bones. Combining ESR isochron enamel analysis with other dating methods has shown that secondary U uptake and leaching have affected the U concentrations in some fossil teeth, and hence, their isochrons (Blackwell et al. 2000, 2001a, 2001e). Blackwell et al. (2001e) postulated several types (Fig. 10):

Random: No pattern exists between the subsamples' U concentrations before and after the second leaching or uptake event. This will not produce viable isochrons.

Uniform (Equal):   All subsamples lost (for leaching) or gained (for secondary uptake) an equal amount of U, i.e.

$$U_i(t_1) = U_i(t_m) + x \qquad (3)$$

where t_m = the time at which the initial uptake event ends (see Fig. 7c),

$U_i(t)$ = the U concentration in tissue i or subsample i at time t,

x = the amount of U added to or removed from the tissue (ppm).

In teeth, each tissue may lose a different amount, but that amount is the same for each tissue.

Proportional: All subsamples lost (for leaching) or gained (for secondary uptake) a constant percentage of their preleaching U concentrations. For example,

$$U_i(t_1) = U_i(t_m) + \varphi \star U_i(t_m) \qquad (4)$$

where φ = the percentage of U added to, or removed from, the tissue (%).

Diffusional (Equilibrative, Dispersive, "Hot-atom", or "Hole-filling"): Not all subsamples absorb or lose the same amount or percentage of U. After equilibrative ("levelling", "hole-filling") diffusional leaching, uptake, or redistribution, the U concentrations show less spread than did those before the secondary event. After dispersive ("disequilibrative") diffusional or U leaching, uptake, or redistribution, the subsamples show more variation in their U concentrations than before the secondary event. "Hot-atom" diffusional leaching or uptake affects those subsamples with higher initial U concentrations more strongly than the less uraniferous subsamples (i.e., more uraniferous subsamples gain or lose more U secondarily). In "hole-filling" diffusional uptake, the subsamples with lower initial U concentrations gain, while those higher initial

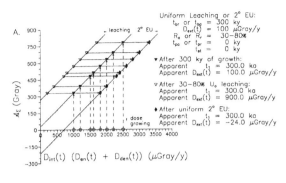

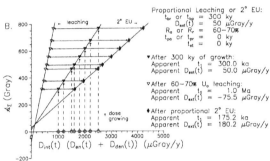

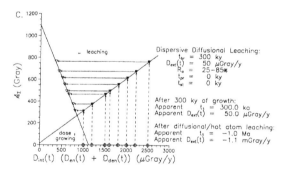

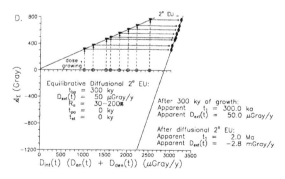

Figure 10. Simulated isochrons following secondary U uptake and leaching events. In the simulations here using data sets from real teeth, the initial uptake event allowed the tooth to accumulate dose for 300 ky, yielding initial isochron analyses which each would have accurately determined the age, t_1, external accumulated dose, $\mathcal{A}_{ext}$, and time-averaged external dose rate, $\overline{D}^I_{ext}(t)$, had another event not occurred. The secondary ("2°") uptake or leaching event was assumed to have occurred geologically instantaneously (EU or EL model, i.e., $t_r \cong 0$ ky), and very recently (i.e., $t_{pr} \cong 0$ ky), with no time gap between the end of the first uptake event and the secondary event (i.e., $t_{st} = 0$ ky, and $t_r = t_m \cong t_1$; adapted from Blackwell et al., 2001e). (a) Uniform (equal) U leaching and secondary EU: Each subsample gained or lost an equal amount of U (i.e., x ppm from the enamel, y ppm from the dentine, and z ppm from the cementum). All the isochrons, both before and after the secondary event, still accurately predict the tooth's age. The post-leaching isochron overestimated $\mathcal{A}_{ext}$ and $\overline{D}^I_{ext}(t)$, but the post-2° EU isochron underestimated $\mathcal{A}_{ext}$ and $\overline{D}^I_{ext}(t)$. (b) Proportional U leaching or secondary EU: Each subsample gained or lost an almost equal proportion of its initial U, here 60–70% from the associated tissues. The post-leaching isochron overestimated the tooth's age significantly, while underestimating $\mathcal{A}_{ext}$ and $\overline{D}^I_{ext}(t)$, whereas the post-2° EU isochron underestimated ages significantly, while overestimating $\mathcal{A}_{ext}$ and $\overline{D}^I_{ext}(t)$. Had exactly the same U amount been removed from all the subsamples, all three lines would have converged on the correct $\mathcal{A}_{ext}$. In real teeth, however, slight deviations from completely proportional loss or gain usually occur, because each subsample's tissues contain different U concentrations, and not all subsamples react to leaching or uptake at exactly the same rate. Such minor differences in the final U amounts gained or lost cause slight to moderate inaccuracies in the $\mathcal{A}_{ext}$ and $\overline{D}^I_{ext}(t)$ predicted by the isochron. If leaching causes slightly less U to be removed from the less uraniferous subsamples, the isochron's $\mathcal{A}_{ext}$ and $\overline{D}^I_{ext}(t)$ will underestimate the actual values, and may even be negative. Here, a 1% deviation in the leaching percentages produced a $\overline{D}^I_{ext}(t)$ that was low by 0.125 mGray/y. If the disproportionality were reversed, the leached isochron would slightly overestimate the actual values for $\mathcal{A}_{ext}$ and $\overline{D}^I_{ext}(t)$. If the secondary uptake causes the more uraniferous subsamples to gain slightly more U, the isochron will overestimate $\mathcal{A}_{ext}$ and $\overline{D}^I_{ext}(t)$. Here, a 6% variation in the uptake percentages caused $\overline{D}^I_{ext}(t)$ to be 0.130 mGray/y too high, whereas reversing the disproportionality causes slightly underestimated and possibly negative values for $\mathcal{A}_{ext}$ and $\overline{D}^I_{ext}(t)$. (c) Equilibrative diffusional ("hot atom") U leaching: Leaching selectively removed more U from subsamples with higher U concentrations. Since the apparent post-leaching $D_{int,i}(t)$ significantly underestimated the preleaching $D_{int,i}(t)$, the final isochron yielded a negative age and $\overline{D}^I_{ext}(t)$, while overestimating $\mathcal{A}_{ext}$ by > 1000 Grays. (d) Equilibrative diffusional ("hole-filling") secondary EU: The secondary uptake event added proportionately more U to the less uraniferous subsamples, decreasing the U concentration variation. Since the apparent post-leaching $D_{int,i}(t)$ overestimated the preleaching $D_{int,i}(t)$, the post-leaching isochron overestimated the age significantly, while giving a negative $\mathcal{A}_{ext}$ and $\overline{D}^I_{ext}(t)$.

U concentrations may gain little or no U. In "hole-filling" redistribution, those with initially higher U concentrations lose U to the less uraniferous subsamples.

For all such events, any of the uptake models (EU, LU, RU, or anything between) or leaching models (EL, LL, RL, or an intermediary model) might pertain (Fig. 7c). If analyzed by isochrons, a tooth which has experienced uniform, proportional, or diffusional leaching for its secondary event will still produce a regular isochron (Figs. 9d and 10). Although such an isochron may be precise, it will not likely be fully accurate. That more than 75% of teeth produce geologically reasonable isochrons suggests that most teeth experience non-random uptake, even if those uptake (and loss) events occur in several discrete episodes, and even if the rates of uptake (or leaching) vary from event to event (Blackwell et al., 2001a).

Because the isochron method averages $D_{ext}(t)$ over the entire burial history, isochron analysis automatically corrects for changes in $D_{ext}(t)$, which may have occurred. By greatly reducing the need to measure $D_{ext}(t)$ *in situ* or to assume that it has remained constant, it can date samples from environments where $D_{ext}(t)$ are likely to have changed in response to complex sedimentological changes, such as lacustrine environments. Isochrons can also date samples from sites that have been destroyed or are otherwise inaccessible, especially samples in museum collections (e.g., Blackwell, 1994).

If an independent method (e.g., TL or γ dosimetry) calculates $D_{\text{ext}}(t)$, and if it can be shown to have been constant throughout time at the site by geological investigations or by an independent date, the isochron method instead can determine the U uptake history (eg. Blackwell et al., 2001e). Since the isochron calculation gives $\mathcal{A}_{\text{ext}}$, which must equal the product of the age, t_1, with $\overline{D}_{\text{ext}}(t)$ the isochron's slope that matches this age represents the "correct" isochron and uptake model for the sample.

ESR microscopy and other new techniques

In ESR microscopy, an ESR spectrometer has been modified to scan across the surface of a solid material to determine the spin concentrations for a preset signal at various points on the surface. With specialized analytical programs, 2D, 3D, and 4D ESR imaging is now possible, some of which are combined with other systems, such as NMR (nuclear magnetic resonance) and CT (computerized tomography; e.g., Hara & Ikeya, 1989; Miki, 1989; Eaton et al., 1989; Miyamura & Ikeya, 1993; Sato et al., 1997; Sueki et al., 1993; Yamamoto & Ikeya, 1994; Miki et al., 1996; Oka et al., 1997). ESR microscopy is still being explored to understand its full potential, but it shows great promise in studying fossil diagenesis (e.g., Hochi et al., 1993; Ikeya, 1993a, 1993b), mapping crystal growth and defects (e.g., Omura & Ikeya, 1995), among other applications. Currently, it works best for materials with very strong ESR signals, such as tooth enamel (e.g., Hochi et al., 1993; Oka et al., 1997), bone (Schauer et al., 1996), coral (Ikeda et al., 1992b), gypsum (Omura & Ikeya, 1995), mollusc shells (Ikeya, 1994a, 1994b), aragonite, barite, and gypsum (Furusawa et al., 1991).

Portable ESR dosimeters and spectrometers can be used in assessing nuclear radiation accidents on site (e.g., Ishii & Ikeya, 1990; Ikeya & Furusawa, 1989; Yamanaka et al., 1991, 1993; Oka et al., 1996). Geoscientists can also use them in the field, where such technology would help to recognize reworked fossils, to aid in selecting the best samples for dating, and to assess the effect of site inhomogeneity on the samples. Eventually, such technology may even allow preliminary age estimates while still in the field.

Applications and datable materials in limnological settings

Within limnological settings, ESR can date materials that might provide valuable insight into a lake's history. Dating volcanic ash, authigenic carbonates or salts can delineate depositional histories and rates. Dates on authigenic cements may date diagenetic events or hydrological changes. Dating fossils, such as molluscs, diatoms, teeth, and phytoliths, in turn can date changes in biological diversity and water chemistry. Dating burnt flints or hearth sands from archaeological sites or fossils from associated basin terrestrial deposits can indicate the age for associated geomorphic surfaces and hint at lake geochemical histories. All fossils, especially loose teeth and fish scales, however, may have been reworked into younger stratigraphic units (Fig. 11; Blackwell, 1994).

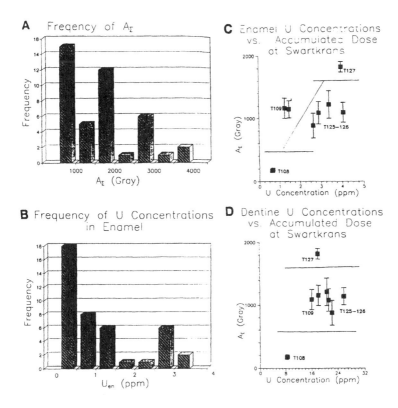

Figure 11. Tests to check for reworked fossils. For teeth from Swartkrans, South Africa: (a) The accumulated dose ($\mathcal{A}_\Sigma$) histogram clearly reveals at least three different populations of teeth. (b) The enamel U concentration histogram shows at least two populations. (c) Plotting $\mathcal{A}_\Sigma$ vs. enamel U concentration reveals four distinct populations. (d) Plotting $\mathcal{A}_\Sigma$ vs. dentinal U concentration shows three different populations well separated from each other. Such plots delineate populations of teeth that have experienced different environmental conditions, one indication for reworking among samples from the same units (after Blackwell, 1994).

Molluscs, ratite eggs shells, ostracodes, and other carbonate fossils

Dated mollusc shells (Table III) can provide diverse information for Quaternary lacustrine studies, as can dated ostracodes, foraminifera, and ratite eggshells which can occur in association with lake basin sediment. Mollusc shells, however, act as open systems for U (Blackwell & Schwarcz, 1995), although the moderate discordance between measured $^{230}Th/^{234}U$ and $^{231}Pa/^{235}U$ ratios suggests that most U uptake accompanies initial sedimentation or very early diagenesis.

Aragonitic mollusc shells normally show five ESR peaks (Fig. 12), but calcitic molluscs have more complex spectra. For the peaks at $g = 2.0018$, 1.9976, and 2.0007, trap density is related to Mg/Ca ratios (Mudelsee et al., 1992), which can change as diagenesis, secondary mineralization, and fossilization progresses, making them unsuitable for dating

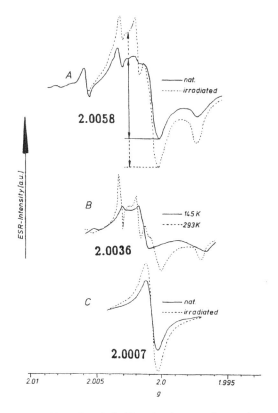

Figure 12. ESR spectra in aragonitic mollusc shells. Three signals commonly occur in aragonitic mollusc shells (modified from Blackwell, 1995a): (a) The signal at $g = 2.0058$ before and after irradiation measured at room temperature. (b) The signal at $g = 2.0036$ measured at room temperature (293 °K) and at 145 °K. (c) The signal at $g = 2.0007$ before and after irradiation measured at room temperature.

some species. Secondary mineralization in the fossils can cause interference problems that affect $\mathscr{A}_\Sigma$ measurement and age calculation. Signal lifetimes vary significantly depending on the peak and species (e.g., Barabas et al., 1992a, 1992b; Table 1 in Blackwell, 1995a). Generally, either the peaks at $g = 2.0014$ and 2.0007 in calcitic shells and the peak at $g = 2.0007$ in aragonitic shells are the most reliable, but that must be tested for each species individually, because complex peaks can occur (e.g., Brumby & Yoshida, 1994a) and peaks other than that at $g = 2.0007$ may be light sensitive (Bartoll et al., 2000). Some species show inflection points in their growth curves, making it difficult to select an appropriate set of added doses for measuring $\mathscr{A}_\Sigma$. Schellmann & Radtke (2001) advocated using a plateau technique with 40–60 irradiation steps to maximize accuracy in the growth curves.

Petrographic or geochemical analysis should accompany any ESR date to avoid remineralized and recrystallized samples. Contamination from Mn peaks often requires overmodulation to discriminate the dating peaks. Due to U uptake, modelling is required for samples that cannot be analyzed by coupled ESR-^{230}Th/^{234}U dating. In some fresh and hypersaline systems, the $(^{234}U/^{238}U)_0$ ratio may also need to be measured or modelled. For each species and signal, the α efficiency factor, k_α, must be measured (Lyons & Brennan, 1991; Skinner & Weicker, 1992; Grün & Katzenberger-Apel, 1994). Long-term signal fading may also need to be considered, depending on the peak and its thermal stability (Molodkov, 1989).

Specimens found in life position give the most reliable results, although that does not guarantee that reworking has not occurred. Larger species are preferred so that each subsample represents a single individual (Table II), but several shells can be combined from a smaller species, assuming that none have been reworked. Fragmentary samples still need to be speciated. Since species effects do occur, submitting two or three different species from each unit can increase dating precision and accuracy. For untested species, ~100 g of pristine shell are needed to perform the necessary signal stability and calibration tests (e.g., Brumby & Yoshida, 1995).

Applications in lacustrine systems are rare as yet, but terrestrial and freshwater molluscs do give reasonable ESR ages. For example, Skinner & Mirecki (1993) found good agreement between ESR, TL, ^{14}C, and AAR (amino acid racemization) ages for *Hendersonia* and *Allogona* using $g = 2.0007$. Molodkov (1996) measured ESR ages in *Lymnaea baltica* from Ancylus Lake, Estonia, and the brackish *Cerastoderma glaucum* using the $g = 2.0012$ peak. Molodkov (1993) reported that thermal stabilities in *Monauha caucaicala* significantly exceeded those in marine molluscs.

In ratite egg shells, Kai et al. (1988) identified two signals with good sensitivity. Wieser et al. (1994b) detected doses as small as 0.3 Grays, suggesting that this technique may prove very sensitive for ostrich, emu, and other ratite shells, which are often found associated with lacustrine sediments in Africa and Australia. Regular application must await calibration tests against other dating methods, but Schwenninger & Rhodes (1999) dated extinct birds in Madagascar.

In theory, ostracodes and stromatolites should also be datable, assuming that trace contaminants do not obscure the dating peaks. No one has yet attempted these, but preliminary tests on echinoderms (Yamamoto & Ikeya, 1994), *Halimeda* (Schramm & Rossi, 1996), and success with foraminifera (e.g., Barabas et al., 1992a, 1992b) suggest that most carbonate fossils have datable peaks waiting to be utilized.

Authigenic carbonates: Travertine, calcrete, caliche

Travertine, which is commonly precipitated in swamps, shallow hypersaline lakes, and around springs, contains calcite or aragonite, but also high organic concentrations that can add organic interference peaks. Nonetheless, several ESR dates have been attempted (Table I). Travertine and other authigenic carbonates often contain proxy data regarding specific climatic conditions (Hennig et al., 1983c; Martinez Tudela et al., 1986), and may coincide with prehistoric human activities.

How post-sedimentary processes affect the ESR signals in authigenic carbonates (Table I in Blackwell, 1995a) is not well understood. Although most travertine spectra (Fig. 13) resemble those for speleothems, which have been extensively studied, other peaks do occur. The humic acid signal at $g = 2.0040$ does not appear accurate (Grün, 1989a, 1989b). In Mn-rich samples, the peak at $g = 2.0022$ yielded reliable ages (Grün, 1989a, 1989b), but needs

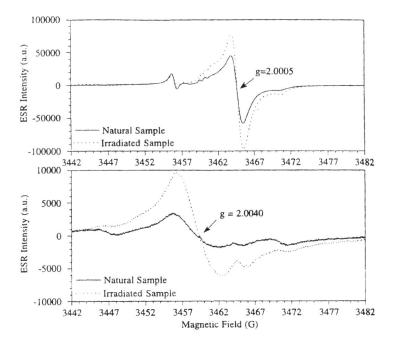

Figure 13. ESR spectra in tufa and travertine. In tufa and other slowly precipitated carbonates, the ESR spectra can vary dramatically, often due to interference signals from included organic matter, contaminant minerals, and trace elements. The signal at $g = 2.0005$ has yielded reliable dates. The broad peak at $g = 2.0040$ hides interference signals that make its use for dating unreliable (modified from Grün, 1997).

more tests for annealing behaviour and replicability before general application. The most reliable peak occurs at $g = 2.0007$ (Huang et al., 1989; Idrissi et al., 1996), but peaks other than that at $g = 2.0036$ may show light sensitivity (Bartoll et al., 2000). Although many authigenic carbonates lack the peak at $g = 2.0007$, carefully sampling densely crystallized calcite can increase the success rate (Grün et al., 1988b). For some calcrete and caliche, contamination causes complex interference signals that affect accuracy, but preannealing samples before analysis may improve the results (Chen et al., 1989a, 1989b, 1994).

Because authigenic carbonates often have experienced remineralization, secondary mineralization or cementation, petrographic, SEM, XRD, or similar analyses should complement the ESR dating analysis to ensure viable geological conclusions. Otherwise, sample preparation is fast, requiring only powdering and a dilute acid leach to remove any transitory peaks induced by the grinding.

Few ESR studies have systematically examined lacustrine travertines or authigenic carbonates. ESR ages for lacustrine travertine from the hominid site at Weimar-Ehringsdorf (Schwarcz et al., 1988b) compared well with U series ages, confirming that the sequence was probably precipitated during Isotope Stages 5 to 7. At Beceite, Spain, ESR ages agreed well with U series ages, if samples were selected to minimize contaminant mineral concentrations (Grün et al., 1988b). For Jordanian travertine, Wieser et al. (1993) found that ESR and TL accumulated doses did not agree, but the fault could lie within either method, as was true for the lack of agreement between TL and ESR ages for some Turkish pisolites (Özer et al., 1989). Although measuring the signals can be challenging, ESR dating in travertines and other authigenic minerals shows great potential and should provide valuable information if tested more extensively. Modern signal subtraction and Q-band studies might solve some of the enigmas.

Hydroxyapatite (HAP): Vertebrate fossils and crustacean chitin

ESR analysis can date hydroxyapatite (HAP) because a single radiation-sensitive ESR signal occurs at $g = 2.0018$ in fossil, but not modern enamel (Fig. 14a; Table I and II). Currently, most labs use mammal enamel. Theoretically, reptilian enamel should work, but this needs to be verified by extensive testing before general applicability can be assumed. Bones, dentine, some fish scales, and crustacean chitin also show the same signal (Fig. 14b) which grows similarly to that in tooth enamel (Stuglik & Sadlo, 1995). In tissues other than enamel, the signals do not fade (Stuglik et al., 1994), but their low sensitivity causes very low signal intensity unless the sample age approaches 0.8–1 Ma. Since diagenetic alteration in bone also complicates its use, bone dating has largely been abandoned in favour of enamel. Analytical protocols for fish scales have been developed (Blackwell et al., 1994a), but need further testing. Unfortunately, only enameloid scales, such as those from gar (*Lepisosteus*), can be dated. In addition to interference problems, other scales do not appear to give sufficiently large signals for accurate dates. ESR dates must consider U uptake models and ingrowth by U daughters for HAP, as well as possible Rn loss and U leaching (Fig. 7).

In HAP, the long ESR signal lifetime, $\tau \sim 10^{19}$ y (Skinner et al., 2000), guarantees its utility. In mammals, its radiation-sensitivity does not depend on species (Serezhenkov et al., 1996), but does depend on the crystallinity which is affected by the animal's age and health (Brik et al., 1996; Lau et al., 1997; Skinner et al., 2001a). In deciduous teeth (i.e., "milk" teeth), poorly crystallized HAP causes analytical problems. Although signal saturation depends on the sample's U concentration, saturation in enamel generally does not occur before the tooth is ~5–10 Ma. Teeth as old as 4.0 Ma have been dated successfully (e.g., Blackwell, 1994; Huang et al., 1995; G. Chen et al., 1996). Although some teeth as young as 8–10 ka have been dated, dosimetry experiments suggest teeth with doses of ~0.05–0.1 Gray may be datable (Egersdörfer et al., 1996b; Haskell et al., 1997b). Currently few attempts have made to date sites younger than ~25–30 ka (~2–5 Gray), because ^{14}C dating is usually used instead.

The standard ESR method (i.e., not isochrons) for tooth enamel has now been tested extensively against other dating methods for sites in the age range 30–300 ka (see Table I in Blackwell, 1995a), but for teeth >300–400 ka, only a few calibration tests have been attempted (eg. Skinner et al., 2001b). Archaeological applications have been extensive (e.g., Table I). Despite recent calls for much more complex measurement protocols (e.g., Grün, 1998a, 1998b; Jonas, 1995, 1997; Jonas & Grün, 1997; Fattibene et al., 2000; Vanhaelewyn et al., 2000), tests using the Q-band spectrum indicate that, although the peak is complex, it grows uniformly and can be accurately measured by a simple peak height measurement (Skinner et al., 2000). Human dosimetry experiments (Table I) have hinted

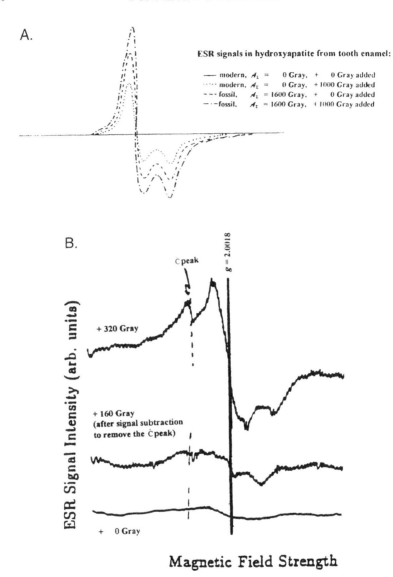

Figure 14. The hydroxyapatite (HAP) ESR spectrum in vertebrate tissues. In most modern tissues, the HAP ESR signal has zero intensity. The exceptions are those that have experienced a nuclear accident. (a) If modern tooth enamel experiences irradiation, a measureable signal will appear after 0.01 Grays exposure, making it a useful signal for monitoring dose exposure during nuclear accidents. In a fossil tooth, a measureable signal is present after 10–20 ka, depending on the total dose rate that the tooth experienced. When a fossil tooth experiences irradiation, the signal will grow larger. (b) In these *Lepisosteus platostomus* (gar fish) scales from the Sangamonian lake at Hopwood Farm, IL, low signal intensity in the natural sample (lower) makes the signal difficult to discern, but artificial irradiation reveals the distinctive hydroxyapatite signal at $g = 2.0018$, along with a carbon radical signal that partially interferes with the dating signal (after Blackwell et al., 2001d).

at possible problems with interference, temperature sensitivity, and signals induced by grinding and UV light exposure. Several researchers have suggested complex preparation techniques to compensate for these problems (e.g., Onori et al., 2000; Ramanyukha & Regulla, 1996; Wieser et al., 2000a), but their effect on teeth older than 10 ka must be minimal or the ESR age would not agree with those from other dating methods. While standard ESR can still be improved methodologically, such as by fully understanding U uptake, this does not hamper its application to many geological sites. In several sites, the dental U concentrations have been so low that all the model ages were statistically identical (Fig. 8a; e.g., Lau et al., 1997; Rink et al., 1995, 1996a, 1996b; Blackwell et al., 2001b).

For the isochron method in enamel, calibration tests have been completed against ^{230}Th/^{234}U and standard ESR (e.g., Blackwell et al., 1994b, 1994c, 2001a, 2001e). Currently, calibration tests between isochrons against ^{40}Ar/^{39}Ar are in progress (Skinner et al., 2001b). More than 80% teeth yield isochrons with good regression statistics, of which only 20% definitively indicate secondary U uptake or leaching. A lack of agreement between standard ESR and isochrons may not necessarily imply problems with the isochron method, but changes in $D_{ext}(t)$ or the U uptake rate (Fig. 9, 10; Blackwell et al., 2000, 2001a, 2001c, 2001e).

For enamel dating, large, relatively intact molars and premolars from large herbivores make the best specimens, because the isochron method can be applied as well as standard ESR. Smaller teeth are analyzed with the conventional ESR technique or the ramped dosing technique, but the enamel must be completely separated from the dentine manually. For small teeth, several teeth from the same jaw can be attempted for isochron analysis. ESR dating does not require taxonomically identified teeth. Fragmentary teeth are fine, providing enough enamel and dentine remains for analysis (Table II). For example, one or two mammoth molar plates usually provide enough enamel for an isochron.

For bones, dentine, ivory (mixed dentine and enamel), and antler, the ESR dating method is more difficult to apply and has not been particularly successful in most attempts. Their low signal sensitivity causes, if nothing else, a much higher minimum age limit. For dentine, tests suggest that sensitivity problems might be overcome by using it to date teeth > 500 ka. Diagenetic minerals in dentine cause few problems, except further lowering sensitivity (Skinner et al., 2000). In bone, tusk, and antler, contaminants and secondary mineralization can also complicate the signal measurement. Since all can absorb significant U, uptake modelling becomes even more essential in determining accurate dates. Crustacean shell chitin shows a typical HAP signal (e.g., Desrosiers, 1996; Stewart & Gray, 1996), but the method needs development to determine if it might be applicable to other chitinous species. More research is needed before these will become routine applications.

Blackwell et al. (1993) dated teeth from a tufa mound associated with a hot spring and small pond at Longola, Zambia (Fig. 9b). Using gar fish scales, Blackwell et al. (1994a) reported ages of ~200 ka for gyttja deposits in a Sangamonian lake deposit at Hopwood Farm, IL, and ages of ~80 ka for an mastodon molar from the deposits associated with the lake's final evaporation. For the type Cromerian freshwater beds at Runton, Rink et al. (1996c) got reliable $\mathscr{A}_\Sigma$, but not surprisingly, had trouble assessing changes in water concentrations over time, since they did not use isochron analysis. ESR dates for a tooth from the paleolake at Üröm, Hungary, date the Templomhegy mammal substage to 670 ± 110 ka (Fig. 6a; Berman et al., 2000).

Heated silica: Volcanic ash, volcanic rocks, baked sediment, burnt flint and chert

Volcanic eruptions and meteorite impacts can form lakes, while their basins may preserve volcanic ash, tektites, and baked sediment. Volcanoes are associated with many rift valley lakes, but few ESR applications have been attempted. Heated chert and flint artefacts occur in archaeological sites associated with Late Pleistocene and Holocene lakes.

Quartz and silica exhibit several radiation-sensitive ESR signals (Fig. 15; Porat & Schwarcz, 1991; Skinner & Rudolph, 1996a). Due to the Ti and Ge signals' low sensitivity, fast saturation, and propensity to bleach (e.g., Buhay et al., 1988; Ulusoy & Apaydín, 1996; Woda et al., 1999), most studies use the OHC (oxygen hole centre), E', or Al signals. Some samples do require signal subtraction to remove trace contaminant interference signals (Fig. 6b; Skinner & Rudolph, 1996a). Because quartz does not absorb U over time, their age calculations do not require modelling for U uptake like tooth enamel. To provide meaningful dates, any preexisting geological signals, however, must have been zeroed completely during the limnologically related event (Fig. 2b). In some flint, an unbleachable component may survive typical heating (Skinner, 2000a). Porat & Schwarcz (1995) measured $\tau \leq 100$ y for the E' and Al signals, but heated flints show much longer lifetimes, suggesting that the signals' kinetics may change on heating. A short-lived interference signal, E'_1 with $\tau = 40$ y, can interfere with the E' signal for some heated quartz samples (Toyoda & Schwarcz, 1996, 1997a, 1997b), complicating dating for volcanic rocks and impact craters.

For burnt flint, chert, and sand, the ESR method is still in its infancy (Tables I and II). More investigations are needed, including calibration tests against other methods. The precision for $\mathcal{A}_{\Sigma}$ values from ESR compares well with those obtained from TL on the same materials (Aitken, 1985). Flints and cherts as young as 10–20 ka may be datable (Skinner & Rudolph, 1996b), but the maximum dating limit, which depends on the flint type, has not yet been well established. Schwarcz & Rink (2001) suggested a new ESR-TL technique for flint that may more effective than either technique individually. Applications to dating burnt sand and volcanic ash are even less advanced, but theoretically feasible.

Strained quartz and feldspar: Fault gouge, mylonite

In many lacustrine systems, faults bound the lake basin. ESR can date the most recent, and sometimes several earlier, fault movements (Fig. 16), allowing complex basin histories to be unravelled. In Japan, the technique has been widely applied to numerous faults (Table I), but few directly associated with lake basins.

In dating gouge, strain zeroes the signals in the gouge minerals (Fig. 3). Several grain sizes must be tested to ensure that the signals have been completely reset (e.g., Lee & Schwarcz, 1994a, 1994b, 1996). Most researchers use the E', OHC, or Al signals in quartz (Fig. 15) or occasionally feldspar, but the grains must be selected by hand after heavy mineral separation and HF leaching to ensure that only gouge minerals with no secondary overgrowths are used. Lee (2001) advocates using at least two signals to ensure accuracy.

Quartz zeroed by light: Beach sand, loess, fluvial sediment

If a radiation-sensitive ESR signal found in quartz can be completely zeroed by exposure to strong light, as can the Ge signal (Fig. 2a), then its deposition in a subaerial, shallow

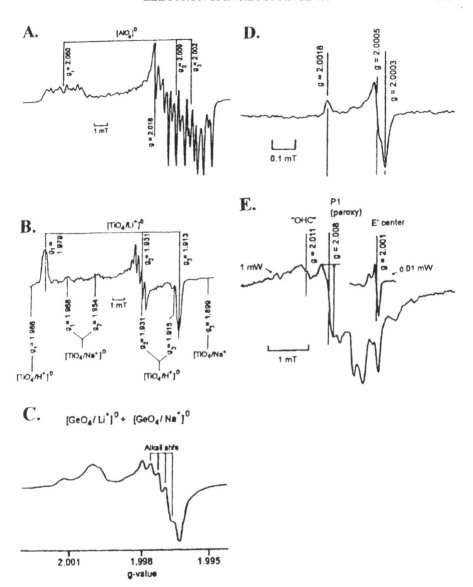

Figure 15. ESR signals in quartz. Several signals occur in quartz, flint, and fault gouge minerals (modified from Rink, 1997): (a) The aluminum (Al) signal, often used for dating fault gouge, must be measured at 70 °K. An $(AlO_4)^0$ defect causes this signal. (b) The titanium (Ti) signal, which has not been used often for dating arises from $(TiO_4/H^+)^0$, $(TiO_4/Li^+)^0$, $(TiO_4/Na^+)^0$ defects. (c) Because the germanium (Ge) signal is more easily bleached than most other signals in many quartz samples, it is used for dating quartz sediment. This complex signal arises from overlapping $(GeO_4/Li^+)^0$ and $(GeO_4/Na^+)^0$ defects. (d) The E' signal at $g = 2.0005$ is easily measured at room temperature to date quartz, flint, and fault gouge, but a secondary interference signal can occur at approximately the same g value. (e) The complex oxygen hole centre (OHC) signal and the P1 peroxy signal are also measured at room temperature. OHC has been used to date quartz, flint, and fault gouge.

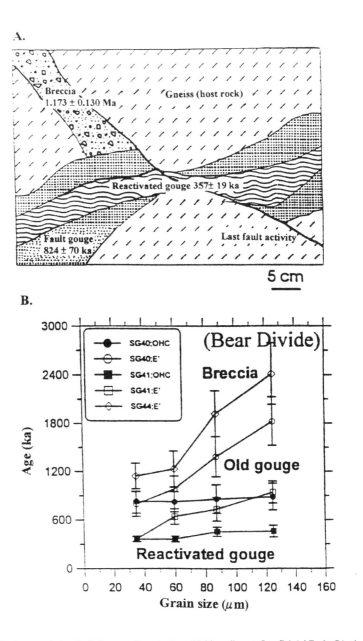

Figure 16. Fault gouge dating. In fault gouge from the Bear Divide sediment, San Gabriel Fault, CA, the gouge records several periods of activity (after Lee & Schwarcz, 1994a): (a) At least three earthquakes occurred in this outcrop at 357 ± 19, 824 ± 70, and 1173 ± 130 ka. (b) Plotting the ESR ages vs. grain size shows different plateaux in old and reactivated fault gouge.

fluvial or marine environment can be dated. As yet, it remains controversial whether any signal is completely zeroed during natural deposition (e.g., Tanaka et al., 1995, 1997). If sediment does not bleach completely, then any ages become maximum ages. Since most applications attempted thus far have used dubious analytical techniques (e.g., Huang et al., 1988; Monnier et al., 1994; Table I), it is difficult to assess if the results are fortuitous or genuine. Although general use of these techniques awaits several necessary theoretical studies, the recent successes with TL and OSL using similar sediment suggest that the potential exists here for many applications.

Authigenic quartz: Phytoliths, diatoms, cement, laterite, and silcrete

Both diatoms and phytoliths, which occur as fossils in lacustrine sediment, theoretically should be datable by ESR. Ikeya & Golson (1985) completed preliminary studies using phytoliths, but no one has yet followed these up. Inherently, diatoms should also have radiation sensitive signals similar to those in other quartz samples (Schwarcz, 1994). In both, the ESR signals should be zero when the crystals form, thereby eliminating the problem of incomplete zeroing seen in other quartz applications.

Were one able to date laterite and silcrete in a lacustrine basin, much geomorphic information might be discovered. In Goa, Nambi & Sankaran (1985) used the signal at $g = 2.0028$ to estimate a geologically reasonable average laterization age, but Radtke & Brückner (1991) found the growth curve was non-linear for silcrete samples, suggesting a complex signal. Diagenetic alteration and secondary cementation may complicate these applications. Nonetheless, all these have potential that should be developed further.

Clay minerals

Several clay minerals have viable ESR signals (Angel et al., 1974; J. P. E. Jones et al., 1974; Muller & Callas, 1989). Both kaolinite and montmorillonite have an OHC signal associated with their silicate layers. In the latter, the stability, $\tau = 10^7$ years at surface temperatures (Fukuchi, 1996a), suggests that its applicability for dating should include at least the Middle and Late Quaternary. Montmorillonite also has a radiation-sensitive carbonate signal (Fukuchi, 1996a), but with even lower stability. Radionuclides located in associated Fe-oxides caused the signals in kaolinite, which Muller & Callas (1989) used to fingerprint and source the clay. Fukuchi (2001) has tried using the OHC signal in montmorillonite to date Japanese faults.

Other salts: Dolomite, gypsum, gypcrete, halite, sulphates

Dating salts can provide detailed information about basin filling and remineralization. Since salts frequently experience remineralization, secondary mineralization or cementation, they require petrographic or geochemical checks to ensure accurate ages.

Several salts have strong ESR signals (Alcalá, 1996; Table I in Blackwell, 1995a). Strong radiation-sensitive signals in other carbonates, sulphates, and phosphates all show potential to be developed into viable techniques. Useful signals may also exist in rare salts with analagous geochemical formulae, but few have been examined. Success may hinge on the salts' purity, since the organic radicals common in some subaerially precipitated salts tend to interfere with dating signals.

As yet, ESR dating has been attempted only for gypsum, anhydrite, halite, monohydro-calcite, dolomite, and barite, but not with unqualified success. In gypsum, the $g = 2.0082$ signal gives the best results (Ikeda & Ikeya, 1992; Y. J. Chen et al., 1989b). Y. J. Chen et al. (1988a) dated Chinese gypcrete by using the known ^{14}C ages to determine the α efficiency factor, k_α. Preliminary results on salt deposits indicate that signal intensities increase with sampling depth (Ikeya & Kai, 1988), but agreement with other dating methods has been poor. In dolomite, contamination, possibly from humic acids, may complicate the signal measurement and affect accuracy (Radtke et al., 1988b; Y. J. Chen et al., 1989b, 1992), while interference signals in monohydrocalcite have been attributed to organic radicals (Debuyst et al., 1999). Attempts to use gypcrete were hampered by the difficulties in obtaining sufficient sample for adequate growth curves (e.g., Y. J. Chen et al., 1988a), but Kohno et al. (1996) measured an accumulated dose in a barite desert rose. Once the idiosyncrasies in sample preparation have been standardized, these applications should provide interesting details about lacustrine systems.

Other applications

Other applications include using ESR imaging systems to explore mineral (e.g., Gotze & Plotze, 1997) and fossil growth and diagenesis (e.g., Grün et al., 1997c; Tsukamoto & Heikoop, 1996). Omura & Ikeya (1995) used ESR microscopy to map gypsum crystal growth. Similar techniques could theoretically be applied to other salts (Ikeya, 1994a, 1994b). In a rather simplistic approach, Yugo et al. (1998) proposed a model for paleowind patterns based on the provenance of aeolian quartz dust as determined by the ESR intensity.

Summary

Electron spin resonance (ESR) dating has been developed for many materials, including hydroxyapatites in enamel, bone and some fish scales, aragonite and calcite in travertine, molluscs, and calcrete, and quartz from ash, which have many potential applications in lacustrine settings. Although the complexity of the signals in some materials has hampered routine application, research is solving these problems to make the method more widely applicable. When tested against other dating techniques, age agreement has usually been excellent. Generally, the most reliable applications seem to be tooth enamel, quartz, some mollusc species and calcite deposits.

ESR dating uses signals resulting from trapped charges created by radiation in crystalline solids. Ages are calculated by comparing the accumulated dose in the dating sample with the internal and external radiation dose rates produced by natural radiation in and around the sample. For fossils and authigenic minerals, no zeroing is necessary to obtain accurate ages. In sediment which contain reworked mineral clasts, such as quartz, ESR signals must have been zeroed in order to give the correct age for sedimentation. High pressure, heating, and from some minerals, light exposure and grinding can zero an ESR signal. For materials that absorb uranium (U) during their burial history, such as teeth, bones, or mollusc shells, the age calculation considers their U uptake by cross-calibrating with U series or U/Pb dating or by assuming different uptake models. Some difficulties in calculating the external dose rate can be overcome by applying the ESR isochron method, in which the sample acts as its own dosimeter. In lacustrine environments, changes in the external dose rate due to altered sediment cover, sedimentary water concentrations and mineralogy also need to be considered. While ESR applications in lacustrine settings have been few, many uses are currently available. The next few years should see development and testing for several more applications.

In lacustrine settings, one must expect that changing sedimentary water concentrations, secondary leaching or addition of U or Th in the sediment, and changing cosmic dose rates in response to basin infilling will affect the external dose rates. Therefore, accurate dates must consider these phenomena carefully. While this complicates the age calculations, ESR can still provide accurate dates for many materials found in or near lacustrine environments, including teeth, egg shells, mollusc shells, burnt flint, fault gouge, and theoretically for foraminifera, phytoliths, diatoms, and ostracodes.

Acknowledgements

Some examples cited herein were analyzed thanks to support from the National Science Foundation. Over the years, Barry Brennan, Jack Rink, Bill Buhay, Rainer Grün, Michel Barabas, Darren Curnoe, Eddie Rhodes, Ed Haskell, Anotoly Molodkov, Albrecht Wieser, Ulrich Radtke, Galena Hütt, Glen Berger, Anne Wintle, Martin Aitken, Gerd Hennig, John Dennison, Andrew Pike, Christien Falguères, Shin Toyoda, Mimi Divjak, Hee Kwon Lee, Daniel Richter, Hélène Valladas, Ruth Lyons, Nomi Porat, and especially Henry Schwarcz have provided valuable insights and discussions about ESR dating. Anne Skinner and Joel Blickstein not only stimulated many discussions on ESR dating, they provided many useful comments on this manuscript. L. Provencher, T. Sani, H. Leung, S. Berman, and J. Blickstein assisted with manuscript preparation. Bill Last, John Smol, and an anonymous reviewer provided excellent suggestions to improve the work.

References

Abeyratne, M., N. A. Spooner, R. Grün & J. Head, 1997. Multi-dating studies of Batadomba Cave, Sri Lanka. Quat. Sci. Rev. (Quat. Geochron.) 16: 243–255.

Abragam, A. & B. Bleaney, 1986. Electron Paramagnetic Resonance of Transition Ions. Dover, New York, 911 pp.

Adamiec, G. & M. Aitken, 1998. Dose rate conversion factors: Update. Ancient TL 16: 37–50.

Aitken, M. J., 1985. Thermoluminescence Dating. Academic Press, London, 359 pp.

Aitken, M. J., 1992. Optical dating. Quat. Sci. Rev. 11: 127–131.

Albarrán, G., G. H. Collins & K. E. Collins, 1987. Formation of organic products in self-radiolyzed calcium carbonate. J. Molec. Evol. 25: 12–14.

Alcalá, R., 1996. Centers and mechanisms in ionic materials. Appl. Rad. Isot. 47: 1471–1477.

Aldrich, J. E. & B. Pass, 1988. Determining radiation exposure from nuclear accidents and atomic tests using dental enamel. Health Phys. 54: 469–471.

Angel, B. R., J. P. E. Jones & P. L. Hall, 1974. Electron spin resonance studies of dopes synthetic kaolinite I. Clay Minerals 10: 247–255.

Apers, D., R. Debuyst, P. de Cannière, F. Dejehet & E. Lombard, 1981. Critique de la datation par résonance paramagnêtique électronique de planchers stalagmitiques de la Caune de l'Arago à

Tautavel. In de Lumley, H. & J. Labeyrie (eds.) Datations Absolus et les Analyses Isotopiques en Préhistoire: Méthodes et Limites. Prétirage. CNRS, Paris: 533–550.

Arakawa, T. & N. Hori, 1989. ESR dating of carbonate speleothem rings and late Quaternary climate changes in the Ryukyu Islands, Japan. Appl. Rad. Isot. 40: 1143–1146.

Ariyama, T., 1985. Conditions of resetting the ESR clock during faulting. In Ikeya, M. & T. Miki (eds.) ESR Dating and Dosimetry. Ionics, Tokyo: 249–256.

Baffa, O. & S. Mascarenhas, 1985a. ESR dating of shells from Sambaquis (Brazilian shell mounds). In Ikeya, M. & T. Miki (eds.) ESR Dating and Dosimetry. Ionics, Toyko: 139–143.

Baffa, O. & S. Mascarenhas, 1985b. Radiation dependence of ESR dating of bones and shells. In Ikeya, M. & T. Miki (eds.) ESR Dating and Dosimetry. Ionics, Toyko: 369–372.

Baffa, O., A. Brunetti, I. Karmann & C. M. Dias Neto, 2000. ESR dating of a Toxodon tooth from a Brazilian karstic cave. Appl. Rad. Isot. 52: 1345–1349.

Bahain, J. J., Y. Yokoyama, C. Falguères & M. N. Sarcia, 1992. ESR dating of tooth enamel: A comparison with K/Ar dating. Quat. Sci. Rev. 11: 245–250.

Bahain, J. J., M. N. Sarcia, C. Falguères & Y. Yokoyama, 1993. Attempt at ESR dating of tooth enamel of French Middle Pleistocene sites. Appl. Rad. Isot. 44: 267–272.

Bahain, J. J., Y. Yokoyama, C. Falguères & R. Bibron, 1994a. Choice of the useful signal in electron spin resonance (ESR) dating of Quaternary stalagmitic calcites. C. r. Acad. Sci., Paris 318: 375–379.

Bahain, J. J., Y. Yokoyama, H. Masaoudi, C. Falguères & M. Laurent, 1994b. Thermal behavior of ESR signals observed in various natural carbonates. Quat. Sci. Rev. (Quat. Geochron.) 13: 671–674.

Bahain, J. J., Y. Yokoyama, C. Falguères & R. Bibron, 1995. Datation par résonance de spin électronique (ESR) de carbonates marins Quaternaires. Quaternaire 6: 13–19.

Baietto, V., G. Villeneuve & M. Schvoerer, 1999. The perinaphthenyl radical: A potential probe for EPR dating. LED 99: 9th International Conference on Luminescence and Electron Spin Resonance Dating, Abstracts, Rome: 164.

Baker, J. M., D. J. Twitchen & M. E. Newton, 1997. Electron paramagnetic resonance data on the defect R_1 in irradiated diamond. Phil. Mag. Lett. 76: 57–62.

Baltakmens, T., 1975. Energy loss of β particles on backscattering. Nucl. Instrum. Meth. 125: 169–171.

Barabas, M., 1991. The nature of the paramagnetic centres at $g = 2.0057$ and $g = 2.0031$ in marine carbonates. Nucl. Tr. Rad. Meas. 20: 453–464.

Barabas, M., A. Bach & A. Mangini, 1988a. An analytical model for the growth of ESR signals. Nucl. Tr. Rad. Meas. 14: 231–235.

Barabas, M., A. Mangini, M. Sarnthein & H. E. Stremme, 1988b. The age of the Holstein Interglaciation: Reply. Quat. Res. 29: 80–84.

Barabas, M., A. Bach, M. Mudelsee & A. Mangini, 1989. Influence of the Mg-content on ESR-signals in synthetic calcium carbonate. Appl. Rad. Isot. 40: 1105–1111.

Barabas, M., A. Bach, M. Mudelsee & A. Mangini, 1992a. General properties of the paramagnetic centre at $g = 2.0006$. Quat. Sci. Rev. 11: 165–171.

Barabas, M., M. Mudelsee, R. Walther & A. Mangini, 1992b. Dose response and thermal behaviour of the ESR signal at $g = 2.0006$. Quat. Sci. Rev. 11: 173–179.

Barabas, M., R. Walther, A. Wieser, U. Radtke & R. Grün, 1993. 2nd interlaboratory comparison project on ESR dating. Appl. Rad. Isot. 44: 119–129.

Bartoll, J. & M. Ikeya, 1997. ESR dating of pottery: A trial. Appl. Rad. Isot. 48: 981–984.

Bartoll, J., R. Stößer & M. Nofz, 2000. Generation and conversion of electronic defects in calcium carbonates by UV/VIS light Appl. Rad. Isot. 52: 1099–1105.

Bella, F., 1971. Applicazione della technica ESR alla datazione geologia. Rendiconti della Societa Italiana de Mineralogia e Petrologia 27: 525–528.

Berger, G. W., 1986. Dating volcanic ash by electron spin resonance: Letter. Nature 319: 795–796.

Berman, S. S., B. A. B. Blackwell, J. I. P. Blickstein, A. R. Skinner, L. Kordos & H. P. Schwarcz, 2000. Electron spin resonance (ESR) dating for the Tarkő and Templomhegy Mammalian Substages in Hungary. Geol. Soc. Am. Abst. 32: A20

Bershov, L. V., M. D. Krylova & A. V. Speranskiy, 1975. O^-, Al^{3+} holes and Ti^{3+} electron centers in quartz as indicators of temperature of regional metamorphism. AN SSSR Izvestiya, Ser. Geol. 1975 (10): 113–117.

Bershov, L. V., A. S. Marfunin & A. V. Speranskiy, 1978. A new stable radiation center in quartz. Int. Geolog. Rev. 22: 1225–1233.

Blackwell, B. A., 1989. Laboratory Procedures for ESR Dating of Tooth Enamel. McMaster University Department of Geology Technical Memo 89.2, 234 pp.

Blackwell, B. A., 1994. Problems associated with reworked teeth in electron spin resonance dating. Quat. Sci. Rev. (Quat. Geochron.) 13: 651–660.

Blackwell, B. A., 1995a. Electron spin resonance dating. In Rutter, N. W., N. R. Catto (eds.) Dating Methods for Quaternary Deposits. Geological Association of Canada, St. John's, Geotext 2: 209–251.

Blackwell, B. A., 1995b. New correction methods for electron spin resonance (ESR) dating rates: Examples from Sterkfontein and Swartkrans, South Africa. J. World Archaeology 1 (2).

Blackwell, B. A. B., 1999. Laboratory Procedures for Performing ^{60}Co Irradiations Used in ESR Dating. Williams College Department of Chemistry Technical Memo 99.4, 28 pp.

Blackwell, B. A. B. & J. I. B. Blickstein, 2000. Considering sedimentary U uptake in external dose rate determinations for ESR and luminescent dating. Quat. Int. 68–71: 329–343.

Blackwell, B. A. & H. P. Schwarcz, 1993a. ESR isochron dating for teeth: A brief demonstration in solving the external dose calculation problem. Appl. Rad. Isot. 44: 243–252.

Blackwell, B. A. & H. P. Schwarcz, 1993b. Archaeochronology and scale. In Stein, J. K. & A. Linse (eds.) The Effects of Scale on Archaeological and Geoscientific Perspectives. Geol. Soc. Am. Spec. Pap. 283: 39–58.

Blackwell, B. A. & H. P. Schwarcz, 1995. Uranium series disequilibrium dating. In Rutter, N. W. & N. R. Catto (eds.) Dating Methods for Quaternary Deposits. Geological Association of Canada, St. John's, GEOtext 2: 167–208.

Blackwell, B. A. B. & A. F. R. Skinner, 1999a. Collecting Samples for ESR Dating. ESR Dating Laboratory Williams College Technical Memo 99.1, 16 pp.

Blackwell, B. A. & A. F. R. Skinner, 1999b. Assessing results from conventional ESR (electron spin resonance) dating: Subalyuk, Hungary, vs. Tsagaan Agui, Mongolia. In Mitchell, P. & H. M. Ambrose (eds.) Research in Archaeometry. Buffalo U. Press, Buffalo: 1–18.

Blackwell, B. A., N. Porat, H. P. Schwarcz & A. Debénath, 1992. ESR dating of tooth enamel: Comparison with $^{230}Th/^{234}U$ speleothem dates at La Chaise-de-Vouthon (Charente), France. Quat. Sci. Rev. 11: 231–244.

Blackwell, B. A., H. P. Schwarcz, K. Schick & N. Toth, 1993. ESR dating tooth enamel from the Paleolithic site at Longola, Zambia. Appl. Rad. Isot. 44: 253–260.

Blackwell, B. A., N. Porat & H. P. Schwarcz, 1994a. The potential for ESR dating at archaeological sites in the Great Lakes Region: Mammal teeth, gar scales, burnt flint hearth sand. In MacDonald, R. I. (ed.) Great Lakes Archaeology and Paleoecology, Exploring Interdisciplinary Initiatives for the Nineties: Symposium Proceedings. Waterloo U. Press, Waterloo: 321–366.

Blackwell, B. A., H. P. Schwarcz & W. R. Farrand, submitted, 1994b. Electron spin resonance (ESR) and $^{230}Th/^{234}U$ dating at Yarımburgaz, Turkey. In Howell, W. C. (ed.) The Yarımburgaz Pebble Culture Site.

Blackwell, B. A., H. P. Schwarcz & A. Debénath, submitted, 1994c. La datation des materiaux archéologiques à La Chaise-de-Vouthon, Charente. In Debénath, A. (ed.) La Gisement Archéologique à la Chaise-de-Vouthon, Charente. CNRS.

Blackwell, B. A. B, H. P. Schwarcz & J. F. Thackeray, submitted, 1998. The challenges in electron spin resonance (ESR) dating tooth enamel from the Australopithecine site Swartkrans, South Africa: A progress report. In Kuman, K. (ed.) Proceedings of the 1998 Dual Congress on Human Biology & Paleontology, South Africa.

Blackwell, B. A. B., H. Leung, A. R. Skinner, H. P. Schwarcz, S. Lebel, H. Valladas, J. I. B. Blickstein & M. Divjak, 2000. External dose rate determinations for ESR dating at Bau de l'Aubesier, Provence. Quat. Int. 68–71: 345–361.

Blackwell, B. A. B., A. R. Skinner & J. I. B. Blickstein, 2001a. ESR isochron exercises: How accurately do modern dose rate measurements reflect paleodose rates? Quat. Sci. Rev. (Quat. Geochron.) 20: 1031–1039.

Blackwell, B. A. B., J. W. Olsen, A. P. Derevianko, D. S. Tseveendorj, A. F. R. Skinner & M. Dwyer, 2001b. ESR dating the Paleolithic site at Tsagaan Agui, Mongolia. Proceedings of the International Archaeometry Congress, Budapest, 1998. Archaeolingua Press/B.A.R., Budapest/London, in press.

Blackwell, B. A. B., A. R. Skinner & J. F. Thackeray, submitted, 2001c. Electron Spin Resonance (ESR) Isochrons in Tooth Enamel from the Australopithecine Site. Swartkrans, South Africa. Geoarchaeology.

Blackwell, B. A. B., B. B. Curry, H. P. Schwarcz, J. J. Saunders & N. Woodman, submitted, 2001d. Dating the Sangamon: ESR dating using mammalian teeth and gar fish scales at Hopwood Farm, Illinois. Quat. Res.

Blackwell, B. A. B., A. R. Skinner, J. I. B. Blickstein, S. Lebel & H. Y. M. Leung, 2001e. ESR isochron analyses at Bau de l'Aubesier, Provence: Clues to U uptake in fossil teeth. Geoarchaeology 16: 719–761.

Bluszcz, A., T. Goslar, H. Hercman, M. F. Pazdur & A. Walanus, 1988. A comparison of TL, ESR and ^{14}C dates of speleothems. Quat. Sci. Rev. 7: 417–421.

Borziac, I. A., P. Allsworth-Jones, C. A. I. French, S. I. Medyanik & W. J. Rink, 1997. The Upper Paleolithic site of Cuintu on the Middle Pruth: A multidisciplinary reinterpretation. Prehist. Soc., Proc. 16: 1–17.

Boslough, M. B., T. J. Ahrens., J. Vizgirda, R. H. Becker & S. Epstein, 1982. Shock induced devolatilization of calcite. Earth & Planet. Sci. Lett. 61: 166–170.

Bouchez, R., E. Cox, E. Lopez Carranza, J. L. Ma, M. Piboule, G. Poupeau, P. Rey & A. Herve, 1988. Q-band studies of fossil teeth: Consequences for ESR dating. Quat. Sci. Rev. 7: 497–501.

Breen, S. L. & J. J. Battista, 1995. Radiation dosimetry in human bone using electron paramagnetic resonance. Phys. Med. Biol. 40: 2065–2077.

Brennan, B. J., 2000. Systematic underestimation of the age of samples with saturating exponential behaviour and inhomogeneous dose distribution. Rad. Meas. 32: 731–734.

Brennan, B. J., R. G. Lyons & S. W. Phillips, 1991. Attenuation of α-particle track dose for spherical grains. Nucl. Tr. Rad. Meas. 18: 249–253.

Brennan, B. J., H. P. Schwarcz & W. J. Rink, 1997a. Simulation of the γ radiation field in lumpy environments. Rad. Meas. 27: 299–305.

Brennan, B. J., W. J. Rink, E. L. McGuirl & H. P. Schwarcz, 1997b. β doses in tooth enamel by "one-group" theory and the Rosy ESR dating software. Rad. Meas. 27: 307–314.

Brennan, B. J., W. V. Prestwich, W. J. Rink, R. E. Marsh & H. P. Schwarcz, 2000. α and β dose gradients in tooth enamel. Rad. Meas. 32: 759–765.

Brik, A., V. Radchuk, O. Scherbina, M. Matyash & O. Gaver, 1996. Metamorphic modifications and EPR dosimetry in tooth enamel. Appl. Rad. Isot. 47: 1317–1319.

Brik, A., E. J. Haskell, V. Brik, O. Scherbina & O. Atamenko, 2000a. Anisotropy effects of EPR signals and mechanisms of mass transfer in tooth enamel and bones. Appl. Rad. Isot. 52: 1077–1083.

Brik, A., V. Baraboy, O. Atamenko, Y. Shevchenko & V. Brik, 2000b. Metabolism in tooth enamel and reliability of retrospective dosimetry. Appl. Rad. Isot. 52: 1305–1310.

Brillibit, M. G., G. A. Klevezal, P. I. Mordivintoev, L. I. Sukhovskaya, B. A. Rzhenkov, N. V. Boevodskya & A. V. Vanin, 1990. On the use of dental enamel as an *in vivo* dosimeter to γ rays. Hematologia 35: 11–15.

Brooks, A. D., D. M. Halgren, J. S. Cramer, A. Franklin, W. Hornyak, J. M. Keating, R. G. Klein, W. J. Rink, H. P. Schwarcz, J. N. L. Smith, K. Stewart, N. E. Todd, J. Verniers & J. E. Yellen, 1995. Dating and context of three Middle Stone Age sties with bone points in the Upper Semliki Valley, Zaire. Science 268: 548–553.

Brückner, H. & U. Radtke, 1985. Neue Erkenntnisse zum marinen Quartäran spaniens Mittelmeerküste. Kieler Geographische Schriften 62: 49–71.

Brückner, H. & U. Radtke, 1986. Paleoclimatic implications derived from profiles along the Spanish Mediterranean coast. Symposium on Climatic Fluctuations During the Quaternary in the Western Mediterranean, Proceedings: 467–486.

Brunetti, A., R. Giola, W. Avelar, F. Mantellato, A. Françosco & O. Baffa, 1999. ESR dating of subfossil shells from Couve Island (Ubatuba-Brazil). LED 99: 9th International Conference on Luminescence and Electron Spin Resonance Dating, Abstracts, Rome: 183.

Brumby, S. & H. Yoshida, 1994a. ESR dating of molluscs shell: Investigations with modern shell of four species. Quat. Sci. Rev. 13: 157–162.

Brumby, S. & H. Yoshida, 1994b. An investigation of the effect of sunlight on the ESR spectra of quartz centres: Implications for dating. Quat. Sci. Rev. (Quat. Geochron.) 13: 615–618.

Brumby, S. & H. Yoshida, 1995. The annealing kinetics of ESR signals due to paramagnetic centers in mollusk shell. Rad. Meas. 24: 255–263.

Brunnacker, K., K. D. Jager, G. J. Hennig, J. Preuss & R. Grün, 1983. Radiometrische Untersuchungen zur Datierung mitteleuropäischer Travertinvorkornmen. Ethnographisch-Archäologische Zeitschrift 24: 217–266.

Buhay, W. M., 1987. A Theoretical Study of ESR Dating of Geological Faults in Southern California. Unpub. M.Sc. thesis, McMaster University, Hamilton, 136 pp.

Buhay, W. M., 1991. Increase of radiation sensitivity of ESR centres by faulting and criteria of fault dates: Comment. Earth Planet. Sci. Lett. 105: 574–577.

Buhay, W. M., H. P. Schwarcz & R. Grün, 1988. ESR studies on quartz grains from fault gouge. Quat. Sci. Rev. 7: 515–522.

Buhay, W. M., P. M. Clifford & H. P. Schwarcz, 1992. ESR dating of the Rotoiti Breccia in the Taupo volcanic zone, New Zealand. Quat. Sci. Rev. 11: 267–271.

Bustos, M. E., M. E. Romero, A. Gutiérrez & J. Azorín, 1996. Identification of irradiated mangoes by means of ESR spectroscopy. Appl. Rad. Isot. 47: 1655–1656.

Caddie, D. A., H. J. Hall, D. S. Hunter & P. J. Pomery, 1985. ESR considerations in the dating of Holocene and Late Pleistocene bone material. In Ikeya, M. & T. Miki (eds.) ESR Dating and Dosimetry. Ionics, Toyko: 353–362.

Callens, F. J., E. R. Boesman, F. C. M. Driessens, L. C. Martens, P. F. A. Matthys & R. M. H. Verbeeck, 1986. The electron spin resonance spectrum near $g = 2$ of carbonated calcium apatites synthesized at high temperature. Bull. Soc. chim. Bêlges 95: 589–596.

Callens, F. J., E. R. Boesman, L. C. Martens, P. F. A. Matthys & R. M. H. Verbeeck, 1987. The contribution of CO_3^{3-} and CO_3^{2-} to the electron spin resonance spectrum near $g = 2$ of powdered human tooth enamel. Calc. Tissue Int. 41: 124–129.

Callens, F. J., P. Moëns & R. M. H. Verbeeck, 1995. An EPR study of intact and powdered human tooth enamel dried at 400 °C. Calc. Tissue Int. 56: 543–548.

Caracelli, I., S. Mascarenhas & M. C. Terrile, 1986. Electron spin resonance dosimetric properties of bone. Health Phys. 50: 259–263.

Çetin, O., A. M. Özer & A. Wieser, 1994. ESR dating of tooth enamel from Karaïn excavation (Antalya, Turkey). Quat. Sci. Rev. (Quat. Geochron.) 13: 661–669.

Chen, G., B. A. B. Blackwell, H. P. Schwarcz, L. Kordos & A. F. R. Skinner, 1996. Electron spin resonance (ESR) dating of mammalian teeth from the Villányian site, Kisláng, Hungary. LED 96: 8th International Conference on Luminescence and Electron Spin Resonance Dating, Abstracts, Canberra: 52.

Chen, T. M., Q. Yang & W. En, 1994. Antiquity of *Homo sapiens* in China. Nature 368: 55–56.

Chen, T. M., Q. Yang, Y. Q. Hu, W. B. Bao & T. Y. Li, 1997. ESR dating of tooth enamel from Yunxian *Homo erectus* site, China. Quat. Sci. Rev. (Quat. Geochron.) 16: 455–458.

Chen, T. M., Q. Chen, Q. Yang & Y. Q. Hu, 2001. The problems in ESR dating of tooth enamel of early Pleistocene and the age of the Longgupo hominid, Wushan, China. Quat. Sci. Rev. 20: 1041–1045.

Chen, Y. J., J. F. Lu, J. Head, G. Jacobson & A. V. Arakel, 1988a. ^{14}C and ESR dating of calcrete and gypcrete cores from the Amadeus Basin, Northern Territory, Australia. Quat. Sci. Rev. 7: 447–453.

Chen, Y. J., J. F. Lu, T. M. Cao, R. G. He & R. L. Zhang, 1988b. A comparative examination of in validity of AD value estimated from ESR spectra of various carbonate sediments. Quat. Sci. Rev. 7: 455–460.

Chen, Y. J., A. V. Arakel & J. F. Lu, 1989a. Investigation of sensitive signals due to γ-ray radiation of chemical precipitates. A feasibility study for ESR dating of gypsum, phosphate and calcrete deposits. Appl. Rad. Isot. 40: 1163–1170.

Chen, Y. J., P. Fortin & Q. Li, 1989b. A preliminary study on ESR dating of sediments from Ouritiba Basin, Brazil. Appl. Rad. Isot. 40: 1123–1126.

Chen, Y. J., S. Brumby, G. Jacobson, A. L. J. Beckwith & H. A. Polach, 1992. A novel application of the ESR method: Dating of insular phosphorites and reef limestone. Quat. Sci. Rev. 11: 209–217.

Chen, Y. J., J. Gao & J. Feng, 1993. ESR dating of geyserites from intermittent geyser sites on the Tibetan plateau. Appl. Rad. Isot. 44: 207–213.

Chen, Y. J., S. Brumby & J. Gao, 1994. Observations on the suitability of the signal at $g = 2.0040$ for ESR dating of secondary carbonates. Quat. Sci. Rev. (Quat. Geochron.) 13: 675–678.

Chen, Y. J., J. Feng, J. Gao, L. Taylor & R. Grün, 1997a. Observations on the microtexture and ESR spectra of quartz from fault gouge. Quat. Sci. Rev. (Quat. Geochron.) 16: 487–493.

Chen, Y. J., J. Feng, J. Gao & R. Grün, 1997b. Investigation of the potential use of ESR signals in quartz for paleothermometry. Quat. Sci. Rev. (Quat. Geochron.) 16: 495–499.

Chong, T. S., T. Iida & K. Ieda, 1985. ESR studies in Naumann elephant teeth and natural minerals. In Ikeya, M. & T. Miki (eds.) ESR Dating and Dosimetry. Ionics, Toyko: 335–340.

Chong, T. S., H. Ohta, Y. Nakashima, T. Iida, K. Ieda & H. Saisho, 1989. ESR dating of elephant teeth and radiation dose rate estimation in soil. Appl. Rad. Isot. 40: 1199–1202.

Chumak, V. V., J. Pavlenko & S. V. Sholom, 1996. An approach to the assessment of overall uncertainty of determination of dose using an ESR technique. Appl. Rad. Isot. 47: 1287–1291.

Ciochón, R., V. T. Long, R. Larick, L. Gonzalez, R. Grün, J. Devos, C. Yonge, L. Taylor, H. Yoshida & M. Reagan, 1996. Dated co-occurrence of *Homo erectus* and *Gigantopithecus* from Tham-Khuyen Cave, Vietnam. Proc. Natl. Acad. Sci., USA 93: 3016–3020.

Copeland, J. F., K. P. Gall, S. Y. Lee & G. E. Chabot, 1996. Proton dosimetry in bone using electron spin resonance. Appl. Rad. Isot. 47: 1533–1538.

Cowan, D. L., V. Priest & S. S. Levy, 1993. ESR dating of quartz from Exile Hill, Nevada. Appl. Rad. Isot. 44: 1035–1039.

Debuyst, R., F. Dejehet, R. Grün, D. Apers & P. de Cannière, 1984. Possibility of ESR dating without determination of annual dose. J. Radioanalyt. Nucl. Chem. 86: 399–410.

Debuyst, R., P. de Cannière & F. Dejehet, 1990. Axial CO_2^- in a particle irradiated calcite: Potential use in ESR dating. Nucl. Tr. Rad. Meas. 17: 525–530.

Debuyst, R., M. Bichamambu & F. Dejehet, 1991. An EPR study of γ-irradiated synthetic powdered calcite labelled with ^{13}C. Nucl. Tr. Rad. Meas. 18: 193–201.

Debuyst, R., F. Dejehet & S. Irdrissi, 1993. Paramagnetic centres in γ-irradiated synthetic monohydrocalcite. Appl. Rad. Isot. 44: 293–297.

Debuyst, R., M. Frenchen & S. Irdrissi, 2000a. Problems encountered in a TL and ESR study of natural monohydrocalcite. Rad. Meas. 32: 725–729.

Debuyst, R., F. Callens, M. Frenchen & F. Dejehet, 2000b. ESR study of tooth enamel from the Kärlich-Seeufer Site in Germany. Appl. Rad. Isot. 52: 1327–1336.

de Cannière, P., T. Joppard, R. Debuyst, F. Dejehet & D. Apers, 1985. ESR dating: A study of humic acids incorporated in synthetic calcite. Nucl. Tr. Rad. Meas.10: 853–863.

de Cannière, P., R. Debuyst, F. Dejehet, D. Apers & R. Grün, 1986. ESR dating: A study of ^{210}Po-coated geological and synthetic samples. Nucl. Tr. Rad. Meas. 11: 211–220.

de Cannière, P., R. Debuyst, F. Dejehet & D. Apers, 1988. Electron spin resonance study of internally α-irradiated (^{210}Po-nitrate-doped) calcite single crystal. Nucl. Tr. Rad. Meas. 14: 267–273.

de Jesus, E. F. O., A. M. Rossi & R. T. Lopes, 1996. Influence of sample treatment on the ESR signal of irradiated citrus. Appl. Rad. Isot. 47: 1647–1653.

de Jesus, E. F. O., A. M. Rossi & R. T. Lopes, 1996. Identification and dose determination using ESR measurements in the flesh of irradiated vegetable products. Appl. Rad. Isot. 52: 1375–1383.

Dennison, K. J. & B. M. Peake, 1992. ESR bone dating in New Zealand. Quat. Sci. Rev. 11: 251–255.

Dennison, K. J., R. Houghton, B. F. Leach & B. M. Peake, 1985. ESR considerations in the dating of Holocene and late Pleistocene bone material. In Ikeya, M. & T. Miki (eds.) ESR Dating and Dosimetry. Ionics, Toyko: 341–352.

Dennison, K. J., A. D. Oduwale & K. D. Sales, 1997. The anomalous ESR dating signal intensity observed for human remains from the Namu burial site on the Island of Taumako, Solomon Islands. Quat. Sci. Rev. (Quat. Geochron.) 16: 459–464.

Desrosiers, M. F., 1996. Current status of the EPR method to detect irradiated food. Appl. Rad. Isot. 47: 1621–1628.

Desrosiers, M. F. & W. L. McLaughlin, 1989. Examination of γ-irradiated fruits and vegetables by electron spin resonance spectroscopy. Rad. Phys. Chem. 34: 895–898.

Desrosiers, M. F. & M. G. Simic, 1988. Post-irradiation dosimetry of meat by electron spin resonance spectroscopy of bones. J. Agric. Food Chem. 36: 601–603.

Desrosiers, M. F., M. G. Simic, F. C. Eichmiller, A. D. Johnston & R. L. Bowen, 1989. Mechanically-induced generation of radicals in tooth enamel. Appl. Rad. Isot. 40: 1195–1197.

Dodd, N. J. F., J. S. Lea, A. J. Swallow, 1988. ESR detection of irradiated food. Nature 334: 387.

Dolo, J. M., N. Lecerf, V. Mihajlovic, C. Falguères & J. J. Bahain, 1996. Contribution of ESR dosimetry for irradiation of geological and archaeological samples with a ^{60}Co panoramic source. Appl. Rad. Isot. 47: 1419–1421.

Duarte, C. L., A. L. C. H. Villa Vicencio, N. L. del Mastro, F. M. Wiendl, 1995. Detection of irradiated chicken by ESR spectroscopy of bone. Rad. Phys. Chem. 46: 689–692.

Duchesne, J., J. Deprieux & J. M. van der Kaa, 1961. Origin of radicals in carbonaceous rocks. Geoch. Cosmoch. Acta 23: 209–218.

Duliu, O. G., 2000. Electron paramagnetic resonance identification of irradiated cuttlefish (*Sepia officinalis* L.). Appl. Rad. Isot. 52: 1385–1390.

Eaton, G. R., S. S. Eaton & M. M. Maltempo, 1989. Three approaches to spectral-spatial ESR imaging. Appl. Rad. Isot. 40: 1227–1231.

Egersdörfer, S., A. Wieser, A. Müller, 1996a. Tooth enamel as a detector material for restrospective EPR dosimetry. Appl. Rad. Isot. 47: 1299–1303.

Egersdörfer, S., A. Wieser, A. Müller, 1996b. Restrospective EPR dosimetry with tooth enamel: A comparison of different dose estimation methods. Prog. Biophys. Molec. Biol. 65: PH512.

Ehlermann, D. A. E., 1996. Comparison of ESR/TLD analyses for quartz as a routine dosimeter. Appl. Rad. Isot. 47: 1547–1550.

Faïn, J., S. Soumana, M. Montret, D. Miallier, T. Pilleyre & S. Sanzelle, 1998. Luminescence and ESR dating: β dose attenuation for various grain shapes calculated by a Monte Carlo method. Quat. Sci. Rev. 18: 231–234.

Falguères, C., Y. Yokoyama & R. Bibron, 1990. Electron spin resonance (ESR) dating of hominid-bearing deposits in the Caverna delle Fate, Ligure, Italy. Quat. Res. 34: 121–128.

Falguères, C., Y. Yokoyama & D. Miallier, 1991. Stability of some centers in quartz. Nucl. Tr. Rad. Meas. 18: 155–161.

Falguères, C., D. Miallier, S. Sanzelle, J. Faïn, M. Laurent, M. Montret, T. Pilleyre & J. J. Bahain, 1994. Potential use of the E' centre as an indicator of initial resetting in TL/ESR dating of volcanic materials. Quat. Sci. Rev. (Quat. Geochron.) 13: 619–623.

Falguères, C., J. J. Bahain & H. Saleki, 1997. U-series and ESR dating of teeth from Acheulean and Mousterian levels at La Micoque (Dordogne, France). J. Arch. Sci. 24: 537–545.

Fattibene, P., D. Aragno, S. Onori & M. C. Pressello, 2000. Thermal-induced EPR signals in tooth enamel. Rad. Meas. 32: 793–798.

Friebele, E. J., D. L. Griscom, M. Stapelbroek & R. A. Weeks, 1979. Fundamental defect centres in glass: The peroxy radical in irradiated, high purity silica. Phys. Rev. Lett. 42: 1346–1349.

Fukuchi, T., 1988. Applicability of ESR dating using multiple centers to fault movement: The case of the Itoigawa-Shizuoka Tectonic Line, a major fault in Japan. Quat. Sci. Rev. 7: 509–514.

Fukuchi, T., 1989a. Theoretical study on frictional heat by faulting using ESR. Appl. Rad. Isot. 40: 1181–1193.

Fukuchi, T., 1989b. Increase of radiation sensitivity of ESR centers by faulting and criteria of fault dates. Earth & Planet. Sci. Lett. 94: 109–122.

Fukuchi, T., 1991. The Itoigawa-Shizuoka Tectonic Line at the western edge of the south Fossa Magna, Japan. Mod. Geol. 15: 347–366.

Fukuchi, T., 1992a. Increase of radiation sensitivity of ESR centres by faulting and criteria of fault dates: Reply. Earth & Planet. Sci. Lett. 114: 211–213.

Fukuchi, T., 1992b. ESR studies for absolute dating of fault movements. J. Geolog. Soc., London 149: 265–272.

Fukuchi, T., 1993. Vacancy-associated type ESR centres observed in natural silica and their application to geology. Appl. Rad. Isot. 44: 179–184.

Fukuchi, T., 1996a. Quartet ESR signals detected from natural clay minerals and their applicability to radiation dosimetry and dating. Jap. J. Appl. Phys. 1 35: 1977–1982.

Fukuchi, T., 1996b. Direct ESR dating of fault gouge using clay minerals and the assessment of fault activity. Engin. Geol. 43: 201–211.

Fukuchi, T., 1996c. A mechanism for the formation of E' and peroxy centers in natural deformed quartz. Appl. Rad. Isot. 47: 1509–1521.

Fukuchi, T., 2001. Assessment of fault activity by ESR dating of fault gouge: An example of the 500 m core samples drilled into the Nojima earthquake fault in Japan. Quat. Sci. Rev. 20: 1005–1008.

Fukuchi, T., N. Imai & K. Shimokawa, 1985. Dating of the fault movement by various ESR signals in quartz: Cases on the faults in the South Fossa Magna, Japan. In Ikeya, M. & T. Miki (eds.) ESR Dating and Dosimetry. Ionics, Toyko: 211–217.

Fukuchi, T., N. Imai & K. Shimokawa, 1986. ESR dating of fault movement using various defect centres in quartz; the case in the western South Fossa Magna, Japan. Earth & Planet. Sci. Lett. 78: 121–128.

Furusawa, M. & M. Ikeya, 1993. ESR imaging of irradiated teflon tube using a highly linear field gradient by straight wires inserted in a cavity. Appl. Rad. Isot. 44: 381–384.

Furusawa, M., M. Kasuya, S. Ikeda & M. Ikeya, 1991. ESR Imaging of minerals and its application to dating. Nucl. Tr. Rad. Meas. 18: 185–188.

Gamarra, L. Z., M. Matsuoka, S. Watanabe, T. Nakajima & C. M. Sunta, 1998. thermoluminescence and paramagnetic centers created by radiation in quartz. Rad. Phys. Chem. 51: 529–530.

Garrison, E. G., 1989. Characterization of an ESR geochronological dating centre in flints. Phys. Chem. Minerals 16: 767–773.

Garrison, E. G., M. Rowlett, D. L. Cowan & L. V. Holroyd, 1981. ESR dating of ancient flints. Nature 290: 44–45.

Gerstenhauer, A., U. Radtke & A. Mangini, 1983. Neue Ergebnisse zur quartären Küstenentwicklung der Halbinsel Yucatan, Mexico. Essener Geographische Arbeiten 6: 187–199.

Geyh, M. A. & G. J. Hennig, 1986. Multiple dating of a long flowstone core. Radiocarbon 28: 503–509.

Geyh, M. A. & G. J. Hennig, 1987. Multiple dating of a long flowstone core: Reply. Radiocarbon 29: 153–155.

Ghelawi, M. A., J. S. Moore, N. J. F. Dodd, 1996. Use of ESR for the detection of irradiated dates (Phoenix dactylifera L.). Appl. Rad. Isot. 47: 1641–1645.

Ginsbourg, S. F., T. A. Babushkina, L. B. Basova & T. P. Klimova, 1996. ESR-spectroscopy of building materials as a dosimetry technique. Appl. Rad. Isot. 47: 1369–1374.

Goede, A. & J. L. Bada, 1985. Electron spin resonance dating of Quaternary bone material from Tasmanian caves: A comparison with ages determined by aspartic acid racemization and ^{14}C. Austr. J. Earth Sci. 32: 155–162.

Goede, A. & M. A. Hitchman, 1987. Electron spin resonance analysis of marine gastropods from coastal archaeological sites in South Africa. Archaeometry 29: 163–174.

Göksu, H. Y., A. Wieser, A. Waibel, A. Vogenauer & D. F. Regulla, 1989. Comparing measurements of free radicals, optical density and thermoluminescence in solids for high level dosimetry. Appl. Rad. Isot. 40: 905–909.

Göksu, H. Y., A. Wieser, D. Stoneham, I. K. Bailiff & M. Figel, 1996. EPR, OSL, TL, and spectral studies of porcelain. Appl. Rad. Isot. 47: 1369–1374.

Gorter, C. J., 1936a. Paramagnetic relaxation. Physica 3: 503–513.

Gorter, C. J., 1936b. Paramagnetic relaxation in a transversal magnetic field. Physica 3: 1006–1008.

Gorter, C. J., Kronig, R. deL., 1936. On the theory of absorption and dispersion in paramagnetic and dielectric media. Physica 3: 1009–1020.

Goslar, T. & H. Hercman, 1988. TL and ESR dating of speleothems and radioactive disequilibrium in the uranium series. Quat. Sci. Rev. 7: 423–427

Gotze, J. & M. Plotze, 1997. Investigation of trace element distribution in detrital quartz by electron paramagnetic resonance (EPR). Eur. J. Mineralogy 9: 529–537.

Gray, R., M. H. Stevenson & D. J. Kilpatrick, 1990. The effect of radiation dose and age of bird on the ESR signal in irradiated chicken drumsticks. Rad. Phys. Chem. 35: 284–287.

Gray, S. C., J. R. Hein, R. Hausmann & U. Radtke, 1992. Geochronology and subsurface stratigraphy of Pukapuka and Rakahanga Atolls, Cook Islands: Late Quaternary reef growth and sea level history. Palaeogeog. Palaeoclim. Palaeoecol. 91: 377–394.

Griffiths, D. R., G. V. Robins, H. Chandra, D. A. C. McNeil & M. C. R. Symons, 1982. Trapped methyl radicals in chert. Nature 300: 435–436.

Griffiths, D. R., N. J. Seeley, H. Chandra & M. C. R. Symons, 1983. ESR dating of heated chert. PACT 9: 399–409.

Griffiths, J. H. E. J. Owen & I. M. Ward, 1954. Paramagnetic resonance in neutron irradiated diamond and smoky quartz. Nature 173: 439–442.

Grün, R., 1985a. ESR-dating speleothem records: Limits of the method. In Ikeya, M. & T. Miki (eds.) ESR Dating and Dosimetry. Ionics, Toyko: 61–72.

Grün, R., 1985b. ESR dating without determination of annual dose: A first application on dating mollusc shells. In Ikeya, M. & T. Miki (eds.) ESR Dating and Dosimetry. Ionics, Toyko: 115–123.

Grün, R., 1985c. Beiträge zur ESR-Datierung. Sonderveruoffentlichungen des Geologischen Instituts der Universität zu Köln 59: 1–157.

Grün, R., 1986. ESR-dating of a flowstone core from Cava de Sa Bassa Blanca (Mallorca, Spain). ENDINS 12: 19–23.

Grün, R., 1987. Die ESR Datierung an Höhlensintern. Laichinger Höhlenfreund 21: 13–24.

Grün, R., 1988. The potential of ESR dating of tooth enamel. Colloque International l'Homme de Néanderthal, Proceedings 1: 37–46.

Grün, R., 1989a. Die ESR-Alterbestimmungs-methode. Springer, Berlin, 132 pp.

Grün, R., 1989b. Electron spin resonance (ESR) dating. Quat. Int. 1: 65–109.

Grün, R., 1989c. Present status of ESR-dating. Appl. Rad. Isot. 40: 1045–1055.

Grün, R., 1989d. ESR dating for the early earth. Nature 338: 543–544.

Grün, R., 1991a. Potential and problems of ESR dating. Nucl. Tr. Rad. Meas. 18: 143–153.

Grün, R., 1991b. Electron spin resonance age estimates on elephant teeth from the Balderton Sand and Gravel. In Brandon, A., M. G. Sumbler (eds.) The Balderton Sand and Gravel: Pre-Ipswichian Cold Stage Fluvial Deposits Near Lincoln, England. Oxford U Press, Oxford: 135–137.

Grün, R., 1992. Remarks on ESR dating of fault movements. J. Geolog. Soc., London 149: 261–264.

Grün, R., 1993. Status and present problems in ESR dating. Appl. Rad. Isot. 40: 1045–1055.

Grün, R., 1995. Semi-non-destructive single aliquot ESR dating. Ancient TL 13: 3–7.

Grün, R., 1996a. A re-analysis of electron spin resonance dating results associated with the Petralona hominid. J. Human Evol. 30: 227–241.

Grün, R., 1996b. Errors in dose assessment introduced by the use of the "linear part" of a saturating dose response curve. Rad. Meas. 18: 143–152.

Grün, R., 1997. Electron spin resonance dating. In Taylor, R. E. & M. J. Aitkin (eds.) Chronometric Dating in Archaeology. Plenum, New York: 217–260.

Grün, R., 1998a. Reproducibility measurements for ESR signal intensity and dose determination: High precision but doubtful accuracy. Rad. Meas. 29: 177–193.

Grün, R., 1998b. Dose determination on fossil tooth enamel using spectrum deconvolution with Gaussian and Lorentzian peak shapes. Ancient TL 16: 3–7.

Grün, R., 2000. Methods of dose determination using ESR spectra of tooth enamel. Rad. Meas. 32: 767–772.

Grün, R. & S. Brumby, 1994. The assessment of errors in past radiation doses extrapolated from ESR/TL dose response data. Rad. Meas. 23: 307–315.

Grün, R. & K. Brunnacker, 1987. ESR Datierungen eines Mammuth-Zahnes aus dem Travertinbruch "Lauster", StuttgartBad Canstatt. Jahrbuch des Geologischen Landesarrdes Baden-Würtemberg 28: 43–47.

Grün, R. & R. Clapp, 1997. An automated sample changer for Brücker ES spectrometers. Ancient TL 14: 1–6.

Grün, R. & P. de Cannière, 1984. ESR dating: Problems encountered in the evaluation of the naturally accumulated dose (AD) of secondary carbonates. J. Radioanalyt. Nucl. Chem. 85: 213–226.

Grün, R. & C. Invernati, 1985. Uranium accumulation in teeth and its effect on ESR dating: Detailed study of a mammoth tooth. Nucl. Tr. Rad. Meas. 10: 869–878.

Grün, R. & M. Jonas, 1996. Plateau tests and spectrum deconvolution for ESR dose determination in tooth enamel. Rad. Meas. 26: 621–629.

Grün, R. & O. Katzenberger-Apel, 1994. An alpha irradiator for ESR dating. Ancient TL 12: 35–38.

Grün, R. & P. D. M. MacDonald, 1989. Nonlinear fitting of TL/ESR dose-response curves. Appl. Rad. Isot. 40: 1077–1080.

Grün, R. & F. McDermott, 1994. Open system modelling for U-series and ESR dating of teeth. Quat. Sci. Rev. 13: 121–125.

Grün, R. & E. J. Rhodes, 1991. On the selection of dose points for saturating exponential ESR/TL dose response curves: Weight of intensity by inverse variance. Ancient TL 10: 50–56.

Grün, R. & E. J. Rhodes, 1992. Simulations of saturating exponential ESR/TL dose response curves: Weighting of intensity by inverse variance. Ancient TL 9: 40–46.

Grün, R. & H. P. Schwarcz, 1987a. Comments on multiple dating of a long flowstone core. Radiocarbon 29: 148–152.

Grün, R. & H. P. Schwarcz, 1987b. Some Remarks on "ESR Dating of Bones". Ancient TL, vol. 5, p. 1–9.

Grün, R. & C. B. Stringer, 1991. Electron spin resonance dating and the evolution of modern humans. Archaeometry 33: 153–199.

Grün, R. & L. Taylor, 1996. Uranium and thorium in the constituents of fossil teeth. Ancient TL 14: 21–26.

Grün, R. & A. Thorne, 1997. Comment: Dating the Ngangdong humans. Science 276: 1575.

Grün, R., H. P. Schwarcz & S. Zymela, 1987. ESR dating of tooth enamel. Can. J. Earth Sci. 24: 1022–1037.

Grün, R., H. P. Schwarcz & J. Chadham, 1988a. Electron spin resonance dating of tooth enamel: Coupled correction for U uptake and U-series disequilibrium. Nucl. Tr. Rad. Meas. 14: 237–241.

Grün, R., H. P. Schwarcz, D. C. Ford & B. Hentzsch, 1988b. ESR dating of spring deposited travertines. Quat. Sci. Rev. 7: 429–432.

Grün, R., P. B. Beaumont & C. B. Stringer, 1990a. ESR dating evidence for Early Modem Humans at Border Cave in South Africa. Nature 344: 537–539.

Grün, R., N. J. Shackleton & H. J. Deacon, 1990b. Electron spin resonance dating of tooth enamel from Klasies River Mouth Cave. Curr. Anthro. 31: 427–432.

Grün, R., C. B. Stringer & H. P. Schwarcz, 1991a. ESR dating of teeth from Garrod's Tabun Cave collection. J. Human Evol. 20: 231–248.

Grün, R., P. Mellars & H. Laville, 1991b. ESR chronology of a 100,000 year archaeological sequence at Pech de l'Azé II, France. Antiq. 65: 544–551.

Grün, R., U. Radtke & A. Omura, 1992. ESR and U-series analyses on corals from Huon Peninsula, New Guinea. Quat. Sci. Rev. 11: 197–202.

Grün, R., W. J. Rink & P. C. Smalley, 1994. ESR analysis of the E' centre in quartz grains and quartz cement of sandstones. 7th International Specialist Seminar on Luminescence and Electron Spin Resonance Dating, Abstracts, Krems.

Grün, R., J. S. Brink, N. A. Spooner, L. Taylor, C. B. Stringer, R. G. Franciscus & A. S. Murray, 1996a. Direct dating of the Florisbad hominid. Nature 382: 500–501.

Grün, R., M. Abeyratne, J. Head, C. Tuniz & R. E. M. Hedges, 1997a. AMS ^{14}C analysis of teeth from archaeological sites showing anomalous ESR dating results. Quat. Sci. Rev. (Quat. Geochron.) 16: 437–444.

Grün, R., P. H. Huang, X. Zu, C. B. Stringer, A. G. Thorne & M. McCullouch, 1997b. ESR analysis of teeth from the Palaeoanthropological site of Zhoukoudian China. J. Human Evol. 32: 83–91.

Grün, R., H. Kohno, A. Tani, C. Yamanaka, M. Ikeya & H. P. Huang, 1997c. Pulsed ESR measurements on fossil teeth. Rad. Meas. 27: 425–432.

Grün, R., P. H. Huang, W. P. Huang, F. McDermott, A. G. Thorne, C. B. Stringer & G. Yan, 1998. ESR and U-series analysis of teeth from the paleoanthropological site of Hexian, Anhui Province, China. J. Human Evol. 34: 555–564.

Grün, R., A. Tani, A. Gurbanov, D. Koshchug, I. Williams & J. Braun, 1999. Reconstruction of cooling and denudation rates of the Eldzhurtinskiy Granite, Caucasus, using paramagnetic centres in quartz. LED 99: 9th International Conference on Luminescence and Electron Spin Resonance Dating, Abstracts, Rome: 32.

Halliburton, L. E., A. Hofstaetter, A. Scharmann, M. P. Scripsick & G. J. Edwards, 1993. Dose-rate dependence in the production of point defects in quartz. Appl. Rad. Isot. 44: 273–279.

Hannß, C., 1985. The ESR dating of a Würm moraine in the Trieves (French Alps) and its significance with respect to the sequence of glaciation phases during the last ice age. In Ikeya, M. & T. Miki (eds.) ESR Dating and Dosimetry. Ionics, Toyko: 157–163.

Hantoro, W. S., P. A. Pirazzoli, C. Jouannic, H. Faure, C. T. Hoang, U. Radtke, C. Causse, M. B. Best, R. Lafont, S. Bieda & K. Lambeck, 1994. Quaternary uplifted coral reef terraces on Alor Island, East Indonesia. Coral Reef 13: 215–223.

Hara, H. & M. Ikeya, 1989. Frequency sweep ESR spectrometer for dosimetry and dating. Appl. Rad. Isot. 40: 841–843.

Haskell, E. H., G. R. Kenner & R. B. Hayes, 1995. Electron paramagnetic resonance dosimetry of dentin following removal of organic material. Health Phys. 68: 579–584.

Haskell, E. H., R. B. Hayes & G. R. Kenner, 1996a. Plasterboard as an emergency dosimeter. Health Phys. 71: 95.

Haskell, E. H., R. B. Hayes & G. R. Kenner, 1996b. Preparation-induced errors in EPR dosimetry of enamel: Pre- and post-crushing sensitivity. Appl. Rad. Isot. 47: 1305–1310.

Haskell, E. H., R. B. Hayes & G. R. Kenner, 1997a. Improved accuracy of EPR dosimetry using a constant rotation goniometer. Rad. Meas. 27: 325–329.

Haskell, E. H., G. H. Kenner, R. B. Hayes, S. V. Sholom & V. V. Chumak, 1997b. An EPR intercomparison using teeth irradiated prior to crushing. Rad. Meas. 27: 419–424.

Haskell, E. H., R. B. Hayes & G. R. Kenner, 1999. An EPR model for separating internal ^{90}Sr doses from external γ-ray doses in teeth. LED 99: 9th International Conference on Luminescence and Electron Spin Resonance Dating, Abstracts, Rome: 167.

Haskell, E. H., R. B. Hayes, A. R. Romanyukha & G. R. Kenner, 2000. Preliminary report on the development of virtually non-destructive additive dose technique for EPR dosimetry. Appl. Rad. Isot. 52: 1065–1070.

Hataya, R., K. Tanaka & T. Miki, 1997a. Studies on a new ESR signal (R signal) of fault gouges for fault dating. Quat. Sci. Rev. (Quat. Geochron.) 16: 477–481.

Hataya, R., K. Tanaka & T. Miki, 1997b. A new ESR signal (R signal) in quartz grains taken from fault gouges: Its properties and significance for ESR fault dating. Appl. Rad. Isot. 48: 423–429

Hayes, R. B. & E. H. Haskell, 2000. A method for identification of interfering signals in EPR dating of tooth enamel. Rad. Meas. 32: 781–785.

Hayes, R. B., E. H. Haskell & G. R. Kenner, 1997. A mathematical approach to optimal selection of dose values in the additive dose method of EPR dosimetry. Rad. Meas. 27: 315–323.

Hayes, R. B., G. R. Kenner & E. H. Haskell, 1998. ESR dosimetry of Pacific walrus (Odobenus rosmarus divergens) teeth. Rad. Protec. Dosim. 77: 55–63.

Hayes, R. B., E. H. Haskell & J. K. Barrus, 2000. A virtually non-destructive EPR technique accounting for diagnostic X-rays. Rad. Meas. 32: 559–566.

Hennig, G. J., 1982. Notes and comments on the ^{230}Th/^{234}U dating on speleothem samples from the Grotte d'Aldène (Cesseras, Herault, France). Musée d'Anthropologie Préhistorique de Monaco, Bullétin 26: 21–25.

Hennig, G. J. & R. Grün, 1983. ESR dating in Quaternary geology. Quat. Sci. Rev. 2: 157–238.

Hennig, G. J. & F. Hours, 1982, Dates pour les passage entre l'Acheuléen et le Paléolithique Moyen à El Kown (Syrie). Paléorient 8: 81–83.

Hennig, G. J., W. Herr & E. Weber, 1981a. Electron spin resonance (ESR) dating on speleothems (notes on problems and progress). In de Lumley, H. & J. Labeyrie (eds.) Datations Absolus et les Analyses Isotopiques en Préhistoire: Méthodes et Limites. Prétirage. CNRS, Paris: 551–556.

Hennig, G. J., W. Herr, E. Weber & N. I. Xirotiris, 1981b. ESR-dating of the fossil hominid cranium from Petralona Cave, Greece. Nature 292: 533–536.

Hennig, G. J., W. Herr, E. Weber & N. I. Xirotiris, 1982. Petralona cave dating controversy. Nature 299: 280–282.

Hennig, G. J., R. Grün & K. Brunnacker, 1983a. Interlaboratory comparison project of ESR dating, Phase I. PACT 9: 447–452.

Hennig, G. J., R. Grün, K. Brunnacker & M. Pécsi, 1983b. ^{230}Th/^{234}U-Sowie ESR-Altersbestimmungen einiger Travertine in Ungarn. Eiszeitalter und Gegenwart 33: 9–19.

Hennig, G. J., R. Grün & K. Brunnacker, 1983c. Speleothems, travertines, and paleoclimates. Quat. Res. 20: 1–29.

Hennig, G. J., M. A. Geyh & R. Grün, 1985. The interlaboratory comparison project of ESR dating, Phase II. Nucl. Tr. Rad. Meas. 10: 945–952.

Hewgill, F. R., G. W. Kendrick, R. J. Weber & K. H. Tyrwoll, 1983. Routine ESR dating of emergent Pleistocene marine units in Western Australia. Search 14: 215–217.

Hewgill, F. R., G. W. Kendrick, R. J. Weber & K. H. Tyrwoll, 1985. Electron spin resonance dating: Reply. Search 16: 170.

Hirai, M., M. Ikeya, Y. Tsukamoto & C. Yamanaka, 1994. Radiolysis of solid CO_2^- and future electron spin resonance dating on outer planets. Jap. J. Appl. Phys. 2 33: L1453–L1455.

Hochi, A., M. Furusawa & M. Ikeya, 1993. Applications of microwave scanning ESR microscope: Human tooth with metal. Appl. Rad. Isot. 44: 401–405.

Hoffman, D., C. Woda, C. Strobl & A. Mangini, 2001. ESR-dating of the Arctic sediment core PS1535: Dose-response and thermal behaviour of the CO_2^- signal in foraminifera. Quat. Sci. Rev. (Quat. Geochron.) 20: 1009–1014.

Houben, J. L., 1971. Free radicals produced by ionizing radiation in bone and its constituents. Int. J. Rad. Biol. 20: 373–389.

Huang, P. H., Z. C. Peng, S. Z. Jin, R. Y. Liang & Z. R. Wang, 1985. An attempt to determine the archaeological doses of the travertine and the deer horn with ESR. In Ikeya, M. & T. Miki (eds.) ESR Dating and Dosimetry. Ionics, Toyko: 321–324.

Huang, P. H., S. Z. Jin, R. Y. Liang, Z. C. Peng, Y. C. Quan & Z. R. Wang, 1988. ESR dating and trapped electrons lifetime of quartz grains in loess of China. Quat. Sci. Rev. 7: 533–536.

Huang, P. H., R. Y. Liang, S. Z. Jin, Z. C. Peng & N. W. Rutter, 1989. Study on accumulated dose in littoral shells of Argentina. Appl. Rad. Isot. 40: 1119–1122.

Huang, P. H., S. Z. Jin, Z. C. Peng, R. Y. Liang, Z. J. Lu, Z. R. Wang, J. B. Chen & Z. X. Yuan, 1993. ESR dating of tooth enamel: Comparison with U-series, FT, and TL dating at the Peking Man site. Appl. Rad. Isot. 44: 239–242.

Huang, W. P., R. Ciochón, Y. M. Gu, R. Larrick, Q. R. Fang, H. P. Schwarcz, C. Yonge, J. de Vos & W. J. Rink, 1995. Early Homo and associated artefacts from Asia. Nature 378: 275–278.

Hütt, G. & J. Jaek, 1989. Dating accuracy from laboratory reconstruction of palaeodose. Appl. Rad. Isot. 40: 1057–1061.

Hütt, G., A. Molodkov, J. M. Punning & L. Pung, 1983. The first experience in ESR dating of fossil shells in Tallinn. PACT 9: 433–446.

Hütt, G., A. Molodkov, H. Kassel & A. Raukas, 1985. ESR dating of subfossil Holocene shells in Estonia. Nucl. Tr. Rad. Meas. 10: 891–898.

Idrissi, S., F. J. Callens, P. Moëns, R. Debuyst & F. Dejehet, 1996. The electron nuclear double resonance and electron spin resonance study of isotropic CO_2^- and SO_2^- radicals in natural carbonates. Jap. J. Appl. Phys. 1 35: 5331–5332.

Ignatiev, E. A., A. A. Romanyukha, A. A. Koshta & A. Wieser, 1996. Selective saturation method for EPR dosimetry with tooth enamel. Appl. Rad. Isot. 47: 333–337.

Ikeda, S. & M. Ikeya, 1992. Electron spin resonance (ESR) signal in natural and synthetic gypsum: An application of ESR to the age estimation of gypsum precipitates from the San Andreas fault. Jap. J. Appl. Phys. 2 31: L136–L138.

Ikeda, S. & M. Ikeya, 1993. ESR signals in synthetic gypsum-doped with $^{13}CO_3^{2-}$ ions. Appl. Rad. Isot. 44: 321–323.

Ikeda, S., M. Kasuya & M. Ikeya, 1991. ESR ages of Middle Pleistocene corals from the Ryukyu, Islands. Quat. Res. 36: 61–71.

Ikeda, S., D. Neil, M. Ikeya, A. Kai & T. Miki, 1992a. Spatial variation of CO_2^- and SO_3^- radicals in massive coral as environmental indicator. Jap. J. Appl. Phys. 2 31: L1644–L1646.

Ikeda, S., M. Kasuya & M. Ikeya, 1992b. ESR dating of corals and preannealing effects on ESR signals. Quat. Sci. Rev. 11: 203–207.

Ikeya, M., 1975. Dating a stalactite by electron paramagnetic resonance. Nature 255: 48–50.

Ikeya, M., 1976. Natural radiation dose in Akiyoshi Cavern and on a karst plateau. Health Phys. 31: 76–78.

Ikeya, M., 1977. Electron spin resonance dating and fission track detection of Petralona stalagmite. Anthropos 4: 152–166.

Ikeya, M., 1978a. Electron spin resonance as a method of dating. Archaeometry 20: 147–158.

Ikeya, M., 1978b. Spin-resonance ages of brown rings in cave deposits. Naturwissenschaften 65: 489.

Ikeya, M., 1980a. ESR dating of carbonates at Petralona Cave. Anthropos 7: 143–150.

Ikeya, M., 1980b. Paramagnetic alanine molecular radicals in fossil shells and bones. Naturwissenschaften 67: 474.

Ikeya, M., 1982a. A model of linear uranium accumulation for ESR age of Heidelberg (Mauer) and Tautavel bones. Jap. J. Appl. Phys. 2 21: L690–L692.

Ikeya, M., 1982b. Electron spin resonance of petrified woods for geological age assessment. Jap. J. Appl. Phys. 2 21: L128–L132.

Ikeya, M., 1982c. Petralona Cave dating controversy: Comment. Nature 299: 281–282.

Ikeya, M., 1983a. A trip to interdisciplinary world: Electron spin resonance (ESR) dating in archaeology and geology. Geology News 19A: 26–30.

Ikeya, M., 1983b. ESR studies of geothermal boring core at Hachobara Power Station. Jap. J. Appl. Phys. 2 22: L763–L765.

Ikeya, M., 1984a. Age limitation of ESR dating for carbonate fossils. Naturwissenschaften 71: 421–423.

Ikeya, M., 1984b. ESR dating of organic materials. Archeometry 84, Smithsonian, Washington.

Ikeya, M., 1985a. Electron spin resonance. In Rutter, N. W. (ed.) Dating Methods of Pleistocene Deposits and their Problems. Geological Association of Canada, St. John's, Geoscience Canada Reprint Series 2: 73–87.

Ikeya, M., 1985b. ESR ages of bones in paleoanthropology: Uranium and fluorine accumulation. In Ikeya, M. & T. Miki (eds.) ESR Dating and Dosimetry. Ionics, Toyko: 373–379.

Ikeya, M., 1985c. Car mileage determination with ESR signal of engine oil: A case of organic ESR dating. In Ikeya, M. & T. Miki (eds.) ESR Dating and Dosimetry. Ionics, Toyko: 453–456.

Ikeya, M., 1988. Dating and radiation dosimetry with electron spin resonance (ESR). Magn. Reson. Rev. 13: 237–255.

Ikeya, M., 1991. Electron spin resonance (ESR) microscopy in materials science. Ann. Rev. Materials Sci. 21: 45–63.

Ikeya, M., 1992. A theoretical growth curve of defect formation for electron spin resonance and thermoluminescence dating. Jap. J. Appl. Phys. 2 31: L1618–L1620.

Ikeya, M., 1993a. From earth to space: ESR dosimetry moves toward the 21st century. Appl. Rad. Isot. 44: 1–6.

Ikeya, M., 1993b. New Applications of Electron Spin Resonance: Dating, Dosimetry, and Microscopy. World Scientific, Singapore.

Ikeya, M., 1994a. ESR dating, dosimetry and microscopy: A trip to interdisciplinary fields as a wanderer from physics of defects. Nucl. Instrum. Meth. Phys. Res. B 91: 43–51.

Ikeya, M., 1994b. ESR and ESR microscopy in geosciences and radiation dosimetry. Appl. Magn. Reson. 7: 237–255.

Ikeya, M., 1996. Radiation effects in organics and inorganics for ESR dosimetry. Appl. Rad. Isot. 47: 1479–1481.

Ikeya, M. & M. Furusawa, 1989. A portable spectrometer for ESR microscopy, dosimetry, and dating. Appl. Rad. Isot. 40: 841–850.

Ikeya, M. & J. Golson, 1985. ESR dating of phytoliths (plant opal) in sediments: A preliminary study. In Ikeya, M. & T. Miki (eds.) ESR Dating and Dosimetry. Ionics, Toyko: 281–286.

Ikeya, M. & H. Ishii, 1989. Atomic bomb and accident dosimetry with ESR: Natural rocks and human tooth *in vivo* spectrometer. Appl. Rad. Isot. 40: 1021–1027.

Ikeya, M. & A. Kai, 1988. ESR dating of saline sediments using $NAHCO_3$ and NaCl. Quat. Sci. Rev. 7: 471–475.

Ikeya, M. & T. Miki, 1980a. A new dating method with a digital ESR. Naturwissenschaften 67: 191–192.

Ikeya, M. & T. Miki, 1980b. Electron spin resonance dating of animal and human bones. Science 207: 977–979.

Ikeya, M. & T. Miki, 1981a. Archaeological dose of Arago and Choukoutien materials with electron spin resonance (ESR). In de Lumley, H. & J. Labeyrie (eds.) Datations Absolus et les Analyses Isotopiques en Préhistoire: Méthodes et Limites. Prétirage. CNRS, Paris: 493–505.

Ikeya, M. & T. Miki, 1981b. Archaeological radiation dose of bones at archaeologically important sites with electron spin resonance. Union International des Sciences Préhistoriques et Protohistoriques, X Congress, Mexico, Proceedings: 39–53.

Ikeya, M. & T. Miki, 1985a. Electron spin resonance dating of organic materials from potato chips to a dead body. Nucl. Tr. Rad. Meas. 10: 909–912.

Ikeya, M. & T. Miki, 1985b. ESR age of cave deposits and natural radiation in Akiyoshi cave. In Ikeya, M. & T. Miki (eds.) ESR Dating and Dosimetry. Ionics, Toyko: 493–497.

Ikeya, M. & T. Miki, 1985c. ESR dating and preservation of papers. Naturwissenschaften 72: 32–33.

Ikeya, M. & K. Ohmura, 1981. Dating of fossil shells with electron spin resonance. J. Geol. 89: 247–251.

Ikeya, M. & K. Ohmura, 1983. Comparison of ESR ages of corals from marine terraces with ^{14}C and $^{230}Th/^{234}U$ ages. Earth & Planet. Sci. Lett. 65: 34–38.

Ikeya, M. & K. Ohmura, 1984. ESR age of Pleistocene shells measured by radiation assessment. Geoch. J. 18: 1147.

Ikeya, M. & A. N. Poulianos, 1979. ESR age of the trace of fire at Petralona. Anthropos 6: 44–47.

Ikeya, M. & S. Toyoda, 1991. Thermal effect in metamorphic rock around an intrusion zone with ESR studies. Appl. Magn. Reson. 2: 69–81.

Ikeya, M., T. Miki & R. Gospodaric, 1982a. ESR dating of Postojna Cave stalactite. Acta Carstologia 11: 117–130.

Ikeya, M., T. Miki & K. Tanaka, 1982b. Dating of fault by electron spin resonance on intrafault materials. Science 215: 1392–1393.

Ikeya, M., T. Miki, K. Tanaka, Y. Sakuramoto & K. Ohmura, 1983. ESR dating of faults at Rokko and Atotsugawa. PACT 9: 411–419.

Ikeya, M., J. Miyajima & S. Okajima, 1984. ESR dosimetry for atomic bomb survivors using shell buttons and tooth enamel. Jap. J. Appl. Phys. 2 23: L699–L701.

Ikeya, M., Y. Miúra, T. Tanosaki & H. Miúra, 1985. ESR spectroscopy and a possible method of Mn^{2+} dating for minerals and fossils. In Ikeya, M. & T. Miki (eds.) ESR Dating and Dosimetry. Ionics, Toyko: 477–484.

Ikeya, M., S. D. Devine, N. E. Whitehead & J. Hedenquist, 1986. Detection of methane in geothermal quartz by ESR. Chem. Geol. 56: 185–192.

Ikeya, M., F. O. Baffa & S. Mascarenhas, 1989. Quality assessment of coffee beans with ESR and γ-ray irradiation. Appl. Rad. Isot. 40: 1219-1222.

Ikeya, M., H. Kohno, S. Toyoda & Y. Mizuta, 1992. Spin-spin relaxation time of E' centers in neutron-irradiated quartz (SiO_2) and in fault gouge. Jap. J. Appl. Phys. 2 31: L1539–L1541.

Ikeya, M., A. Tani & C. Yamanaka, 1995. Electron spin resonance isochrone (sic) dating of fracture age: Grain-size dependence of the dose rates for fault gouge. Jap. J. Appl. Phys. 2 34: L334–L337.

Ikeya, M., H. Ochiai, A. tani, 1997a. Total dose (Td) formula for uranium saturation-uptake model for ESR and TL dating. Rad. Meas. 27: 339–343.

Ikeya, M., H. Sasaoka, H. Toda, K. Kanosue & M. Hirai, 1997b. Future ESR and optical dating of outer planet icy materials. Quat. Sci. Rev. 16: 431–435.

Ildefonse, P., J. P. Muller, B. Clozel & G. Callas, 1990. The study of two alteration systems as natural analogues for radionuclide release and migration. Engin. Geol. 29: 413–439.

Imai, N. & K. Shimokawa, 1985. Dating of volcanic ash by electron spin resonance using aluminum and titanium centers in plagioclase. In Ikeya, M. & T. Miki (eds.) ESR Dating and Dosimetry. Ionics, Toyko: 187–190.

Imai, N. & K. Shimokawa, 1988. ESR dating of Quaternary tephra from Mt. Osore-Zan using Al and Ti centers in quartz. Quat. Sci. Rev. 7: 523–527.

Imai, N. & K. Shimokawa, 1989. ESR dating of the tephra 'crystal ash' distributed in Shinshu, central Japan. Appl. Rad. Isot. 40: 1177–1180.

Imai, N. & K. Shimokawa, 1993. ESR ages and trace elements in a fossil mollusc shell. Appl. Rad. Isot. 44: 161–165.

Imai, N., K. Shimokawa & M. Hirota, 1985. ESR dating of volcanic ash. Nature 314: 81–83.

Imai, N., K. Shimokawa & M. Hirota, 1986. Dating volcanic ash by electron spin resonance: Reply. Nature 319: 796.

Imai, N., K. Shimokawa, K. Sakaguchi & M. Takada, 1992. ESR dates and thermal behavior of Al and Ti centers in quartz for the tephra and welded tuff in Japan. Quat. Sci. Rev. 11: 257–265.

Imai, N., K. Shimokawa & M. Yamamoto, 1994. ESR study of radiation centres and thermal-behavior in chert. Quat. Sci. Rev. (Quat. Geochron.) 13: 641–645.

Ishii, H. & M. Ikeya, 1990. An electron spin resonance system for *in vivo* human tooth dosimetry. Jap. J. Appl. Phys. 1 29: 871–875.

Ishii, H. & M. Ikeya, 1993. Defects in synthesized apatite powder and sintered material. Appl. Rad. Isot. 44: 95–100.

Ishii, H., M. Ikeya & M. Okano, 1990. ESR dosimetry of teeth of residents close to Chernobyl reactor accident. J. Nucl. Sci. Techn. 27: 1153–1155.

Ishii, H., M. Ikeya, M. Kasuya & M. Furusawa, 1991. ESR, TL, and FT measurements of a natural apatite. Nucl. Tr. Rad. Meas. 18: 189–192.

Ito, T. & S. Sawada, 1985. Reliable criteria for the selection of sampling points for ESR fault dating. In Ikeya, M. & T. Miki (eds.) ESR Dating and Dosimetry. Ionics, Toyko: 229–237.

Ivannikov, A. I., V. G. Skvortsov, V. F. Stepanenko, A. F. Tsyb, L. G. Kamidova, D. D. Tikunov, 2000. Tooth enamel EPR dosimetry: Sources of error and their correction. Appl. Rad. Isot. 52: 1291–1296.

Ivanovich, M., C. Vita-Finzi & G. J. Hennig, 1983. Uranium-series dating of molluscs from uplifted Holocene beaches in the Persian Gulf. Nature 302: 408–410.

Jacobi, R. M., P. W. Rowe, M. A. Gilmour, R. Grün & T. C. Atkinson, 1998. Radiometric dating of the Middle Paleolithic tool industry and associated fauna of Pin Hole Cave, Creswell Crags, England. J. Quat. Sci. 13: 29–42.

Jacobs, C., P. de Cannière, R. Debuyst, F. Dejehet & D. Apers, 1989. ESR study of γ-ray-irradiated synthetic calcium carbonates. Appl. Rad. Isot. 40: 1147–1152.

Jacobson, G., Y. J. Chen & A. V. Arakel, 1988. The central Australian groundwater discharge zone: Evolution of associated calcrete and gypcrete deposits. Austr. J. Earth Sci. 35: 549–565.

Jaek, I., G. Hütt, V. Seeman, L. Brodski, 1995. Luminescence and ESR of the natural alkali feldspars extracted from sediments doped by probe impurities. Rad. Meas. 24: 557–563.

Jani, M. G., R. B. Bossoli, L. E. Halliburton, 1983. Further characterization of the E_1' center in crystalline SiO_2. Phys. Rev. B 27: 2285–2293.

Jin, S. Z., Z. Deng & P. H. Huang, 1993. A comparative study on optical effects of E' centre in quartz grains from loess and faults. Appl. Rad. Isot. 44: 175–178.

Johnson, D. L., M. A. Glassow, K. B. Graettinger, B. A. Blackwell, T. Morgan, D. R. Muhs, J. A. Parsons & K. A. Rockwell, 1991. A field guide to the geoarchaeology of the Vandenberg-Lompoc-Point Conception Area, Santa Barbara County, California. In Walawender, M. J. & B. B. Hanan (eds.) Geological Excursions in Southern California and Mexico. Geological Society of America, Boulder: 244–271.

Jonas, M., 1995. Spectral deconvolution of the ESR dating signal in fossil tooth enamel. Quat. Sci. Rev. (Quat. Geochron.) 14: 431–438.

Jonas, M., 1997. Concepts and methods of ESR dating. Rad. Meas. 27: 943–973.

Jonas, M. & R. Grün, 1997. Q-band studies of fossil tooth enamel: Implications for spectrum deconvolution and dating. Rad. Meas. 27: 49–58.

Jonas, M. & E. Marseglia, 1997. The case for the use of integrated spectrum deconvolution in ESR dating: A numerically generated examples. Rad. Meas. 27: 359–363.

Jonas, M., L. P. Zhou, E. Marseglia & P. Mellars, 1994. New analysis of EPR spectra of fossil tooth enamel. Camb. Archaeol. J. 4: 139–146.

Jones, A. T., B. A. Blackwell & H. P. Schwarcz, 1991. ESR age of emerged marine fossil corals from Molokai, Hawaii. Geol. Soc. Am., Abstr. 23: A459.

Jones, A. T., B. A. Blackwell & H. P. Schwarcz, 1993. Annealing and etching of corals for ESR dating. Appl. Rad. Isot. 44: 153–156.

Jones, J. P. E., B. R. Angel & P. L. Hall, 1974. Electron spin resonance studies of dopes synthetic kaolinite II. Clay Minerals 10: 257–270.

Joppard, T., 1984. Contribution à la Datation des Concretions calcaires par RPE: Etude de la Calcite dopée aux Acides humiques. Unpub. Lic. Sci. thèse, Université de Louvain-la-Neuve, 59 pp.

Kai, A., 1996. ESR study of radiation-induced organic radicals in $CaCO_3$. Appl. Rad. Isot. 47: 1483–1487.

Kai, A. & M. Ikeya, 1989. ESR study of fossil shells in sediments at Hamana Lake. Appl. Rad. Isot. 40: 1139–1142.

Kai, A. & T. Miki, 1989. Electron spin resonance of organic radicals derived from amino acids in calcified fossils. Jap. J. Appl. Phys. 1 28: 2277–2282.

Kai, A. & T. Miki, 1991. Sulfite radicals in irradiated calcite. Jap. J. Appl. Phys. 1 30 : 1109–1110.

Kai, A. & T. Miki, 1992. Electron spin resonance of sulfite radicals in irradiated calcite and aragonite. Rad. Phys. Chem. 40: 469–476.

Kai, A., T. Miki & M. Ikeya, 1988. ESR dating of teeth, bones and eggshells excavated at a Paleolithic site of Douara Cave, Syria. Quat. Sci. Rev. 7: 503–507.

Kai, A., T. Murata & T. Miki, 1993. Effects of trivalent metal impurities on radiation-induced radicals in calcite. Appl. Rad. Isot. 44: 311–314.

Kalefezra, J. & Y. S. Horowitz, 1979. Electron backscattering corrections for β dose rate estimations in archaeological objects. PACT 2/3: 428–438.

Kanaori, Y., K. Tanaka & K. Miyakoshi, 1985. Further studies on the use of quartz grains from fault gouges to establish the age of faulting. Engin. Geol. 21: 175–194.

Kanosue, K., M. Hirai & M. Ikeya, 1996. Preliminary study for future ESR dating of solid SO_2. Appl. Rad. Isot. 47: 1433–1436.

Kanosue, K., H. Toda, M. Hirai, H. Kanamori & M. Ikeya, 1997. Thermoluminescence (TL) and ESR study of γ-irradiated SO_2 frost for future dating in outer planets. Rad. Meas. 27: 399–403.

Karakostanoglou, I. & H. P. Schwarcz, 1983. ESR isochron dating. PACT 9: 391–398.

Kasuya, M., S. Brumby & J. Chappell, 1991. ESR signals from natural gypsum single crystals: Implications for ESR dating. Nucl. Tr. Rad. Meas. 18: 329–333.

Kasuya, M., M. Furusawa & M. Ikeya, 1990. Distributions of paramagnetic centres and α-emitters in a zircon single crystal. Nucl. Tr. Rad. Meas. 17: 563–568.

Katzenberger, O., 1985. Untersuchungen zur ESR Datierung von dünnschlaligen Molluskengehäusen. Unpub. Ph. D. thesis, University of Köln, Köln, 39 pp.

Katzenberger, O. & R. Grün, 1985. ESR dating of circumarctic molluscs. Nucl. Tr. Rad. Meas. 10: 885–890.

Katzenberger, O. & N. Willems, 1988. Interferences encountered in the determination of AD of mollusc samples. Quat. Sci. Rev. 7: 485–489.

Katzenberger, O., R. Debuyst, P. de Cannière, F. Dejehet, D. Apers & M. Barabas, 1989. Temperature experiments on mollusc samples: An approach to ESR signal identification. Appl. Rad. Isot. 40: 1113–1118.

Kenner, G. H., E. H. Haskell, R. B. Hayes, A. Baig & W. I. Higuchi, 1998. ESR properties of synthetic apatites, deorganified dentin, and enamel. Calc. Tissue Int. 62: 443–446.

Kislyakov, Y. M., B. M. Moiseyev, L. T. Rakov & E. G. Kugalin, 1975. Distribution of concentrations of E-centers in minerals of rock enclosing uranium mineralization. Geologiya Rudnnykh Mestorozhdeniy 1975 (3): 86–92.

Koba, M., M. Ikeya, T. Miki & T. Nakata, 1985. ESR ages of the Pleistocene coral reef limestones in the Ryukyu Islands, Japan. In Ikeya, M. & T. Miki (eds.) ESR Dating and Dosimetry. Ionics, Toyko: 93–104.

Koba, M., M. Tamura, M. Ikeya, T. Kiagara, T. Nakashima & H. Kan, 1987. Quaternary shorelines and crustal movements on Minamidaito-jima, northwestern Pacific. In Qin, Y. & S. Zhao (eds.) Late Quaternary Sea-level Changes. China Ocean Press, Beijing: 187–198.

Kobiyashi, T. & M. Suhara, 1985. Single crystal ESR dating: Method and apparatus. In Ikeya, M. & T. Miki (eds.) ESR Dating and Dosimetry. Ionics, Toyko: 293–298.

Kohno, H., C. Yamanaka & M. Ikeya, 1994a. Pulsed and CW-ESR studies of CO_2^- in $CaCO_3$ irradiated by 1.6 MeV He^+ and α-rays for ESR dating. Jap. J. Appl. Phys. 1 33: 5743–5746.

Kohno, H., C. Yamanaka, M. Ikeya, S. Ikeda & Y. Horino, 1994b. An ESR study of radicals in $CaCO_3$ produced by 1.6 MeV He^+ irradiation and γ irradiation. Nucl. Instrum. Meth. in Phys. Res. B 91: 366–369.

Kohno, H., C. Yamanaka & M. Ikeya, 1996. Effects of α-irradiation and pulsed ESR measurements of evaporites. Appl. Rad. Isot. 47: 1459–1463.

Komura, K. & M. Sakanoue, 1985. γ-ray spectroscopy for ESR dating. In Ikeya, M. & T. Miki (eds.) ESR Dating and Dosimetry. Ionics, Toyko: 9–18.

Korolov, K. G., B. M. Moiseyev & Y. M. Schmariovich, 1977. Allowance for solubility of mineral components when determining absolute age of uranium ores. AN SSSR Izvestiya, Ser. Geol. 1977 (4): 74–85.

Kosaka, K. & S. Sawada, 1985. Fault gouge analysis and ESR dating of the Tsurukawa Fault, west of Tokyo: Significance of minute sampling. In Ikeya, M. & T. Miki (eds.) ESR Dating and Dosimetry. Ionics, Toyko: 257–266.

Koshta, A. A., A. Wieser, E. A. Igantiev, S. Bayankin, A. A. Romanyukha & M. O. Degteva, 2000. A new computer proceure for routine EPR-dosimetry on tooth enamel: Description and verification. Appl. Rad. Isot. 52: 1287–1290.

Lau, B., B. A. B. Blackwell, H. P. Schwarcz, I. Turk & J. I. B. Blickstein, 1997. Dating a Flautist? Using ESR (electron spin resonance) in the Mousterian Cave Deposits at Divje Babe I, Slovenia. Geoarchaeology 12: 507–536.

Laurent, M., C. Falguères, J. J. Bahain & Y. Yokoyama, 1994. Géochronologie du système de terraces fluviatiles quaternaires du bassin de la Somme par datation RPE sur quartz, déséquilibres des familles de l'uranium et magnétostratigraphie. C. r. Acad. Sci., Paris 318: 521–526.

Laurent, M., C. Falguères, J. J. Bahain, L. Rousseau & B. Van Vliet Lanoé, 1998. ESR dating of quartz extracted from Quaternary and Neogene sediments: Method, potential and actual limits. Quat. Sci. Rev. 17: 1057–1062.

Lea, J. S., N. J. F. Dodd, A. J. Swallow, 1988. A method for testing for irradiation of poultry. J. Food Sci. Techn. 23: 34–40.

Lea-Wilson, M. A., J. N. Lomer, 1996. Electron spin resonance of a new defect *R*13 and additional data on the defect R_1 in irradiated diamond. Phil. Mag. A 74: 685–695.

Lee, H. K., 1994. ESR Dating of Fault Rocks. Unpub. Ph.D. thesis, McMaster University, Hamilton.

Lee, H. K., 2001. ESR dating of the subside faults in the Yangsan fault system, South Korea. Quat. Sci. Rev. (Quat. Geochron.) 999–1003.

Lee, H. K. & H. P. Schwarcz, 1993. An experimental study of shear-induced zeroing of ESR signals in quartz. Appl. Rad. Isot. 44: 191–195.

Lee, H. K. & H. P. Schwarcz, 1994a. Criteria for complete zeroing of ESR signals during faulting of the San Gabriel Fault Zone, Southern California. Tectonophys. 235: 317–337.

Lee, H. K. & H. P. Schwarcz, 1994b. ESR plateau dating of fault gouge. Quat. Sci. Rev. (Quat. Geochron.) 13: 629–634.

Lee, H. K. & H. P. Schwarcz, 1995. Fractal clustering of fault activity in California. Geology 23: 377–380.

Lee, H. K. & H. P. Schwarcz, 1996. Electron spin resonance plateau dating of periodicity of activity on the San Gabriel Fault Zone, southern California. Geol. Soc. Am. Bull. 108: 735–746.

Lee, H. K., W. J. RINK & H. P. Schwarcz, 1997. Comparison of ESR signal dose-responses in modern and fossil tooth enamels. Rad. Meas. 27: 405–411.

Li, D. S., H. Gao, Z. Z. Ding, X. T. Zhao & B. M. Yang, 1993a. On the estimation of depositional ages of loess by ESR. Appl. Rad. Isot. 44: 197–201.

Li, D. S., X. T. Zhao, Z. Z. Ding, B. M. Yang & H. Gao, 1993b. A study of the clock zero of sedimentary loess for ESR dating. Appl. Rad. Isot. 44: 203–206.

Li, H. H., 1988. Properties of thermoluminescence (TL) and electron spin resonance (ESR) in deposited carbonates. Nucl. Tr. Rad. Meas. 14: 259–266.

Li, S. H., 1999. Detection of insufficiently bleached sediments and selection of suitable grains in optical dating of quartz. LED 99: 9th International Conference on Luminescence and Electron Spin Resonance Dating, Abstracts, Rome: 86.

Liang, R. Y, Z. C. Peng, S. H. Jin & P. H. Huang, 1989. Estimation of the influence of experimental conditions on ESR dating results. Appl. Rad. Isot. 40: 1071–1075.

Libin, L., Z. Yiyun, L. Lizhong & L. Tiechen (sic), 1996. The Mn valence variation induced by neutron irradiation in sapphire: Mn. Appl. Rad. Isot. 47: 1523–1526.

Liidja, G., 1999. UV light induced paramagnetism in tooth enamel. LED 99: 9th International Conference on Luminescence and Electron Spin Resonance Dating, Abstracts, Rome: 170.

Liidja, G., J. Past. J. Puskar, E. Lippmaa, 1996. Paramagnetic resonance in tooth enamel created by ultraviolet light. Appl. Rad. Isot. 47: 785–788.

Lin, Z. R., M. E. Yang, Q. C. Fan & M. Ikeya, 1985. ESR dating of the age of the fault movement at San-Jiang fault zone in West China. In Ikeya, M. & T. Miki (eds.) ESR Dating and Dosimetry. Ionics, Toyko: 205–210.

Linke, G., O. Katzenberger & R. Grün, 1985. Description and ESR dating of the Holstein interglaciation. Quat. Sci. Rev. 4: 319–331.

Liritzis, Y., 1982. Petralona cave dating controversy: Comment. Nature 299: 280–282.

Liritzis, Y. & Y. Maniatis, 1989. ESR experiments on Quaternary calcites and bones for dating purposes. J. Radioanalyt. Nucl. Chem. 129: 3–21.

Lloyd, R. V. & D. N. Lumsden, 1987. The influence of temperature on the radiation damage line in electron spin resonance spectra of metamorphic dolomites: A potential paleothermometer. Chem. Geol. 64: 103–108.

Lyons, R. G., 1988a. Determination of α effectiveness in electron spin resonance dating using nuclear accelerator techniques: Methods and energy dependence. Nucl. Tr. Rad. Meas. 14: 275–280.

Lyons, R. G., 1988b. A simple statistical method for the comparison of accumulated dose in electron spin resonance dating. Nucl. Tr. Rad. Meas. 14: 361–363.

Lyons, R. G., 1993. That uncertain age: A multilevel calibration exercise. Appl. Rad. Isot. 44: 107–118.

Lyons, R. G., 1996. Back to basics: Qualitative spectral analysis as an investigatory tool, using calcite as a case study. Appl. Rad. Isot. 47: 1385–1391.

Lyons, R. G., 1997. The difference in integrating. Rad. Meas. 27: 345–350.

Lyons, R. G. & B. J. Brennan, 1989. α-particle effectiveness in ESR dating: Energy dependence and implications for dose rate calculations. Appl. Rad. Isot. 40: 1063–1070.

Lyons, R. G. & B. J. Brennan, 1991. a/γ effectiveness ratios of calcite speleothems. Nucl. Tr. Rad. Meas. 18: 223–227.

Lyons, R. G. & S. M. Tan, 2000. Differentials or integrals: Pluses and minuses in their application to additive dose techniques. Appl. Rad. Isot. 52: 1051–1057.

Lyons, R. G., W. B. Wood & P. W. Williams, 1985. Determination of α efficiency in speleothem calcite by nuclear accelerator techniques. In Ikeya, M. & T. Miki (eds.) ESR Dating and Dosimetry. Ionics, Toyko: 39–48.

Lyons, R. G., G. A. Bowmaker & C. J. O'Connor, 1988. Dependence of accumulated dose in ESR dating on microwave power: A contra-indication to the routine use of low power levels. Nucl. Tr. Rad. Meas. 14: 243–251.

Lyons, R. G., P. C. Crossley, R. G. Ditchburn, W. C. McCabe & N. Whitehead, 1989. Radon escape from New Zealand speleothems. Appl. Rad. Isot. 40: 1153–1158.

Lyons, R. G., G. A. Bowmaker & C. J. O'Connor, 1992. High or low: The optimal microwave power for determination of accumulated dose in ESR dating of speleothem calcite. Appl. Rad. Isot. 43: 825–827.

Lyons, R. G., B. J. Brennan & M. Redhead, 1993. Games with γ's: Problems in environmental γ dose determination. Appl. Rad. Isot. 44: 131–138.

Mackey, J. H., 1963. EPR study of impurity-colour centres in germanium-doped quartz. J. Chem. Phys. 39: 74–83.

MacPhee, R. D. E., D. A. McFarlane & D. C. Ford, 1989. Pre-Wisconsinan mammals from Jamaica and models of Late Quaternary extinction in the Greater Antilles. Quat. Res. 31: 94–106.

Malmberg, R. & U. Radtke, 2000. The α-efficiency and its importance for ESR dating. Rad. Meas. 32: 747–750.

Mangini, A., M. Segl & W. Schmitz, 1983. ESR studies on $CaCO_3$ of deep-sea sediments. PACT 9: 439–446.

Marfunin, A. S., 1979. Spectroscopy, Luminescence, and Rad. Centers in Minerals. Springer Verlag, Berlin, 352 pp.

Marks, A. E., Demidenko, E. Yu, K. Monigal, V. I. Usik, C. R. Ferring, A. Burke, W. J. Rink & C. McKinney, 1997. Starsosele: New excavations, new and different results. Curr. Anthro. 37: 112–123.

Marsh, R. E. & W. J. Rink, 2000. β-gradient isochron dating of thin tooth enamel layers using ESR. Rad. Meas. 32: 567–570.

Martinez Tudela, A., F. Robles Cuenca, C. Santisteban Bove, R. Grün & B. Hentzsch, 1986. Los travertinos del Rio Matarraña, Beceite (Teruel), como indicatores paleoclimiaticos del Quaternario. Symposium on Climatic Fluctuations during the Quaternary in the Western Mediterranean, Proceedings: 307–324.

Masaoudi, H., C. Falguères, J. J. Bahain & M. H. Moncel, 1997. Dating of the Middle Paleolithic site of Payre (Ardèche): New radiometric data (U-series and ESR methods). C. r. Acad. Sci., Paris 324: 149–156.

Mascarenhas, S., F. O. Baffa & M. Ikeya, 1982. Electron spin resonance dating of human bones from Brazilian shell mounds (Sambaquis). Am. J. Phys. Anthro. 59: 413–417.

Matasuoka, M., S. H. Tatumi, S. Watanabe, K. Inabe & T. Nikajima, 1993. ESR and TL in quartz from a Brazilian sediment. Appl. Rad. Isot. 44: 185–189.

Matsuki, K., Y. Shimoyawa & H. Watari, 1985. ESR dating of coal and carbonization. In Ikeya, M. & T. Miki (eds.) ESR Dating and Dosimetry. Ionics, Toyko: 439–446.

McDermott, F., R. Grün, C. B. Stringer & C. J. Hawkesworth, 1993a. Mass-spectrometric U-series dates for Israeli Neanderthal/early modern hominid sites. Nature 363: 252–255.

McDermott, F., C. J. Hawkesworth, R. Grün & C. B. Stringer, 1993b. Reply: Dating hominid remains. Nature 366: 415.

McDermott, F., C. B. Stringer, R. Grün, C. T. Williams, V. K. Din & C. J. Hawkesworth, 1996. New late Pleistocene U/Th and ESR dates for the Singa hominid (Sudan). J. Human Evol. 31: 507–516.

McMorris, D. W., 1969. Trapped electron dating: ESR studies. Nature 222: 870–871.

McMorris, D. W., 1970. ESR detection of fossil α damage in quartz. Nature 226: 146–148.

McMorris, D. W., 1971. Impurity color centers in quartz and trapped electron dating: Electron spin resonance, thermoluminescence studies. J. Geophys. Res. 76: 7875–7887.

Meguro, K. & M. Ikeya, 1992. Stabilization of radicals by doping from aqueous solutions into crystals of hydroxyapatite. Jap. J. Appl. Phys. 1 31: 1353–1357.

Mellars, P. A. & R. Grün, 1991. A comparison of the electron spin resonance and thermoluminescence dating methods: The results of the ESR dating at Le Moustier (France). Camb. Arch. J. 1: 269–276.

Mellars, P. A., L. P. Zhou & E. A. Marseglia, 1997. Compositional inhomogeneity of sediments and its potential effects on dose rate estimation for electron spin resonance dating of tooth enamel. Archaeometry 39: 169–176.

Miallier, D., S. Sanzelle, J. Faïn, M. Montret, T. Pilleyre, S. Soumana & C. Falguères, 1994a. Attempts at dating pumice deposits around 580 ka by use of red TL and ESR of xenolithic quartz inclusions. Rad. Meas. 23: 143–153.

Miallier, D., J. Faïn, S. Sanzelle, T. Pilleyre, M. Montret, S. Soumana & C. Falguères, 1994b. Intercomparisons of red TL and ESR signals from heated quartz grains. Rad. Meas. 23: 399–404.

Miallier, D., S. Sanzelle, C. Falguères, J. Faïn, T. Pilleyre & P. M. Vincent, 1997. TL and ESR of quartz from the astrobleme, Aorunga (Sahara of Chad). Quat. Sci. Rev. 16: 265–274.

Michel, V., C. Falguères & J. M. Dolo, 1998. ESR signal behavior study at $g \sim 2.002$ of modern and fossil bones for heating paleotemperature assessments. Rad. Meas. 29: 95–103.

Miki, T., 1989. ESR spatial dosimetry using localized magnetic field modulation. Appl. Rad. Isot. 40: 1243–1246.

Miki, T. & M. Ikeya, 1978. Thermoluminescence and ESR dating of Akiyoshi stalactite. Jap. J. Appl. Phys. 17: 1703–1704.

Miki, T. & M. Ikeya, 1982. Physical basis of fault dating with ESR. Naturwissenschaften 69: 90–91.

Miki, T. & M. Ikeya, 1985. A plateau method for total dose evaluation with digital data processing. Nucl. Tr. Rad. Meas. 10: 913–919.

Miki, T. & A. Kai, 1991. Thermal annealing of radicals in aragonitic $CaCO_3$ and $CaHPO_4 \cdot 2H_2O$. Jap. J. Appl. Phys. 1 30: 404–410.

Miki, T., T. Yahagi & M. Ikeya, 1985a. Dendrochronological ESR dating. In Ikeya, M. & T. Miki (eds.) ESR Dating and Dosimetry. Ionics, Toyko: 315–320.

Miki, T., T. Yahagi, M. Ikeya, N. Sugawara & J. Furuno, 1985b. ESR dating of organic substances: Corpse for forensic medicine. In Ikeya, M. & T. Miki (eds.) ESR Dating and Dosimetry. Ionics, Toyko: 447–452.

Miki, T., M. Ikeya & A. Kai, 1987a. ESR dating utilizing change in valency or ligand of transition metals. Jap. J. Appl. Phys. 1 26: 972–973.

Miki, T., M. Ikeya & A. Kai, 1987b. Electron spin resonance of bloodstains and its application to the estimation of the time after bleeding. Forensic Sci. Int. 35: 149–158.

Miki, T., A. Kai & M. Ikeya, 1988. Electron spin resonance dating of organic substances utilizing paramagnetic degradation products. Nucl. Tr. Rad. Meas. 14: 253–258.

Miki, T., A. Kai & Murata, 1993. Radiation-induced radicals in sulfite-doped $CaCO_3$. Appl. Rad. Isot. 44: 315–319.

Miki, T., T. Murata, H. Kumai & A. Yamashiro, 1996. A high resolution EPR-CT microscope using cavity-resonators equipped with small field gradient coils. Appl. Rad. Isot. 47: 1599–1603.

Millard, A. R. & R. E. M. Hedges, 1996. A diffusion-adsorption model of uranium uptake by archaeological bone. Geoch. Cosmoch. Acta 60: 2139–2152.

Millard, A. R. & A. W. G. Pike, 1999. Uranium series dating of the Tabun Neanderthal: A cautionary note. J. Human Evol. 36: 581–585.

Mirecki, J., J. F. Wehmiller & A. R. Skinner, 1995. Geochronology of Quaternary coast plain deposits, southeastern Virginia, USA. J. Coastal Res. 11: 1135–1144.

Mitchell, S. F., R. K. Pickerill, B. A. B. Blackwell & A. R. Skinner, 2000. The age of the Port Morant Formation, southeastern, Jamaica. Carib. J. Earth Sci. 34: 1–4.

Miúra, Y., Y. Ohkura, M. Ikeya, T. Miki, J. Rucklidge, N. Takaoka & T. D. F. Nielsen, 1985. ESR data of Danish calcite at Cretaceous and Tertiary boundary. In Ikeya, M. & T. Miki (eds.) ESR Dating and Dosimetry. Ionics, Toyko: 469–476.

Miyakawa, K. & K. Tanaka, 1985. An ESR study on the geothermal histories of the Kurobe, Central Japan. In Ikeya, M. & T. Miki (eds.) ESR Dating and Dosimetry. Ionics, Toyko: 165–174.

Miyake, M., K. J. Liu, T. M. Walczak & H. M. Swartz, 2000. In vivo dosimetry of accidental exposures to radiation: Experimental results indicating the feasibility of practical use in human subjects. Appl. Rad. Isot. 52: 1031–1038.

Miyamaru, H. & M. Ikeya, 1993. One dimensional scanning ESR microscope using microwire array. Appl. Rad. Isot. 44: 397–400.

Moëns, R. D. W., F. J. Callens, R. M. H. Verbeek & D. E. Naessens, 1993. An EPR spectrum decomposition study of precipitated carbonated apatites (NCAP) at 25 °C: Adsorption of molecules from the atmosphere on the apatite powders. Appl. Rad. Isot. 44: 279–285.

Moiseyev, B. M., 1980. Paleodosimetric method for determining the age of deposits of radioactive elements. Doklady Acadamii, Nauk SSSR 233: 1227–1229.

Moiseyev, B. M. & L. T. Rakov, 1977. Paleodosimetry properties of E' centers in quartz. Doklady Acadamii, Nauk SSSR 233: 679–683.

Moiseyev, B. M., T. M. Petropavlov, N. P. ShmariovAch, N. P. Strelianov & V. M. Rekharskaya, 1982. Paleodosimetric age determination for uranium deposits. Geologiya Rudnykh Mestorozhdeniy 1982 (3): 61–70.

Molodkov, A., 1986. Application of ESR to the dating of subfossil shells from marine deposits. Ancient TL 4: 49–54.

Molodkov, A., 1988. ESR dating of Quaternary mollusc shells: Recent advances. Quat. Sci. Rev. 7: 477–484.

Molodkov, A., 1989. The problem of long-term fading of absorbed palaeodose on ESR dating of Quaternary mollusc shells. Appl. Rad. Isot. 40: 1087–1093.

Molodkov, A., 1993. ESR dating of nonmarine mollusc shells. Appl. Rad. Isot. 44: 145–148.

Molodkov, A., 1996. ESR dating of Lymnaea baltica and Cerastoderma glaucum from Low Ancylus Level and Transgress Littorina Sea deposits. Appl. Rad. Isot. 47: 1427–1432.

Molodkov, A., 2001. Early Paleolithic cave-site in the Northern Caucasus: ESR-evidence for early man as derived from terrestrial mollusc shells. Quat. Sci. Rev. (Quat. Geochron.) 20: 1051–10.

Molodkov, A. & G. Hütt, 1985. ESR dating of subfossil shells: some refinements. In Ikeya, M. & T. Miki (eds.) ESR Dating and Dosimetry. Ionics, Toyko: 145–155.

Molodkov, A. & A. Raukas, 1985. The age of upper Pleistocene marine deposits of the boreal transgression on the basis of ESR (electron spin resonance) dating of subfossil mollusk shells. Boreas 17: 267–272.

Molodkov, A., A. Dreimanis, O. Aboltins & A. Raukas, 1998. The ESR age of Portlandia arctica shells from glacial deposits of central Latvia: An answer to a controversy on the age and genesis of their enclosing sediments. Quat. Sci. Rev. 17: 1077–1094.

Monigal, K., A. E. Marks, Demidenko, E. Yu, W. J. Rink, H. P. Schwarcz, C. R. Ferring & C. McKinney, 1997. Nouvelles découvertes des restes humains au site Paléolithique Moyen de Starosele, Crimée (Ukraine). Préhistoire Européene 11.

Monnier, J. L., B. Hallégouet, S. Hinguant, M. Laurent, P. Auguste, J. J. Bahain, C. Falguères, A. Gebhardt, D. Marguerie, N. Molines, H. Morzadec & Y. Yokoyama, 1994. A new regional group of the lower Paleolithic in Brittany (France), recently dated by electron spin resonance. C. r. Acad. Sci., Paris 319: 155–160.

Morency, M., P. L. Emond & P. H. von Bitter, 1969. Dating conodonts using electron spin resonance. Kansas Geol. Survey, Bull. 199: 17–19.

Morinaga, H., H. Inokuchi, K. Yasakawa, M. Ikeya, T. Miki & M. Kusakabe, 1985. Paleomagnetism, paleoclimatology, and ESR stalagmite deposits. In Ikeya, M. & T. Miki (eds.) ESR Dating and Dosimetry. Ionics, Toyko: 31–37.

Mudelsee, M., M. Barabas & A. Mangini, 1992. ESR dating of the Quaternary deep sea sediment core RC17–177. Quat. Sci. Rev. 11: 181–189.

Muller, J. P. & G. Callas, 1989. Tracing kaolinites through their defect centers: Kaolinite paragenesis in a laterite (Cameroon). Econ. Geol. 84: 694–707.

Muñoz, R. E., E. Adem, G. Burillo, V. R. Gleason & S. H. Murrieta, 1994. ESR studies of irradiated ground corn as a dosimeter. Rad. Phys. Chem. 43: 311–313.

Murata, T., A. Kai & T. Miki, 1993a. Electron spin resonance of radicals from amino acids in synthetic calcified tissues. Appl. Rad. Isot. 44: 299–303.

Murata, T., A. Kai & T. Miki, 1993b. Hydration effects on CO_2 radicals in calcium carbonates and hydroxyapatite. Appl. Rad. Isot. 44: 305–309.

Murata, T., K. Shiraishi, Y. Ebina & T. Miki, 1996. An ESR study of defects in irradiated hydroxyapatite. Appl. Rad. Isot. 47: 1527–1531.

Murray, A. S., S. G. E. Bowman & M. J. Aitken, 1979. Evaluation of the γ dose rate contribution. PACT 2: 84–96.

Murray-Wallace, C. V. & A. Goede, 1995. Aminostratigraphy and electron spin resonance dating of Quaternary coastal neotectonism in Tasmania and the Bass Straight Islands. Austr. J. Earth Sci. 42: 51–67.

Murrieta, H., E. Muñoz, P. E. Adem, G. Barillo, M. Vasquez & E. Cabrera B., 1996. Effect of irradiation dose, storage time, and temperature on the ESR signal in irradiation oat, corn, and wheat. Appl. Rad. Isot. 47: 1657–1661.

Nagy, V., 2000. Accuracy considerations in EPR dosimetry. Appl. Rad. Isot. 52: 1039–1050 .

Nakajima, T., T. Osuki & H. Hara, 1993a. Radiation accident dosimetry using shells. Appl. Rad. Isot. 44: 91–94.

Nakajima, T., T. Osuki, M. Aonuma, J. Satoti & M. Shouji, 1993b. Dating of shellmound (sic) at Chiba-city in Japan. Appl. Rad. Isot. 44: 157–161.

Nakamura, N. & C. Miyazawa, 1997. Alkaline denaturation of dentin: A simple way to isolate human tooth enamel for electron spin resonance dosimetry. J. Rad. Res. 38: 173–177.

Nakazato, H., K. Shimokawa & N. Imai, 1993. ESR dating for Pleistocene shell fossils and value of annual dose. Appl. Rad. Isot. 44: 167–173.

Nambi, K. S. V., 1982. ESR and TL dating studies on some marine gypsum crystals. PACT 6: 314–321.

Nambi, K. S. V., 1985. Scope of ESR studies in thermally stimulated luminescence studies and chronological applications. Nucl. Tr. Rad. Meas. 10: 113–131.

Nambi, K. S. V. & M.J. Aitken, 1986. Annual dose conversion factors for TL and ESR dating. Archaeometry 28: 202–205.

Nambi, K. S. V. & A. V. Sankaran, 1985. ESR dating of laterite of basaltic origin. In Ikeya, M. & T. Miki (eds.) ESR Dating and Dosimetry. Ionics, Toyko: 175–180.

Ninagawa, K., I. Yamamoto, Y. Yamashita, T. Wada, H. Sakai & T. Fujii, 1985. Comparison of ESR with TL for fossil calcite shells. In Ikeya, M. & T. Miki (eds.) ESR Dating and Dosimetry. Ionics, Toyko: 105–114.

Nishiwaki, Y., T. Shimano, 1990. Uncertainties in dose estimation under emergency conditions and ESR dosimetry with human teeth. Rad. Protec. Dosim. 34: 295–297.

Norizawa, K., K. Kansoue & M. Ikeya, 2000. Radiation effects in dry ice: Models for a peak on the Arrhenius curve. Appl. Rad. Isot. 52: 1259–1263.

O'Brien, M. C., 1955. The structure of colour centres in smoky quartz. R. Soc. London, Proc. A231: 404–414.

Odom, A. L. & W. J. Rink, 1989. Natural accumulation of Schottky-Frenkel defects: Implications for a quartz geochronometer. Geology 17: 55–58.

Oduwole, A. D. & K. D. Sales, 1991. ESR signals in bones: Interference from Fe^{3+} ions and a new method of dating. Nucl. Tr. Rad. Meas. 18: 213–221.

Oduwale, A. D. & K. D. Sales, 1994. Transient ESR signals induced by γ-irradiation in tooth enamel and bone. Quat. Sci. Rev. (Quat. Geochron.) 13: 647–650.

Oduwale, A. D., K. D. Sales & J. K. Dennison, 1993. Some ESR observations on bone, tooth enamel and egg shell. Appl. Rad. Isot. 44: 261–266.

Ogoh, K., S. Toyoda, S. Ikeda, M. Ikeya & F. Goff, 1993. Cooling history of the Valles Caldera, New Mexico using ESR dating method. Appl. Rad. Isot. 44: 233–237.

Ogoh, K., C. Yamanaka, S. Toyoda, M. Ikeya & E. Ito, 1994. ESR studies on radiation-induced defects in high pressure phase SiO_2. Nucl. Instrum. Meth. Phys. Res. B 91: 331–333.

Oka, T., M. Ikeya, N. Sugawara, A. Nakanishi, 1996. A high-sensitivity portable spectrometer for ESR dosimetry. Appl. Rad. Isot. 47: 1589–1594.

Oka, T., R. Grün, A. Tani, C. Yamanaka, M. Ikeya & H. P. Huang, 1997. ESR microscopy of fossil teeth. Rad. Meas. 27: 331–337.

Oliveiri, L. M., A. M. Rossi & R. T. Lopes, 2000. γ-dose response of synthetic A-type carbonated apatie in comparison with the resonse of tooth enamel. Appl. Rad. Isot. 52: 1093–1097.

Omura, T., M. Ikeya, 1995. Evaluation of the ambient environment of mineral gypsum ($CaSO_4 \cdot H_2O$) growth by ESR microscope. Geoch. J. 29: 317–324.

Onori, S., M. Pantaloni, S. Baccaro & P. G. Fuochi, 1996. Influencing factors on ESR bone dosimetry. Applied Rad. Isot. 47: 1637–1640.

Onori, S., D. Aragno, P. Fattibene, E. Petetti & M. C. Pressello, 2000. ISS protocol for EPR tooth enamel dosimetry. Rad. Meas. 32: 787–792.

Ostrowski, K., A. Dziedzic-Goclawska, W. Stachowicz & J. Michalik, 1994. Accuracy, sensitivity, and specificity of electron spin resonance analysis of mineral constituents of irradiated tissues. Ann. NY Acad. Sci. 238: 186–200.

Ostrowski, K., A. Dziedzic-Goclawska, J. Michalik & W. Stachowicz, 1996. Radiation induced paramagnetic centers in bone and other mineralized tissues. J. Chimie Physique et de Physico-chemie Biologique 93: 174–181.

Ostrowski, K., A. Dziedzic-Goclawska, W. Stachowicz, J. Michalik, G. Burlinska & J. Sadlo, 1997. Detection limits of absorbed dose of ionizing radiation on molluscan shells as determined by EPR spectrometry. J. Chimie Physique et de Physico-chemie Biologique 94: 382–389.

Ottey, J. M., R. G. Roberts & A. S. Murray, 1997. Disequilibria in the uranium series decays series in the sedimentary deposits at Allen's Cave, Nullabor Plain, Australia: Implications for dose rate determinations. Rad. Meas. 27: 433–443.

Özer, A. M., 1985. Electron spin resonance studies of ancient flints in Turkey. Nucl. Tr. Rad. Meas. 10: 899–908.

Özer, A. M., A. Wieser, H. Y. Göksu, P. Müller, D. F. Regulla & O. Erol, 1989. ESR and TL age determination of caliche nodules. Appl. Rad. Isot. 40: 1159–1162.

Pass, B. & J. E. Aldrich, 1985. Dental enamel as an *in vivo* dosimeter. Medical Phys. 12: 305.

Papastefanou, C. & S. Charalambous, 1978. ^{226}Ra leaching in fossil bones. Nucl. Instrum. Meth. 151: 599–601.

Papastefanou, C., S. Charalambous, M. Manolopoulou & E. Savvides, 1986. Natural radiation dose in Petralona Cave. Health Phys. 50: 281–286.

Peng, Z. C., S. H. Jin, R. Y. Liang, P. H. Huang, Y. C. Quan & M. Ikeya, 1989. Study on comparison of ESR dating of coral and shells with ^{230}Th/^{234}U and ^{14}C methods. Appl. Rad. Isot. 40: 1127–1131.

Pick, H. & M. Ikeya, 1980. Wie alt sind Tropfsteine. Umschau 80: 472–474.

Pike, A. W. G. & R. E. Hedges, 2001. Sample geometry and U uptake in archaeological teeth: Implications for U-series and ESR dating. Quat. Sci. Rev. (Quat. Geochron.) 20: 1021–1025.

Pilbeam, D. & O. Bar-Yosef, 1993. Comment: Dating hominid remains. Nature 366: 415.

Pilbrow, J. R., 1996. ESR fundamentals. Appl. Rad. Isot. 47: 1465–1470.

Pilbrow, J. R., 1997. Subtleties in EPR. Rad. Meas. 27: 413–417.

Pirazzoli, P. A., U. Radtke, W. S. Hantoro, C. Jouannic, C. T. Hoang, C. Causse & M. B. Best, 1991. Quaternary raised coral, reef terraces on Sumba Island, Indonesia. Science 252: 1834–1836.

Pirazzoli, P. A., U. Radtke, W. S. Hantoro, C. Jouannic, C. T. Hoang, C. Causse & M.B. Best, 1993. A one million year long sequence of marine terraces on Sumba Island, Indonesia. Mar. Geol. 109: 221–236.

Pivovarov, S., A. Rukhin & T. Seredavina, 2000. ESR of environmental objects from Semipalatinsk Nuclear Test Site. Appl. Rad. Isot. 52: 1255–1258.

Polyakov, V., E. Haskell, G. Kenner, G. Hütt, R. Hayes, 1995. Effect of mechanically induced background signal on EPR dosimetry of tooth enamel. Rad. Meas. 24: 249–254.

Porat, N. & H. P. Schwarcz, 1991. Use of signal subtraction methods in ESR dating of burned flint. Nucl. Tr. Rad. Meas. 18: 203–212.

Porat, N. & H. P. Schwarcz, 1994. ESR dating of tooth enamel: A universal growth curve. In Corrucini, R. & R. Ciochón (eds.) Integrative Paths to the Past: Paleoanthropological Advances in Honor of F.C. Howell. Prentice-Hall, Englewood Cliffs: 521–530.

Porat, N. & H. P. Schwarcz, 1995. Problems in determining lifetimes of ESR signals in natural and burned flint by isothermal annealing. Rad. Meas. 24: 161–167.

Porat, N., B. A. Blackwell, H. P. Schwarcz & F. C. Howell, 1990. Late uranium adsorption in teeth: Indications from electron spin resonance (ESR) and U-series dating of enamel and calcite nodules from Torralba and Ambrona, Spain. CANQUA/AMQUA Annual Meeting, Abstracts, Waterloo.

Porat, N., B. A. Blackwell & H. P. Schwarcz, 1991. ESR dating of tusks: Preliminary results of an early Holocene mammoth tusk from Dawson, Yukon. Geol. Assoc. Can. - Mineral. Assoc. Can. Abstr. 16: A101.

Porat, N., R. Amit & E. Zilberman, 1994a. Electron spin resonance dating of carbonate nodules from Shivta, western Negev: Preliminary results. Curr. Res. Geol. Survey Israel 9: 41–46.

Porat, N., H. P. Schwarcz, H. Valladas, O. Bar-Yosef & B. Vandermeersch, 1994b. Electron spin resonance dating of burnt flint from Kebara Cave, Israel. Geoarchaeology 9: 393–407.

Poupeau, G. & A. M. Rossi, 1989. Electron spin resonance dating. In Roth, E. & E. Poty (eds.) Nuclear Methods of Dating. CEA, Paris: 275–293.

Poulianos, A. N., 1982. Petralona cave dating controversy: Comment. Nature 299: 280–282.

Prescott, J. R. & J. T. Hutton, 1994. Cosmic ray contributions to dose rates for luminescence and ESR dating: Large depths and long-term time variations. Rad. Meas. 23: 497–500.

Prescott, J. R. & L. G. Stephen, 1982. The contribution of cosmic radiation to the environmental dose for thermoluminescence dating: Latitude, altitude, and depth dependencies. PACT 6: 17–25.

Radtke, U., 1983. Elektronenspin-Resonanz-Altersdatierungen fossiler Mollusken in "Beachrock-Generatiorien" Latinums, Mittelitalien. Essener Geographische Arbeiten 6: 201–215.

Radtke, U., 1985a. Küstenmorphologische Untersuchungen auf den Äolischen Insein und Ustica (Sizilien). Berliner Geographische Studien 16: 71–89.

Radtke, U., 1985b. Chronostratigraphie und Neotektonik mariner Terassen in Nord-und Mittelchile: Erste Ergebnisse. IV Congresso Geologico Chilerio, Antofagasta, Proceedings: 4.436–4.457.

Radtke, U., 1985c. Untersuchungen zur zeitlichen Stellung mariner Terassen und Kalkkrusten auf Fuerteventura (Kanadsche Inseln, Spanien). Kieler Geographische Schriften 62: 73–95.

Radtke, U., 1985d. Electron spin resonance dating of shells. Search 16: 169–170.

Radtke, U., 1986. Value and risks of radiometric dating of shorelines: Geomorphological and geochronological investigations in central Italy, Eolian Islands, and Ustica (Sicily). Zeitschrift für Geomorphologie, Supplementband 62: 167–181.

Radtke, U., 1987. Marine terraces in Chile (22 °–32 °S): Geomorphology, chronostratigraphy, and neotectonics. In The Quaternary of South America and Antarctica Peninsula.

Radtke, U., 1988. How to avoid "useless" radiocarbon dating. Nature 333: 304–308.

Radtke, U., H. Brückner, 1991. Investigation on age and genesis of silcretes in Queensland (Australia): Preliminary results. Earth Surf. Process. Landf. 16: 547–554.

Radtke, U. & R. Grün, 1988. ESR dating of corals. Quat. Sci. Rev. 7: 465–470.

Radtke, U., G. J. Hennig, W. Linke & J. Mungersdorf, 1981. $^{230}Th/^{234}U$ and ESR dating problems of fossil shells in Pleistocene marine terraces (Northern Latinum, Central Italy). Quaternaria 23: 37–50.

Radtke, U., G. J. Hennig & A. Mangini, 1982. Untersuchungen zur Chronostratigraphie mariner Terrassen in Mittelitalien $^{230}Th/^{234}U$ und ESR-Datierungen an fossilen Mollusken. Eiszeitalter und Gegenwart 32: 49–55.

Radtke, U., A. Mangini & R. Grün, 1985. Nuclear ESR dating of fossil marine shells. Nucl. Tr. Rad. Meas. 10: 879–884.

Radtke, U., R. Hausmann & B. Hentzsch, 1986. The travertine complex of Vulci (Central Italy): An indicator of Quaternary climatic change. Symposium on Climatic Fluctuations during the Quaternary in the Western Mediterranean, Proceedings: 273–292.

Radtke, U., R. Grün & H. P. Schwarcz, 1988a. New results from ESR dating of Pleistocene coral reef tracts of Barbados (W.I.). Quat. Res. 29: 197–215.

Radtke, U., H. Brückner, A. Mangini & R. Hausmann, 1988b. Problems encountered with absolute dating (U-series, ESR) of Spanish calcretes. Quat. Sci. Rev. 7: 439–445.

Radtke, U., R. Grün, A. Omura & A. Mangini, 1996. The Quaternary coral reef tracts of Hateruma Ryukyu Islands, Japan. Quat. Int. 31: 61–70.

Raffi, J., J. C. Evans, J. P. Agnel, C. C. Rowlands & G. Lesgards, 1989. ESR analysis of irradiated frogs' legs and fishes. Appl. Rad. Isot. 40: 1215–1218.

Raffi, J., C. Hasbany, G. Lesgards & D. Ochin, 1996. ESR detection of irradiated shells. Appl. Rad. Isot. 47: 1633–1636.

Rakov, L. T., N. D. Milovodova, K. A. Kuvshinova & B. M. Moiyseyev, 1985. An ESR study of Ge centers in natural polycrystalline quartz. Geokhimica 9: 1339–1344.

Rakov, L. T., N. D. Milovodova, K. A. Kuvshinova & B. M. Moiyseyev, 1986. An ESR study of Ge centers in natural polycrystalline quartz. Geoch. Int. 23: 61–66.

Rambaud, X., J. Faïn, D. Miallier, T. Pilleyre & S. Sanzelle, 2000. Annual dose assessment for luminescence and ESR dating: Evaluation of disequilibrium kinetics using a pedologic approach. Rad. Meas. 32: 741–746.

Rao, T. K. G., S. S. Shinde, B. C. Bhatt, J. K. Srivastava & K. S. V. Nambi, 1996. ESR, TL, and fluorescence correlation in $BaSO_4$-Eu,P thermoluminescent phosphor. Physica Status Solidi A 157: 173–179.

Regulla, D. F., 2000. From dating to biophysics: 20 years of progress in applied ESR spectroscopy. Appl. Rad. Isot. 52: 1023–1030.

Regulla, D. F., A. Wieser & H. Y. Göksu, 1985. Effects of sample preparation on the ESR of calcite, bone and volcanic material. Nucl. Tr. Rad. Meas. 10: 825–830.

Regulla, D. F., A. Bartolotta, U. Deffner, S. Onoci, M. Pantaloni & A. Wieser, 1993. Calibration network based on alanine ESR dosimetry. Appl. Rad. Isot. 44: 23–31.

Regulla, D. F., H. Y. Göksu, A. Vogenauer & A. Wieser, 1994. Retrospective dosimetry based on egg shells. Appl. Rad. Isot. 45: 371–373.

Rhodes, E. J., J. P. Haynai, D. Geraads & F. Z. Sbihialaoui, 1994. First ESR dates for the Acheulean of Atlantic Morocco (Rhinoceros Cave, Casablanca). C. r. Acad. Sci., Paris 319: 1109–1115.

Rink, W. J., 1997. Electron spin resonance (ESR) dating and ESR applications in Quaternary science and archaeometry. Rad. Meas. 27: 975–1025.

Rink, W. J. & V. A. Hunter, 1998. Density Variation among Fossil and Modern Dental Tissues. Ancient TL.

Rink, W. J. & A. L. Odom, 1991. Natural α recoil particle radiation and ionizing radiation in quartz detected with EPR: Implications for geochronometry. Nucl. Tr. Rad. Meas. 18: 163–173.

Rink, W. J. & H. P. Schwarcz, 1994. Dose response of ESR signals in tooth enamel. Rad. Meas. 23: 481–484.

Rink, W. J., H. P. Schwarcz, 1995. Tests for diagenesis in tooth enamel: ESR dating signals and carbonate contents. J. Arch. Sci. 22: 251–255.

Rink, W. J., H. P. Schwarcz, R. Grün, I. Yalçinkaya, H. Taskiran, M. Otte, H. Valladas, N. Mercier, O. Bar-Yosef & J. Kowlowski, 1994. ESR dating of the Last Interglacial Mousterian at Karaïn Cave, southern Turkey. J. Arch. Sci. 21: 839–849.

Rink, W. J., H. P. Schwarcz, F. H. Smith & J. Radovcic, 1995. ESR ages for the Krapina hominids. Nature 378: 24.

Rink, W. J., H. P. Schwarcz & H. K. Lee, 1996a. ESR dating of tooth enamel: Comparison with AMS ^{14}C at El Castillo cave, Spain. J. Arch. Sci. 23: 945–951.

Rink, W. J., R. Grün, I. Yalçinkaya, M. Otte, H. Taskiran, H. Valladas, N. Mercier, J. Kowlowski & H. P. Schwarcz, 1996b. ESR dating of Middle Paleolithic materials and warm climate phases at Karaïn Cave, Southern Turkey. J. Arch. Sci. 23: 839–849.

Rink, W. J., H. P. Schwarcz, A. J. Stuart, A. M. Lister, E. Marselgia & B. J. Brennan, 1996c. ESR dating of the type Cromerian Freshwater Bed at West Runton, U.K. Quat. Sci. Rev. (Quat. Geochron.) 15: 727–738.

Rink, W. J., H. P. Schwarcz, K. Valoch, L. Seitl & C. B. Stringer, 1996d. ESR dating of the Micoquian industry and Neanderthal remains at Külna Cave, Czech Republic. J. Arch. Sci. 23: 889–901.

Rink, W. J., H. P. Schwarcz, H. K. Lee, V. Cabrera Valdés, F. Barnald de Quirós & M. Hoyos, 1997. ESR dating of the Mousterian levels at El Castillo Cave, Cantabria, Spain. J. Arch. Sci. 24: 593–600.

Rink, W. J., H. K. Lee, J. Rees-Jones & K. Goodger, 1998. Electron spin resonance (ESR) and mass-spectrometric U-series dating of teeth in Crimean Middle Palaeolithic sites: Starosele, Kobazi II, and Kobazi V. In Marks, A. E. & V. P. Chaboi (eds.) Palaeolithic of the Crimea, vol. 1. ERAUL, Liège: 323–340.

Robertson, S. & R. Grün, 2000. Dose determination on tooth fragments from two human fossils. Rad. Meas. 32: 773–779.

Robins, D., K. Sales & A. D. Oduwole, 1985a. ESR dating of bone: The effect of contemporary treatment upon measured age. In Ikeya, M. & T. Miki (eds.) ESR Dating and Dosimetry. Ionics, Toyko: 363–367.

Robins, D., K. Sales & A. D. Oduwole, 1985b. Electron trapping in plant resins: An assessment of ESR dating possibilities. In Ikeya, M. & T. Miki (eds.) ESR Dating and Dosimetry. Ionics, Toyko: 435–438.

Robins, D., A. D. Oduwole & K. Sales, 1987. Electron spin resonance studies of bones from the Lower Paleolithic site at Choukoutien, China. Am. Chem. Soc., Abstr. 193: 19.

Robins, D., K. Sales & A. D. Oduwole, 1988. Electron spin resonance dating of human bones from the Kansas River at Bonner Springs, northeastern Kansas. Am. J. Phys. Anthro. 75: 262.

Robins, G. V., N. J. Seeley, M. C. R. Symons & S. G. E. Bowman, 1982. The Institute of Archaeology (London) ESR flint project. PACT 6: 322–331.

Robins, G. V., N. J. Seeley, D. A. C. McNeil & M. C. R. Symons, 1978. Identification of ancient heat treatment in flint artefacts by ESR spectroscopy. Nature 276: 703–704.

Rodas, J. E. D., H. Panzeri & S. Mascarenhas, 1985. EPR dosimetry of irradiated human teeth. In Ikeya, M. & T. Miki (eds.) ESR Dating and Dosimetry. Ionics, Toyko: 391–396.

Romanyukha, A. A. & D. Regulla, 1996. Aspects of retrospective ESR dosimetry. Appl. Rad. Isot. 47: 1293–1297.

Romanyukha, A. A., D. Regulla, E. Vasilenko & A. Wieser, 1994. South Ural nuclear workers: Comparison of individual doses from restrospective EPR dosimetry and operational personal monitoring. Appl. Rad. Isot. 45: 1195–1199.

Romanyukha, A. A., E. A. Ignatiev, M. O. Degtiev & V. P. Kozheurov, 1996. Radiation doses from the Ural region. Nature 381: 199–200.

Romanyukha, A. A., M. F. Desrosier & D. F. Regulla, 2000. Current issues on EPR dose reconstruction in tooth enamel. Appl. Rad. Isot. 52: 1265–1273.

Rossi, A. M. & G. Poupeau, 1989. Radiation-induced paramagnetic species in natural calcite speleothems. Appl. Rad. Isot. 40: 1133–1137.

Rossi, A. M. & G. Poupeau, 1990. Radiation damage in bioapatites: The ESR spectrum of irradiated dental enamel revisited. Nucl. Tr. Rad. Meas. 17: 537–545.

Rossi, A. M., G. Poupeau, O. Chaix, J. Raffi, J.P. Angel & A. Heunet, 1992. Paramagnetic species induced in bioapatites by foodstuff irradiation. In Catoire, B. (ed.) Electron Spin Resonance (ESR) Applications in Organic and Bioorganic Materials. Springer-Verlag, Berlin: 151–156.

Rossi, A. M., C. C. Wafcheck, E. F. de Jesus & F. Pelegrini, 2000. Electron spin resonance doismetry of teeth of Goiania radiation accident victims. Appl. Rad. Isot. 52: 1297–1303.

Sakurai, H., K. Okada, K. Tuchiya & Y. Fujita, 1989. Dating of human blood by electron spin resonance spectroscopy. Naturwissenschaften 76: 24–25.

Sales, K. D., A. D. Oduwole, G. V. Robins & S. Olsen, 1985. The radiation and thermal dependence of electron spin resonance signals in ancient and modern bones. Nucl. Tr. Rad. Meas. 10: 845–851.

Sales, K. D., A. D. Oduwole & G. V. Robins, 1989. Electron spin resonance study of bones from the Paleolithic site at Zhoukoudian, China. American Chemical Society, Advances in Chemistry 220: 353–368.

Sarnthein, M., H. E. Stremme & A. Mangini, 1986. The Holstein interglaciation: time, stratigraphic position, and correlation to stable isotope stratigraphy of deep-sea sediments. Quat. Res. 26: 283–298.

Sasaoka, H., C. Yamanaka, M. Ikeya, 1996. Is the quartet due to $\cdot CH_3$ and $\cdot C_2H_5$ or $\cdot NH_3^+$ in alkali feldspars? Appl. Rad. Isot. 47: 1415–1417.

Sato, T., 1981. Electron spin resonance dating of calcareous microfossils in deep-sea sediment. Rock Magn. Paleogeophys. 8: 85–88.

Sato, T., 1982. ESR dating of planktonic foraminifera. Nature 300: 518.

Sato, T., 1983a. ESR studies of planktonic foraminifera: Reply. Nature 305: 162.

Sato, T., 1983b. Bleaching of ESR signals from planktonic foraminifera. Rock Magn. Paleogeophys. 10: 7–8.

Sato, T., K. Suito & Y. Ichikawa, 1985. Characteristics of ESR and TL signals on quartz from fault regions. In Ikeya, M. & T. Miki (eds.) ESR Dating and Dosimetry. Ionics, Toyko: 267–274.

Sato, T., K. Oikawa, H. Ohyanishigushi & H. Kamada, 1997. Development of and L-band electron spin resonance proton nuclear magnetic resonance imaging instrument. Rev. Sci. Instrum. 68: 2076–2081.

Schauer, D. A., M. F. Desrosiers, P. Kuppusamy & J. L. Zweier, 1996. Radiation dosimetry of an accidental overexposure using EPR spectrometry and imaging of human bone. Appl. Rad. Isot. 47: 1345–1350.

Schellmann, G. & U. Radtke, 1997. Electron spin resonance (ESR) techniques applied to mollusc shells from South America (Chile, Argentina) and implications for the palaeo-sea-level curve. Quat. Sci. Rev. (Quat. Geochron.) 16: 465–475.

Schellmann, G. & U. Radtke, 1999. Problems encountered in the determination of dose and dose rate in ESR dating of mollusc shells. Quat. Sci. Rev. 18: 1515–1527.

Schellmann, G. & U. Radtke, 2000. ESR dating of stratigraphically well constrained marine terraces along the Patagonian Atlantic coast (Argentina). Quat. Int. 68: 261–273.

Schellmann, G. & U. Radtke, 2001. Progress in ESR dating of pleistocene corals: A new approach for D_E determination. Quat. Sci. Rev. (Quat. Geochron.) 20: 1015–1020.

Scherbina, O. & A. Brik, 2000. Temperature stability of carbonate groups in tooth enamel. Appl. Rad. Isot. 52: 1071–1075.

Schramm, D. U. & A. M. Rossi, 1996. Electron spin resonance (ESR) and electron nuclear double resonance (ENDOR) and general triple resonance of irradiated biocarbonates. Appl. Rad. Isot. 47: 1443–1455.

Schramm, D. U. & A. M. Rossi, 2000. Electron spin resonance (ESR) studies of CO_2^- radicals in irradiated A- and B-type carbonate-containing apatites. Appl. Rad. Isot. 52: 1085–1091.

Schwarcz, H. P., 1985. Electron spin resonance studies of tooth enamel. Nucl. Tr. Rad. Meas. 10: 865–867.

Schwarcz, H. P., 1986. Geochronology and isotope geochemistry of speleothem. In Fontes, J. C. & S. Fritz (eds.) Handbook of Environmental Geochemistry. The Terrestrial Environment. Elsevier, Amsterdam: 271–303.

Schwarcz, H. P., 1994. Current challenges to ESR dating. Quat. Sci. Rev. (Quat. Geochron.) 13: 601–605.

Schwarcz, H. P., 1999. ESR dating comes of age. LED 99: 9th International Conference on Luminescence and Electron Spin Resonance Dating, Abstracts, Rome: 31.

Schwarcz, H. P., B. A. Blackwell, 1992. Uranium series dating of archaeological sites. In Ivanovich, M., R. S. Harmon (eds.) Uranium Series Disequilibrium: Application to Environmental Problems, 2nd ed. Clarendon, Oxford: 513–552.

Schwarcz, H. P. & R. Grün, 1988a. Comment on: M. Sarnthein, H. E. Stremme, and A. Mangini, The Holstein Interglaciation: Time-stratigraphic position and correlation to stable-isotope stratigraphy of deep-sea sediments. Quat. Res. 29: 75–79.

Schwarcz, H. P. & R. Grün, 1988b. ESR dating of level L 2/3 at La Micoque (Dordogne), France: Excavations of Debénath and Rigaud. Geoarchaeology 3: 293–296.

Schwarcz, H. P. & R. Grün, 1989. ESR dating of tooth enamel from prehistoric archaeological sites. Appl. Geoch. 4: 329–330.

Schwarcz, H. P. & R. Grün, 1992. Electron spin resonance (ESR) dating of the origin of modern man. R. Soc. London, Phil. Trans. B337: 145–148.

Schwarcz, H. P. & R. Grün, 1993a. Electron spin resonance (ESR) dating of the Lower Industry. In Singer, R., J. J. Wymer & B. G. Gladfelter (eds.) The Lower Palaeolithic Site at Hoxne, England. University of Chicago Press, Chicago: 207–217.

Schwarcz, H. P. & R. Grün, 1993b. Electron spin resonance dating of tooth enamel from Bir Tarfawi. In Wendorf, F., R. Schild & A. Close (eds.) Egypt During the Last Interglacial. Plenum, New York: 234–237.

Schwarcz, H. P. & W. J. Rink, 2001. Skinflint dating. Quat. Sci. Rev. (Quat. Geochron.) 20: 1047–1050.

Schwarcz, H. P. & Zymela, S., 1985. ESR dating of Pleistocene teeth from Alberta, Canada. In Ikeya, M. & T. Miki (eds.) ESR Dating and Dosimetry. Ionics, Toyko: 325–333.

Schwarcz, H. P., W. M. Buhay & R. Grün, 1987. Electron spin resonance (ESR) dating of fault gouge. In Crone, A. J. & E. Olmdahl (eds.) Directions in Paleoseimology. USGS Open File Report OF 87–673: 50–64.

Schwarcz, H. P., R. Grün, A. G. Latham, D. Mania & K. Brunnacker, 1988a. New evidence for the age of the Bilzingsleben archaeological site. Archaeometry 30: 5–17.

Schwarcz, H. P., R. Grün, B. Vandermeersch, O. Bar-Yosef, H. Valladas & E. Tchernov, 1988b. ESR dates for the hominid burial site of Qafseh in Israel. J. Human Evol. 17: 733–737.

Schwarcz, H. P., W. M. Buhay, R. Grün, H. Valladas, E. Tchernov, O. Bar-Yosef & B. Vandermeersch, 1989. ESR dating of the Neanderthal site, Kebara Cave, Israel. J. Arch. Sci. 16: 653–659.

Schwarcz, H. P., W. M. Bietti, M. C. Stiner, R. Grün & A. Segre, 1991a. On the reexamination of Grotta Guatari: Uranium-series and electron spin resonance dates. Curr. Anth. 32: 313–316.

Schwarcz, H. P., W. M. Buhay, R. Grün, M. C. Stiner, S. Kuhn & G. H. Miller, 1991b. Absolute dating of sites in coastal Lazio. Quaternaria Nova 1: 51–67.

Schwarcz, H. P., R. Grün & P. V. Tobias, 1994. ESR dating studies of the Australopithecine site of Sterkfontein, South Africa. J. Human Evol. 26: 175–181.

Schwenninger, J. & E. J. Rhodes, 1999. ESR extinction dates for the giant elephant bird of Madagascar. LED 99: 9th International Conference on Luminescence and Electron Spin Resonance Dating, Abstracts, Rome: 188.

Seitz, M. G. & R. E. Taylor, 1974. Uranium variations in a dated fossil bone series from Olduvai Gorge, Tanzania. Archaeometry 16: 129–135.

Serzhant, I., e. Yesirkenov, D. Grodzinsky, 1996. Resin and barb of pine (Pinus silvestris) as new objects for respective dosimetry by the EPR method. Appl. Rad. Isot. 47: 1357–1358.

Serezhenkov, V. A., I. A. Moroz, G. A. Klevezal & A. F. Vanin, 1996. Estimation of accumulated dose of radiation by the method of ESR-spectrometry of dental enamel of mammals. Appl. Rad. Isot. 47: 1321–1328.

Sherman, C. E., C. R. Glenn, A. T. Jones, W. C. Burnett & H. P. Schwarcz, 1993. New evidence for two highstands of the sea during the last interglacial, Oxygen Isotope Substage 5e. Geology 21: 1079–1082.

Shilles, T., J. Habermann, 2000. Radioluminescence dating: The IR emission of feldspar. Rad. Meas. 32: 679–683.

Shimano, T., M. Iwasaki, C. Miyazawa, T. Miki, A. Kai & M. Ikeya, 1989. Human tooth dosimetry for γ-rays and dental x-rays using ESR. Appl. Rad. Isot. 40: 1035–1038.

Shimokawa, K. & N. Imai, 1985. ESR dating of quartz in tuff and tephra. In Ikeya, M. & T. Miki (eds.) ESR Dating and Dosimetry. Ionics, Toyko: 181–185.

Shimokawa, K. & N. Imai, 1987. Simultaneous determination of alteration and eruption ages of volcanic rocks by electron spin resonance. Geoch. Cosmoch. Acta 51: 115–119.

Shimokawa, K., N. Imai & M. Hirota, 1984, Dating of volcanic rock by electron spin resonance. Chem. Geol. 46: 365–373.

Shimokawa, K., N. Imai & A. Moriyama, 1988. ESR dating of volcanic and baked rocks. Quat. Sci. Rev. 7: 529–532.

Shimokawa, K., N. Imai, H. Nakazato & K. Mizuno, 1992. ESR dating of fossil shells in the middle to upper Pleistocene strata in Japan. Quat. Sci. Rev. 11: 219–224.

Sholom, S. V. & V. V. Chumak, 1999. Possibility of dosimetry using thermally activated EPR centers of enamel. LED 99: 9th International Conference on Luminescence and Electron Spin Resonance Dating, Abstracts, Rome: 177.

Sholom, S. V., E. H. Haskell, R. B. Hayes, V. V. Chumak & G. H. Kenner, 1998a. Influence of crushing and additive irradiation procedures on EPR dosimetry of tooth enamel. Rad. Meas. 29: 105–111.

Sholom, S. V., E. H. Haskell, R. B. Hayes, V. V. Chumak & G. H. Kenner, 1998b. Properties of light induced EPR signals in enamel and their possible interference with γ-induced signals. Rad. Meas. 29: 113–118.

Sholom, S. V., E. H. Haskell, R. B. Hayes, V. V. Chumak & G. H. Kenner, 2000. EPR-dosimetry with carious teeth. Rad. Meas. 32: 799–803.

Shopov, Y. Y., R. I. Kostov & S. R. Manoushev, 1985. γ-spectroscopy applied to EPR-dating of calcite stalactites from Bulgaria. In Ikeya, M. & T. Miki (eds.) ESR Dating and Dosimetry. Ionics, Toyko: 73–76.

Siegele, R. & A. Mangini, 1985. Progress of ESR studies on $CaCO_3$ of deep sea sediments. Nucl. Tr. Rad. Meas. 10: 937–943.

Skalaric, U., P. Cevc, B. Gaspirc & M. Schara, 1998. Crystallite arrangement of hydroxyapatite microcrystals in human tooth cementum as revealed by electron paramagnetic resonance (EPR). Eur. J. Oral Sci. 106: 365–367.

Skinner, A. R., 1983. ESR dating of stalagmitic calcite: Overestimate due to laboratory heating. Nature 304: 152–154.

Skinner, A. R., 1985. Comparison of ESR and $^{230}Th/^{234}U$ ages in fossil aragonitic corals. In Ikeya, M. & T. Miki (eds.) ESR Dating and Dosimetry. Ionics, Toyko: 135–138.

Skinner, A. R., 1986. Electron spin resonance dating of aragonitic materials. In Olin, J. S. & M. J. Blackmun (eds.) Proceedings of the 24th International Archaeometry Symposium. Smithsonian Institution Press, Washington: 477–480.

Skinner, A. R., 1988. Dating of marine aragonite by electron spin resonance. Quat. Sci. Rev. 7: 461–464.

Skinner, A. R., 1989. ESR dosimetry and dating in aragonitic mollusks. Appl. Rad. Isot. 40: 1081–1085.

Skinner, A. R., 1995. ESR dating of flint artefacts: Hot tips and stone-cold facts. American Chemical Society Abstracts.

Skinner, A. R., 2000a. What can electron spin resonance tell us about chert? In Doud, A. (ed.) Archaeological Chert, in press.

Skinner, A. R., 2000b. ESR dating: Is it still an experimental technique? Appl. Rad. Isot. 52: 1311–1316.

Skinner, A. R. & J. Mirecki, 1993. ESR dating of molluscs: Is it only a shell game? Appl. Rad. Isot. 44: 139–143.

Skinner, A. R. & M. N. Rudolph, 1996a. Dating flint artefacts with electron spin resonance: Problems and perspectives. In Orna, M. V. (ed.) Archaeological Chemistry 5 (American Chemical Society, Advances in Chemistry 625): 37–46.

Skinner, A. R. & M. N. Rudolph, 1996b. The use of the E' signal in flint for ESR dating. Appl. Rad. Isot. 47: 399–1404.

Skinner, A. R. & C. E. Shawl, 1994. ESR dating of terrestrial Quaternary shells. Quat. Sci. Rev. (Quat. Geochron.) 13: 679–684.

Skinner, A. R. & N. Weicker, 1992. ESR dating of *Chione cancellata* and *Chama sinuosa*. Quat. Sci. Rev. 11: 225–229.

Skinner, A. R., B. A. B. Blackwell, D. E. Chasteen, J. M. Shao & S. S. Min, 2000. Improvements in dating tooth enamel by ESR. Appl. Rad. Isot. 52: 1337–1344.

Skinner, A. R., B. A. B. Blackwell, D. E. Chasteen & J. M. Shao, 2001a. Q-band studies of the ESR signal in dating tooth enamel. Quat. Sci. Rev. (Quat. Geochron.) 20: 1027–1030.

Skinner, A. R., B. A. B. Blackwell & V. Lothian, 2001b. Calibrating ESR ages in the 2-Ma range at Olduvai Gorge, Tanzania. J. Human Evol. 40: A22.

Skvortzov, V. G., A. I. Ivannikov & U. Eichhoff, 1995. Assessment of individual accumulated irradiation doses using EPR spectroscopy of tooth enamel. J. Molec. Structure 347: 321–329.

Skvortzov, V. G., A. I. Ivannikov, V. F. Stepanenko, A. F. Tsyb, L. G. Khamidova, A. E. Kondrashov & D. D. Tikunov, 2000. Application of EPR retrospective dosimetry for large scale accidental situation (sic). Appl. Rad. Isot. 52: 1275–1282.

Smart, P. L., B. W. Smith, H. Chandra, J. N. Andrews & M. C. R. Symons, 1988. An intercomparison of ESR and uranium series ages for Quaternary speleothem calcites. Quat. Sci. Rev. 7: 411–416.

Smith, B. W., P. L. Smart, M. C. R. Symons & J. N. Andrews, 1985a. ESR dating of detritally contaminated calcites. In Ikeya, M. & T. Miki (eds.) ESR Dating and Dosimetry. Ionics, Toyko: 49–59.

Smith, B. W., P. L. Smart & M. C. R. Symons, 1985b. ESR signals in a variety of speleothem calcites and their suitability for dating. Nucl. Tr. Rad. Meas. 10: 837–844.

Smith, B. W., P. L. Smart & M. C. R. Symons, 1986. A routine ESR technique for dating calcite speleothems. Rad. Protection Dosimetry 17: 241–245.

Smith, B. W., P. L. Smart, P. J. Fox, J. R. Prescott & M. C. R. Symons, 1989. An investigation of ESR signals and their related TL emission in speleothem calcite. Appl. Rad. Isot. 40: 1095–1104.

Smolyanskiy, P. L. & V. L. Masaytis, 1979. Reconstruction of paleotemperature anomalies of old rocks from radiation-induced defects in quartz. Doklady Akademii, Nauk, SSSR 248: 1428–1431.

Stachowicz, W., J. Michalik, G. Burlinska, J. Sadlo, A. Dziedzic-Goclawska & K. Ostrowski, 1994. Detection limits of absorbed dose of ionizing radiation in molluscan shells as determined by EPR spectroscopy. Appl. Rad. Isot. 46: 1047–1052.

Stewart, E. M. & R. Gray, 1996. A study of the effect of irradiation dose and storage on the ESR signal in the cuticle of pink shrimp (Pandalus montagui) from different geographical regions. Appl. Rad. Isot. 47: 1629–1632.

Stößer, R., J. Bartoll, L. Schirrmeister, R. Ernst & R. Lueck, 1996. ESR of trapped holes and electrons in natural and synthetic carbonates, silicates, and aluminosilicates. Appl. Rad. Isot. 47: 1489–1496.

Stringer, C. B. & R. Grün, 1991. Paleoanthropology: Time for the last Neanderthals. Nature 351: 701–702.

Stringer, C. B., R. Grün, H. P. Schwarcz & P. Goldberg, 1989. ESR dates for the hominid burial site of Es Skhul in Israel. Nature 338: 756–758.

Stuglik, Z., J. Sadlo, 1995. Latent tracks generated in microcrystalline α-L-alanine and standard bone powder by ^{59}Co ion beams as investigated by the EPR method. Rad. Meas. 25: 95–98.

Stuglik, Z., J. Michalik, W. Stachowicz, K. Ostrowski, I. Zvara & A. Dziedzic-Goclawska, 1994. Bone powder exposed to the action of ^{12}C and ^{25}Mg ion beams as investigated by electron paramagnetic resonance spectroscopy. Appl. Rad. Isot. 45: 1181–1187.

Sueki, M., S. S. Eaton & G. R. Eaton, 1993. Spectral spatial EPR imaging of irradiated silicon dioxide. Appl. Rad. Isot. 44: 377–380.

Sueki, M., S. S. Eaton & G. R. Eaton, 1996. EPR imaging of irradiated silicon dioxide: Increased concentrations of E' defects near the surface. Appl. Rad. Isot. 47: 1595–1598.

Swisher, C. C., III, W. J. Rink, S. C. Antón, H. P. Schwarcz, G. H. Curtis, A. Suprijo & Widiasmoro, 1996. Latest Homo erectus of Java: Potential contemporaneity with Homo sapiens in Southeast Asia. Science 274: 1870–1874.

Swisher, C. C., III, W. J. Rink, H. P. Schwarcz & S. C. Anthony, 1997. Technical reply: Dating the Ngangdong humans. Science 276: 1575–1576.

Taguchi, S., M. Harayama & M. Hayashi, 1985. ESR signal of zircon and geologic age. In Ikeya, M. & T. Miki (eds.) ESR Dating and Dosimetry. Ionics, Toyko: 191–196.

Takano, M. & Y. Fukao, 1994. ESR dating of Pleistocene fossil shells of the Atsumi group, Central Honshu, Japan: On the discrepancy in TD value among different ESR peaks. Appl. Rad. Isot. 45: 49–56.

Takeuchi, A. & R. Saeki, 1985. Electron spin resonance (ESR) signals of pelagic sediments in the Southern Pacific. In Ikeya, M. & T. Miki (eds.) ESR Dating and Dosimetry. Ionics, Toyko: 125–133.

Tanaka, K. & T. Shidahara, 1985. Fracturing, crushing, and grinding effects on ESR signal of quartz. In Ikeya, M. & T. Miki (eds.) ESR Dating and Dosimetry. Ionics, Toyko: 239–248.

Tanaka, K., S. Sawada & T. Ito, 1985. ESR dating of late Pleistocene near-shore and terrace sands in southern Kanto, Japan. In Ikeya, M. & T. Miki (eds.) ESR Dating and Dosimetry. Ionics, Toyko: 275–280.

Tanaka, K., M. N. Machette, A. J. Crone & J. R. Bowman, 1995. ESR dating of aeolian sand near Tennant Creek, Northern Territory, Australia. Quat. Sci. Rev. 14: 385–393.

Tanaka, K., R. Hataya, N. Spooner, D. Questiauz, Y. Saito & T. Hashimoto, 1997. Dating of marine terrace sediments by ESR, TL, and OSL methods and their applications. Quat. Sci. Rev. (Quat. Geochron.) 16: 257–264.

Tani, A., M. Ikeya & C. Yamanaka, 1996. ESR dating of a geological fault with a new isochrone (sic) method: Granite fractures on the earthquake in 1995. Appl. Rad. Isot. 47: 1423–1426.

Tani, A., J. Bartoll, M. Ikeya, K. Komura, H. Kajiwara, S. Fujimura, T. Kamada & Y. Yokoyama, 1998. ESR study of thermal history and dating of a stone tool. Appl. Magn. Reson. 13: 561–569.

Tatsumi-Miyajima, J., 1987. Electron spin resonance dosimetry for atomic bomb survivors and radiologic technologists. Nucl. Instrum. Meth. Phys. A257: 417–422.

Toyoda, S. & F. Goff, 1996. Quartz in post-caldera rhyolites of Valles Caldera New Mexico: ESR fingerprinting and discussion of ESR ages. In New Mexico Geological Society, 4th Field Conference Guidebook: 303–309.

Toyoda, S. & S. Hattori, 2000. Formation and decay of the E_1' center and of its precursor. Appl. Rad. Isot. 52: 1351–1356.

Toyoda, S. & M. Ikeya, 1989. ESR as a paleothermometer of volcanic materials. Appl. Rad. Isot. 40: 1171–1175.

Toyoda, S. & M. Ikeya, 1991a. ESR dating of quartz and plagioclase from volcanic ashes using E', Al, and Ti centres. Nucl. Tr. Rad. Meas. 18: 179–184.

Toyoda, S. & M. Ikeya, 1991b. Thermal stabilities of paramagnetic defect and impurity centers in quartz: Basis for ESR dating of thermal history. Geoch. J. 25: 437–445.

Toyoda, S. & M. Ikeya, 1994a. Formation of oxygen vacancies in quartz and its application to dating. Quat. Sci. Rev. (Quat. Geochron.) 13: 607–609.

Toyoda, S. & M. Ikeya, 1994b. ESR dating of quartz with stable component of impurity centres. Quat. Sci. Rev. (Quat. Geochron.) 13: 625–628.

Toyoda, S. & H. P. Schwarcz, 1996. The spatial distribution of ESR signals in fault gouge revealed by abrading technique. Appl. Rad. Isot. 47: 1409–1413.

Toyoda, S. & H. P. Schwarcz, 1997a. Counterfeit E_1' signals in quartz. Rad. Meas. 27: 59–66.

Toyoda, S. & H. P. Schwarcz, 1997b. The hazard of the counterfeit E_1' signal in quartz to the ESR dating of fault movements. Quat. Sci. Rev. (Quat. Geochron.) 16: 483–487.

Toyoda, S., M. Ikeya, J. Morikawa & T. Nagatomo, 1992. Enhancement of oxygen vacancies in quartz by natural external α, β, γ ray dose: A possible ESR geochronometer of Ma–Ga range. Geoch. J. 26: 111–115.

Toyoda, S., H. Kohno & M. Ikeya, 1993a. Distorted E' centres in crystalline quartz: An application to ESR dating of fault movements. Appl. Rad. Isot. 44: 215–220.

Toyoda, S., M. Ikeya, R. C. Dunnell & P. T. McCutcheon, 1993b. The use of electron spin resonance (ESR) for the determination of prehistoric lithic heat treatment. Appl. Rad. Isot. 44: 227–231.

Toyoda, S., F. Goff, S. Ikeda & M. Ikeya, 1995a. ESR dating of quartz phenocrysts in the El Cajete and Battleship Rock Members of Valles Rhyolite, Valles Caldera, New Mexico. J. Volc. Geotherm. Res. 67: 29–40.

Toyoda, S., D. L. Kirschner, W. J. Rink & H. P. Schwarcz, 1995b. ESR dating of quartz vein. Advanced ESR Applications 11: 8–13.

Toyoda, S., W. J. Rink, H. P. Schwarcz & M. Ikeya, 1996. Formation of E_1' precursors in quartz: Applications to dosimetry and dating. Appl. Rad. Isot. 47: 1393–1398.

Toyoda, S., C. Falguères & J. M. Dolo, 1999. The method to represent the ESR signal intensity of the aluminum hole center in quartz for the purpose of dating. LED 99: 9th International Conference on Luminescence and Electron Spin Resonance Dating, Abstracts, Rome: 179.

Toyoda, S., P. Voinchet, C. Falguères, J. M. Dolo & M. Laurent, 2000. Bleaching of ESR signals by sunlight: A laboratory experiment for estabilishing the ESR dating of sediments. Appl. Rad. Isot. 52: 1357–1362.

Toyoda, S., W. J. Rink, C. Yonezawa, H. Matsue & T. Kagami, 2001. *In situ* production of α particles and α recoil particles in quartz applied to ESR oxygen vacancies. Quat. Sci. Rev. (Quat. Geochron.) 20: 1057–1061.

Toyokura, I., Y. Sakuramoto, K. Ohmura, E. Iwasaki & M. Ishiguchi, 1985. Determination of the age of fault movement the Rokko Fault. In Ikeya, M. & T. Miki (eds.) ESR Dating and Dosimetry. Ionics, Toyko: 219–228.

Tsay, F. D., S. I. Chan & L. M. Stanley, 1972. Electron paramagnetic resonance of radiation damage in lunar rocks. Nature 237: 121–122.

Tsuji, Y., Y. Sakuramoto, E. Iwasaki, M. Ishigushi & K. Ohmura, 1985. Ages of pelecipod shells of the Last Interglacial Shimosueyoshi stage by electron spin resonance. In Ikeya, M. & T. Miki (eds.) ESR Dating and Dosimetry. Ionics, Toyko: 87–92.

Tsukamoto, Y. & J. Heikoop, 1996. Sulfite radicals in banded coral. Appl. Rad. Isot. 47: 1437–1441.

Tsukamoto, Y., M. Ikeya & C. Yamanaka, 1993. Fundamental study on ESR dating of outer planets and their satellites. Appl. Rad. Isot. 44: 221–225.

Übersfeld, J., A. Etienne & J. Combrisson, 1954. Paramagnetic resonance: A new property of coal-like materials. Nature 174: 614.

Ulusoy, Ü. L. K., 2000. ESR dating of a quartz single crystal from the Menderes Massif, Turkey. Appl. Rad. Isot. 52: 1363–1370.

Ulusoy, Ü. L. K. & F. Apaydìn, 1996. ESR studies and ESR dating of quartz collected from Kapadokya, Turkey. Appl. Rad. Isot. 47: 1405–1407.

Usatyi, A. F., N. V. Verein, 1996. EPR-based dosimetry of large dimensional radiation fields (Chernobyl experience and new approaches). Appl. Rad. Isot. 47: 1351–1356.

Vainshtein, D. J. & H. W. den Hartog, 1996. EPR study of Na-colloids in heavily irradiated NaCl: Size effects and interprecipitate interactions. Appl. Rad. Isot. 47: 1503–1507.

Valladas, H., G. Valladas, 1987. TL dating of burnt flint and quartz: Comparative results. Archaeometry 29: 214–220.

Valladas, G., H. Valladas & J. C. Messot, 1981. Thermoluminescence and electron spin resonance dating of some stalagmite deposits of the Caune de l'Arago at Tautavel. In de Lumley, H. & J. Labeyrie (eds.) Datations Absolus et les Analyses Isotopiques en Préhistoire: Méthodes et Limites. Prétirage. CNRS, Paris: 391–401.

Vanhaelewyn, G., F. Callens & R. Grün, 2000. ESR spectrum deconvolution and dose assessment of fossil tooth enamel using maximum likelihood common factor analysis. Appl. Rad. Isot. 52: 1317–1326.

Vanmoort, J. C. & D. W. Russell, 1987. Electron spin resonance of auriferous and barren quartz at Beaconsfield, Northern Tasmania. J. Geoch. Explor. 27: 227–237.

Vizgirda, J., T. J. Ahrens & F. D. Tsay, 1980. Shock-induced effects in calcite from Cactus Crater Geoch. Cosmoch. Acta 44: 1059–1069.

Volterra, V., W. J. Rink & H. P. Schwarcz, 1999. Dating of Iberian Mousterian layers by ESR. LED 99: 9th International Conference on Luminescence and Electron Spin Resonance Dating, Abstracts, Rome: 189.

Vugman, N. V., A. M. Rossi & S. E. J. Rigby, 1995. EPR dating CO_2^- sites in tooth enamel apatites by ENDOR and triple resonance. Appl. Rad. Isot. 46: 311–315.

Walther, R. & D. Zilles, 1994a. ESR studies on bleached sedimentation quartz. Quat. Sci. Rev. (Quat. Geochron.) 13: 611–614.

Walther, R. & D. Zilles, 1994b. ESR studies on flint with a difference-spectrum method. Quat. Sci. Rev. (Quat. Geochron.) 13: 635–640.

Walther, R., M. Barabas & A. Mangini, 1992. Basic ESR studies on recent corals. Quat. Sci. Rev. 11: 191–196.

Weil, J. A., 1984. A review of electron spin resonance spectroscopy and its application to the study of paramagnetic defects in crystalline quartz. Phys. Chem. Minerals 10: 149–165.

Wendorf, F., A. E. Close, R. Schild, A. Gauthier, H. P. Schwarcz, G. H. Miller, K. Kowalski, H. Krolik, A. Bluszcz, D. Robins, R. Grün & C. McKinney, 1987. Chronology and stratigraphy of the Middle Paleolithic at Bir Tarfawi, Eqypt. In Clark, J. d. (ed.) Culutral Beginnings. Hablet, Bonn: 197–208.

Wertz, J. E., J. R. Bolton, 1986. Electron Spin Resonance: Elementary Theory and Practical Applications, 2nd ed. Chapman Hall, London.

Whitehead, N. E., S. D. Devine & B. F. Leach, 1986. Electron spin resonance dating of human teeth from the Namu burial ground, Taumako, Solomon Islands. New Zeal. J. Geol. Geophys. 29: 359–361.

Wieser, A. & D. F. Regulla, 1989. ESR dosimetry in the "gigarad" range. Appl. Rad. Isot. 40: 911–913.

Wieser, A., H. Y. Göksu & D. F. Regulla, 1985. Characteristics of γ-induced ESR spectra in various calcites. Nucl. Tr. Rad. Meas. 10: 831–836.

Wieser, A., H. Y. Göksu, D. F. Regulla & A. Vogenauer, 1988. Dose rate assessment in tooth enamel. Quat. Sci. Rev. 7: 491–495.

Wieser, A., H. Y. Göksu, D. F. Regulla & A. Waibel, 1991. Unexpected superlinear dose dependence of the E' centre in fused Silica. Nucl. Tr. Rad. Meas. 18: 175–178.

Wieser, A., H. Y. Göksu, D. F. Regulla, P. Fritz, A. Vogenauer & I. D. Clark, 1993. ESR and TL dating of travertine from Jordan: Complications in paleodose assessment. Appl. Rad. Isot. 44: 149–152.

Wieser, A., E. Haskell, G. Kenner & F. Bruenger, 1994a. EPR dosimetry of bone gains accuracy by isolation of calcified tissue. Appl. Rad. Isot. 45: 525–526.

Wieser, A., H. Y. Göksu, D. F. Regulla & A. Vogenauer, 1994b. Limits of retrospective accident dosimetry by EPR and TL with natural materials. Rad. Meas. 23: 509–514.

Wieser, A., K. Mehta & EPR Working Group, 2000a. The 2nd international intercomparison on EPR tooth dosimetry. Rad. Meas. 32: 549–557.

Wieser, A., S. Onori, D. Aragno, P. Fattibene, A. Romanyukha, E. Ignatiev, A. Koshta, V. Skvortzov, A. Ivannikov, V. Stepanenko, V. Chumak, S. Sholom, E. Haskell, R. Hayes & G. Kenner, 2000b. Comparison of sample preparation and signal evaulatuion methods for EPR analysis of tooth enamel. Appl. Rad. Isot. 52: 1059–1064.

Wintle, A. G. & D. J. Huntley, 1983. ESR studies of planktonic foraminifera: Comment. Nature 305: 161–162.

Woda, C., A. Mangini & G. A. Wagner, 2001. ESR dating of volcanic rocks. Quat. Sci. Rev. (Quat. Geochron.) 20: 993–998.

Wright, P. M., J. A. Weil, T. Buch & J. H. Anderson, 1963. Titanium centres in rose quartz. Nature 197: 246–248.

Xirotiris, N. J., W. Henke & G. J. Hennig, 1982. Die phylogenetische Stellung des Petralona Schädels auf Grund computertomographischer Analysen und der absoluten Datierung mit der ESR-Methode. Humanbiologia Budapestinensis 9: 89–94.

Yamamoto, M. & M. Ikeya, 1994. Electron spin resonance computer tomography with field gradient wires inside the cavity: Images of alanine dosimeters and some materials. Jap. J. Appl. Phys. 1 33: 4887–4890.

Yamanaka, C., M. Ikeya, K. Meguro & A. Nakanishi, 1991. A portable ESR spectrometer using Nd-Fe-B permanent magnets. Nucl. Tr. Rad. Meas. 18: 279–282.

Yamanaka, C., M. Ikeya & H. Hara, 1993. ESR cavities for in vivo dosimetry of tooth enamel. Appl. Rad. Isot. 44: 77–80.

Yamanaka, C., H. Kohno & M. Ikeya, 1996. Pulsed ESR measurements of oxygen deficient type centers in various quartz. Appl. Rad. Isot. 47: 1573–1577.

Ye, Y., J. C. Gao, J. He, S. B. Diao, X. Lui & Y. Du, 1991. ESR and U-series ages of coral reef samples from shallow drill holes in the South China Sea. Acta Oceanol. Sin. 10: 423–431.

Ye, Y., S. B. Diao, J. He & J. C. Gao, 1993a. ESR dating of quartz in sediment. Nucl. Techniques 16: 222–224.

Ye, Y., S. B. Diao, J. He & J. C. Gao, 1993b. Preliminary study on the Late Pleistocene ESR chronology of core QC2 in the Southern Huanghi Sea. Chin. Sci. Bull. 38: 1280–1284.

Ye, Y., S. B. Diao, J. He & J. C. Gao, 1996a. Abnormal response to irradiation dose of E' signal of quartz in coastal eolian sand. Appl. Rad. Isot. 47: 1457–1458.

Ye, Y., S. B. Diao, J. He & J. C. Gao, 1996b. ESR chronology of core QC2 in the South Yellow Sea. Mar. Geol. Mineral Resources 99: 122–129.

Yokoyama, Y., J. P. Quaegebeur, R. Bibron, C. Leger, H. V. Nguyen & G. Poupeau, 1981a. Electron spin resonance (ESR) dating of fossil bones of the Caune de l'Arago at Tautavel. In de Lumley, H. & J. Labeyrie (eds.) Datations Absolus et les Analyses Isotopiques en Préhistoire: Méthodes et Limites. Prétirage. CNRS, Paris: 457–492.

Yokoyama, Y., J. P. Quaegebeur, R. Bibron, C. Leger, H. V. Nguyen & G. Poupeau, 1981b. Electron spin resonance (ESR) dating of stalagmites of the Caune de l'Arago at Tautavel. In de Lumley, H. & J. Labeyrie (eds.) Datations Absolus et les Analyses Isotopiques en Préhistoire: Méthodes et Limites. Prétirage. CNRS, Paris: 507–532.

Yokoyama, Y., J. P. Quaegebeur, R. Bibron, C. Leger, H. V. Nguyen & G. Poupeau, 1982. Datation du site de l'Homme de Tautavel par la résonance de spin électronique (ESR). C. r. Acad. Sci., Paris 294: 759–764.

Yokoyama, Y., J. P. Quaegebeur, R. Bibron, C. Leger, N. Chappaz, C. Michelot, G. J. Chen & H. V. Nguyen, 1983a. ESR dating of stalagmites of the Caune de l'Arago, the Grotte du Lazarat, the Grotte Vallonet and the Abri Pie Lombard: A comparison with the U-Th method. PACT 9: 381–389.

Yokoyama, Y., J. P. Quaegebeur, R. Bibron & C. Leger, 1983b. ESR dating of Paleolithic calcite: Thermal annealing experiment and trapped electron lifetime. PACT 9: 371–379.

Yokoyama, Y., C. Falguères & J. P. Quaegebeur, 1985a. ESR dating of sediment baked by lava flows: Comparison of paleodoses for Al and Ti centers. In Ikeya, M. & T. Miki (eds.) ESR Dating and Dosimetry. Ionics, Toyko: 197–204.

Yokoyama, Y., R. Bibron, C. Leger & J. P. Quaegebeur, 1985b. ESR dating of Paleolithic calcite: Fundamental studies. Nucl. Tr. Rad. Meas. 10: 929–936.

Yokoyama, Y., C. Falguères & J. P. Quaegebeur, 1985c. ESR dating of quartz from Quaternary sediments first attempt. Nucl. Tr. Rad. Meas. 10: 921–928.

Yokoyama, Y., R. Bibron & C. Leger, 1988. ESR dating of Paleolithic calcite: A comparison between powder and monocrystal spectra with thermal annealing. Quat. Sci. Rev. 7: 433–438.

Yoshida, H., S. Brumby, 1999. Comparison of ESR ages of corals using different signals at X- and Q-band: Re-examination of corals from Huon Peninsula, Papua New Guinea. Quat. Sci. Rev. 18: 1529–1536.

Yugo, O., T. Naruse, M. Ikeya, H. Kohno & S. Toyoda, 1998. Origin and derived courses of eolian dust quartz deposited during marine Isotope Stage 2 in east Asia suggested by ESR signal intensity. Glob. Planet. Change 18: 129–135.

Zeller, E. J., 1968. Use of electron spin resonance for measurement of natural radiation damage. In McDougall, D. J. (ed.) Thermoluminescence of Geological Materials, Academic Press, London: 271–279.

Zeller, E. J., G. Dreschhoff & G. V. Robins, 1988. Impact of radioelement distribution in bones on electron spin resonance dating. Am. J. Phys. Anth. 75: 289–290.

Zhoa, X. T., H. Gao & D. Li, 1991. Response of the E' signal in quartz to light exposure. Nucl. Techniques 14: 87.

Zhou, L. P., F. McDermott, E. J. Rhodes, E. A. Marselgia & P. A. Mellars, 1997. ESR and mass-spectrometric uranium-series dating of a mammoth tooth from Stanton Harcourt, Oxfordshire, England. Quat. Sci. Rev. 16: 727–738.

Ziaei, M., H. P. Schwarcz, C. M. Hall & R. Grün, 1990. Radiometric dating of the Mousterian site at Quneitra. In Goren-Inbar, N. (ed.) QEDEM. Monographs of the Institute of Archaeology, Hebrew University, Jerusalem 37: 232–235.

Zymela, S., 1986. ESR Dating of Pleistocene Deposits. Unpub. M.Sc. thesis, McMaster University, Hamilton.

Zymela, S., H. P. Schwarcz, R. Grün, A. M. Stalker & C. S. Churcher, 1988. ESR dating of Pleistocene fossil teeth from Alberta and Saskatchewan. Can. J. Earth Sci. 25: 235–245.

14. USE OF PALEOMAGNETISM IN STUDIES OF LAKE SEDIMENTS

JOHN KING (jking@gso.uri.edu)
Graduate School of Oceanography
University of Rhode Island
Narragansett, RI 02882-1197, USA

JOHN PECK (jpeck@uakron.edu)
Department of Geology
University of Akron
Akron, OH, 44325-4101, USA

Keywords: paleomagnetism, lake sediments, geomagnetic field, secular variation, magnetostratigraphy, excursions

Introduction

The Earth's magnetic field is a vector and can be completely characterized by three elements at any point in time and space. The first element is declination (D), the angle of deviation in the horizontal plane between geographic north and magnetic north. The second element is inclination (I), the dip angle of the magnetic vector below the horizontal plane. The final element is the magnitude of the field or geomagnetic intensity (F). These elements are illustrated in Figure 1.

The variation of magnetic field elements over the Earth's surface on timescales of 10^4 years or greater can be generalized. The general behavior can be modeled as a dipole M or bar magnet located along the rotation axis of the Earth. This model shown in Figure 2 is called the geocentric axial dipole (GAD) model of the Earth's magnetic field. The GAD inclination can be described by the relationship $\tan I = 2 \tan \lambda$, where λ is the site latitude. During positive polarity intervals, like the present one, I is near 0° at the equator, positive in the Northern Hemisphere, +90° at the North magnetic pole, negative in the Southern Hemisphere, and −90° at the South Pole. D averages to 0° at all points in space on timescales of 10^4 years. The geomagnetic intensity has minimum values at the equator and maximum values at the poles.

The Earth's main magnetic field is of internal origin. A discussion of source models of this field is beyond the scope of this paper and we refer the interested reader to Backus et al. (1996) for a recent review of the topic. However, it is useful to conceptualize the field as being produced by the flow pattern of molten Fe-Ni, an excellent electrical conductor, in the Earth's liquid outer core. Some features of the flow pattern have a regional influence at the surface of the Earth, whereas others have a global influence.

W. M. Last & J. P. Smol (eds.), 2001. *Tracking Environmental Change Using Lake Sediments. Volume 1: Basin Analysis, Coring, and Chronological Techniques.* Kluwer Academic Publishers, Dordrecht, The Netherlands.

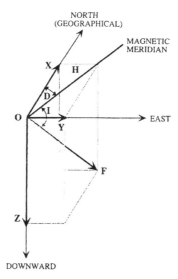

Figure 1. Elements of the geomagnetic field. From Jacobs (1994).

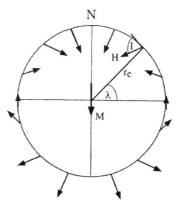

Figure 2. Geocentric axial dipole model. Magnetic dipole M is located at the center of the Earth (Earth radius $= r_e$) aligned with the Earth's axis of rotation. The latitude (λ) and inclination (I), and field vector (H) for one location are shown. The other field vectors drawn are a schematic representation of the variation of I with λ. N is the north geographic pole. From Butler (1992).

The Earth's main magnetic field shows three types of spatial and temporal variation on timescales that are useful for dating and correlation purposes in studies of lake sediments. These three types of variation are geomagnetic secular variation (SV), excursions, and reversals and they are defined by both the amplitude of the change in field direction and intensity and the duration of the change (Jacobs, 1994).

Geomagnetic secular variation (SV) is the "typical" temporal variation of the three components of the Earth's magnetic field between polarity transitions (reversals). SV records of I and F have maximum amplitudes at low latitudes, whereas D records have maximum amplitudes at high latitudes.

Numerous studies indicate that the Holocene secular variation patterns of all three elements are reproducible on a regional scale of 3000–5000 km (e.g., King et al., 1983a, b; Thompson, 1984; Thompson & Oldfield, 1986; Lund, 1996). The SV pattern of all three geomagnetic elements can be used for regional dating and correlation on timescales of 10^2–10^4 years.

The polarity of the Earth's magnetic field has reversed repeatedly in the geological past. Although polarity transitions only last a few thousand years, they are isochronous over the entire Earth and can be used for global dating and correlation. Magnetostratigraphy organizes rocks and sediments according to the site dependent polarity acquired at the time of emplacement. A polarity timescale that is well constrained has been developed for the last 160 million years. The temporal control for this timescale is provided by radiometric dating (e.g., Cox et al., 1964), calculation and extrapolation of the seafloor spreading rates using the marine magnetic anomaly pattern (e.g., Heirtzler et al., 1968), biostratigraphic constraints (e.g., Berggren et al., 1985) and more recently orbital tuning of the polarity timescale using the magnetic record and oxygen isotope stratigraphy of long marine sedimentary sections (e.g., Shackleton et al., 1990). Magnetostratigraphy is useful on timescales of 10^5–10^8 years.

Polarity transitions involve a change in I from the latitude dependent GAD value (see Fig. 2) of one polarity to the comparable value of the opposite polarity (e.g., at 40° N latitude the I will change from $-60°$ to $+60°$ during negative to positive polarity transitions and vice versa). D values change by 180° during a polarity transition, irrespective of site location. Geomagnetic intensity decays from higher values to relatively low values during a polarity transition and returns to higher values upon completion of the transition.

An intermediate type of geomagnetic behavior between SV and polarity transitions is the geomagnetic excursion. A geomagnetic field excursion is a brief (i.e., 10^3 years) but significant departure from the GAD configuration, where the field can approach, but does not necessarily attain, a polarity transition and then returns to its pre-existing polarity. If the field attains a complete polarity transition and the opposite polarity persists for at least 10^4 years, then the field behavior is described as a polarity event (Barbetti & McElhinny, 1976). In practice it is often difficult to distinguish between excursions and short events and the spatial extent of these types of field behavior is controversial. Short events are global, whereas it is unclear if excursions are regional or global. At present the use of excursions and short events to provide stratigraphic control and dating of lake sediments shows considerable promise but requires extensive further study.

Recording fidelity of geomagnetic behavior by sediments

A complete discussion of the recording fidelity of geomagnetic behavior by sediments is beyond the scope of this paper, but several important concepts will be touched on in this section. The first concept involves reliability criteria. Thompson (1984) reviewed the reliability criteria needed to demonstrate that paleomagnetic data from sediments are reliably recording geomagnetic behavior. These criteria (Thompson, 1984) include:

(1) reproducibility of data between samples, cores, and lakes; (2) independence of paleomagnetic data from lithologic variation; (3) comparisons with historical records from geomagnetic observatories; (4) comparisons with lava flow and archeomagnetic data; and (5) laboratory tests of magnetic stability. Paleomagnetic data obtained from sediments that pass all of these criteria are very likely to accurately record geomagnetic variations, particularly large amplitude ones.

A second important concept involves geomagnetic signal smoothing or attenuation in depositional environments with low sedimentation rates and a thin surficial zone of moderate bioturbation. Lund and Keigwin (1994) have shown that SV recorded in fine-grained sediments with sedimentation rates $\geq$ 25 cm/ky show little or no smoothing, whereas sediments with sedimentation rates of ~12.5 cm/ky will smooth SV by 50%. Therefore, in intermediate sedimentation rate environments, SV features will still be recorded but their amplitude will be recorded at 50% of the actual geomagnetic change. Environments with sedimentation rates significantly less than 10 cm/ky and moderate bioturbation are probably not useful for SV studies. On the other hand, these environments will still record very large amplitude changes (i.e., excursions and polarity transitions), albeit with some attenuation of the amplitude.

A final important concept involves the possible effects of post-depositional reductive diagenesis of the original magnetic carrier. Although much more common in marine sediments than lake sediments (Jones & Bowser, 1978), the reductive diagenesis of fine-grained magnetite to either less magnetic pyrrhotite or greigite, or non-magnetic pyrite can alter or destroy that original magnetic record (e.g., Karlin & Levi, 1983). Reductive diagenesis is readily recognized by rock-magnetic studies (e.g., Reynolds & King, 1995), and sediments strongly affected by the process should be avoided when attempting to obtain records of geomagnetic behavior.

Field and laboratory methods

Field methods

Multiple, high-quality, sediment cores must be collected for paleomagnetic studies from each study site to satisfy the first reliability criterion of Thompson (1984) that the geomagnetic record be reproducible. Declination, inclination, and intensity data can be obtained from cores that are either absolutely, or relatively oriented in the horizontal plane (i.e., have either a known, and/or constant azimuth) and penetrate the sediments vertically. On the other hand, inclination and intensity data can still be obtained from unoriented cores that penetrate the sediments vertically (King et al., 1983b). Most commonly used piston and gravity coring and piston drilling tools provide sediments suitable for paleomagnetic studies. In general, large diameter corers (~10 cm) provide the least physically disturbed and therefore best materials for paleomagnetic studies.

Laboratory methods

The trend during the last decade has been for increasing use of automated systems that can rapidly do high-resolution (~1 cm) paleomagnetic and physical property studies of either whole-core, or continuous U-channel samples (Weeks et al., 1993). Two automated

Figure 3. Photograph of a 2-G Enterprises, Mountain View, California, USA, 755R automated, long-core magnetometer system located at the University of Rhode Island, USA. Cores are automatically measured, AF demagnetized, and remeasured with a computer-controlled core handling track that utilizes a stepper motor and controller and optical sensors to provide 1 cm resolution paleomagnetics data. This system is ~8 m in length and can measure cores up to 3 m in length.

systems currently define the state-of-the-art in the field. A 2-G Enterprises pass-through superconducting cryogenic magnetometer and alternating field (AF) demagnetizer system is shown in Figure 3. The computer-controlled system can automatically do measurements of inclination, declination, and magnetization. In addition, the built-in AF demagnetizer provides a reliable method of "magnetic cleaning" (Collinson, 1983) (i.e., removal of secondary magnetic signals that might obscure the original geomagnetic field record). A variety of mineral magnetic properties of lake sediments useful for paleoenvironmental characterization, including concentration, grain size, and mineralogy can also be studied using this system (Weeks et al., 1993; Reynolds & King, 1995).

A second automated system useful for whole- or split-core logging of physical properties (i.e, low-field susceptibility, gamma wet bulk density, compressional-wave velocity, natural gamma, resistivity, digital photography and red-green-blue (RGB) color analysis) is the GEOTEK system shown in Figure 4. These core logs are routinely used for stratigraphic correlation, dating using orbital tuning approaches, and paleoenvironmental interpretation (Reynolds & King, 1995). The combined use of these two systems will produce the data sets described in subsequent sections of this paper.

Holocene SV records

The most powerful approach to obtaining accurate age determinations from lake sediments is the multi-disciplinary use of SV records, radiocarbon, biostratigraphic studies, and, where

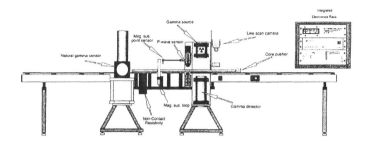

Figure 4. Diagram of GEOTEK, Limited, Surrey, England, multi-sensor core logger with gamma ray porosity evaluator (GRAPE) sensor for wet bulk density measurements, compressional wave (P-wave) velocity sensor, magnetic susceptibility (mag. sus.) loop and point sensors, resistivity sensor, natural gamma sensor, and line-scan digital camera with red-green-blue color analysis capability. This automated system can log core sections up to 1.55 m long with 0.5 cm resolution using a computer-controlled stepper motor and controller and laser sensors for core handling.

feasible, tephrochronology and varve chronology. Numerous studies (e.g., King et al., 1983a,b; Thompson, 1984; Thompson & Oldfield, 1986; Lund, 1996) have shown that radiocarbon dated Holocene SV curves obtained from rapidly deposited lake sediments are reproducible on a regional scale of several thousand kilometers.

We have tested the robustness of using radiocarbon dated Holocene SV records for stratigraphic correlation and dating in the northeastern U.S. Three lakes, shown in Figure 5, separated by 300 km and with very different basin morphometry, sediment characteristics and sedimentation rates, were selected for SV, radiocarbon, and pollen studies. Seneca Lake, New York, one of the Finger lakes, is a large, deep lake, with inorganic sediments (organic C is 0.6–2.2%), and relatively low sedimentation rates (~40 cm/ky). LeBoeuf Lake, Pennsylvania, is a small kettle lake with relatively large inlet and outlet streams, intermediate organic content (organic C is 2.0–7.0%), and a relatively high sedimentation rate (~210 cm/ky). Sandy Lake, Pennsylvania, is a small kettle lake with relatively small inlet and outlet streams, higher organic content (organic C is 2.3–13.6%), and an intermediate sedimentation rate (~130 cm/ky).

Pollen studies (King, 1983) were done to identify two major pollen features of known age within the study area. The *Ambrosia* or ragweed rise at the time of initial European settlement occurs between 1790–1810 A.D. within the study area. Calibration curves (Stuiver, 1982) indicate these calendar ages correspond to a radiocarbon age of 150 years B.P. The *Tsuga* or hemlock decline has an average ^{14}C age of 4634 ± 304 yr. B.P. calculated from 51 sites in northwestern North America by Webb (1982). Both pollen features were observed in Sandy and Seneca lakes, whereas only the *Ambrosia* rise was observed in LeBoeuf Lake indicating that the cores from LeBoeuf Lake were younger than ~4600 years B.P.

Radiocarbon studies of bulk sediment samples were done at each lake. The *Ambrosia* rise was dated from two cores from each lake and the *Tsuga* decline dated from two cores each from Seneca and Sandy lakes. Multiple dates were obtained from Seneca Lake (14 total), Sandy Lake (15 total), and LeBoeuf Lake (12 total), and an independent age model was determined for each lake. An old-carbon correction factor was determined for each lake by comparing the average radiocarbon age determination for the base of

Figure 5. Location map of the study area. Locations are: (1) Seneca Lake, (2) LeBoeuf Lake, and (3) Sandy Lake. Sites are shown in relation to important regional geographic and glacial features.

Ambrosia rise with the known age of 150 years B.P. The correction factors ranged from 402 ± 50 years for Sandy Lake, to 854 ± 50 years for LeBoeuf Lake, to 1577 ± 50 years for Seneca Lake. The magnitude of the correction factor appears to be directly proportional to the carbon content of the sediments from the three lakes. These correction factors were applied uniformly to the independent age models previously determined for each lake.

High-resolution SV studies were done on multiple cores from each lake. Three cores were studied from LeBoeuf Lake, two cores from Sandy Lake, and five cores from Seneca Lake. Composite *D* records were constructed for each site by matching each independent record to a reference core based on correlation of their susceptibility records and stacking the data versus depth. The stacked data are then plotted versus radiocarbon age using the independently determined age models of the reference cores. The declination results are compared in Figures 6 and 7. The corrected radiocarbon ages for the declination and pollen factors are summarized in Table I.

If we assume that the declination and pollen features are isochronous within the study area, then the average radiocarbon age difference observed for the 29 features is 172 years, comparable to radiocarbon counting errors. This result primarily confirms that an excellent old carbon correction can be obtained by radiocarbon dating the *Ambrosia* rise, a feature of known age, calculating the difference in known and measured age, and uniformly applying this correction to Holocene sediments. In addition, this test strongly indicates that regional radiocarbon dated SV curves can be used to date Holocene sediments within that region with centennial scale accuracy.

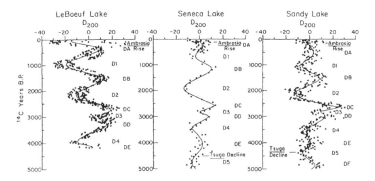

Figure 6. Comparison of composite *D* curves for northeastern U.S. lakes. Radiocarbon age models are independently constructed and ages are corrected for old carbon effects.

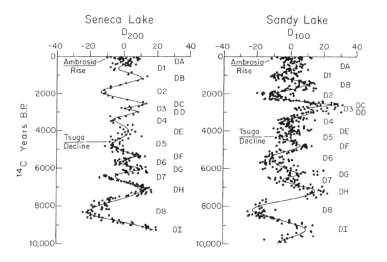

Figure 7. Comparison of composite *D* curves for Seneca and Sandy Lakes. Radiocarbon age models are independently constructed and ages are corrected for old carbon effects. *D* peaks are lettered (i.e. DA) and troughs are numbered (i.e D1). Depths of *Ambrosia* rise and *Tsuga* decline are indicated.

For this reason, we stacked the *D* data from all three lakes and the *I* data from Seneca and LeBoeuf lakes to construct regional SV curves. We selected Seneca Lake as the reference SV curve and used CORPAC (Martinson et al., 1982; 1987), a well-known data stacking program, and the correlation points shown in Figure 8 to construct the regional declination SV curve shown in Figure 9a. The comparable regional inclination SV curve is shown in Figure 9b. These curves can be used to date the SV results obtained from other lakes within North America with centennial accuracy.

Table I. Radiocarbon ages of pollen and declination features.

Feature	Sandy Lake Corrected Age in ^{14}C years (Correction Factor = 402 years)	Seneca Lake Corrected Age in ^{14}C years (Correction Factor = 1577 years)	LeBoeuf Lake Corrected Age in ^{14}C years (Correction Factor = 854 years)
Ambrosia Rise	150	150	150
Declination *Features*			
Peak DA	529	263	386
Trough D1	1029	670	976
Peak DB	1528	1212	1570
Trough D2	2109	1958	2175
Peak Dc	2649	2591	2740
Trough D3	2827	2864	3020
Peak DD	2993	3043	3293
Trough D4	3561	3422	3954
Peak DE	4035	4060	4216
Tsuga Decline	4353	4534	
Declination *Features*			
Trough D5	4389	4787	
Peak Df	4848	5422	
Trough D6	5439	5709	
Peak DG	6346	6064	
Trough D7	6629	6428	
Peak DH	7252	7196	
Trough D8	8210	8317	
Peak DI	9343	9276	

Average Differences in Feature Age = 172 years.

In addition to I and D records, relative paleointensity records can also be used for regional correlation. Sediments must meet rigorous criteria (e.g., King et al., 1983a; Tauxe, 1993) to be suitable for paleointensity studies. Among our northeastern U.S. lakes, only LeBoeuf Lake readily met these criteria. However, the paleointensity curve obtained from LeBoeuf lake is very comparable to the western U.S. lava flow record (Champion, 1980) as shown in Figure 10. These data indicate that SV curves obtained from suitable Holocene

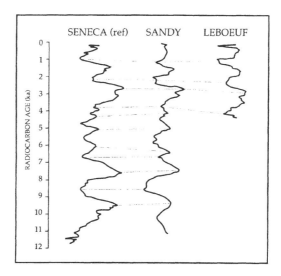

Figure 8. Composite radiocarbon dated *D* curves from three northeastern U.S. lakes used to construct regional composite curve. Tie lines indicate the correlation points used to constrain the CORPAC (Martinson et al., 1987) analysis. Seneca Lake was used as the target curve.

lake sediments can be used for dating and correlation within North America with resolution that approaches the centennial scale.

Magnetostratigraphic studies of Neogene lake sediments

Magnetostratigraphic studies of Neogene lake sediments have increased in importance during the last decade as new deep drilling programs (e.g., the Baikal Drilling Project, BDP) in rift lakes have allowed access to lake sediments old enough to record polarity transitions. The BDP is an international collaboration between Russian, U.S., Japanese, and German scientists to drill sediments in the world's largest and oldest freshwater lake. Drilling on Academician Ridge, an isolated rise, with relatively slow sedimentation rates, has allowed recovery of cores that penetrate a sequence of polarity transitions back to 5 Ma (Williams et al., 1997). It is possible to accurately date these sediments by comparison with the polarity timescale using only *I* records because of the high latitude (53.5°N) of the site as shown in Figure 11. In general, sites with latitudes of greater than 30° (i.e., GAD inclinations of ±45°) can be dated by magnetostratigraphy using their *I* records. On the other hand, sites with latitudes of less than 20° are difficult to date with magnetostratigraphy using only *I* records because changes in polarity can occur as part of the normal SV of the geomagnetic field. Therefore, it is necessary to obtain azimuthally oriented cores at low latitudes because the *D* values change by 180° during polarity transitions, and a variation of this magnitude is easily recognized in an oriented core.

The magnetostratigraphic correlation shown in Figure 11 is used to constrain the overall Baikal age model. However, since polarity transitions only occur every few 10^5 years, the

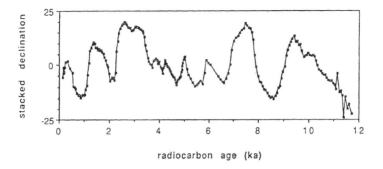

Figure 9a. Regional composite *D* curve for eastern U.S. after CORPAC analysis.

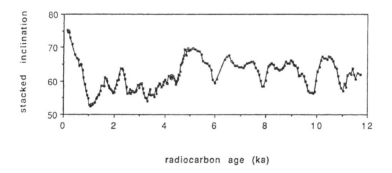

Figure 9b. Regional composite *I* curve for eastern U.S. after CORPAC analysis.

magnetostratigraphic age model is relatively low in resolution. To improve resolution, the Baikal magnetic susceptibility record can be readily correlated to the orbitally tuned $\delta^{18}O$ record obtained from deep-sea sediments (e.g., Tiedemann et al., 1994). The marine record is also constrained by magnetostratigraphy. An example of the marine correlation approach to adjusting the Baikal age model is shown in Figure 12. The marine correlation approach actually shifts the age of the depths of the Baikal polarity transitions by an average of 8 ka from the polarity timescale ages. We view the average difference in polarity age and correlated age as a measure of the accuracy of this approach. For the last 5 million years it is $\sim$10 ka.

An independent approach to dating lake sediments on timescales of 10^6–10^7 years involves the use of the relative paleointensity record. Studies of marine sediments have produced a global paleointensity timescale (e.g., Meynadier et al., 1994) for the last few million years. Use of this approach will be discussed in the next section.

LeBoeuf Lake

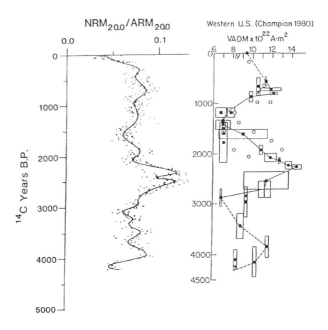

Figure 10. Comparison of the composite relative paleointensity record of LeBoeuf Lake with absolute paleointensity record from western U.S. lava flows of Champion (1980). Relative paleointensity estimates are obtained from natural remanent magnetizations after 20.0 mT demagnetization divided by anhysteretic remanent magnetization after 20.0 mT demagnetization (i.e., $NRM_{20.0}/ARM_{20.0}$).

Excursions, short events and relative paleointensity

Recent compilations of excursions and short events (e.g., Langereis et al., 1997; Nowaczyk & Antonow, 1997) during the last 1 Ma indicate both their potential for obtaining increased resolution in sediment studies on timescales of 10^3-10^5 years and the dating problems that currently limit their use. For example, the compilation of Langereis et al. (1997) is shown in Figure 13. Note that the age of one of the better established excursion/short events, the Jamaica/Pringle Falls/Biwa *I*, is assigned to 205–215 ka, whereas the Nowaczyk & Antonow (1997) compilation for this event shown in Figure 14 indicates a bimodal age distribution, with the dominant mode between 174–186 ka. Similar dating problems exist for all of the best-documented excursions/short events. Intensive global research efforts are underway to document the spatial extent and timing of excursions/short events. We anticipate that an extremely useful and well-documented excursion/short timescale for the last 1 Ma will be available for the study of lake sediments in the near future.

A second approach to dating sediments on timescales of 10^3-10^6 years involves the use of relative paleointensity records. Studies of marine sediments have produced a paleointensity timescale (e.g., Meynadier et al., 1994) for the last 4 Ma. Relative paleointensity results

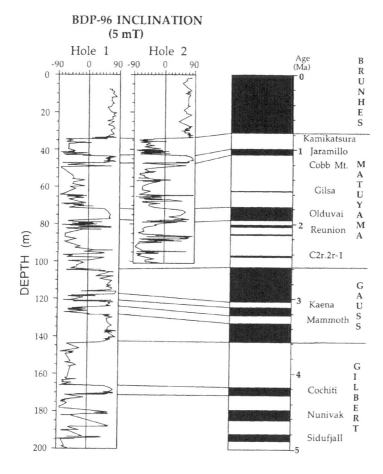

Figure 11. Comparison of Lake Baikal Drilling Project inclination records (BDP-96) from Academician Ridge to the geomagnetic polarity timescale (Shackleton et al., 1995; Cande & Kent, 1995).

obtained from lake sediments can be compared to the marine paleointensity timescale and age estimates for the lake sediments can be obtained. An example of this approach for Lake Baikal sediments is shown in Figure 15 from Peck et al. (1996). Spectral analysis and bandpass filtering of these records indicates high coherence of frequency components greater than 5 ka (Peck et al., 1996). These components were interpreted as the global geomagnetic signal, whereas the frequency component between 3.5–2 kyr was not coherent and was interpreted as the regional geomagnetic signal (Peck et al., 1996). The spectral analysis results indicate that the limit of accuracy of global paleointensity dating is ∼5 ka (Peck et al., 1996).

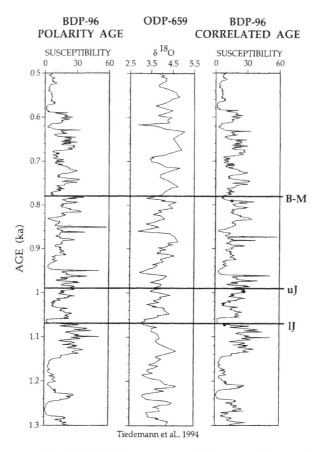

Figure 12. Refinement of Lake Baikal (BDP-96) polarity age model (left column 5) by correlation of the Baikal susceptibility record to the astronomically tuned (Tiedemann et al., 1994) $\delta^{18}O$ marine record from ODP Site 659 (middle column). The right column shows the correlated age model for BDP-96. Position of Brunhes-Matuyama (B-M), upper Jaramillo (UJ) and lower Jaramillo (LJ) polarity transitions are shown. Note the actual positions of the polarity transitions (indicated by dots in the right column) can shift slightly in the correlated age model.

Conclusions

SV and radiocarbon studies of suitable, rapidly deposited (>30 cm/ka) Holocene lake sediments can be used to provide dating and correlation with centennial resolution for regional ($\leq 5 \times 10^3$ km) paleoenvironmental studies.

Magnetostratigraphy studies, in conjunction with correlation of susceptibility records to the orbitally tuned marine $\delta^{18}O$ record, can provide global dating resolution of ~ 10 ka for studies of lake sediments on timescales of 10^6–10^7 years. Paleointensity studies of suitable sediments provide an independent approach to dating sediments on similar timescales that can provide similar resolution.

BRUNHES
Reversal Excursions

kyr		
40-45		**Laschamps**
70-80		Norwegian-Greenland Sea
110-120		**Blake**
155-165		Albuquerque/Fram Strait
205-215		**Jamaica/Pringle Falls**
255-265		Fram Strait/CR0?
315-325		**Calabrian Ridge 1**
360-370		Levantine
400-420		unknown?
515-525		**Calabrian Ridge 2/West Eifel**
560-570		**Emperor/Big Lost/CR3**
778		B/M

Figure 13. Excursions/short events during the Brunhes. Those in bold are better dated and probably global, whereas those in small print are not well dated and may be regional. Age of B/M boundary after Tauxe et al., 1996. From Langereis et al. (1997).

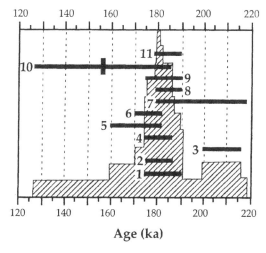

Age (ka)

Figure 14. Age ranges published for the Jamaica/Pringle Falls/Biwa *I* short event. References are: (1) Wollin et al., 1971; (2) Kawai et al., 1972; (3) Ryan, 1972; (4) Bleil & Gard, 1989; (5) Liddicoat, 1990; (6) Nowaczyk & Baumann, 1992; (7) Herrero-Bevera et al., 1994; (8) Negrini et al., 1994; (9) Weeks et al., 1995; (10) Peate et al., 1996; (11) Nowaczyk & Antonow, 1977. The cross-hatched area was obtained by stacking the age ranges with a weighting of 1. The half-width provides an age range 186–174 ka for this short event. From Nowaczyk & Antonow (1997).

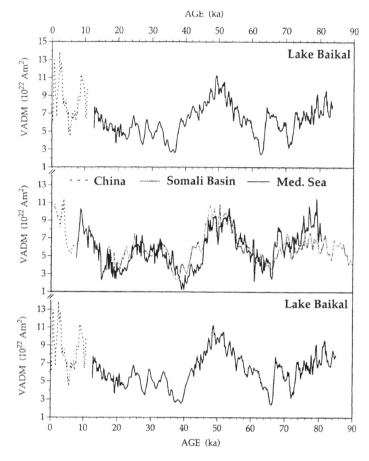

Figure 15. Comparison of relative paleointensity records. (a) Lake Baikal: dotted line = surface cores; solid line = stacked long cores; (b) Mediterranean Sea (Tric et al., 1992), Somali Basin (Meynadier et al., 1992), and NE China (Yang et al., 1993); (c) Lake Baikal record replotted after adjustment to Somali Basin timescale. From Peck et al. (1996).

Studies of excursions/short events hold promise for increasing dating resolution of lake sediments on timescales of 10^5–10^6 years. Intensive study is underway that will clarify the excursion/short event stratigraphy of the last 1 Ma. In the interim, excursion/short event dating should be used with caution.

Summary

SV and magnetostratigraphy studies are well established for dating and correlation of lake sediments. SV and radiocarbon studies of suitable, rapidly deposited (>30 cm/ka) Holocene lake sediments can be used to provide dating and correlation with centennial resolution for regional ($\leq 5 \times 10^3$ km) paleoenvironmental studies.

Magnetostratigraphy studies, in conjunction with correlation of susceptibility records to the orbitally tuned marine $\delta^{18}O$ record, can provide global dating resolution of ~ 10 ka for studies of lake sediments on timescales of 10^6–10^7 years. Paleointensity studies of suitable sediments provide an independent approach to dating sediments on similar timescales that can provide similar resolution.

On the other hand, studies of excursions/short events hold promise for increasing dating resolution of lake sediments on timescales of 10^5–10^6 years. Intensive study is underway that will clarify the excursion/short event stratigraphy of the last 1 Ma. In the interim, excursion/short event dating should be used with caution.

References

Backus, G., R. Parker & C. Constable, 1996. Foundations of Geomagnetism. Cambridge University Press, N.Y., 369 pp.

Barbetti, M. F. & M. W. McElhinny, 1976. The Lake Mungo geomagnetic excursion. Phil. Trans. r. Soc. Lond. 281: 515.

Berggren, A. B., D. V. Kent, J. J. Flynn & J. A. van Couvering, 1985. Cenozoic geochronology. Bull. am. Geol. Soc. 96: 1407.

Bleil, U. & G. Gard, 1989. Chronology and correlation of Quaternary magnetostratigraphy and nannofossil biostratigraphy in Norwegian-Greenland Sea sediments. GEOLOG. Rundschau 78: 1173–1187.

Butler, R. F., 1992. Paleomagnetism: Magnetic Domains to Geologic Terranes. Blackwell Science Publishers, Boston, 319 pp.

Cande, S. C. & D. V. Kent, 1995. Revised calibration of the geomagnetic polarity timescale for the Late Cretaceous and Cenozoic. J. Geophys. Res. 100: 6093–6095.

Collinson, D. W., 1983. Methods in Rock Magnetism and Paleomagnetism: Techniques and Instrumentation. Chapman & Hall, Lond. & N.Y., 503 pp.

Cox, A., R. R. Doell & G. R. Dalrymple, 1964. Reversals of the Earth's magnetic field. Science 144: 1537.

Champion, D., 1980. Holocene Geomagnetic Secular Variation in the Western United States: Implications for the Global Geomagnetic Field. Ph.D. Thesis, California Institute of Technology, Pasadena, 280 pp.

Heirtzler, J. R., G. O. Dickson, E. M. Herron, W. C. Pitman & X. LePichon, 1968. Marine magnetic anomalies, geomagnetic field reversals and motions of the ocean floor and continents. J. Geophys. Res. 73: 2119.

Herrero-Bevera, E., C. E. Helsley, A. M. Sarna-Woycicki, K. R. Lajoie, C. E. Meyer, M. O. McWilliams, R. M. Negrini, B. D. Turrin, J. M. Donnelly Nolan & J. C. Liddicoat, 1994. Age and correlation of a paleomagnetic episode in the western United States by $^{40}Ar/^{39}Ar$ dating and tephrochronology: the Jamaica, Blake, or a new polarity episode? J. Geophys. Res. 99: 24,091–24,103.

Jacobs, J. A., 1994. Reversals of the Earth's Magnetic Field. Cambridge University Press, Cambridge, 346 pp.

Jones, B. F. & C. J. Bowser, 1978. The mineralogy and related chemistry of lake sediments. In Lerman, A. (ed.) Lakes: Chemistry, Geology, Physics. Springer, N.Y.

Karlin, R. & S. Levi, 1983. Diagenesis of magnetic minerals in Recent hemipelagic sediments. Nature 303: 327–330.

Kawai, N., K. Yaskawa, T. Nakajima, M. Torii & S. Horie, 1972. Oscillating geomagnetic field with a recurring reversal discovered from Lake Biwa. Proc. Japan Acad. 48: 186–190.

King, J. W., 1983. Geomagnetic Secular Variation Curves for Northeastern North America for the Last 9,000 Years B.P. Ph.D. Thesis, University of Minnesota, Minneapolis, 195 pp.

King, J. W., S. K. Banerjee & J. Marvin, 1983a. A new rock-magnetic approach to selecting sediments for geomagnetic paleointensity studies: application to paleointensity for the last 4000 years. J. Geophys. Res. 88: 5911–5921.

King, J. W., S. K. Banerjee, J. Marvin & S. Lund, 1983b. Use of small-amplitude paleomagnetic fluctuations for correlation and dating of continental climate changes. Palaeogeogr. Palaeoclimatol. Palaeoecol. 42: 167–183.

Langereis, C. G., M. J. Dekkers, G. J. de Lange, M. Paterne & P. J. M. van Santvoort, 1997. Magnetostratigraphy and astronomical calibration of the last 1.1 Myr from an eastern Mediterranean piston core and dating of short events in the Brunhes. Geophys. J. Int. 129: 75–94.

Liddicoat, J. C., 1990. Aborted reversal of the palaeomagnetic field in the Brunhes Normal Chron in east-central California. Geophys. J. Int. 102: 747–752.

Lund, S. P., 1996. A comparison of Holocene paleomagnetic secular variation records from North America. J. Geophys. Res. 101: 8007–8024.

Lund, S. P. & L. Keigwin, 1994. Measurement of the degree of smoothing in sediment paleomagnetic secular variation records: an example from late Quaternary deep-sea sediments of the Bermuda Rise, western North Atlantic Ocean. Earth Planet. Sci. Lett. 122: 317–330.

Martinson, D. G., W. Menke & P. Stoff, 1982. An inverse approach to signal correlation. J. Geophys. Res. 87: 4807–4818.

Martinson, D. G., N. G. Pisias, J. D. Hay, J. Imbrie, T. C. Moore & N. J. Shackleton, 1987. Age dating and the orbital theory of the ice ages: development of a high-resolution 0 to 300,000 year chronostratigraphy. Quat. Res. 27: 1–29.

Meynadier, L., J.-P. Valet, R. Weeks, N. J. Shackleton & V. L. Hagee, 1992. Relative geomagnetic intensity of the field during the last 140 ka. Earth Planet. Sci. Lett. 114: 39–57.

Meynadier, L., J.-P. Valet, F. C. Bassinot, N. J. Shackleton & Y. Guyodo, 1994. Asymmetrical sawtooth pattern of the geomagnetic field intensity from equatorial sediments in the Pacific and Indian Oceans. Earth Planet. Sci. Lett. 126: 109–127.

Negrini, R. M., D. B. Erbes, A. P. Robertes, K. L. Verosub, A. M. Sarna-Wojcicki & C. E. Meyer, 1994. Repeating waveform initiated by a 180–190 ka geomagnetic excursion in western North America: implications for field behavior during polarity transitions and subsequent secular variation. J. Geophys. Res. 99: 24,105–24,119.

Nowaczyk, N. R. & M. Antonow, 1997. High-resolution magnetostratigraphy of four sediment cores from the Greenland Sea, I. Identification of the mono lake excursion, Laschamp and Biwa I/Jamaica geomagnetic polarity events. Geophys. J. Int. 131: 310–324.

Nowaczyk, N. R. & M. Baumann, 1992. Combined high-resolution magnetostratigraphy and nannofossil biostratigraphy for late Quaternary Arctic Ocean sediments. Deep Sea Res. 39: 567–601.

Peate, D. W., J. H. Chen, G. J. Wasserburg, D. A. Papanastassiou & J. W. Geissman, 1996. ^{238}U-^{230}Th dating of a geomagnetic excursion in Quaternary basalts of the Albuquerque Volcanoes Field, New Mexico (USA). Geophys. Res. Lett. 23: 2271–2274.

Peck, J. A. & J. W. King, 1996. An 84-kyr paleomagnetic record from the sediments of Lake Baikal, Siberia. J. Geophys. Res. 101: 11,365–11,385.

Reynolds, R. L. & J. W. King, 1995. Magnetic records of climate change. Rev. Geophys. Sup., U.S. National Report to International Union of Geodesy and Geophysics 1991–1994: 101–110.

Ryan, W. B. F., 1972. Stratigraphy of late Quaternary sediments in the Eastern Mediterranean. In Stanley, D. J. (ed.) The Mediterranean Sea: A Natural Sedimentation Laboratory. Dowden, Hutchinson & Ross, Inc., Stroudsburg: 149–169.

Shackleton, N. J. A. Berger & W. R. Peltier, 1990. An alternative astronomical calibration of the lower Pleistocene timescale based on ODP Site 677. Trans. R. Acad. Soc. Edinb. 81: 251–261.

Shackleton, J. J., S. Crowhurst, T. Hagelberg, N. G. Pisias & D. A. Schneider, 1995. A new late neogene timescale: application to Leg 138 sites. In Pisias, N. G., L. A. Mayer, T. R. Janecek, A. Palmer-Julson & T. H. van Andel (eds.) Proceedings of the Ocean Drilling Program, Sci. Res. 138: 73–90.

Stuiver, M., 1982. A high-precision calibration of the A.D. radiocarbon timescale. Radiocarbon 24: 1–26.

Tauxe, L., 1993. Sedimentary records of relative paleointensity of the geomagnetic field: theory and practice. Rev. Geophys. 31: 319–354.

Tauxe, L. T. Herbert, N. J. Shackleton & Y. S. Kok, 1996. Astronomical calibration of the Matuyama-Brunhes boundary: consequences for magnetic remanence acquisition in marine carbonates and the Asian loess sequences. Earth Planet. Sci. Lett. 140: 133–146.

Thompson, R., 1984. A global review of paleomagnetic results for wet lake sediments. In Haworth, E. Y. & J. W. G. Lund (eds.) Lake Sediments and Environmental History. Univ. of Minn. Press, Minneapolis: 145–164.

Thompson, R. & F. Oldfield, 1986. Environmental Magnetism. Allen & Unwin, Winchester, 227 pp.

Tiedemann, R., M. Sarnthein & N. J. Shackleton, 1994. Astronomic timescale for the Pliocene Atlantic δ^{18}O and dust flux records of Ocean Drilling Program site 659. Paleoceanography 9: 619–638.

Tric, E., J.-P. Valet, P. Tucholka, M. Paterne, L. Labeyrie, F. Guichard, L. Tauxe & M. Fontugne, 1992. Paleointensity of the geomagnetic field during the last 80 kyr. J. Geophys. Res. 97: 9337–9351.

Webb, T. III, 1982. Temporal resolution in Holocene pollen data. Third North American Paleon. Con. Proc. 2: 569–572.

Weeks, R., C. Laj, L. Endignoux, M. Fuller, A. Roberts, R. Manganne, E. Blanchard & W. Gorce, 1993. Improvements in long-core measurement techniques: applications in paleomagnetism and paleocenaography. Geophys. J. Int. 114: 651–662.

Weeks, R. J., C. Laj, L. Endignoux, A. Mazaud, L. Labeyrie, A. P. Roberts, C. Kissel & E. Blanchard, 1995. Normalized natural remanent magnetization intensity during the last 240,000 years in piston cores from the central North Atlantic Ocean: geomagnetic field intensity or environmental signal? Phys. Earth Planet. Int. 87: 213–229.

Williams, D. F., J. Peck, E. B. Karabanov, A. A. Prokopenko, V. Kravchinsky, J. King & M. I. Kuzmin, 1997. Lake Baikal record of continental climate response to orbital insolation during the past 5 million years. Science 278: 1114–1117.

Wollin, G., D. B. Ericson, W. B. F. Ryan & J. H. Foster, 1971. Magnetism of the earth and climatic changes. Earth Planet. Sci. Lett. 12: 175–183.

Yang, S. J. & Q. Y. Wei, 1993. Tracking a non-dipole geomagnetic anomaly using new archeointensity results from north-east China. Geophys. J. Int. 115: 1189–1196.

15. AMINO ACID RACEMIZATION (AAR) DATING AND ANALYSIS IN LACUSTRINE ENVIRONMENTS

BONNIE A. B. BLACKWELL
(bonnie.a.b.blackwell@williams.edu)
Department of Chemistry
Williams College
Williamstown
MA, 01267
USA

Keywords: Amino acid racemization (AAR) dating, aminostratigraphy, paleolimnology, paleodietary analysis, molluscs, foraminifera, egg shells, bones, teeth, paleoenvironmental analysis.

Introduction

Amino acid racemization (AAR) dating depends on the natural diagenesis of organic components in proteins within fossils and in sediment containing organic matter. Unlike radiometric dating (e.g., ^{14}C, $^{40}Ar/^{39}Ar$, etc.), this process depends on time, temperature, and, for some fossils, other paleoenvironmental conditions near the sample. Therefore, the process must be modelled extensively under various environmental conditions before it can be applied universally. If AAR does not function well as a dating technique for some fossil tissues, this reliance on environmental conditions may permit other applications that can provide important paleoenvironmental data.

This chapter will concentrate on potential limnological applications, illustrated where possible by lacustrine examples. It necessarily omits most applications in non-limnological settings, although Table I does list references for many applications for readers interested in other uses. Other reviews (e.g., Rutter & Blackwell, 1995; Wehmiller & Miller, 1990) offer more general coverage. Two other important references are *The Biogeochemistry of Amino Acids* (Hare et al., 1980) and *Perspectives in Amino Acid and Protein Geochemistry* from the 1998 festschrift for Ed Hare (Goodfriend et al., 2000).

History of the method

Until 1954, few suspected that proteins might have geochronological applications. Pioneering work by Abelson (1954, 1955) and later by Hare (1963, 1969; Hare & Abelson, 1966; Hare & Mitterer, 1965, 1967; Miller & Hare, 1975) using shell and bone demonstrated the potential of the method. With the development of chiral phase chromatographic columns (Hare & Hoering, 1972) and the portable amino acid analyzer, many scientists began to

W. M. Last & J. P. Smol (eds.), 2001. *Tracking Environmental Change Using Lake Sediments. Volume 1: Basin Analysis, Coring, and Chronological Techniques.* Kluwer Academic Publishers, Dordrecht, The Netherlands.

Table I. Quaternary applications for amino acid racemization analyses.

Sample Type -Application	Location	References
Non-marine molluscs		Abbott et al., 1995, 1996; Collins et al., 1999; Ellis & Goodfriend, 1994; Ellis et al., 1996; Goodfriend, 1987a, 1987b; 1989, 1991a, 1991b, 1992a Googfriend, 1992b, 2000; Goodfriend & Meyer, 1991; Goodfriend et al., 1991; Hare et al., 1993; Murray-Wallace, 1993; Qian et al., 1995; Rutter & Blackwell, 1995; Wehmiller & Miller, 1990, 1998.
- relative stratigraphy & regional correlation	L. Bonneville Basin	Bouchard et al., 1998; McCoy, 1981, 1987a; Oviatt et al., 1987; Scott et al., 1983.
	Canadian Prairies	Nielsen et al., 1986; Osborn et al., 1991.
	Caribbean	Goodfriend & Mitterer, 1988, 1993.
	Europe	Oches et. al., 2000.
	Georgia, USA	Goodfriend & Rollins, 1998.
	Hungary	Oches & McCoy, 1995, 1996.
	Madeira	Cook et al., 1993; Goodfriend et al., 1994, 1996b.
	Mississippi Valley	Alford, 1990; Clark et al., 1989, 1990; Forman et al., 1992; B. B. Miller et al., 1987, 1992; Mirecki & Miller, 1994; Oches & McCoy, 1989.
	Namib	Teller et al., 1990.
	Negev Desert	Goodfriend, 1987a, 1987b; Rosen & Goodfriend, 1993.
	Norway	Bowen & Sykes, 1988; Bowen et al., 1989.
	Virginia	Mirecki et al., 1995.
	Texas	Abbott et al., 1995, 1996; Ellis & Goodfriend, 1994; Ellis et al., 1996.
- absolute dates	Caribbean	Goodfriend & Gould, 1996; Goodfriend & Mitterer, 1988.
	Negev Desert	Goodfriend, 1987a, 1987b, 1989, 1992b; Goodfriend & Meyer, 1991; Meyer, 1991, 1992.
- paleoenvironmental analyses	Bonneville Basin	Bouchard et al., 1998.
	Caribbean	Goodfriend & Mitterer, 1988, 1993.
	Georgia, USA	Goodfriend & Rollins, 1998.
	Negev Desert	Goodfriend, 1991a, 1991b, 1992a.
	Nile Delta	Goodfriend & Stanley, 1996; Stanley & Goodfriend, 1997.
- archaeology	Negev Desert	Rosen & Goodfriend, 1993.
	Texas	Abbott et al., 1995, 1996; Ellis & Goodfriend, 1994; Ellis et al., 1996.
- evolutionary rates	Caribbean	Goodfriend & Gould, 1996.
	Madeira	Cook et al., 1993; Goodfriend et al., 1994, 1996b.
- subsidence	Nile Delta	Goodfriend & Stanley, 1996; Stanley & Goodfriend, 1997.
Marine molluscs		Abelson, 1954, 1955; Bada & Schroeder, 1972, 1976a; Bada et al., 1986a; Brigham 1980, 1982, 1983b, 1985; Collins & Riley, 2000; Collins et al., 1999; Corrado et al., 1986; Ernst, 1987; Glavin & Bada, 1998; Goodfriend, 2000; Goodfriend et al., 1995, 1997; Hare, 1963, 1969; 1974b; Hare & Abelson, 1966, 1970; Hare & Mitterer, 1965, 1967; Haugen & Sejrup, 1990, 1992; Hearty & Aharon, 1988; Kaufman & Miller, 1992; Kaufman & Sejrup, 1995; Kimber & Griffin, 1987; Kimber & Milnes, 1984; Kimber et al.,1986; King, 1980a; Kowalewski et al., 1998; Kriasakul & Mitterer, 1980, 1983; Kvenvolden, 1975, 1980; LaJoie et al., 1980a;

Table I. Quaternary applications for amino acid racemization analyses (continued).

Sample Type -Application	Location	References
Marine molluscs (cont'd)		
		Manley et al, 2000; Masters & Bada, 1978; McCoy, 1987b;
		Miller & Brigham-Grette, 1989;
		Miller & Hare, 1975, 1980;
		Mitchell & Curry, 1997; Mitterer, 1993;
		Mitterer & Kriausakul, 1989; Murray-Wallace, 1993;
		Murray-Wallace & Kimber, 1987, 1993;
		Murray-Wallace et al., 1996; Qian et al., 1995;
		Roof, 1997; Rutter & Blackwell, 1995;
		Rutter et al., 1979, 1980, 1989, 1990;
		Schroeder & Bada, 1976; Steinberg & Bada, 1983;
		Wehmiller, 1980, 1981, 1982, 1984a, 1984b, 1993;
		Wehmiller & Miller, 1990, 1998.
		Wehmiller et al., 1977, 1988, 1992;
- relative stratigraphy & regional correlation	Alaska	Brigham, 1985; Brigham-Grette & Carter, 1992; Brigham-Grette & Hopkins, 1995; Carter & Hillhouse, 1991; Kaufman, 1992; Kaufman et al., 1991.
	Antarctica	Pickard et al., 1986, 1988.
	Arctic	Anderson et al., 1991a, 1991b; Brigham-Grette & Carter, 1992; Carter et al., 1986a, 1986b; Goodfriend et al., 1996a; Miller & Brigham-Grette, 1989; Miller & Hare, 1975.
	Australia	Cann et al., 1991; Ferland et al., 1995; Kimber & Griffin, 1987; Kimber & Milnes, 1984; Murray-Wallace, 1991, 2000; Murray-Wallace & Belperio, 1994; Murray-Wallace & Goede, 1995; Murray-Wallace & Kimber, 1987, 1989; Murray-Wallace et al., 1988b, 1991, 1993, 1996; Woodroffe et al., 1995.
	Baffin I.	Andrews et al., 1985a; Brigham, 1980, 1982, 1983a; Miller, 1985; G. H. Miller et al., 1977, 1988; Nielson, 1982; Szabo et al., 1981, 1982.
	Barents Sea	Saettem et al., 1991, 1992.
	Bermuda	Harmon et al., 1983; Hearty et al., 1992.
	Chesapeake Bay	Chen et al., 1995.
	Chile	Clapperton et al., 1995.
	Sea of Cortez	Goodfriend et al., 2000b; Kowalewski et al., 1998.
	N. European coastal plain	Miller & Mangerud, 1985.
	Florida	Mitterer, 1974, 1975.
	Greenland	Andrews et al., 1985a; Boulton et al., 1982; Funder & Simonarson, 1984; Funder et al., 1991.
	Hudson Bay Lowlands	Andrews et al., 1983, 1984b; Sykes, 1984.
	Mediterranean	Dumas et al., 1988; Hearty, 1986; Hearty et al., 1986;

continued ...

Notes:

* Unsuccessful or problematic application

‡ Discusses antler also

† Discusses dentine, enamel, and cement separately

Table I. Quaternary applications for amino acid racemization analyses (continued).

Sample Type -Application	Location	References
Marine molluscs (cont'd)		
-relative stratigraphy		G. H. Miller et al., 1986.
& regional correlation	N. Am. Atlantic coastal plain	Belknap & Wehmiller, 1980; Hollin et al., 1993; Hollin & Hearty, 1990; McCartan et al., 1982 Oldale et al., 1982; Toscano & York, 1992; Wehmiller & Belknap, 1982; Wehmiller, 1982, 1986, 1993; Wehmiller et al., 1988, 1992, 1995; York et al., 1989.
	N. Am. Pacific coastal plain	Atwater et al., 1981; Hicock & Rutter, 1986; Kennedy et al., 1982; Kvenvolden et al., 1979a, 1981; O' Neal et al., 2000; Wehmiller, 1982, 1986, 1993; Wehmiller et al., 1977, 1992; York & Wehmiller, 1992.
	Norway	Boulton et al., 1982; Forman & Miller, 1984; Mangerud et al., 1981; Miller, 1982; G. H. Miller et al., 1983, 1989a; Salvigsen & Nydel, 1981.
	Papua New Guinea	Murray-Wallace, 2000.
	S. Am. Atlantic coastal plain	Rutter et al., 1989, 1990.
	S. Am. Pacific coastal plain	Hsu et al., 1989,
	Spain	Torres et al., 2000a.
	U.K.	Andrews et al., 1979, 1984a; Bowen, 2000; Bowen et al., 1985, 1989; Campbell et al., 1982; Davies, 1983; Jardine et al., 1988; Keen et al., 1981; Miller & Hare, 1980; G. H. Miller et al., 1979, 1987; Sykes, 1988.
	N. Yukon	Rutter et al., 1980.
-absolute dates	Alaska	Brigham-Grette & Carter, 1992; Carter et al., 1986a, 1986b; Carter & Hillhouse, 1991; Kaufman et al., 1991, 1996; Kvenvolden et al., 1979b.
	Antarctica	Pickard et al., 1988.
	Arctic coastal plains	Carter et al., 1986b; Miller & Hare, 1975.
	Australia	Cann et al., 1991; Murray-Wallace & Bourman, 1990.
	W. Barents Sea	Saettem et al., 1991, 1992.
	Bermuda	Harmon et al., 1983; Hearty et al., 1992,
	Caribbean	Hoang & Hearty, 1989.
	Florida	Mitterer, 1974, 1975.
	Italy	Bellumoni & Delitala, 1988; Dumas et al., 1988.
	Norway	G. H. Miller et al., 1983.
	S. Am Atlantic coastal plain	Rutter et al., 1989, 1990.
- absolute dates	N. Am. Atlantic coastal plain, & continental shelf	Groot et al., 1995; Hare & Mitterer, 1967; McCartan et al., 1982; Mirecki & Wehmiller, 1984; Odale et al., 1982; O' Neal et al., 2000; Wehmiller 1982, 1984, 1986, 1993; Wehmiller & Belknap, 1982; Wehmiller et al., 1988, 1992.
	N. Am. Pacific coastal plain	Emerson et al., 1981; Kvenvolden et al., 1979a, 1981; Masters & Bada, 1978a, 1978b; Muhs et al., 1990; Wehmiller, 1982, 1984, 1986, 1993; Wehmiller et al., 1977.

Table I. Quaternary applications for amino acid racemization analyses (continued).

Sample Type -Application	Location	References
Marine molluscs (cont'd)		
- absolute dates	UK	Bowen, 2000; Bowen & Sykes, 1998, 1994.
- tectonic uplift	N. Am. Pacific coastal plain	Muhs et al., 1990.
	S. Am. Pacific coastal plain	Leonard & Wehmiller, 1992.
	Tasmania	Murray-Wallace & Goede, 1995.
- subsidence		Goodfriend & Stanley, 1996; Stanley & Goodfriend, 1997.
- sea level fluctuation	Alaska	Kaufman, 1992.
	Australia	Cann et al., 1991.
	Bermuda	Harmon et al., 1983.
- archaeology	Alaska	Kvenvolden et al., 1979b.
	Australia	Cann et al., 1991.
	California	Masters & Bada, 1978a; Wehmiller, 1977.
	Great Britain	Bowen & Sykes, 1994.
- taxonomy		Andrews et al., 1985b; Kaufman et al., 1992.
- taphonomy		Martin et al., 1996.
- reworked fossil or sediment recognition	Australia	Ferland et al., 1995; Murray-Wallace & Belperio, 1994; Nichol & Murray-Wallace, 1992.
	Sea of Cortez	Kowalewski et al., 1998.
	Nile Delta	Goodfriend & Stanley, 1996.
	N. Am. Atlantic coastal plain	Chen et al., 1995; Wehmiller et al., 1995;
-diagenesis		Akiyama, 1980; Hoering, 1980; Keil et al., 2000; Kvenvolden et al., 1979b; Robbins et al., 2000; Weiner & Lowenstam, 1980.
- paleodiets		Serban et al., 1986.
- forensic analysis		Custer et al., 1989; Powell et al., 1989.
- diagenetic or paleotemperatures	Arctic coastal plains	Miller, 1985; Miller & Brigham-Grette, 1989.
	Scotland	Miller et al., 1987.
	S. Australia	Murray-Wallace et al., 1988a.
Ostracodes		Kaufman, 2000; McCoy, 1988.
relative stratigraphy	W. USA	Kaufman, 2000.
Foraminifera		Bada & Schroeder, 1972; Bada et al., 1970; Collins et al., 1999; Goodfriend, 2000; Harada et al., 1996; Katz & Man, 1980; King, 1977, 1980a; King & Hare, 1972a, 1972b; King & Neville, 1977; Miller & Brigham-Grette, 1989; Rutter & Blackwell, 1995; Rutter et al., 1979; Schroeder & Bada, 1975, 1976, 1977; Wehmiller, 1980; Wehmiller & Hare, 1970, 1971; Wehmiller & Miller, 1990, 1998.
- pelagic sediment stratigraphy	Arctic region	Anderson et al., 1991a, 1991b; Haugen & Sejrup, 1992; Sejrup & Haugen, 1992.

continued . . .

Notes:

* Unsuccessful or problematic application

‡ Discusses antler also

† Discusses dentine, enamel, and cement separately

Table I. Quaternary applications for amino acid racemization analyses (continued).

Sample Type -Application	Location	References
Foraminifera (con't)		
- pelagic sediment stratigraphy	Arctic Ocean	Macko & Aksu, 1986; Sejrup et al., 1984b.
	W. Barents Sea	Haid et al., 1990; Saettem et al., 1991, 1992.
	Caribbean	Fletcher et al., 1991.
	N. Atlantic	Jansen & Sejrup, 1987.
	North Sea	Feyling-Hansen, 1982; Knudsen & Sejrup, 1988; G. H. Miller et al., 1983; Sejrup et al., 1984a, 1987, 1989,1991.
	Norwegian Sea	Haflidason et al., 1991; Lycke et al., 1992; Mienert et al., 1992.
	Pacific	Harada & Handa, 1995.
	S. Australia	Cann & Murray-Wallace, 1986.
- absolute dates	W. Barents Sea	Saettem et al., 1991, 1992.
- taxonomy		Haugen et al., 1989.
- sedimentation rates	Arctic region	Miller & Brigham-Grette, 1989; Wehmiller & Hare, 1970, 1971.
	Arctic Ocean	Macko & Aksu, 1986; Sejrup et al., 1984b.
- diagenetic temperatures		Katz et al., 1983; Keil et al., 2000; Miller & Brigham-Grette, 1989.
- biochemical pathways, paleodiets		Macko et al., 1987; Uhle et al., 1997.
Corals, Anemones		Brinton & Bada, 1995; Collins et al., 1999; Goodfriend, 1997; 2000; Goodfriend & Hare, 1995; Goodfriend et al., 1992; Rutter & Blackwell, 1995; Wehmiller & Hare, 1970.
- growth rates, longevity		Goodfriend, 1997.
Coralline algae	Sea of Cortez	Goodfriend et. al., 2000.
Diatoms		Rutter & Blackwell, 1995.
	estuararies	Sigleo et al., 1983.
	deep ocean	Keil et al., 2000; Warnke et al., 1980.
Radiolaria		Rutter & Blackwell, 1995.
	deep ocean	Keil et al., 2000; King, 1979, 1980a; King & Bada, 1974.
Dinosaur Egg Shells		Rutter & Blackwell, 1995; Shimoyama et al., 1989.
Ratite Egg Shells		Brooks et al., 1990, 1991; Collins et al., 1999; Ernst, 1987, 1989; Goodfriend et al., 2000; Goodfriend et al., 1991; Hare et al., 1984, 1993; Johnson, 1995; Johnson & Miller, 1997; Kaufman & Miller, 1995; Kokis, 1988; Miller & Beaumont, 1989; G. H. Miller et al., 1992, 2000; Murray-Wallace, 1993; Rutter & Blackwell, 1995; Wehmiller & Miller, 1990, 1998.
- archaeology: absolute dates	Africa	Brooks et al., 1990, 1993, 1995; Hare et al., 1984; G. H. Miller et al., 1989b, 1997; Miller & Beaumont, 1989; Wehmiller & Miller, 1990, 1998; Wendorf et al., 1987.
	Levant	Henry & Miller, 1992; G. H. Miller et al., 1992.
- paleontology: absolute dates	Australia	Magee et al., 1995; G. H. Miller et al., 1999, 2000.

Table I. Quaternary applications for amino acid racemization analyses (continued).

Sample Type -Application	Location	References
Ratite Egg Shells (cont'd)		
- extinction causes	Australia	G. H. Miller et al., 1999.
- paleotemperatures	Australia	G. H. Miller et al., 1997, 2000.
	Levant	G. H. Miller et al., 1992.
	South Africa	Johnson et al., 1997; G. H. Miller et al., 1992.
Bones		Abelson, 1954, 1955; Bada, 1972a, 1972b;
		Bada & Shou, 1980; Bada et at, 1974a, 1974b, 1979a, 1979b;
		Blackwell, 1987; Blackwell et al., 1990, 1991, 2000;
		Child, 1996; Child et al., 1993; Collins & Galley, 1998;
		Collins et al., 1998, 1999; Elster et al., 1991;
		El Mansouri et al., 1996; Goodfriend et al., 2000;
		Hare, 1969, 1974a, 1974b, 1980, 1995; Hare et al., 1978;
		Hassan & Hare, 1978; Julg et al., 1987;
		Johnson & Miller, 1997; Kessels & Dungworth, 1980;
		Kimber & Hare, 1992; King, 1980b; King & Bada, 1979;
		Kvenvolden, 1975; Masters, 1986, 1987;
		Murray-Wallace, 1993; Perinet et al., 1975, 1977;
		Prior et al., 1986; Rutter & Blackwell, 1995;
		Rutter et al., 1979; Saint-Martin, 1991;
		Saint-Martin & Julg, 1991; Stafford et al., 1990;
		Taylor et al., 1989, 1995a; Turban-Just & Schramm, 1998;
		van Duin & Collins, 1998; Von Endt, 1979, 1980;
		Wehmiller & Miller, 1990, 1998; Williams & Smith, 1977;
		Wyckoff, 1972; Wyckoff et al., 1963, 1964.
- *archaeology or paleontology: absolute dates	Africa	*Bada, 1974, 1981, 1982, 1985a, 1987;
		*Bada & Deems, 1975; *Bada & Protsch, 1974;
		*Bada & Schroeder, 1975; *Bada et al., 1973, 1974b,
		*Bada et al.,1979a, 1979b; Davis & Treloar, 1977;
		Hare et al., 1978; *King & Bada, 1979;
		*Masters & Bada, 1978b;
		*Turekian & Bada, 1972.
	Australia	*Clarke, 1999; *Goede & Bada, 1985.
* archaeology or paleontology: absolute dates	China	*Bada, 1987a; *Masters & Bada, 1978b.
	Europe	*Bada, 1972a; Bada et al., 1974b, 1979a, 1979b;
		*Dungworth et al., 1974, 1976; *Masters & Bada, 1978b.
	France	†Blackwell, ‡1987; †Blackwell & Rutter, 1985;
		†Blackwell et al., 1990, 1991;
		*de Lumley & Labeyrie, 1981; *de Lumley et al., 1977;
		*El Mansouri et al., 1996; Lafont et al., 1984;
		Rutter et al., 1981; Saint-Martin & Julg, 1991.
	Hungary	*Csapó et al., 1994, 1998.
	Italy	*Bellumoni & Bada, 1985.
	Japan	*Matsu'ura & Ueta, 1980; *Shimoyama & Harada, 1984.
	Middle East	*Bada & Schroeder, 1975; *Bada et al., 1979a, 1979b.

continued . . .

Notes:

* Unsuccessful or problematic application

† Discusses antler also

‡ Discusses dentine, enamel, and cement separately

Table I. Quaternary applications for amino acid racemization analyses (continued).

Sample Type -Application	Location	References
Bones (cont'd)		
	N. Am. "Paleo-indian" sites	*Bada, 1974, 1985a, 1985b; *Bada & Finkel, 1982; *Bada & Helfman, 1975; *Bada & Masters, 1978, 1987, *Bada & Protsch, 1974; *Bada & Schroeder, 1975; *Bada et al., 1974a, 1974b, 1979a, 1979b, 1984; Bender, 1974; Bischoff & Rosenbauer, 1981a, 1981b, 1982; Donahue et al., 1983; Ike et al., 1979; *King & Bada, 1979; LaJoie et al., 1980b; *Masters & Bada, 1978a, 1978b; Pollard & Heron, 1996; Skelton, 1982; Stafford et al., 1990; Taylor, 1983, 1987; Taylor et al., 1983, 1984, 1985a, 1985b; 1989; Von Endt, 1979, 1980; Wehmiller, 1977.
- *sediment: absolute dates		*Bada, 1972a; Dungworth et al., 1974.
- *stratigraphic correlation	France	[‡]Blackwell, 1987; Blackwell & Rutter, 1985; [†]Blackwell et al., 1990, 1991.
- *diagenetic or paleotemperatures		*Bada, 1974; *Bada & Schroeder, 1976b; Bender, 1974; *McCulloch & Smith, 1976;
- diagenesis		*Schroeder & Bada, 1973. Bada, 1985b; Bellumoni & Bada, 1985; [‡]Blackwell, 1987; [†]Blackwell & Rutter, 1985; [†]Blackwell et al., 1988, 1989, [†]Blackwell et al., 1990, 1991, 2000; Burky et al., 2000; DeNiro & Weiner, 1988; Ennis et al., 1986; Hedges & Wallace, 1980; Ho, 1965; Jope, 1980; Kvenvolden & Peterson, 1973; Lowenstam, 1980; Lowenstein, 1980; McMenamin et al., 1982; Tuross et al., 1980; Wyckoff, 1972; Wyckoff & Davidson, 1976; Wyckoff et al., 1963, 1964.
- heated bone recognition		Hare, 1995; Taylor et al., 1995.
- paleodiets		Hare & Estep, 1984; Hare et al., 1987; Millard, 2000.
-monitoring ^{14}C contamination & dating prep.		Burky et al., 1998; Hassan & Hare, 1978.
- monitoring DNA contamination		Krings et al., 1997; Poinar et al., 1996.
Teeth		Bada, 1987; [†]Blackwell, 1987; Collins et al., 1999; Goodfriend, 2000; [†]Helfman, 1976; Ho, 1965; Johnson & Miller, 1997; Rutter & Blackwell, 1995; Saleh et al., 1993; Torres et al., 2000.
- *archaeology or paleontology: absolute dates	Africa Europe France N. Am. Paleoindian sites	*Bada, 1981; *Bada & Helfman, 1975. *Bada & Helfman, 1975. [†]Blackwell, [‡]1987; [†]Blackwell & Rutter, 1985; [†]Blackwell et al., 1990, 1991. *Bada & Helfman, 1975.
- *stratigraphic correlation	France	[‡]Blackwell, 1987; Blackwell & Rutter, 1985; [†]Blackwell et al., 1990, 1991.
- diagenesis		[‡]Blackwell, 1987; [†]Blackwell & Rutter, 1985; [†]Blackwell et al., 1988, 1989, 1990, 1991, 2000.

Table I. Quaternary applications for amino acid racemization analyses (continued).

Sample Type -Application	Location	References
Teeth (cont'd)		
- forensic analysis:		Bada et al., 1983a; Helfman & Bada, 1975;
age at death		Bada & Brown, 1981; Carolan et al., 1997;
		Gillard et al., 1991; [†]Helfman, 1976; Masters, 1982, 1985;
		Ogino & Ogino, 1988; Ohtani & Yamamoto, 1991, [†]1992;
		Ohtani et al., 1995; Ritz et al., 1993.
Otoliths		G. H. Miller & Johnson, 1995.
- absolute dates		G. H. Miller & Johnson, 1995.
Skin		
- parchment		Weiner et al., 1980.
Hair		
-monitoring [14]C		Taylor et al., 1995b.
contamination &		
dating preparation		
Soft Tissues		Bada, 1984.
- forensic analysis		Masters, 1982; Masters et al., 1977; Ritz & Schutz, 1993.
Insects		Bada et al., 1994.
Wood		Hughes et al., 1981; Lee et al., 1976;
		Rutter & Blackwell, 1995; Rutter & Vlahos 1988;
		Rutter et al., 1979, 1980; Zumberge et al., 1980.
- relative stratigraphy	Arctic	Rutter & Crawford, 1984; Rutter & Vlahos, 1988;
& regional correlation		Rutter et al., 1980; Vincent et al., 1983.
	Canadian Prairies	Liverman et al., 1989; Nielson et al., 1986.
	N. Am. Pacific coastal plain	*Blunt et al., 1987, *Hicock & Rutter, 1986.
Seeds	New Mexico	Rutter & Blackwell, 1995; Rutter et al., 1993.
Leaves		Teece et al., 2000.
Packrat Middens		Rutter & Blackwell, 1995; Petit, 1974.
Speleothems		Lauritzen et al., 1994.
Sediment		Friebele et al., 1981; Hedges & Hare, 1987;
		Rutter & Blackwell, 1995; Goodfriend, 2000;
		Milnes et al., 1987; Smith & Evans, 1980;
		Vauchskii et al., 1982.
- lake stratigraphy	L. Bonneville	Colman et al., 1986.
	Clear L., CA	Blunt et al., 1981, 1982; Dungworth, 1982.
	L. Ontario	Dungworth et al., 1977; Schroeder & Bada, 1978.
	L. O'Hara, AB	Reasoner & Rutter, 1988.
- loess	Australia	Milnes et al., 1987.
	New Zealand	Kimber et al., 1994.
- marine stratigraphy	deep ocean	Bada et al., 1970.
- absolute dates	Clear L., CA	*Blunt et al., 1981, 1982.
Soil		Milnes et al., 1987; Murray-Wallace, 1993.
- modern		Colman et al., 1986; Griffin & Kimber, 1988;
		Hedges & Hare, 1987; Kimber & Griffin, 2000;
		Pollock et al., 1977; Rutter & Blackwell, 1995.

continued . . .

Notes:

* Unsuccessful or problematic application

[‡] Discusses antler also

[†] Discusses dentine, enamel, and cement separately

Table I. Quaternary applications for amino acid racemization analyses (continued).

Sample Type -Application	Location	References
Soil (cont'd)		
- paleosols	Australia	Milnes et al., 1987.
	Mt. Kenya	Mahaney & Rutter, 1989; Mahaney et al., 1986, 1991.
	New Zealand	Kimber et al., 1994.
- moraine stratigraphy	China	Mahaney & Rutter, 1992.
Meterorites		Bada, 1997; Bada & McDonald, 1995;
		Bada et al., 1983b, 1986a, 1986b; 1998
		Brinton et al., 1998; Cohen & Chyba, 1998;
		Cronin & Pizzarello, 2000; Cronin et al., 1980;
		Engel & Hare, 1997; Engel & Macko, 1997;
		Engel & Nagy, 1982, 1983; Googfriend, 2000;
		Harada & Hare, 1980; Kevenvolden et al., 2000;
		Rutter & Blackwell, 1995; Zhao & Bada, 1989.
Submarine Vents		Armend & Shock, 2000; Miller & Bada, 1988.
Precambrian &		Brandes et al., 2000; Dungworth & Schwartz, 1974;
Paleozoic Rocks		Rutter & Blackwell, 1995; Towe, 1980.

Notes:

* Unsuccessful or problematic application

‡ Discusses antler also

† Discusses dentine, enamel, and cement separately

research AAR applications. By the mid-1980's, some 30 laboratories were actively pursuing AAR research into many fossils and sediment types.

Unfortunately, the reputation of AAR as a credible dating method suffered a serious setback when it was used extensively to "date" bones and teeth from famous archaeological sites (e.g., Bada, 1972a; references in Table I). The controversy intensified over several Californian "Paleoindian" sites (e.g., Bada, 1974; references in Table I), whose very old AAR ages suggested early North American colonization during or before the Late Wisconsinan glacial advance. By 1987, AMS ^{14}C dating resolved the problem by demonstrating that all the "Paleoindian" bones were Holocene, and, that some were even younger than suggested originally by the standard ^{14}C ages (e.g., Taylor et al., 1983; Donahue et al., 1983), thereby confirming the archaeological associations for some sites which had suggested Archaic and later cultural affinities. For archaeologists, however, the damage had been done: They will now rarely consider AAR as a viable technique even for materials with much greater reliability, such as ratite egg shells (G. H. Miller, pers. comm., 1995).

To some extent, the "bone fiasco" also tarnished AAR's reputation with geologists and paleontologists. While AAR dating and aminostratigraphy are extensively used in the marine and coastal zones, they have not been extensively applied in either the terrestrial or lacustrine realms until recently (see Goodfriend, 2000, for a more extensive history).

Racemization and epimerization

All amino acids commonly found in living tissues, with the exception of glycine (Gly), possess at least one asymmetric carbon atom (Fig. 1; Table I). Isoleucine (Ile; Fig. 2),

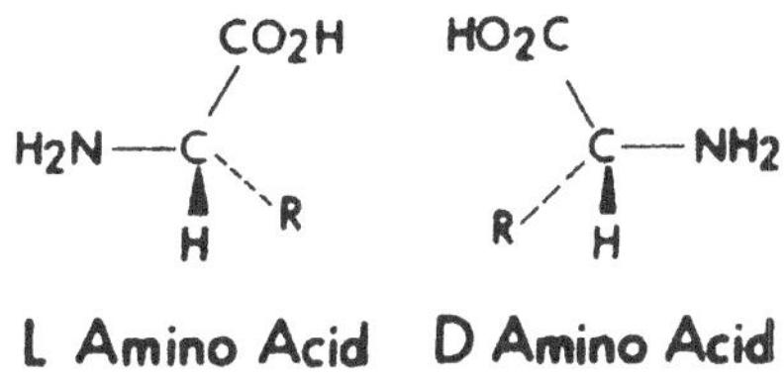

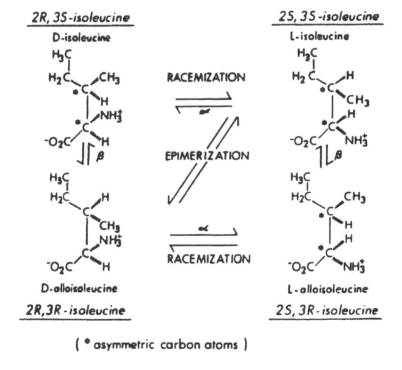

Figure 1. Asymmetric carbon atoms in amino acids. Except glycine, all amino acids contain at least one asymmetric carbon atom. Different R group compounds result in the different amino acids. Isoleucine, threonine, hydroxylysine, and cystine all contain two asymmetric centres (after Rutter & Blackwell, 1995).

Figure 2. Enantiomerization in multi-asymmetric amino acids. In multi-asymmetric amino acids, such as isoleucine, threonine, hydroxylysine, and cystine, either the α or β asymmetric centre can theoretically experience enantiomerization. In nature, only epimerization occurs. Both the common and R-S system names are listed (after Rutter & Blackwell, 1995).

hydroxyproline, hydroxylysine, and threonine, however, have two asymmetric centres. Asymmetric carbon atoms cause polarized light passing through the molecule to be rotated either dextrally (d) or levrally (l). Most biologically produced amino acids are L forms, but a few organisms, including some bacteria, worms, insects, molluscs, frogs, and antibiotics contain D amino acids in some specialized cell wall membranes (Bada et al., 1973; Soper et al., 1977; Nakamiya et al., 1976; Heck et al., 1996; Friedman, 1999). Once a living organism containing L amino acids dies or its tissue ceases to be regenerated regularly, D amino acids form as the proteins degrade and decay.

In inorganically produced amino acid mixtures, amino acids with one asymmetric centre will occur in a racemic concentration, i.e., $D = L$ or $D/L = 1.00$. For multi-asymmetric amino acids, the equilibrium ratio may differ significantly from 1.00, since it depends on the ratio between the forward and backward reaction rate constants, k_1 and k_2:

$$\text{L-isoleucine (L-Ile)} \underset{k_2}{\overset{k_1}{\rightleftharpoons}} \text{D-alloisoleucine (D-alle)} \tag{1}$$

where $k_1 =$ the rate constant for the forward reaction,

             $k_2 =$ the rate constant for the backward reaction.

In *Mercenaria* shells, for example, $k' = k_1/k_2 = 1.25$ for Ile (Hare & Mitterer, 1967). In non-racemic mixtures of uni-asymmetric amino acids, the reaction proceeds so that the more abundant form decreases in concentration until a racemic mixture is achieved with $D/L = 1.00$. In multi-asymmetric amino acid, only one asymmetric centre experiences conversion, a process termed epimerization (Equation 1; Fig. 2; Hare & Abelson, 1966). For simplicity's sake in this article, *racemization* will be used to refer to racemization (*sensu stricto*) for uni-asymmetric amino acids, and epimerization in multi-asymmetric amino acids.

Generally, racemization is complete (i.e., a racemic amino acid mixture occurs) in most Miocene fossils (Hare & Abelson, 1970). Although most racemization occurs in amino acids for which at least one peptide bond has been broken, racemization can occur for amino acids bound into proteins (Hare & Abelson, 1966; Hare & Mitterer, 1967; Hare, 1980). After racemization, a D amino acid in an intraprotein position distorts the adjacent protein lattice, thereby making it more susceptible to peptide bond cleavage at the racemized amino acid, and more prone to leaching (Helfman et al., 1977).

Because AAR has been shown in some tissues to approximate first order reversible linear kinetics with respect to time and temperature, the degree of AAR can be used, in principle, to date the tissue. The AAR rate, however, depends on many factors, including:

1. the amino acid in question

2. its location, whether free in solution or bound into a protein

3. the protein's size and structure

4. the amino acid's positions within the protein

5. the microenvironment at those locations within the protein

6. the degree of structural breakdown near those positions

7. the group at the N-terminal position

8. the ability of the R-group to stabilize the carbanion intermediate

9. metal chelation by the protein or peptide

10. the ambient temperature of the system

11. the pH of the system

12. any incident ionizing radiation in the system

13. the amount of water available for reactions

14. the inorganic mineral matrix in the fossil and minerals in the nearby sediment

15. the species of animal involved

16. the subsampling location within the fossil

Many other factors likely also affect racemization. Luckily, for many tissues, few of these effects are significant enough to produce deviation from first order reversible linear kinetics.

For all amino acids in all fossils or sediment, temperature is a major control on racemization rate. A 2 °C ambient temperature difference can cause a 50% difference in the time required to complete racemization. For example, Ile racemization can be completed in 220 ky at 20 °C, but requires 580 ky at 15 °C (Bada, 1972b). For samples from archaeological contexts, burning or boiling the fossils not only affects the racemization rate by causing variable racemization rates across egg or mollusc shells during burning, but also causes amino acid and protein degradation which further enhances long-term racemization rate differences (see below; Abbott et al., 1996; Brooks et al., 1991; Hare et al., 1993). Therefore, temperature histories can become extremely important for AAR dating and aminostratigraphic analysis.

Geochemical effects on racemization rates

In fossils, not all amino acids racemize at the same rate. Aspartic acid is usually among the fastest, while isoleucine is usually among the slowest. This variation allows AAR to be used to date fossils deriving from much of the Quaternary, and for some Pliocene sites from polar regions. Furthermore, the variation can potentially provide two or more somewhat independent chronometers that can be measured using only one analytical preparation. Unfortunately until now, AAR techniques have focused mainly on a few amino acids, Ile, aspartic acid (Asp), and occasionally leucine (Leu) or glutamic acid (Glu). Alanine (Ala), valine (Val), proline (Pro), and phenylalanine (Phe) are also readily measureable and occur in sufficient abundance in most fossils to be useful. Other less common amino acids, including hydroxylysine and hydroxyproline, while abundant in some fossil materials, are not sufficiently concentrated to be generally useful.

The protein structure and specific location in which the amino acids occur strongly influences the rate at which the amino acids racemize. For example, the Ile racemization rate is inversely proportional to peptide size. N-terminal Ile racemizes faster than that at the C-terminal, but the converse is true for Ala, Leu, Phe, and Asp (Kriausakul & Mitterer, 1980; Mitterer & Kriausakul, 1984; Moir, 1985; Moir & Crawford, 1988). Because peptide chains are held coiled in position with the protein by hydrogen bonding, sulphur bridges, and van der Waal's forces, each amino acid has its own unique microchemical environment and racemization rate, slightly different from other amino acids of the same composition in the protein. Although some exceptions have been reported (Smith & Evans, 1980), amino acids deep within the quaternary protein structure are usually less susceptible to racemization than those on the termini (Bada & Schroeder, 1976a; van Duin & Collins, 1998; Collins et al., 1998, 1999). In collagen, for example, Asp racemization may be completely suppressed when the triple helix is intact (van Duin & Collins, 1998). The racemization rate measured for the whole protein integrates all these variables, and hence, tends to be species specific for many shorter chained proteins (King, 1980a; Wehmiller, 1980).

As the protein undergoes diagenesis, the racemization rate also changes. During diagenesis, hydrolysis cleaves the protein into ever shorter peptide chains (Fig. 3a), which produce an increasingly complex protein-peptide-free amino acid mixture (Fig. 3b), causing amino acids therein to experience a range of different racemization rates (see Collins & Riley, 2000, for an in depth discussion). In general, the shorter the peptide chain, the more unpredictable the racemization rate will be. Some amino acids also decompose into other amino acids or organic compounds with time. For example,

$$\text{aspartic acid (Asp)} \longrightarrow \beta\text{-alanine } (\beta\text{-Ala}) \tag{2}$$

$$\text{glutamic acid (Glu)} \longrightarrow \alpha\text{-amino butyric acid} \tag{3}$$

$$\text{arginine (Arg)} \longrightarrow \text{ornithine (Orn)} \tag{4}$$

(Hare & Mitterer, 1965; Schroeder & Bada, 1975; Akiyama, 1980; Steinberg & Bada, 1981, 1983). Both hydrolysis and protein degradation increase the solubility of proteins, peptides, and their amino acids, making it more likely they will be leached from the fossil if sufficient water is available. Not all proteins, however, have equal likelihoods for being preserved (Masters, 1987), making it necessary in some fossils to select particular proteins to ensure a predictable degradation and AAR kinetics. Therefore, for most amino acids in fossil proteins, a finite limit exists to the D/L range over which any amino acid's racemization will approximate the first-order reversible linear or parabolic kinetics required to permit AAR dating or aminostratigraphy (Fig. 4).

Metallic ions, polymers, clays, and radiation all influence racemization rates. For example, Fe^{3+}, Al^{3+}, Cu^{2+}, and Mg^{2+} in or near the protein can catalyze racemization, whereas Ni^{2+} can retard it (e.g., Liardon & Jost, 1981; Smith et al., 1978, 1982; Smith & Reddy, 1986; Vauchskii et al., 1982). Many polymers and clay surfaces can act as catalysts, but amino acids trapped in clay interlayer positions tend to have their racemization inhibited (e.g., Friebele et al., 1981; Hedges & Hare, 1987; Smith & Evans, 1980). Protein radiolysis and racemization rates depend also radiation (Bonner et al., 1980; Baratova et al., 1981). In fossils that absorb U, such as bones, teeth, and peat, high U concentrations may cause non-linear kinetics (Rutter & Blackwell, 1995).

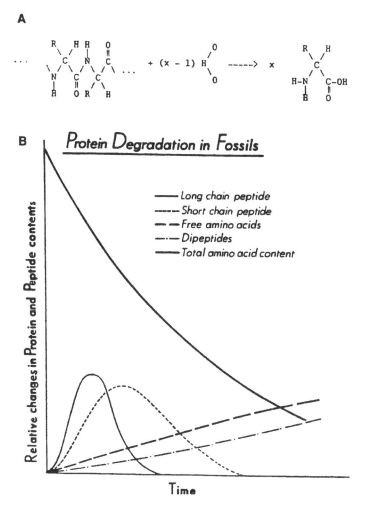

Figure 3. Protein diagenesis in fossils. (a) During hydrolysis, a water molecule is used for each peptide bond broken (after Blackwell, 1987). (b) As the protein ages and the fossil experiences diagenesis, the proteins degrade into long chained peptides, then short chained peptides, and finally free amino acids. In each component, the racemization rates will differ for each amino acid (after Rutter & Blackwell, 1995).

In solution, most amino acids' racemization depends critically on pH (Fig. 5; Bada & Shou, 1980; Frank et al., 1981; Smith & Sivakua, 1983; Steinberg et al., 1984; Baum & Smith, 1986; Blackwell, 1987). Partly, the effect arises from the decomposition of amino acids under strongly basic conditions (Hare, 1969). In buffered solutions, racemization increases as the buffer concentration increases (Helfman & Bada, 1975; Kriausakul & Mitterer, 1980), but phosphates and carbonates in fossils tend to moderate the pH effect (Masters & Bada, 1978b).

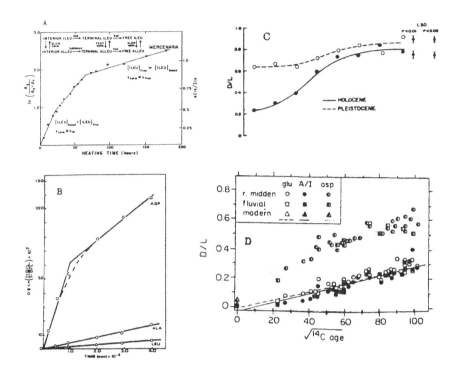

Figure 4. Linear and nonlinear racemization kinetics. Some fossils show linear racemization kinetics over part of their racemization range, while others only display non-linear kinetics throughout: (a) Isoluecine epimerization in *Mercenaria* at 152 °C: Three linear segments occur with two distinct changes in epimerization rate. For the range $0 < D/L < 0.3$, *Mercenaria* can be used for dating using Equation 8, but for $0.3 < D/L < 0.5$, a parabolic equation becomes necessary to calculate ages. For $0.5 < D/L$, dates are not reliable (after Kriausakul & Mitterer, 1980). (b) Aspartic acid, alanine, and leucine racemization in collagen at 122.5 °C, pH $\leq$ 8.0, in 0.05 M phosphate-buffered solution: Aspartic acid racemization shows two linear segments, whereas alanine and leucine appear linear for the range shown (after Smith & Evans, 1980). (c) Aspartic acid in peptides > 1000 Daltons from *Ostrea angasi* at 110 °C: No linear segments occur, suggesting that aspartic acid in this mollusc can not be used for dating (modified from Kimber et al., 1986). (d) Isoleucine, aspartic and glutamic acids all show parabolic kinetics in the Holocene Israeli land snail *Trochoidea seetzeni* (modified from Goodfriend, 1991a).

Hydrolysis and protein diagenesis (Fig. 3) requires water. In controlled kinetic experiments with bone, water availability was critical to racemization (Fig. 6; Hare, 1974b, 1980). In geologic environments, groundwater or soil moisture can leach amino acids from fossils, depositing them into the soil, while increasing racemization (Kessels & Dungworth, 1980; Perinet et al., 1975, 1977). Porous fossils may absorb leached amino acids and peptides, or they may be trapped in mineral inclusions during secondary mineralization. The leaching problem, however, is not limited to bones (e.g., Roof, 1997). Leaching effects can be minimized by using the intracrystalline fraction (e.g., Sykes et al., 1995), but this method can not be used for bones or teeth (Hare, 1980). The most suitable fossils for AAR studies

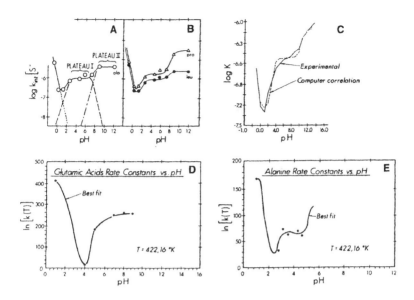

Figure 5. pH effects on racemization. Despite a relatively constant rate at 5 < pH < 8 for several amino acids at high temperatures, racemization rates do depend strongly on pH: (a) Alanine at 142 °C (after Bada & Shou, 1980). (b) Proline and leucine at 142 °C (after Bada & Shou, 1980). (c) Alanine at 120 °C (after Smith & Sivakua, 1983). (d) Glutamic acid at 150 °C (after Blackwell, 1987). (e) Alanine at 150 °C (after Blackwell, 1987). Unfortunately, the extremely slow racemization rates preclude determining pH dependence at Earth surface temperatures.

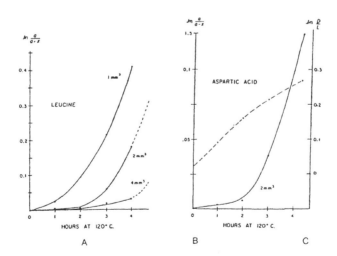

Figure 6. The effect of leaching. At 120 °C, the hydrolysis rates increase, but the racemization rate slows somewhat, with the leaching time (after Von Endt, 1980): (a) Leucine hydrolysis is also a function of bone fragment size. (b) Aspartic acid hydrolysis (solid line, left axis) increases asymptotically with leaching. (c) Aspartic acid racemization (dotted line, right axis) slows moderately.

have proven to be ostrich egg shells. These shells, which Bushmen use to store water for long periods, are impervious to water. Therefore, one test necessary before a species can be effectively used for AAR dating or aminostratigraphy involves checking its potential for leaching and secondary contamination (e.g., Roof, 1997). In molluscs, foraminifera, and egg shells, mineralogical alteration can hint at leaching (Wehmiller & Miller, 1990). In molluscs, plotting aIle/Ile versus Gly/Ala can indicate leaching (Murray-Wallace & Kimber, 1987), whereas high serine or threonine concentrations can indicate secondary contamination (Miller & Brigham-Grette, 1989).

Biogeochemical effects on racemization rates

Due to species effects, most fossils must be speciated before AAR analysis, and hence, any analytical samples must be sufficiently large to be taxonomically identifiable. In molluscs for example, significant differences occur between species at the taxonomic family level, and in some cases, even between subfamilies (LaJoie et al., 1980a; Kimber & Milnes, 1984; Wehmiller, 1980, 1984a). AAR rates in foraminifera also depend strongly on the species (King & Neville, 1977; Miller & Brigham-Grette, 1989), as do rates in wood (Vlahos, 1985). Rat dentine racemizes more rapidly than human dentine (Ohtani et al., 1995), but most species effects in bones, teeth, and antler are overshadowed by other effects arising from geochemical differences or secondary infestations in the fossils during early diagenesis (Blackwell, 1987; Blackwell et al., 2000; Child et al., 1993; Turban-Just & Schramm, 1998).

In bivalves and gastropods, significant differences can exist between AAR ratios seen at different locations on the same shell (Hare, 1963; Brigham, 1983b, 1985; Rutter et al., 1989; Goodfriend et al., 1997). Therefore, it becomes essential to systematically collect molluscs that allow consistent subsampling locations on the shells to ensure good AAR dates or aminostratigraphy. If similar effects occur in foraminifera, the current sampling techniques preclude their detection. For fossil bones and teeth, different tissues yield significantly different AAR rates (Blackwell et al., 1990, 1991, 2000), making tissue separation mandatory.

Sampling protocols

Because amino acids are so ubiquitous in the environment, amino acid analysis demands that sampling protocols minimize external contamination. Consequently, the following precautions need to occur:

1. Samples should be collected using gloves to prevent contamination by hand oils which contain modern amino acids. If the samples are cleaned properly, such contamination theoretically should not present a problem, but to be certain, it seems more sensible to avoid touching samples intended for AAR.

2. All sampling equipment should be as free from potential sources of contaminant amino acids as is possible.

3. To prevent the growth of algae, fungi, and bacteria that can affect the samples if not analyzed rapidly, all samples, especially sediments, should be bagged in air-tight

containers and frozen or refrigerated as soon after collection as possible. Ideally, storage should occur at temperatures below −4 °C.

4. Water samples should be stabilized for pH as quickly as possible.

5. Samples, excepting water samples, should be air-dried in a dessicator or other contaminant-free environment if possible, but not heated.

6. When selecting subsamples from calcified tissues for analysis, avoid using drills or other equipment that will cause localized heating, and which may promote significant racemization.

7. For molluscs, a consistent area on the shell, preferably the umbo, should be subsampled for all samples to be used in an aminostratigraphic study.

8. Any water that contacts the sample after its removal from the sediment should be deionized and doubly distilled.

9. Sample numbering uses inks and paints that can add contaminant organic compounds if applied directly to the sample.

10. Allowing clay samples to dry necessitates extensive grinding which can cause racemization during sample preparation.

11. Packing samples for transport with materials such as old newspapers, dyed paper, etc., can cause trace organic contamination if they contact the sample.

12. Samples should be processed as rapidly as possible after their collection to reduce contamination.

Samples stored in museums often have been exposed to preservatives or other treatments that make AAR analyses for such samples highly unreliable. Most casting procedures add contaminants that preclude casting fossils before AAR analyses.

Samples that have experienced diagenesis, especially bones and teeth, may have suffered abnormal AAR processes (e.g., Blackwell et al., 2000). Since fossils can be easily reworked into younger depositional units, any sampling program should collect at least 8–10 discrete samples from each stratigraphic unit to increase the chance that the samples analyzed provide dates related to the event of interest. To ensure good aminostratigraphies, at least 3–4 stratigraphic units at each exposure must be analyzed. To ensure that samples have not been exposed to extremely high temperatures, and hence, have experienced high racemization rates for short times, select samples on north facing outcrops or samples that are 5–10 cm below the surface. Although this does not guarantee that samples have never been exposed to high surface temperatures, selecting facies that have been rapidly buried make it less likely. For very old samples, this source of error will likely be small compared to the total racemization from all temperatures. A related problem arises with archaeological bones, teeth, molluscs and ratite eggshells that may have been burnt (e.g., Brooks et al., 1991; Kvenvolden et al., 1979; Taylor et al., 1995), which should be avoided for dating, but can provide data for hominid behavioral studies.

Although the required sample weight depends on the analytical method and the concentration of amino acids present in the sample, to analyze at least 2–3 subsamples for both Ile and Asp, the two most commonly used amino acids, requires a minimum of 20–30 g for most sample types. For samples being tested for Ile alone, collecting 2–5 g of tissue usually suffices. For some materials, especially those prone to diagenesis, it is necessary to check for secondary mineralization and remineralization, which can signal amino acid leaching and reworking. This necessitates collecting another 1–2 g for XRD or similar supplementary analyses. For samples from middens, tests to check for burning are necessary. For AAR dating soil or sediment, pristine sample blocks of ∼ 0.5 kg cut from thick or extensive units provide the best results, if available. For very small samples (100–200 mg), samples can only be analyzed with the automatic amino acid analyzer (AAAA) for Ile.

Because amino acids tend to leach out of the fossils, especially with older fossils, sampling the adjacent sediment and even the groundwater or associated surface water, can help to identify amino acid leaching processes, and to determine what, if any, modern sedimentary contaminants may have affected the fossils. In addition to measuring the ground temperatures at the time of collection, some estimate of mean annual tempreraure is vital. To test for effective diagenetic temperature, it is often useful to emplace temperature probes, capsules containing quickly racemizing sugars, to measure the modern mean annual temperature. The water temperature of springs which tap deep groundwater will also provide a reasonably accurate analog for mean annual temperature, as does the air temperature in a cave, if it is measured below the active freeze zone in an area not affected by wind from outside sources.

AAR analysis

AAR analysis uses primarily two methods, gas chromatography (GC) and high precision liquid chromatography (HPLC). The AAAA is essentially a specially modified HPLC capable of detecting amino acids at femtomole concentrations (Hare, 1986). GC systems are now routinely hooked to mass spectrometers to couple AAR analysis with stable isotope analysis for proteins and amino acids in paleodietary and paleoenvironmental studies (e.g., Macko et al., 1997). GC systems provide more precise analysis for the uni-asymmetric amino acids, whereas the AAAA is faster and easier for Ile and other multi-asymmetric amino acids. AAAA columns have been improving steadily and may soon provide the resolution for uni-asymmetric amino acids that the GC currently provides. Although new preparations have been suggested recently (e.g., Glavin & Bada, 1998), most laboratories continue to use the methods perfected in the 1980's (e.g., Hare et al., 1985; Engel & Hare, 1985; Murray-Wallace, 1993).

For both analytical systems, samples first need to be cleaned, subsampled, dissolved, and hydrolyzed. Sonically cleaning fossils with a weak $HCl_{(aq)}$ solution followed by 20 sonicated cleanings with doubly distilled deionized water will remove most surface contaminants without affecting the protein or mineral phase, but may not remove tissue filaments from boring infaunal contaminants, such as algae and fungi. Generally, GC analysis requires 1–10 g of dry sample, whereas the AAAA requires < 0.1–1 g, or ∼ 50 foraminiferal tests. Usually, 2–5 subsamples are prepared from each sample to allow replicability to be established. Prior to chromatography, samples are dissolved, but for some protocols, a particular size fraction may also be selected. Hydrolysis varies depending on the tissues involved, but tends to be either 15 minutes at 150 °C or 24 hours at 105–110 °C in heated acidic solutions, usually 1 M $HCl_{(aq)}$ or $HNO_{3(aq)}$.

In AAAA chromatography, after sample hydrolysis, the machine performs all the liquid chemical preparation as well as the chromatography. In GC analysis, chiral columns are used to separate the two optically active phases, but the hydrolyzed amino acid solutions must first be desalted on a cation exchange column and esterified with a compound such as TFA (trifluoroacetone) or PFPA (pentafluoroproprionic anhydride) before being injected into the GC. During chromatography, a ramped heating program in the GC or a program to alter the buffer concentrations in the AAAA at first traps and then releases the amino acids, thereby separating each D and L form. A integrated peak area analysis, often with some form of baseline correction program, is used to measure the relative peak heights for the D and L peaks, from which the D/L or Ile/alle ratio is calculated. If a standard amino acid has been added as a spike, actual concentrations can be determined. Otherwise, interacid ratios normalized against glycine are usually calculated (Blackwell, 1987). Normally, several aliquots of the prepared solution will be analyzed to establish the precision for the various ratios.

Several interlaboratory calibrations and replicability studies have demonstrated that the best precision occurs for Ala, Leu, Asp, and Glu (e.g., Wehmiller, 1984b; Goodfriend, 1987a; Blackwell et al., 1990). Accuracy for Ile, unfortunately, seems to depend on preparation technique (Brigham, 1982; Wehmiller, 1984b). Accuracy and precision in the analysis depend strongly on the spectral peak separation. Although modern statistical programs to separate peaks, compensate for interferences, and calculate ratios have dramatically improved the analytical precision, chromatographic analytical programs still must maximize peak separation and minimize interferences (e.g., Fig. 7). Total analytical precision and accuracy also depend on uncertainties arising from the many other factors noted above (Fig. 8).

Aminozones: Defining units

Aminozones are widely used to correlate marine, lacustrine, and loess sequences using molluscs, foraminifera, and occasionally other fossils (e.g., Fig. 9b). Rarely are aminozones defined using a single amino acid in a single phase, and any thus defined lack precision and accuracy. Although often appearing more art than science, modern reproducible aminozone definition is usually based on clustering, discriminant, or factor analysis. In selecting the amino acids to use in aminozone definition, the more rapidly racemizing amino acids, such as Asp or Glu, usually offer the best discrimination (Fig. 9a), especially for younger Holocene or Late Quaternary sections or those from cold climates. To discriminate older sections or those from warm terrestrial regions, more slowly racemizing amino acids, such as Ile, may be necessary. Common aminozone definitions plot Ile/alle in free vs. combined fractions, or Glu vs. Asp.

Rigid criteria are usually applied to ensure that the type aminozones have wide applicability:

1. An aminozone cannot be defined using less than 10 independent AAR analyses for a single species.

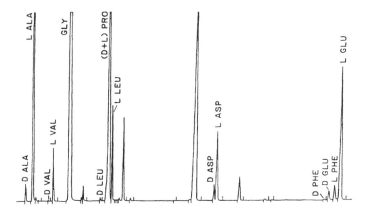

Figure 7. A gas chromatographic analysis for a bone. Gas chromatography will analyze D/L ratios for the uni-asymmetric amino acids, including the ones shown here. Analytical precision depends on the degree of peak separation. Here, separation is ideal for Ala, Val, Phe, and Glu. Resolution is poorer for D- and L-Asp, which overlap slightly, and L-Leu, which overlaps L-Pro. The proline D/L ratio for this analysis will have a large uncertainty due to its large degree of overlap.

Error Propogation in Amino Acid Analysis

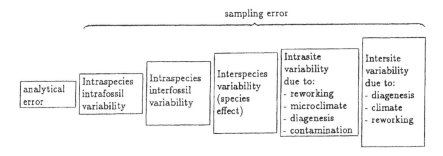

Figure 8. Relative uncertainties and variation in AAR ratio determinations. Many factors add uncertainty or variation, few of which can be analytically controlled, and hence, become difficult to assess numerically (modified from Murray-Wallace & Kimber, 1987).

2. Each independent analysis should comprise a single mollusc shell or individually picked unispecific foraminifera or diatoms.

3. Each sample must be speciated. Mixed populations cannot be used, and fragmentary shells should be avoided (Campbell et al., 1982).

4. Samples with minimal diagenetic alteration must be selected (Bowen et al., 1989).

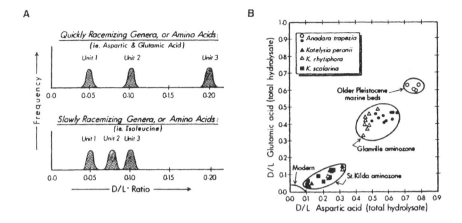

Figure 9. Defining aminozones. Aminozone definitions depend on the fossils available, the temporal range, and the stratigraphic problem to be resolved: (a) For young units (i.e., typically those < 200 ka roughly) and those in cold climates, rapidly racemizing amino acids will discriminate more effectively between the stratigraphic units than more slowly racemizing amino acids. For older units or those in warm climates, slowly racemizing amino acids may be necessary to distinguish units (adapted from Miller & Brigham-Grette, 1989). (b) Aminozones are often defined by plotting D/L ratios for two different amino acids or the same amino acid from two different proteins or peptide components. Here, plotting D/L ratios for glutamic *versus* aspartic acid in the total fraction effectively defines four aminozones for mid and late Quaternary South Australian molluscs (after Murray-Wallace & Kimber, 1987).

5. Aminozones cannot be defined using fossils that have been reworked or transported (Goodfriend, 1987b).

6. Only amino acid ratios from the D/L or aIle/Ile range known to be exhibiting linear reversible or parabolic kinetics can be used (Wehmiller & Miller, 1990; Goodfriend et al., 1996a; e.g., Fig. 4a, 4b, 4d, 10b).

7. Although an aminozone definition can incorporate more than one species, D/L or Ile/aIle ratios from two or more species cannot be combined to produce an "average" AAR ratio for the unit.

8. AAR ratios must clearly discriminate each aminozone from its neighbours.

Once the type aminozones have been defined, other sections can be compared to the type to develop regional correlations (e.g., Fig. 11).

AAR dating

Assuming linear first order kinetics, namely that time and temperature are the only two factors affecting the AAR ratios, then

$$\frac{1 - \frac{D}{L}}{1 + \frac{D}{L}} = e^{-2k_i t} \qquad (5)$$

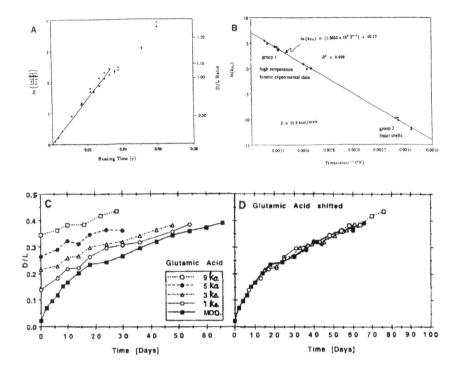

Figure 10. Tests for the suitability of racemization kinetics for dating. Both high temperature kinetic experiments and calibration tests were used to test the suitability of *Struthio* (ostrich) egg shells and *Trochoidea seetzeni* (Israeli land snail) for AAR dating: (a) At 143 °C, isoleucine epimerization in *Struthio* egg shells is a first order linear reaction until $D/L > 1.0$, and deviates only slightly from linearity for $D/L \leq 1.2$ (modified from Wehmiller & Miller, 1990). (b) In an Arrhenius plot of the isoleucine epimerization rate constant versus temperature, the regression line is linear with temperature for both the high temperature kinetic experimental data (data group 1) and calibration test data for fossil shells (data group 2) for *Struthio* egg shells (modified from Wehmiller & Miller, 1990). (c) For glutamic acid in *Trochoidea*, both fossil and modern shells show similar racemization patterns when heated at 106.5 °C (modified from Goodfriend & Meyer, 1991). (d) When the glutamic acid in fossil *Trochoidea* shells are corrected for their initial racemization ratio, their curves overly that seen in the modern shells, indicating the racemization at Earth surface temperatures mimics that created in the heating experiment (modified from Goodfriend & Meyer, 1991).

where $\dfrac{D}{L} = $ the measured AAR ratio,

$k_i = $ the racemization rate constant at temperature, T_i,

$t = $ time.

In performing kinetic experiments, t becomes the heating time. For calculating ages, t is the sample's age. The rate constant, k_i, is itself a function of temperature,

$$k_i = A e^{\frac{-E_a}{R T_i}}$$ (6)

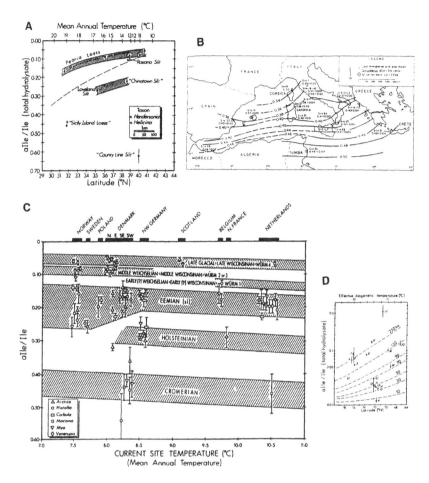

Figure 11. Stratigraphic correlation using molluscs. Stratigraphic correlation requires first building aminozones at individual sites. Then, the correlations can be expanded regionally for relative dating or paleotemperature analysis. (a) Using isoleucine in *Helicia* and *Hendersonia* shells, Clark et al. (1989) defined four aminozones, two of which could be traced regionally to correlate loess units in the Mississippi Valley. (b) Using isoleucine in Last Interglacial *Glycymeris* shells, Hearty et al. (1986) define isopleths of constant alle/Ile ratios, and hence, constant paleotemperature, across the Mediterranean region. (c) Using isoleucine from six genera, Miller and Mangerud (1985) defined six regional correlative aminozones, at least four of which equate with glacial stages. (d) For Arctic sites, Andrews et al. (1985a) plotted alle/Ile ratios versus effective diagenetic temperature to reveal several regionally correlative interglacial events.

where A = the frequency factor, a constant,

E_a = the activation energy at temperature, T_i,

R = Boltzman's constant, 1.9872 cal/mole-°K,

T_i = temperature (°K).

Before using AAR ratios in fossils for dating, several tests need to be completed to ensure that the racemization reaction meets the criteria for the first order linear reversible or parabolic kinetics needed for dating (Fig. 4). The two pronged approach using both modern kinetic studies and calibrated tests on fossils applied to develop AAR dating for ostrich egg shells is an excellent example.

In high temperature kinetic experiments, the fossil species to be dated is heated at several different temperatures for various times. After using the Arrhenius equation (Equation 6) to determine a rate constant for each temperature tested (e.g., Fig. 10a), the rate constants are plotted versus temperature (e.g., Fig. 10b) and extrapolated to Earth surface temperatures. If this method is used in isolation, then the curve must be assumed to extrapolate linearly to lower temperatures, an assumption that cannot be guaranteed without calibration at low temperature (Blackwell, 1987; Rutter & Blackwell, 1995).

Calibration tests use several fossils from several diverse sites that have been accurately and precisely dated by another technique, such as ^{14}C, $^{230}Th/^{234}U$, $^{39}Ar/^{40}Ar$, ESR, etc. The AAR ratios and known ages are used to solve Equation 5 to determine k_i for each site. If used in isolation without the high temperature tests, the calibration method can fail miserably if the fossils' dates (e.g., the Paleoindian site "dates", references in Table I) or the sites' ambient temperature are imprecisely known. After measuring or calculating the sites' ambient temperatures, k_i is plotted *versus* temperature on an Arrhenius plot (e.g., Fig. 10b). If the regression line measured for the high temperature experiments and that for the geological sites coincide, then the fossil can serve as a reliable AAR dating tool. Ideally, the calibration tests should use a wide range of fossil species encompassing the useful age range for the chronometer, and include fossils from several paleoenvironments to ensure that other bio- and geochemical factors (discussed above) have minimal influence on the chronometer's AAR rate.

The preferred way to perform essentially the same test involves using high temperature experiments on both fossil and modern shells. For species without modern congeneric representatives, fossil shells calibrated by ^{14}C dating can be used, but preferably the calibration samples should be either Holocene or approximately 30 ka, and hence, have experienced half their history in warmer and half in colder Pleistocene temperature regimes (Hare et al., 1997). The racemization ratios for the fossil shells are corrected to account for their initial racemization ratios (e.g., Bada & Protsch, 1974; Mitterer, 1975). If the resulting curves all show the same AAR ratio growth pattern (Fig. 10c, 10d), then the racemization obeys identical kinetics at high temperature and Earth surface temperatures (e.g., Goodfriend & Meyer, 1991).

With the combined calibration-kinetic approach, its only limitation arises from imprecision in the ambient temperature. For sites where the ambient temperature may have varied significantly over time, diagenetic temperatures are often calculated:

$$T_d = \frac{E_a}{R} \ln \frac{At(1 + k_i')}{\ln\left[\frac{1+\left(\frac{D}{L}\right)_t}{1-k_i'\left(\frac{D}{L}\right)_t}\right] - \ln\left[\frac{1+\left(\frac{D}{L}\right)_0}{1-k_i'\left(\frac{D}{L}\right)_0}\right]} \qquad (7)$$

where T_d = the diagenetic temperature,

$k_i' = k_1/k_2$, the racemization rate constant at temperature i,

$$\left(\frac{D}{L}\right)_t = \text{the measured AAR ratio in the fossil species,}$$

$$\left(\frac{D}{L}\right)_0 = \text{the measured AAR ratio in the same modern species, essentially the}$$

modern sample background.

For uni-asymmetric acids at Earth surface temperatures, $k_i' = 1.00$, simplifying the equation somewhat. Uncertainties in this calculation arise from poorly known values for E_a, k_i', and A (Miller, 1985; McCoy, 1987b; G. H. Miller et al., 1987). This calculation also requires an accurate site age, which precludes its use where other dating methods are not applicable. Where fossils ages are well known, diagenetic temperature analysis can assess paleotemperatures (e.g., Fig. 11d).

If a species has proven a good chronometer, then AAR ages are calculated from Equation 8:

$$\ln\left[\frac{1+\left(\frac{D}{L}\right)_t}{1-k'\left(\frac{D}{L}\right)_t}\right] - \ln\left[\frac{1+\left(\frac{D}{L}\right)_0}{1-k'\left(\frac{D}{L}\right)_0}\right] = -2k_i t \qquad (8)$$

Here, the second term, often termed the modern sample background, is added to compensate for the racemization induced in the AAR analysis. Mitterer & Kriausakul (1989) proposed a parabolic model for Ile to approximate ages for samples in which $0.3 <$ alle/Ile < 0.5. If alle/Ile $vs.$ $\sqrt{t}$ gives a linear regression, then

$$\sqrt{t} = \frac{\left(\frac{\text{alle}}{\text{Ile}}\right)_s}{m_c} \qquad (9)$$

where $\left(\frac{\text{alle}}{\text{Ile}}\right)_s$ = the measured AAR ratio in the unknown sample,

m_c = the slope of the regression line.

Recent aminostratigraphic and chronometric applications are beginning to use species showing parabolic kinetics as well as those with first order linear kinetics (e.g., Fig. 4d; Murray-Wallace & Kimber, 1993; Goodfriend et al., 1996a).

Dating and aminostratigraphic correlation in lacustrine environments

The most commonly used fossils for AAR studies in lacustrine environments are freshwater and terrestrial molluscs (Table I; Fig. 11). For example, the lake level history of Lake Bonneville was reinterpreted using D/L ratios in *Lymnaea* and *Amnicola* (Scott et al., 1983; McCoy, 1987a). In the Mississippi Valley, B. B. Miller et al. (1987) used the terrestrial gastropods *Catinella*, *Stenostrema*, and *Hendersonia* to define four aminozones for regional correlation, while Clark et al. (1989) used alle/Ile ratios in terrestrial molluscs to identify five distinct loess units (Fig. 11a). Goodfriend et al. (1996b) used land snails to develop Holocene stratigraphies for Madeira. N. W. Rutter (unpublished data) used Asp ratios in *Pisidium*, *Valvata*, *Sphaerium*, *Bulimus*, and *Amnicola* in the Old Crow Basin for correlating alluvial sedimentary units. In northern Europe, marine molluscs incorporated into glaciolacustrine, till, or glaciomarine deposits by glaciers scouring the North or Baltic

Basins have been used to correlate regionally (e.g., Fig. 11c; Miller & Mangerud, 1985). In central Europe, especially in the Hungarian loess lacustrine travertine sequence, loess and paleosol correlations have been developed using the terrestrial snails, *Succinea*, *Trichnia*, and *Pupilla* (Oches & McCoy, 1995, 1996; Oches et al., 2000).

As noted above, however, each mollusc species must be tested for applicability before its use. Most molluscs have an applicable range for $D/L < 0.3$–0.5 using first-order linear kinetics (e.g., Fig. 4a and 4b), making them inapplicable for pre-Quaternary units generally or for early Quaternary units in warmer environments. Using parabolic kinetics, however, the range can be extended to $D/L < 0.5$–1.0 (e.g., Fig. 4d), which increases not only the time and environmental ranges, but also the number of applicable species. Using aspartic and glutamic acids for example, Goodfriend (1991, 1992a) recognized very rapid racemization in tropical landsnails making AAR applicable to the last 500 years where ^{14}C dating can be problematic. In many mollusc species from lacustrine or terrestrial deposits, particularly saline and hypersaline lakes, the systematics have not been extensively tested yet. Therefore, a few hidden pitfalls may await discovery before molluscan AAR ratios are widely applicable in these environments.

AAR analyses of corals, radiolaria, diatoms, coralline algae, and foraminifera all may have limited applications in lacustrine units that are intercalated with marine deposits (see references in Table I). AAR can also be used with extremely light marine fossils that can be carried far inland by strong winds. These invertebrates have applicable D/L ranges similar to molluscs and the same caveats on their utility. The potential to analyze ostracodes or diatoms from saline or hypersaline lake deposits has not been fully exploited as yet, but diatoms have been used in estuarine sediment (e.g., Sigleo et al., 1983). McCoy (1988) suggested several uses for AAR ostracode analyses. Work by Kaufman (2000) suggests that Asp and Glu show first order linear kinetics in the ostracode *Candona* and give consistent ages in calibration tests against other dating methods. More applications, however, will need to be tested before assuming universal applicability.

Ratite egg shell aIle/Ile ratios are the best AAR chronometer yet found, with an applicable range, $0 < D/L \lesssim 1.0$ (Fig. 10a; Miller & Beaumont, 1989; Miller et al., 1989b, 2000; Brooks et al., 1990). While leaching is minimal (Ernst, 1987), excess water in high temperature experiments does not affect their AAR ratios (Hare et al., 1984; Kokis, 1988). Secondary contamination is rare (Wehmiller & Miller, 1990), and the shells approximate a closed system better than other species (Miller et al., 2000). Miller et al. (1999) used eggshells from emu and the extinct ratite, *Genyornis*, found on the shores of Lake Eyre to estimate the timing of some Australian megafauna extinctions at ~ 50 ka and to link them with human colonization. Magee et al. (1995) also used *Genyornis* shell to date and correlate lacustrine deposits at Lake Eyre.

Bones, teeth, and antler have proven extremely unreliable for both dating and stratigraphic correlation in all environments (e.g., Blackwell et al., 1990, 1991; Hare et al., 1978; Taylor et al., 1983; Table 1). Diagenetic alteration, particularly mineralogical changes, infaunal infestation, and amino acid leaching and enrichment in these fossils dramatically affects racemization rates (Fig. 12, 13; Turban-Just & Schramm, 1998; Blackwell et al., 2000). Using only high molecular weight or insoluble fractions (e.g., Julg et al., 1987; Elster et al., 1991; El Mansouri et al., 1996) may improve reliability somewhat, but condensation reactions or other processes that cause denatured peptides to remain in the insoluble residue may still preclude accuracy (Collins et al., 1992, 1999; Waite & Collins, 2000).

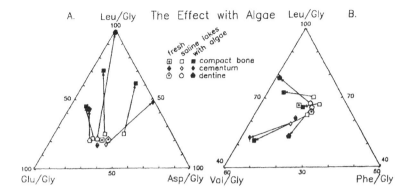

Figure 12. Effects from algal infestation on amino acid ratios in Recent Australian mammal tissues. In subsamples from adjacent regions within individual tissues, the amino acid concentration ratios changed significantly when algal infestation occurred. Arrows point from the subsample *without* the algae to the subsample *with* algae present selected from the same tissue sample (after Blackwell et al., 2000).

Consequently, all dates or stratigraphic correlations based on bones, teeth, or antler (e.g., Csapó et al., 1994, 1998) must be viewed with great skepticism.

Using AAR ratios in wood, seeds, and peat has been slow to develop. Wood shows definite species effects (Vlahos, 1985; Rutter & Crawford, 1984; Rutter & Vlahos, 1988), but further studies to ascertain its kinetics are still needed, as are tests for potential effects related to leaching, pH dependence, secondary mineralization, and contamination. For example, failure to obtain useful stratigraphic information from *Abies* was blamed on leaching or bacterial contamination (Hicock & Rutter, 1986). Wood from lacustrine and till sections, however, has provided some useful stratigraphic data (e.g., Neilsen et al., 1986; Liverman et al., 1989). Rutter et al. (1993) used AAR ratios in seeds, whereas Petit (1974) tried material from pack-rat middens. Some laboratories have apparently tried to use AAR ratios in peat, but little has appeared in the literature, suggesting that the application may be geochemically challenging.

Equally problematic has been the use of AAR ratios in soil and sediment profiles. Significant controversies have arisen over what exactly the soil and sediment AAR ratios are measuring, which could include amino acids from bacterial or other infaunal residents or humic acid contaminants to amino acids leached from adjacent fossils or dissolved in the groundwater (Pollock et al., 1977; Griffin & Kimber, 1988; Hedges & Hare, 1987). Colman et al. (1986) suggested using serine concentrations to monitor contamination. Nonetheless, several studies have reported increased D/L ratios with depth in lake sediment (e.g., Dungworth et al., 1977; Blunt et al., 1981, 1982; Reasoner & Rutter, 1988), and in soils on Mt. Kenya (Mahaney et al., 1986, 1991). Reason and Rutter (1988) used abnormal AAR ratios to recognize reworked sediment in Lake O'Hara.

Other potential chronometers are in development. For example, G. H. Miller & Rosewater (1995) have been testing fish otoliths. Peat, ostracodes, diatoms, and several other fossil species may also prove useful when fully developed.

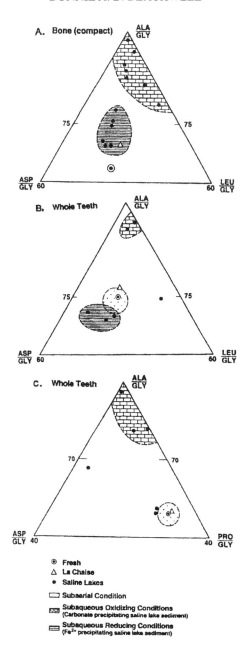

Figure 13. Variations in amino acid geochemistry in Australian mammal tissues compared to sediment chemistry. Plotting relative amino acid compositions on ternary plots clearly differentiated the sediment chemistries from which the Australian samples were collected. The amino acid chemistry, especially the Ala/Gly ratio, directly correlated with the oxidation conditions (after Blackwell et al., 2000).

Paleoenvironmental analyses in lacustrine environments

Several different paleoenvironmental applications have been developed using AAR ratios and amino acid geochemistry, ranging from paleotemperature and paleosedimentological analysis to species identification and paleodietary analysis.

Because AAR depends so strongly on temperature, diagenetic temperature analysis can provide a time-integrated paleotemperature for the site where the fossil occurred. Most diagenetic temperature analysis has used marine molluscs or foraminifera (e.g., Andrews et al., 1985a), but the method could equally well be used with lacustrine molluscs or ostracodes if the species kinetics are well delineated. Amino acid compositions in bone can be used to recognize archaeologically heated bone (Hare, 1995) ostrich egg and mollusc shells (e.g., Brooks et al., 1991). Using emu shells from the Lake Eyre Basin, Miller et al. (1997) determined the extent of cooling in the southern hemisphere during the Wisconsinan. Using terrestrial and freshwater molluscs, Oches & McCoy (1989) estimated paleotemperatures in the Mississippi Valley at $-8\,°C$ to $-12\,°C$ at 18–20 ka.

Molluscan AAR ratios and amino acid concentrations can also be used to speciate fragmentary molluscs (e.g., Fig. 14; Andrews et al., 1985b). This technique is theoretically applicable to any groups showing strong taxonomic dependence in their AAR rates. When more nonmarine species have been well studied, this technique should be fully applicable to lacustrine and terrestrial molluscs, and probably other groups. Recently, paleodietary analysis has used amino acid coupled with stable isotopic analysis to establish paleodiets for terrestrial mammals, foraminifera and marine molluscs (e.g., Hare et al., 1987; Uhle et al., 1997; Powell et al., 1989). Similar studies should be expected soon for lacustrine species. AAR ratios in dentine from modern animals can theoretically be used to determine the organism's age at death (e.g., Fig. 15; Ohtani et al., 1995) provided it has not been dead and exposed for more than a few days. Rapid infestation and diagenesis can radically alter the AAR ratios in recent fossils (Fig. 12, 13; Blackwell et al., 2000). Asp D/L ratios in dentine from historical animals, for example, appear independent of age (Gillard et al., 1991; Carolan et al., 1997). Goodfriend et al. (1995) used AAR ratios to determine growth rates and longevity for marine molluscs and anemones, but similar techniques should be applicable to lacustrine molluscs or other species. If Asp $D/L > 0.08$ in DNA studies, intact DNA will likely not be retrieved (e.g., Poinar et al., 1996; Krings et al., 1997). Using D/L ratios calibrated by ^{14}C ages, Goodfriend & Gould (1996) determined the hybridization rates for the land snail *Cerion* (Fig. 16).

Using Sr isotopes and AAR ratios from *Valvata*, *Tryonia*, *Lymnaea*, and *Sphaerium*, Bouchard et al. (1998) measured the contribution from river water to Lake Bonneville through several lake phases. Goodfriend & Rollins (1998) measured beach barrier retreat and marsh exhumation with D/L Asp ratios in the mussel *Geukensia*. Abnormal AAR ratios have been used to recognize reworked fossils in lacustrine, estuarine, and marine settings (e.g., Kowalewski et al., 1998; Mirecki et al., 1995; Nichol & Murray-Wallace, 1992). When combined with ^{18}O analyses, AAR ratios can be used to map water sources or recognize marine incursions in deltaic lake sequences (e.g., Goodfriend & Stanley, 1996).

Goodfriend & Mitterer (1988) reported a correlation between alle/Ile ratios in Israeli land snails and the amount of secondarily precipitated carbonate. In bone, dentine, and whole teeth from Australian saline lakes, Blackwell et al. (2000) reported correlations between their AAR ratios, amino acid concentrations, and secondary mineral precipitates.

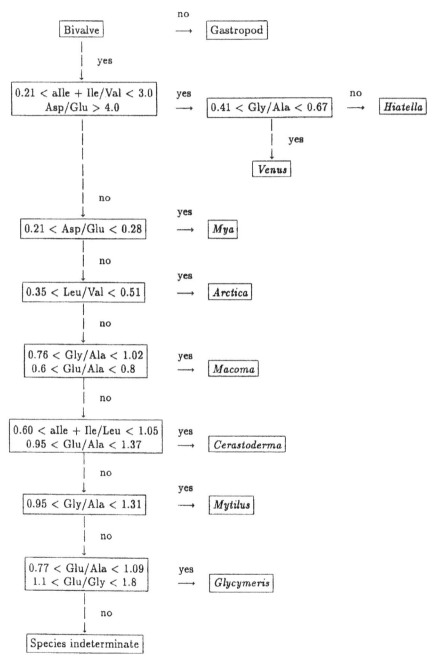

Figure 14. Mollusc identification flow chart. Using interacid or concentration ratios, small mollusc fragments can be identified by following the flow chart (modified from Andrews et al., 1985b).

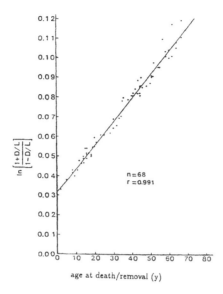

Figure 15. Aspartic acid AAR ratios versus age at removal/death in human teeth. Theoretically, Asp AAR ratios in dentine will yield the age at death with considerable accuracy (after Ogino & Ogino, 1988).

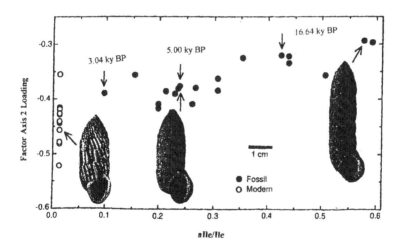

Figure 16. Evolutionary change in *Cerion* during the late Quaternary. Easily discernable anatomical changes in *Cerion* occurred over a time scale in which measureable changes in alle/Ile ratios occurred. The ^{14}C ages used here were corrected for dead carbon effects by using the alle/Ile ratio in the shells (modified from Goodfriend & Gould, 1996).

Moreover, they also reported correlations with algal infestation (Fig. 12), sedimentary geochemistry and mineralogy (Fig. 13), as well as evidence for bacterial infestation. With further development and testing, these intriguing results all show the potential to become biomarkers for paleoenvironments in which the tissues undergo fossilization. In lakes, where many generations of minerals may be precipitated and later completely removed by subsequent hydrological changes, AAR analysis may present the only method to identify these paleoenvironmental conditions.

Summary

All living protein, with a few exceptions, consists of purely L optically active amino acids. After an organism dies or when the tissue is not being regenerated regularly, the L amino acids slowly racemize into D amino acids until a stable equilibrium is achieved. Because racemization in some fossils can be modelled by assuming reversible first-order linear kinetics, in which the rate varies only with time and temperature, proteinaceous fossils can be dated by amino acid racemization (AAR) analysis. Dating also becomes possible in some special cases of nonlinear parabolic AAR kinetics. To calculate an age, diagenetic temperatures must be measured or modelled.

In limnological settings, AAR dating and aminostratigraphy works well for ratite egg shells, as well as many mollusc and diatom species, but not bones or teeth. Aminostratigraphy has proven a powerful tool for Quaternary and late Pliocene correlation, but has been mainly applied in marine settings thus far. Development continues for AAR analysis and dating of fish otoliths, peat, sediment, soil, wood and other plant remains. AAR analysis can be used to determine paleotemperatures for fossils dated by other methods. Other paleoenvironmental applications include faunal identification, longevity, growth, and evolutionary rate studies. When coupled with Sr or O isotopes, AAR analysis can monitor sediment transport patterns, deposition and subsidence rates in lacustrine, glacial, fluvial, and estuarine systems. Combined with N, O, and H isotopes, AAR analysis can yield paleodietary information. AAR analysis is used routinely to screen DNA samples and ^{14}C preparations for contamination. AAR compositions in bones and teeth correlate well with the sediment in which they fossilized and may be developed as a tool for future paleoenvironmental analysis.

Although AAR analysis has not been used as extensively in paleolimnological research as in marine settings, AAR analysis can provide several important tools for geochronometry, aminostratigraphic correlation, and paleoenvironmental analysis. Although some theoretical biogeochemical research must occur to facilitate AAR application in lacustrine settings, it will occur simultaneously as limnologists begin to require AAR analyses more routinely. The next few years should see the development of amino acid analysis for several other fossil types, such as insects (e.g., Bada et al., 1994), that may have uses in lacustrine settings, as well as new applications undreamt at the moment.

Acknowledgements

I thank John Wehmiller, Gif Miller, Jeff Bada, Keith Kvenvolden, Ron Kimber, Colin Murray-Wallace, Julie Brigham-Grette, Paul Ennis, Donald Bowen, Lucy McCartney, Steve

Forman, Dick Mitterer, Glen Sykes, Linda York, Maggie Toscano, Grant Gil Smith, Mike Moir, Bob Crawford, Donald Belknap, Simon Clarke, Darryl Kaufman, June Mirecki, and expecially Nat Rutter, Ed Hare, and Glenn Goodfriend for extensive discussions over many years on the topics presented herein. Had Bill Last and Pat De Deckker not introduced me to the intricacies of saline lake studies, I might have remained frustrated by AAR in bones and teeth forever. T. Sani, H. Leung, L. Provencher, S. Berman, and J. I. B. Blickstein assisted with manuscript preparation. Bill Last, John Smol, and two anonymous reviewers provided excellent suggestions to improve the manuscript.

References

Abbott, J. T., G. L. Ellis & G. A. Goodfriend, 1995. Chronometric and integrity analyses using land snails. In Abbott, J. T. & W. N. Trierweiler (eds.) NRHP Significance Testing of 57 Prehistoric Archaeological Sites on Fort Hood, Texas, Vol. 2. United States Army Fort Hood Arcaheological Resource Management Series Research Report 34: 801–814.

Abbott, J. T., G. A. Goodfriend & G. L. Ellis, 1996. Landsnail investigations. In Trierweiler, W. N. (ed.) Archaeological Testing at Fort Hood, 1994–95, Vol. 2. United States Army Fort Hood Arcaheological Resource Management Series Research Report 35: 619–636.

Abelson, P. H., 1954. Organic constituents of fossils. Carnegie Inst. Washington Yrbk. 53: 97–101.

Abelson, P. H., 1955. Organic constituents of fossils. Carnegie Inst. Washington Yrbk. 54: 107–109.

Akiyama, M., 1980. Diagenetic decomposition of peptide-linked serine residues in fossil scallop shells. In Hare, P. E., T. C. Hoering & K. King Jr. (eds.) Biogeochemistry of Amino Acids. Wiley, New York: 115–120.

Alford, J. J., 1990. Quaternary aminostratigraphy of Mississippi Valley loess: Discussion. Geol. Soc. Am. Bull. 102: 1136–1138.

Anderson, P. & 26 others, 1991a. Report of the 1st discussion group: The Last Interglacial in the high latitudes of the Northern Hemisphere: Terrestrial and marine evidence. Quat. Int. 10–12: 9–28.

Anderson, P. & 26 others, 1991b. Report of the 2nd discussion group: Inter-relationships and linkages between the land, atmosphere, and oceans during the Last Interglacial. Quat. Int. 10–12: 29–48.

Andrews, J. T., D. Q. Bowen & C. Kidson, 1979. Amino acid ratios and the correlation of raised beaches in southwest England and Wales. Nature 281: 556–558.

Andrews, J. T., G. H. Miller & W. W. Shilts, 1983. Multiple deglaciations of the Hudson Bay Lowlands, Canada, since deposition of the Missinaibi (last interglacial questionable) Formation. Quat. Res. 19: 18–37.

Andrews, J. T., D. D. Gilbertson & A. B. Hawkins, 1984a. The Pleistocene succession of the Severn Estuary: A revised model based upon amino acid racemization studies. J. Geol. Soc. 141: 967–974.

Andrews, J. T., G. H. Miller & W. W. Shilts, 1984b. Multiple deglaciations of the Hudson Bay Lowlands, Canada, since deposition of the Missinaibi (Last Intergiacial) Formation: Reply. Quat. Res. 22: 253–258.

Andrews, J. T., A. Aksu, M. Kelly, R. Klassen, G. H. Miller, W. N. Mode & P. Mudie, 1985a. Land-ocean correlations during the last interglacial-glacial transition, Baffin Bay, northwestern North Atlantic: A review. Quat. Sci. Rev. 4: 333–355.

Andrews, J. T., K. H. Davies, D. C. Davies & G. H. Miller, 1985b. Generic identification of fragmentary Quaternary mollusks by amino acid chromatography: A tool for Quaternary and paleontological research. Geol. J. 20: 1–20.

Armend, J. P. & E. L. Shock, 2000. Thermodynamics of amino acid synthesis in hydrothermal systems on early Earth. In Goodfriend, G. A., M. J. Collins, M. L. Fogel, S. A. Macko & J. F. Wehmiller (eds.) Perspectives in Amino Acid and Protein Geochemistry. Oxford University Press, New York: 23–40.

Atwater, B. F., B. E. Ross & J. F. Wehmiller, 1981. Stratigraphy of Late Quaternary estuarine deposits and amino acid stereochemistry of oyster shells beneath San Francisco Bay, California. Quat. Res. 16: 181–200.

Bada, J. L., 1972a. The dating of fossil bones using the racemization of isoleucine. Earth Planet. Sci. Lett. 15: 223–231.

Bada, J. L., 1972b. Kinetics of racemization of amino acids as a function of pH. Am. Chem. Soc. J. 94: 1371–1373.

Bada, J. L., 1974. Reliability of amino acid racemization dating and paleotemperature analysis on bones: Reply. Nature 252: 379–381.

Bada, J. L., 1981. Racemization of amino acids in fossil bones and teeth from the Olduvai Gorge region, Tanzania, East Africa. Earth & Planet. Sci. Lett. 55: 292–298.

Bada, J. L., 1982. The amino acid dating of African fossil bones. Natl. Geog. Soc., Res. Rep. 14: 23–30.

Bada, J. L., 1984. *In vivo* racemization in mammalian proteins. Meth. Enzym. 106: 98–115.

Bada, J. L., 1985a. Amino acid racemization dating of fossil bones. Ann. Rev. Earth Planet. Sci. 13: 241–268.

Bada, J. L., 1985b. Amino acids in California Paleoindian skeletons and their use in age and plaeodietary determinations. Geol. Soc. Am., Abstr. 17: 517.

Bada, J. L., 1987. Paleoanthropological applications of amino acid racemization dating of fossil bones and teeth. Anthropologisch Anzeiger 45: 1–8.

Bada, J. L., 1997. Meteoritics: Extraterrestrial handedness. Science 275: 942–943.

Bada, J. L. & S. E. Brown, 1981. The Amino Acid Aging Method: Dolphin Studies. NOAA (US) Report LJ-81-06C, 6 pp.

Bada, J. L. & M. Deems, 1975. Accuracy of dates beyond the C-14 dating limit using aspartic-acid racemization. Nature 255: 218–219.

Bada, J. L. & R. Finkel, 1982. Uranium series ages of the Del Mar Man and Sunnyvale skeletons: Comment. Science 217: 755–756.

Bada, J. L. & P. M. Helfman, 1975. Amino acid racemization dating of fossil bones. World Archaeol. 7: 160–173.

Bada, J. L. & P. M. Masters, 1978. The antiquity of human beings in the Americas: Evidence from amino acid racemization dating of Paleoindian skeletons. California Archaeology, Occasional Papers in Method & Theory 2: 17–24.

Bada, J. L. & P. M. Masters, 1987. Amino acids in the Del Mar Man skeleton. Abstr. Am. Chem. Soc. 193: Hist 16.

Bada, J. L., G. D. McDonald, 1995. Amino acid racemization on Mars: Implications for the preservation of biomlecules from an extinct Martian biota. Icarus 114: 139–143.

Bada, J. L. & R. Protsch, 1974. Racemization reaction of aspartic-acid and its use in dating fossil bones. Proc. Natl. Acad. Sci. USA 70: 1331–1334.

Bada, J. L. & R. A. Schroeder, 1972. Racemization of isoleucine in calcareous marine sediments. Earth Planet. Sci. Lett. 15: 1–11.

Bada, J. L. & R. A. Schroeder, 1975. Correction in the postglacial temperature difference computed from amino-acid racemization. Science 191: 103–105.

Bada, J. L. & R. A. Schroeder, 1976a. Amino-acid racemization reactions and their geochemical implications. Naturwissenschaften 62: 71–79.

Bada, J. L. & R. A. Schroeder, 1976b. Correction in the glacial-postglacial temperature difference computed from amino acid racemization: Reply. Science 191: 103.

Bada, J. L. & M. Y. Shou, 1980. Kinetics and mechanisms of amino acid racemization in aqueous solution and in bones. In Hare, P. E., T. C. Hoering & K. King Jr. (eds.) Biogeochemistry of Amino Acids. Wiley, New York: 235–255.

Bada, J. L., B. P. Luyendyk & J. B. Maynard, 1970. Marine sediments: dating by racemization of amino acids. Science 170: 730–732.

Bada, J. L., K. A. Kvenvolden & E. Peterson, 1973. Racemization of amino acids in bones. Nature 245: 308–310.

Bada, J. L., R. A. Schroeder & G. F. Carter, 1974a. New evidence for the antiquity of man in North America deduced from aspartic acid racemization. Science 184: 791–793.

Bada, J. L., R. A. Schroeder, R. Protsch & R. Berger, 1974b. Concordance of collagen-based radiocarbon and aspartic acid racemization ages. Proc. Natl. Acad. Sci. USA 71: 914–917.

Bada, J. L., P. M. Masters, E. Hoopes & D. Darling, 1979a. The dating of fossil bones using amino acid racemization. In Berger, R. & H. Suess (eds.) Radiocarbon and Other Dating Methods. University of California Press, Berkeley: 740–756.

Bada, J. L., E. Hoopes, D. Darling, G. Dungworth, H. J. Kessels, K. A. Kvenvolden & D. J. Blunt, 1979b. Amino acid racemization dating of fossil bones, 1. Interlaboratory comparison of racemization measurements. Earth Planet. Sci. Lett. 43: 265–268.

Bada, J. L., B. Kemper & E. Mitchell, 1983a. Aspartic acid racemization in narwhal teeth. Nature 303: 418–420.

Bada, J. L., J. R. Cronin, M. S. Ho, K. A. Kvenvoiden, J. G. Lawless, S. L. Miller, J. Oro & S. M. Steinberg, 1983b. On the reported optical activity of amino acids in the Murchison Meteorite: Comment. Nature 301: 494–496.

Bada, J. L., R. Gillespie, J. A. J. Gowlett & R. E. M. Hedges, 1984. Accelerator mass spectormetry ages of amino acid extracts from California paleoindian skeletons. Nature 312: 442–444.

Bada, J. L., M. X. Zhao, S. M. Steinberg & E. Ruth, 1986a. Isoleucine stereoisomers on the earth. Nature 319: 314–316.

Bada, J. L., N. Lee & M. X. Zhao, 1986b. Did extraterrestrial impactors supply the organics necessary for the origin of terrestrial life: Amino acid evidence in Cretaceous-Tertiary boundary sediments. Orig. Life Evol. Biosph. 16: 185–185.

Bada, J. L., X. S. Wang, H. N. Poinar, S. Paabo & G. O. Poinar, 1994. Amino acid racemization in amber-entombed insects: Implications for DNA preservation. Geoch. Cosmoch. Acta 58: 3131–3135.

Bada, J. L., D. P. Glavin, G. D. MacDonald & L. Becker, 1998. A search for endogenous amino acids in Martian Meteorite Alh84001. Science 279: 362–365.

Baratova, L. A., Y. M. Rumyantsev, A. V. Shishkov, E. F. Simonov, V. A. Tsyryapkin & M. S. Unukovich, 1981. Reactions of atomic tritium with amino acids racemization of L-alanine. High Energy Chem. 15: 284–287.

Baum, R. & G. G. Smith, 1986. Systematic pH study on the acid-catalyzed and base-catalyzed racemization of free amino acids to determine the 6 constants, one for each of the 3 ionic species. J. Am. Chem. Soc. 108: 7325–7327.

Belknap, D. F. & J. F. Wehmiller, 1980. Amino acid racemization in Quaternary mollusks: Examples from Delaware, Maryland, and Virginia. In Hare, P. E., T. C. Hoering & K. King Jr. (eds.) Biogeochemistry of Amino Acids. Wiley, New York: 401–414.

Belluomini, G. & J. L. Bada, 1985. Isoleucine epimerization ages of the dwarf elephants of Sicily. Geology 13: 451–452.

Belluomini, G. & L. Delitala, 1988. Amino acid racemization dating of Quaternary deposits of Central and Southern Italy. Org. Geoch. 13: 735–740.

Bender, M. L., 1974. Reliability of amino acid racemization dating and paleotemperature analysis on bones: Comment. Nature 252: 378–381.

Bischoff, J. L. & J. F. Rosenbauer, 1981a. Uranium series dating of human skeletal remains from the Del Mar and Sunnyvale sites, California. Science 213: 1003–1005.

Bischoff, J. L. & J. F. Rosenbauer, 1981b. Uranium series dating of bones and carbonate deposits of the Caune de L'Arago at Tautauel. In de Lumley, H. & J. Labeyrie (eds.) Datations Absolus et les Analyses Isotopiques en Préhistoire: Méthodes et Limites, Prétirage. CNRS, Paris: 327–347.

Bischoff, J. L. & J. F. Rosenbauer, 1982. Uranium series ages of the Del Mar Man and Sunnyvale skeletons: Reply. Science 217: 756–757.

Blackwell, B. A., 1987. Problems of Amino Acid Analyses in Mammalian Bones and Teeth from the Archaeological Sites Lachaise and Montgaudier (Charente), France. Unpublished Ph.D. thesis, University of Alberta, Edmonton, 650 pp.

Blackwell, B. A. & N. W. Rutter, 1985. Concentration variation of amino acids in mammalian fossils: Effects of diagenesis and the implications for amino acid racemization analysis. Geol. Soc. Am. Abstr. 17: A524–525.

Blackwell, B. A., W. M. Last & N. W. Rutter, 1988. Evidence for rapid mineralogical and organic diagenesis in mammalian bones and teeth from saline lakes in Saskatchewan and Australia. Geol. Soc. Am. Abstr. 20: A392.

Blackwell, B. A., N. W. Rutter & W. M. Last, 1989. Effects of fossilization on amino acid racemization in Recent mammalian bones and teeth from saline lakes, Australia and Saskatchewan. Geol. Soc. Am. Abstr. 21: A210.

Blackwell, B. A., N. W. Rutter & A. Debénath, 1990. Amino acid racamization analysis of mammalian bones and teeth from La Chaise-de-Vouthon (Charente), France. Geoarchaeology 5: 121–147.

Blackwell, B. A., N. W. Rutter & A. Debénath, 1991. Amino acid racamization analysis of mammalian bones and teeth from La Chaise-de-Vouthon (Charente), France. Paleoanthropology Annuals (1990) 1: 245–270.

Blackwell, B. A., N. W. Rutter & W. M. Last, 2000. Biogeochemical diagenesis in Recent mammalian bones from saline lakes in western Victoria, Australia. In Goodfriend, G. A., M. J. Collins, M. L. Fogel, S. A. Macko & J. F. Wehmiller (eds.) Perspectives in Amino Acid and Protein Geochemistry. Oxford University Press, New York: 88–119.

Blunt, D. J., K. A. Kvenvolden & J. D. Sims, 1981. Geochemistry of amino acids in sediments from Clear Lake, California. Geology 9: 378–382.

Blunt, D. J., K. A. Kvenvolden & J. D. Sims, 1982. Geochemistry of amino acids in sediments from Clear Lake, California: Reply. Geology 10: 124–125.

Blunt, D. J., D. J. Easterbrook & N. W. Rutter, 1987. Chronology of Pleistocene sediments in the Puget Lowland, Washington. In Schuster, J. E. (ed.) Selected Papers on the Geology of Washington. Washington Div. Earth Resources Bull. 77: 321–353.

Bonner, W. A., N. E. Blair & R. M. Lemmon, 1980. The radioracemization of amino acids by ionizing radiation, geochemical and cosmochemical implications. In Hare, P. E., T. C. Hoering & K. King Jr. (eds.) Biogeochemistry of Amino Acids. Wiley, New York: 357–374.

Bouchard, D. P., D. S. Kaufman, A. Hochberg & J. Quade, 1998. Quaternary history of the Thatcher Basin, Idaho, reconstructed from $^{87}Sr/^{86}Sr$ and amino acid composition of lacustrine fossils: Implications for the diversion of the Bear River into the Bonneville Basin. Palaeogeog. Palaeoclim. Palaeoecol. 141: 95–114.

Boulton, G. S., C. T. Baldwin, R. N. Chroston, T. E. Day, N. Eyles, R. Gibbard, P. E. Hare, B. Horsefield, J. Jarvis, A. M. Mctabe, G. H. Miller, J. D. Peacock, V. Vonbrunn & P. Worsley, 1982. A glacio-isostatic facies model and amino acid stratigraphy for Late Quaternary events in Spitsbergen and the Arctic. Nature 298: 437–441.

Bowen, D. Q., 2000. Revised aminostratigraphy for land-sea correlations from the northeastern North Atlantic margin. In Goodfriend, G. A., M. J. Collins, M. L. Fogel, S. A. Macko & J. F. Wehmiller (eds.) Perspectives in Amino Acid and Protein Geochemistry. Oxford University Press, New York: 253–262.

Bowen, D. Q. & G. A. Sykes, 1989. Correlations of marine events and glaciations on the northeast Atlantic margin. Trans. R. Soc. London 318B: 619–635.

Bowen, D. Q. & G. A. Sykes, 1994. How old is "Boxgrove Man"? Nature 371: 751.

Bowen, D. Q., G. H. Miller, G. A. Sykes & S. Hughes, 1989. Land-sea correlations in the Pleistocene based on isoleucine epimerization in non-marine mollusks. Nature 340: 49–51.

Bowen, D. Q., G. A. Sykes, A. Reeves, G. H. Miller, J. T. Andrews, J. S. Brew & P. E. Hare, 1985. Amino acid geochronology of raised beaches in Southwest England. Quat. Sci. Rev. 4: 279–318.

Brandes, J. A., R. M. Hazen, H. S. Yoder, Jr. & G. D. Cady, 2000. Early pre- and post-biotic synthesis of alanine: An alternative to the Strecker synthesis. In Goodfriend, G. A., M. J. Collins, M. L. Fogel, S. A. Macko & J. F. Wehmiller (eds.) Perspectives in Amino Acid and Protein Geochemistry. Oxford University Press, New York: 41–47.

Brigham, J. K., 1980. Stratigraphy, Amino Acid Geochronology, and Genesis of Quaternary Sediments, Broughton Island, E. Baffin Island, Canada. Unpublished M.Sc. thesis, University of Colorado, Boulder, 224 pp.

Brigham, J. K., 1982. Comparison of uranium series, radiocarbon, and amino acid data from marine mollusks, Baffin Island, Arctic Canada: Comment. Geology 10: 215.

Brigham, J. K., 1983a. Stratigraphy, amino acid geochronology, and correlation of Quaternary sea level and glacial events, Broughton Island, Arctic Canada. Can. J. Earth Sci. 20: 577–598.

Brigham, J. K., 1983b. Intershell variations in amino acid concentrations and isoleucine epimerization in fossil *Hiatella arctica*. Geology 11: 509–513.

Brigham, J. K., 1985. Marine Stratigraphy and Amino Acid Geochronology of the Gubik Formation, Western Arctic Coastal Plain, Alaska. Unpublished Ph.D. thesis, University of Colorado, Boulder, 336 pp.

Brigham-Grette, J. K. & L. D. Carter, 1992. Pliocene marine transgressions of Northern Alaska: Circumarctic correlations and paleoclimatic interpretations. Arctic 45: 74–89.

Brigham-Grette, J. K. & D. M. Hopkins, 1995. Emergent marine record and paleoclimate of the Last Interglaciation along the northwest Alaskan coast. Quat. Res. 43: 159–173.

Brinton, K. L. F. & J. L. Bada, 1995. Comment on "Aspartic acid racemization and protein diagenesis in corals over the last 350 years". Geoch. Cosmoch. Acta 59: 415–416.

Brinton, K. L. F., E. Engrand, D. P. Glavin, J. L. Bada & M. Maurette, 1998. A search for extraterrestrial amino acids in carbonaceous Antarctic micrometeorites. Orig. Life Evol. Biosph. 28: 413–424.

Brooks, A. S., P. E. Hare, J. E. Kokis, G. H. Miller, R. D. Ernst & F. Wendorf, 1990. Dating Pleistocene archaeological sites by protein diagenesis in ostrich eggshell. Science 248: 60–64.

Brooks, A. S., P. E. Hare, J. E. Kokis & K. Durana, 1991. A burning question: Differences between laboratory-induced and natural diagenesis in ostrich eggshell proteins. Annual Report of the Director, Geophysical Laboratory, Carnegie Institution of Washington, 1990–1991. Geophysical Laboratory, Carnegie Institution of Washington, Washington: 176–179.

Brooks, A. S., P. E. Hare & J. E. Kokis, 1993. Age of early anatomically modern human fossils from the cave of Klasies River Mouth, South Africa. Carnegie Inst. Washington Yrbk 92: 95–96.

Brooks, A. D., D. M. Halgren, J. S. Cramer, A. Franklin, W. Hornyak, J. M. Keating, R. G. Klein, W. J. Rink, H. P. Schwarcz, J. N. L. Smith, K. Stewart, N. E. Todd, J. Verniers & J. E. Yellen, 1995. Dating and context of three Middle Stone Age sties with bone points in the Upper Semliki Valley, Zaire. Science 268: 548–553.

Burky, R. R., D. L. Kirner, R. E. Taylor, P. E. Hare & J. R. Southon, 1998. ^{14}C dating of bone using γ-carboxygluatmic acid and α-carboxyglycine (aminomalonate). Radiocarbon 40: 11–20.

Burky, R. R., D. L. Kirner, R. E. Taylor, P. E. Hare & J. R. Southon, 2000. Isotopic integrity of α-carboxyglycine (aminomalonate) in fossil bone based on ^{14}C data. In Goodfriend, G. A., M. J. Collins, M. L. Fogel, S. A. Macko & J. F. Wehmiller (eds.) Perspectives in Amino Acid and Protein Geochemistry. Oxford University Press, New York: 60–66.

Campbell, S., J. T. Andrews & R. A. Shakesby, 1982. Amino acid evidence for Devensian ice, West Gower, South Wales. Nature 300: 249–251.

Cann, J. H. & C. V. Murray-Wallace, 1986. Holocene distribution and amino acid racemization of the benthic foraminifera Massilina milletti, northern Spencer Gulf, South Australia. Alcheringa 10: 45–54.

Cann, J. H., P. De Deckker & C. V. Murray-Wallace, 1991. Coastal aboriginal shell middens and their palaeoenvironmental significance, Robe Range, South Australia. Trans. R. Soc. S. Austr. 115: 161–175.

Carolan, V. A., M. L. G. Gardener, D. Lucy & A. M. Pollard, 1997. Some considerations regarding the use of amino acid racemization in human dentine as an indicator of age at death. J. Forensic Sci. 42: 10–16.

Carter, L. D. & J. W. Hillhouse, 1991. Age of the late Cenozoic Big Bendian marine transgression of the Arctic Coastal Plain: Significance for permafrost history and paleoclimate. US Geol. Survey, Bull. 1999: 44–51.

Carter, L. D., J. K. Brigham-Grette, L. Marincovich Jr., V. L. Pease & J. W. Hillhouse, 1986. Late Cenozoic Arctic Ocean sea ice and terrestrial paleoclimate. Geology 14: 675–678.

Chen, Z. Q., C. H. Hobbs, J. F. Wehmiller & S. M. Kimball, 1995. Late Quaternary paleochannel systems on the continental shelf, south of the Chesapeake Bay entrance. J. Coastal Res. 11: 605–614.

Child, A. M., 1996. Amino acid racemization and the effects of microbial diagenesis. In Orna, M. V. (ed.) Archaeological Chemistry 5 (American Chemical Society, Advances in Chemistry 625): 366–377.

Child, A. M., D. R. Gillard & A. M. Pollard, 1993. Microbially indcuced promotion of amino acid racemization in bone: Isolation of the microorganisms and detection of their enzmes. J. Arch. Sci. 20: 159–168.

Clapperton, C. M., D. E. Sugden, D. S. Kaufman & R. D. McCulloch, 1995. The last glaciation in central Magellan Strait, southernmost Chile. Quat. Res. 44: 133–146.

Clark, P. U., D. K. Barnes, W. D. McCoy, B. B. Miller & A. R. Nelson, 1989. Quaternary aminostratigraphy of Mississippi Valley loess. Geol. Soc. Am. Bull. 101: 918–926.

Clark, P. U., W. D. McCoy, E. A. Oches, A. R. Nelson & B. B. Miller, 1990. Quaternary aminostratigraphy of Mississippi Valley loess: Reply. Geol. Soc. Am. Bull. 102: 1136–1138.

Clarke, S. J., 1999. The Application of Amino Acid Racemization Geochronological Techniques to Late Pleistocene Fossil Teeth from the Australian Megafauna Locality, Cuddie Springs. Unpub. B.Sc. thesis, University of Wollangong, Wollangong.

Cohen, B. A. & C. F. Chyba, 1998. Racemization of meteoritic amino acids. Meteoritics & Planet. Sci. Abstr. 33: 33.

Collins, M. J. & P. Galley, 1998. Towards an optimal method of archaeological collagen extraction: The influence of pH and grinding. Ancient Biomol. 2: 209–222.

Collins, M. J. & M. S. Riley, 2000. Amino acid racemization in biominerals: The impact of protein degradation and loss. In Goodfriend, G. A., M. J. Collins, M. L. Fogel, S . A. Macko & J. F. Wehmiller (eds.) Perspectives in Amino Acid and Protein Geochemistry. Oxford University Press, New York: 120–142.

Collins, M. J., D. Walton & A. King, 1998. The geochemical fate of proteins. In Stankiewicz, B. A. & P. F. van Bergen (eds.) Nitrogen-containing Macromolecules in the Biosphere and Geosphere. American Chemical Society, Advances in Chemistry 707: 74–86.

Collins, M. J., E. R. Waite & A. C. T. van Duin, 1999. Predicting protein decomposition: The case of aspartic-acid racemization kinetics. Phil. Trans. R. Soc. London B 354: 51–64.

Colman, S. M., A. F. Choquette, D. J. Huntley, G. H. Miller & J. N. Rosholt, 1986. Dating the Upper Cenozoic sediments in Fisher Valley, southeastern Utah. Geol. Soc. Am. Bull. 97: 1422–1431.

Cook, L. M., G. A. Goodfriend & R. A. D. Cameron, 1993. Changes in the land snail fauna of eastern Madeira during the Quaternary. Phil. Trans. R. Soc. London B339: 83–103.

Corrado, J. C., R. E. Weems, P. E. Hare & K. K. Bambach, 1986. Capabilities and limitations of applied aminostratigraphy, as illustrated by analyses of *Mulinia lateralis* from the late Cenozoic marine beds near Charleston, South Carolina. South Carolina Geology 30: 19–46.

Cronin, J. R. & S. Pizzarello, 2000. Chirality of meteoritic organic matter: A brief review. In Goodfriend, G. A., M. J. Collins, M. L. Fogel, S. A. Macko & J. F. Wehmiller (eds.) Perspectives in Amino Acid and Protein Geochemistry. Oxford University Press, New York: 15–23.

Cronin, J. R., W. E. Grandy & S. Pizzarello, 1980. Amino acids of the Murchison meteorite. In Hare, P. E., T. C. Hoering & K. King Jr. (eds.) Biogeochemistry of Amino Acids. Wiley, New York: 153–168.

Csapó, J., Z. Csapó-Kiss, S. Némenthy, S. Folestad, A. Tivesten & T. G. Martin, 1994. Age determination based on amino acids racemization: A new possibility. Amino Acids 7: 317–325.

Csapó, J., Z. Csapó-Kiss & J. Csapó, 1998. Use of amino acids and their racemization for age determination in archaeometry. Trends in Analyt. Chem. 17: 140–148.

Custer, J. F., J. C. Kraft & J. F. Wehmiller, 1989. The Holly Oak Shell. Science 243: 151.

Davies, K. H., 1983. Amino acid analysis of Pleistocene marine mollusks from the Gower Peninsula. Nature 302: 137–139.

Davis, W. E. & F. E. Treloar, 1977. The application of racemisation dating in archaeology: A critical review. The Artifact 2: 63–94.

de Lumley, H., M. A. de Lumley, J. L. Bada & K. K. Turekian, 1977. The dating of pre-Neandertal remains at Caune de l'Arago, Tautavel, Pyrénées-Orientales, France. J. Human Evol. 6: 223–224.

de Lumley, H. & J. Labeyrie (eds.), 1981. Datations Absolues et Analyses Isotopiques en Préhistoire: Méthodes et Limites, Datation du Remplisage de la Caune de l'Arago à Tautavel, Prétirage. CNRS, Paris, 447 pp.

DeNiro, M. J. & S. Weiner, 1988. Chemical, enzymatic, and sprectroscopic characterization of "collagen" and other organic fractions from prehistoric bones. Geoch. Cosmoch. Acta 52: 2197–2206.

Donahue, D. J., T. H. Zabel, A. J. T. Jull & P. E. Damon, 1983. Results of tests and measurements from the NSF regional accelerator facility for radioisotope dating. Radiocarbon 25: 719–728.

Dumas, B., P. Gueremy, P. J. Hearty, R. Lhenaff & J. Raffy, 1988. Morphometric analysis and amino acid geochronology of uplifted shorelines in a tectonic region near Reggio-Calabria, South Italy. Palaeogeog. Palaeoclim. Palaeoecol. 68: 273–289.

Dungworth, G., 1982. Geochemistry of amino acids in sediments from Clear Lake, California: Comment. Geology 10: 124–125.

Dungworth, G. & A. W. Schwartz, 1974. Organic matter and trace elements in Precambrian rocks from South Africa. Chem. Geol. 14: 163–172.

Dungworth, G., A. W. Schwartz, L. Vandelee, 1976. Composition and racemization of amino acids in mammoth collagen determined by gas and liquid chromatography. Comp. Bioch. Physiol. 53: 473–480.

Dungworth, G., M. Thijssens, J. Zuurveld, W. van der Velden & A. W. Schwartz, 1977. Distributions of amino acids, amino sugars, purines, and pyrimidines in a Lake Ontario sediment core. Chem. Geol. 19: 295–308.

Dungworth, G., N. J. Vincken & A. W. Schwartz, 1974. Composition of fossil collagen: Analysis by gas-liquid chromatography. Comp. Bioch. Physiol. 47b: 391–399.

El Mansouri, A., A. El Fouikar & B. Saint-Martin, 1996. Correlation between ^{14}C ages and aspartic acid racemization at the Upper Paelaeolithic site of Abri Pataud (Dordogne, France). J. Arch. Sci. 23: 803–809.

Ellis, G. L., G. A. Goodfriend, J. T. Abbott, P. E. Hare & D. W. Von Endt, 1996. Assessment of integrity and geochronology of archaeological sites using amino acid racemization in land snail shells: Examples from central Texas. Geoarchaeology 11: 189–213.

Ellis, G. L. & G. A. Goodfriend, 1994. Chronometric and site formation studies using land snail shells: Preliminary results. In Abbott, J. T. & W. N. Trierweiler (eds.) Arcaheological Investigations on 571 Prehistoric Sites at Fort Hood and Coryell Conties, Texas. United States Army Fort Hood Arcaheological Resource Management Series Research Report 31: 183–201.

Elster, H., E. Gil-Av & S. Weiner, 1991. Amino acid racemization of fossil bone. J. Arch. Sci. 18: 605–617.

Emerson, W. K., G. L. Kennedy, J. F. Wehmiller & E. Keenan, 1981. Age relations and zoogeographic implications of Late Pleistocene marine invertebrate faunas from Turtle Bay, Baja California Sur, Mexico. Nautilus 95: 105–116.

Engel, M. H. & P. E. Hare, 1985. Gas liquid chromatographic separation of amino acids and their derivatives. In Barrett, G. C. (ed.) Chemistry and Biochemistry of Amino Acids. Chapman Hall, London: 462–479.

Engel, M. H. & S. A. Macko, 1997. Isotopic evidence for extraterrestrial non-racemic amino acids in the Murchison Meteorite. Nature 389: 265–268.

Engel, M. H. & B. Nagy, 1982. Distribution and enantiomeric composition of amino acids in the Murchison Meteorite. Nature 296: 837–840.

Engel, M. H. & B. Nagy, 1983. On the reported optical activity of amino acids in the Murchison Meteorite: Reply. Nature 301: 496–497.

Ennis, P. J., P. E. Hare, E. A. Noltmann, L. A. Payen, C. A. Prior, P. J. Slota & R. E. Taylor, 1986. Use of AMS ^{14}C analysis in the study of problems in aspartic acid racemization deduced age estimates on bone. Radiocarbon 28: 539–546.

Ernst, R. D., 1987. Reaction Kinetics of Protein and Amino Acid Degradation and Isoleucine Epimerization in Molluscan Shells, *Mya truncata* and *Patella vulgata*, and Eggshells of the African Ostrich, *Struthio camelus*. Unpublished B.Sc. thesis, University of Colorado, Boulder.

Ernst, R. D., 1989. Reaction Kinetics of Protein Hydrolysis, Amino Acid Decomposition, and Isoleucine Epimerization in Eggshells of the African Ostrich, *Struthio camelus*. Unpublished M.Sc. thesis, University of Arizona, Tuscon, 151 pp.

Ferland, M. A., P. S. Roy & C. V. Murray-Wallace, 1995. Glacial lowstand deposits on the outer contineal shelf of southeastern Australia. Quat. Res. 44: 294–299.

Feyling-Hanssen, R. W., 1982. Foraminiferal zonation of a boring in Quaternary deposits of the northern North Sea. Bull. Geol. Soc. Denm. 29: 175–184.

Fletcher, R. R., J. F. Wehmiller, R. E. Martin & B. J. Johnson, 1991. Comparative planktonic foraminifereal stratigraphy of the Colombia Basin and the northeast Gulf of Mexico. Am. Assoc. Petrol. Geol. Bull. 75: 575.

Forman, S. L. & G. H. Miller, 1984. Time dependent soil morphologies and pedogenic processes on raised beaches, Broggerhalvoya, Spitsbergen, Svalbard Archipelago. Arct. Alp. Res. 16: 381–394.

Forman, S. L., E. A. Bettis, T. J. Kemmis & B. B. Miller, 1992. Chronologic evidence for multiple periods of loess deposition during the late Pleistocene in the Missouri & Mississippi Valley, United States: Implications for the activity of the Laurentide Ice sheet. Palaeogeog. Palaeoclim. Palaeoecol. 93: 71–83.

Frank, H., E. Bayer, G. Nicholson & W. Woiwode, 1981. Determination of the rate of acidic catalyzed racemization of protein amino acids. Liebigs Annalen der Chemie 1981: 354–365.

Friedman, M., 1999. Chemistry, nutrition, and microbiology of D-amino acids. J. Agric. Food Chem. 47: 3457–3479.

Friebele, E., P. E. Hare, C. Ponnamperuma & A. Shimoyama, 1981. Adsorption of amino acid entantiomers by Na-montmorillonite. Orig. Life Evol. Biosph. 11: 173–184.

Funder, S. & L. A. Simonarson, 1984. Biostratigraphy and aminostratigraphy of some Quaternary marine deposits in West Greenland. Can. J. Earth Sci. 21: 843–852.

Funder, S., C. Hjort & M. Kelly, 1991. Isotope Stage 5 (130–74 ka) in Greenland: A review. Quat. Int. 10–12: 107–122.

Gillard, R. D., S. M. Hardman, A. M. Pollard, P. A. Sutton & D. K. Whittacker, 1991. Determinations of age at death in archaeological populations using the *D/L* ratio of aspartic acid in dental collagen. Archaeometry 32: 637–644.

Glavin, D. P. & J. L. Bada, 1998. Isolation of amino acids from anturals samples using sublimation. Analyt. Chem. 70: 3119–3122.

Goede, A. & J. L. Bada, 1985. Electron spin resonance dating of Quaternary bone material from Tasmanian caves: A comparison with ages determined by aspartic acid racemization and [14]C. Austr. J. Earth Sci. 32: 155–162.

Goodfriend, G. A., 1987a. Evaluation of amino acid racemization-epimerization dating using radiocarbon-dated fossil land snails. Geology 15: 698–700.

Goodfriend, G. A., 1987b. Chronostratigraphic studies of sediments in the Negev Desert, using amino acid epimerization analysis of land snail shells. Quat. Res. 28: 374–392.

Goodfriend, G. A., 1989. Complementary use of amino acid epimerization and radiocarbon analysis for dating of mixed age fossil assemblages. Radiocarbon 31: 1041–1047.

Goodfriend, G. A., 1991a. Patterns of racemization and epimerization of amino acids in land snail shells over the course of the Holocene. Geoch. Cosmoch. Acta 55: 293–302.

Goodfriend, G. A., 1991b. Holocene trends in 0–18 in land snail shells from the Negev Desert and their implications for changes in rainfall source areas. Quat. Res. 35: 417–426.

Goodfriend, G. A., 1992a. The use of land snail shells in paleoenvironmental reconstruction. Quat. Sci. Rev. 11: 665–685.

Goodfriend, G. A., 1992b. Rapid racemization of aspartic acid in mollusc shells and potential for dating over recent centuries. Nature 357: 399–401.

Goodfriend, G. A., 1997. Aspartic acid racemization and amino acid composition of the organic endoskeleton of the deep-water colonial anemone *Gerardia*: Determination of longevity from kinetic experiments. Geoch. Cosmoch. Acta 61: 1931–1939.

Goodfriend, G. A., 2000. Introduction. In Goodfriend, G. A., M. J. Collins, M. L. Fogel, S. A. Macko & J. F. Wehmiller (eds.) Perspectives in Amino Acid and Protein Geochemistry. Oxford University Press, New York: 1–4.

Goodfriend, G. A. & S. J. Gould, 1996. Paleontology and chronology of two evolutionary transitions by hybridization in the Bahamian land snail *Cerion*. Science 274: 1894–1897.

Goodfriend, G. A. & P. E. Hare, 1995. Reply to the comment by K. L. F. Brinton and J. L. Bada on "Aspartic acid racemization and protein diagenesis in corals over the last 350 years". Geoch. Cosmoch. Acta 59: 417–418.

Goodfriend, G. A. & V. R. Meyer, 1991. A comparative study of the kinetics of amino acid racemization and epimerization in fossil and modern mollusc shells. Geoch. Cosmoch. Acta 55: 3355–3367.

Goodfriend, G. A. & R. M. Mitterer, 1988. Late Quaternary land snails from the north coast of Jamaica: Local extinctions and climatic change. Palaeogeog. Palaeoclim. Palaeoecol. 63: 293–311.

Goodfriend, G. A. & R. M. Mitterer, 1993. A 45,000-yr record of a tropical lowland biota: The land snails fauna from cave sediments at Coco Ree Jamica. Geol. Soc. Am., Bull. 105: 18–29.

Goodfriend, G. A. & H. B. Rollins, 1998. Recent barrier beach retreat in Georgia: Dating exhumed salt marshes by aspartic acid racemization and post-Bomb radiocarbon. J. Coastal Res. 14: 960–969.

Goodfriend, G. A. & D. J. Stanley, 1996. Reworking and discontinuities in Holocene sedimentation in the Nile Delta: Documentation from amino acid racemization and stable isotopes in mollusk shells. Mar. Geol. 129: 271–283.

Goodfriend, G. A., D. W. Von Endt & P. E. Hare, 1991. Rapid Fossilization of Aspartic Acid in Mollusk and Ostrich Egg Shells: A New Method for Dating on a Decadal Time Scale. Annual Report of the Director, Geophysical Laboratory, Carnegie Institution of Washington, 1990–91. Geophysical Laboratory, Carnegie Institution of Washington, Washington: 172–176.

Goodfriend, G. A., P. E. Hare & E. R. M. Druffel, 1992. Aspartic acid racemization and protein diagenesis in corals over the last 350 years. Geoch. Cosmoch. Acta 56: 3847–3850.

Goodfriend, G. A., R. A. D. Cameron & L. M. Cook, 1994. Fossil evidence of recent human impact on the land snail fauna of Madeira. J. Biogeog. 21: 703–715.

Goodfriend, G. A., M. Kashgarian & M. G. Harasewych, 1995. Use of aspartic acid racemization and post-Bomb ^{14}C to reconstruct growth rates and longevity of the deep water slit shell *Entemnotrochus adamsonianus*. Geoch. Cosmoch. Acta 59: 1125–1129.

Goodfriend, G. A., J. K. Brigham-Grette & G. H. Miller, 1996a. Enhanced age resolution of the marine Quaternary record in the Arctic using aspartic acid racemization dating of bivalve shells. Quat. Res. 45: 176–187.

Goodfriend, G. A., R. A. D. Cameron, L. M. Cook, M. A. Courtney, N. Federoff, E. Livett & J. Tallis, 1996b. The Quaternary eolian sequence of Madeira: Stratigraphy, chronology, and paleoenvironmental interpretation. Palaeogeog. Palaeoclim. Palaeoecol. 120: 195–234.

Goodfriend, G. A., K. W. Flessa & P. E. Hare, 1997. Variation in amino acid epimerization rates and amino acid composition among shell layers in the bivalve *Chione* from the Gulf of California. Geoch. Cosmoch. Acta 61: 1487–1493.

Goodfriend, G. A., M. J. Collins, M. L. Fogel, S. A. Macko & J. F. Wehmiller (eds.), 2000a. Perspectives in Amino Acid and Protein Geochemistry. Oxford University Press, New York, 366 pp.

Goodfriend, G. A., J. Halfar & L. Godinez-Orta, 2000b. Chronostratigraphy of sediments in the southern Gulf of California, based on amino acid racemization analysis of mollusks and coralline algae. In Goodfriend, G. A., M. J. Collins, M. L. Fogel, S. A. Macko & J. F. Wehmiller (eds.) Perspectives in Amino Acid and Protein Geochemistry. Oxford University Press, New York: 320–329.

Gribben, J. & H. H. Lamb, 1978. Climatic change in historic times. In Gribben, J. J. (ed.) Climatic Change. Cambridge University Press, Cambridge: 68–82.

Griffin, C. V. & R. W. L. Kimber, 1988. Racemization of amino acids in agricultural soils: An age effect. Austr. J. Soil Res. 26: 531–536.

Groot, J. J., R. N. Benson & J. F. Wehmiller, 1995. Palynological, foraminiferal, and aminostratigraphic studies of Quaternary sediments from the US Middle Atlantic upper continental slope, continental shelf, and coastal plain. Quat. Sci. Rev. 14: 17–49.

Hald, M., J. Saettem & E. Nesse, 1990. Middle and Late Weichselian stratigraphy in shallow drillings from the southwestern Barents Sea: Foraminiferal, amino acid, and radiocarbon evidence. Norsk Geologisk Tidsskrift 70: 241–257.

Haflidason, H., I. Aarseth, J. E. Haugen, H. P. Sejrup, R. Lovlie & E. Reither, 1991. Quaternary stratigraphy of the Draugen area, Mid-Norwegian Shelf. Mar. Geol. 101: 125–146.

Harada, K. & N. Handa, 1995. Amino acid chronology in the fossil planktonic foraminifera, *Pulleniatina obliquiloculata* from the Pacific Ocean. Geophys. Res. Lett. 22: 2353–2356.

Harada, K. & P. E. Hare, 1980. Analyses of the amino acids from the Allende meteorite. In Hare, P. E., T. C. Hoering & K. King Jr. (eds.) Biogeochemistry of Amino Acids. Wiley, New York: 169–181.

Harada, K., N. Handa, M. Ito, T. Oba & E. Matsumoto, 1996. Chronology of marine sediments by the racemization reaction of aspartic acid in planktonic foraminifera. Org. Geoch. 24: 921–930.

Hare, P. E., 1963. Amino acids in the proteins from aragonite and calcite in shells of *Mytilus californianus*. Science 139: 216–217.

Hare, P. E., 1969. Geochemistry of proteins, peptides, and amino acids. In Eglinton, G. & M. T. J. Murphy (eds.) Organic Geochemistry. Springer-Verlag, Berlin: 438–463.

Hare, P. E., 1974a. Amino acid dating: A history and an evaluation. Museum Appl. Sci. Center for Archaeol. Newsletter 10: 4–7.

Hare, P. E., 1974b. Amino acid dating of bone: The influence of water. Carnegie Inst. Washington Yrbk 73: 576–581.

Hare, P. E., 1980. Organic geochemistry of bones, and its relation to the survival of bone in the natural environment. In Behrensmeyer, A. K. & A. P. Hill (eds.) Fossils in the Making. Chicago University Press, Chicago: 208–219.

Hare, P. E., 1986. Detection limits for amino acids in environmental samples. Abstr. Am. Chem. Soc. 191: Anyl 94.

Hare, P. E., 1995. Identifying thermally altered bone from archaeological and paleoanthropological contexts: Biochemical criteria. Abstr. Am. Chem. Soc. 209: Hist 13.

Hare, P. E. & P. H. Abelson, 1966. Racemization of amino acids in fossil shells. Carnegie Inst. Washington Yrbk. 66: 526–528.

Hare, P. E. & P. H. Abelson, 1970. Dynamics of nitrogenous organic matter in marine sediments. Geol. Soc. Am., Abstr. 2: 569.

Hare, P. E. & M. L. S. Estep, 1984. Nitrogen and carbon stable isotope ratios in amino acids of modern and fossil collagens. Abstr. Am. Chem. Soc. 187: Geoc 45.

Hare, P. E. & T. C. Hoering, 1972. Separation of amino acid optical isomers by gas chromatography. Carnegie Inst. Washington Yrbk. 72: 690–692.

Hare, P. E. & R. M. Mitterer, 1965. Nonprotein amino acids in fossil shells. Carnegie Inst. Washington Yrbk. 65: 362–364.

Hare, P. E. & R. M. Mitterer, 1967. Laboratory simulation of amino acid diagenesis in fossils. Carnegie Inst. Washington Yrbk. 67: 205–207.

Hare, P. E., H. F. Turnbill & R. E. Taylor, 1978. Amino acid dating of Pleistocene fossil materials: Olduvai Gorge, Tanzania. In Freeman, L. G. (ed.) Views of the Past: Essays in Old World Prehistory and Paleoanthropology. Mouton, Den Haag, 445 pp.

Hare, P. E., T. C. Hoering & K. King Jr. (eds.), 1980. Biogeochemistry of Amino Acids. Wiley, New York, 558 pp.

Hare, P. E., A. S. Brooks, D. M. Helgren, J. E. Kokis & K. Kuman, 1984. Aminostratigraphy: The use of ostrich eggshell in dating the Middle Stone Age at ≠Gi, Botswana. Geol. Soc. Am., Abstr. 16: A529.

Hare, P. E., P. A. St. John & M. H. Engel, 1985. Ionic exchange separation of amino acids. In Barrett, G. C. (ed.) Chemistry and Biochemistry of Amino Acids. Chapman Hall, London: 415–425.

Hare, P. E., G. A. Goodfriend, A. S. Brooks, J. E. Kokis & D. W. Von Endt, 1993. Chemical clocks and thermometers: Diagenetic reactions of amino acids in fossils. Carnegie Inst. Washington Yrbk. 92: 80–85.

Hare, P. E., D. W. Von Endt & J. E. Kokis, 1997. Protein and amino acid diagenesis dating. In Taylor, R. E. & M. J. Aitkin (eds.) Chronometric Dating in Archaeology. Plenum, New York: 261–296.

Hare, P. E., M. L. Fogel, T. C. Hoering, A. D. Mitchell & T. W. Stafford, 1987. Stable isotopes in amino acids from fossil bones and their relationship to ancient diets. Abstr. Am. Chem. Soc. 193: Hist 17.

Harmon, R. S., R. Garrett, N. Kriausakul, L. S. Land, G. J. Larson, R. M. Mitterer, M. Rowe, H. P. Schwarcz & H. L. Vacher, 1983. U-series and amino acid racemization geochronology of Bermuda: Implications for eustatic sea level fluctuation over the past 250,000 years. Palaeogeog. Palaeoclim. Palaeoecol. 44: 41–70.

Hassan, A. A. & P. E. Hare, 1978. Amino acid analysis in radiocarbon dating of bone collagen. Archaeological Chemistry 2 (American Chemical Society, Advances in Chemistry 171): 109–116.

Haugen, J. E. & H. P. Sejrup, 1992. Isoleucine epimerization kinetics in the shell of *Arctica islandica*. Norsk Geologisk Tidsskrift 72: 171–180.

Haugen, J. E., N. B. Vogt & H. P. Sejrup, 1989. Chemotaxonomy of Quaternary benthic foraminifera using amino acids. J. Foram. Res. 19: 38–51.

Hearty, P. J., 1986. An inventory of Last Interglacial (*sensu lato*) age deposits from the Mediterranean Basin: A study of isoleucine epimerization and U-series dating. Zeitschrift für Geomorphologie 62: 51–69.

Hearty, P. J. & P. Aharon, 1988. Amino acid chronostratigraphy of Late Quaternary coral reefs: Huon Peninsula, New Guinea, and the Great Barrier Reef, Australia. Geology 16: 579–583.

Hearty, P. J., G. H. Miller, C. E. Stearns & B. J. Szabo, 1986. Aminostratigraphy of Quaternary shorelines in the Mediterranean Basin. Geol. Soc. Am. Bull. 97: 850–858.

Hearty, P. J., H. L. Vacher & R. M. Mitterer, 1992. Aminostratigraphy and ages of Pleistocene limestones of Bermuda. Geol. Soc. Am. Bull. 104: 471–480.

Heck, S. D., W. S. Faracii, P. R. Kelbaugh, N. A. Saccomano, P. F. Thadeio & R. A. Volkmann, 1996. Post-translational amino acid epiemerization: Enzyme catalyzed isomerization of amino acid residues in peoptide chains. Proc. Natl. Acad. Sci. USA 93: 4036–4039.

Hedges, J. I. & P. E. Hare, 1987. Amino acid adsorption by clay minerals in distilled water. Geoch. Cosmoch. Acta 51: 255–259.

Hedges, R. E. M. & C. J. A. Wallace, 1980. The survival of protein in bone. In Hare, P. E., T. C. Hoering & K. King Jr. (eds.) Biogeochemistry of Amino Acids. Wiley, New York: 35–40.

Helfman, P. M., 1976. Aspartic acid racemization in dentine as a measure of aging. Nature 262: 279–281.

Helfman, P. M. & J. L. Bada, 1975. Aspartic acid racemization in tooth enamel from living humans. Proc. Natl. Acad. Sci. USA 72: 2891–2894.

Helfman, P. M., J. L. Bada & M. Y. Shou, 1977. Considerations on the role of aspartic acid racemization in the aging process. Gerontology 23: 419–425.

Henry, D. O. & G. H. Miller, 1992. The implications of amino acid racemization dates of Levantine Mousterian deposits in southern Jordan. Paleorient 18: 45–52.

Hicock, S. R. & N. W. Rutter, 1986. Pleistocene aminostratigraphy of the Georgia Depression, southwest British Columbia. Can. J. Earth Sci. 23: 383–392.

Ho, T. Y., 1965. The amino acid composition of bone and tooth proteins in the late Pleistocene. Proc. Natl. Acad. Sci. USA 54: 25–31.

Hoang, C. T. & P. J. Hearty, 1989. A comparison of U-series disequilibrium dates and amino acid epimerization ratios between corals and marine mollusks of Pleistocene age. Chem. Geol. 79: 317–323.

Hoering, T. C., 1980. The organic constituents in fossil mollusc shells. In Hare, P. E., T. C. Hoering & K. King Jr. (eds.) Biogeochemistry of Amino Acids. Wiley, New York: 193–201.

Hollin, J. T. & P. J. Hearty, 1990. South Carolina interglacial sites and Stage 5 sea levels. Quat. Res. 33: 1–17.

Hollin, J. T., F. L. Smith, J. T. Renouf & D. G. Jenkins, 1993. Sea-cave temperature measurements and amino acid geochronology of British late Pleistocene sea stands. J. Quat. Sci. 8: 359–364.

Hsu, J. T., E. M. Leonard & J. F. Wehmiller, 1989. Aminostratigraphy of Peruvian and Chilean Quaternary marine terraces. Quat. Sci. Rev. 8: 255–262.

Hughes, O. L., C. R. Harrington, J. A. Janssens, J. V. Matthews, R. E. Morlan, N. W. Rutter & C. E. Schweger, 1981. Upper Pleistocene stratigraphy, paleoecology, and archaeology of the northern Yukon interior, Eastern Beringia. 1. Bonnet Plume Basin. Arctic 34: 329–365.

Ike, D., J. L. Bada, P. M. Masters, G. Kennedy & J. C. Vogel, 1979. Aspartic acid racemizition dating of an early Milling Stone horizon burial in California. Am. Antiq. 44: 524–530.

Jansen, E. & H. P. Sejrup, 1987. Stable isotope stratigraphy and amino acid epimerization for the last 2.4 m.y. at Site 610, Hole 610 and Hole 610A. Init. Rep. Deep Sea Drilling Proj. 94: 879–888.

Jardine, W. B., D. D. Harkness, P. D. W. Haughton, D. Q. Bowen, J. H. Dickson & G. A. Sykes, 1988. A Late Devensian Interstadial site at Sourlie, near Irvine, Strathclyde. Scot. J. Geol. 24: 288–295.

Jope, E. M., 1980. Ancient bone and plant proteins: the molecular state of preservation. In Hare, P. E., T. C. Hoering & K. King Jr. (eds.) Biogeochemistry of Amino Acids. Wiley, New York: 23–33.

Johnson, B. J., 1995. Stable Isotope Biogeochemistry of Ostrich Eggshell and its Application to Late Quaternary Paleoenvironmental Reconstruction in South Africa. Unpubl. Ph.D. thesis, University of Colorado, Boulder.

Johnson, B. J. & G. H. Miller, 1997. Archaeoligcal applications of amino acid racemization. Archaeometry 39: 265–267.

Johnson, B. J., G. H. Miller, M. L. Fogel & P. B. Beaumont, 1997. The determination of late Quaternary paleoenvironments at Equus Cave, South Africa, using stable isotopes and amino acid racemization in ostrich eggshell. Palaeogeog. Palaeoclim. Palaeoecol. 136: 121–137.

Julg, A., R. Lafont & G. Perinet, 1987. Mechanisms of collagen racemizaion in fossil bones: Applications to absolute dating. Quat. Sci. Rev. 6: 25–28.

Katz, B. J. & E. H. Man, 1980. The effects and implications of ultrasonic cleaning on the amino acid geochemistry of foraminifera. In Hare, P. E., T. C. Hoering & K. King Jr. (eds.) Biogeochemistry of Amino Acids. Wiley, New York: 215–222.

Katz, B. J., C. G. A. Harrison & E. H. Man, 1983. Geothermal and other effects on amino acid racemization in selected Deep Sea Drilling Project cores. Org. Geoch. 5: 151–156.

Kaufman, D. S., 1992. Aminostratigraphy of Pliocene-Pleistocene high-sea-level deposits, Nome coastal plain and adjacent nearshore area, Alaska. Geol. Soc. Am., Bull. 104: 40–52.

Kaufman, D. S., 2000. Amino acid racemization in ostracodes. In Goodfriend, G. A., M. J. Collins, M. L. Fogel, S. A. Macko & J. F. Wehmiller (eds.) Perspectives in Amino Acid and Protein Geochemistry. Oxford University Press, New York: 145–160.

Kaufman, D. S. & G. H. Miller, 1992. Overview of amino acid geochronology. Comp. Bioch. Physiol. B102: 199–204.

Kaufman, D. S. & G. H. Miller, 1995. Isoleucine epimerization and amino acid composition in molecular weight separations Pleistocene *Genyornis* eggshell. Geoch. Cosmoch. Acta 59: 2757–2765.

Kaufman, D. S. & H. P. Sejrup, 1995. Isoleucine epimerization in the high molecular-weight fraction of Pleistocene *Arctica*. Quat. Sci. Rev. 14: 337–350.

Kaufman, D. S., G. H. Miller & J. T. Andrews, 1992. Amino acid composition as a taxonomic tool for molluscan fossils: An example from Pliocene-Pleistocene Arctic marine deposits. Geoch. Cosmoch. Acta 56: 2445–2453.

Kaufman, D. S., R. C. Walter, J. K. Brigham-Grette & D. M. Hopkins, 1991. Middle Pleistocene age of the Nome River Glaciation, northwestern Alaska. Quat. Res. 36: 277–293.

Kaufman, D. S., S. L. Forman, P. D. Lea & Wobus, C. W., 1996. Age of pre-Late Wisconsinan glacial estuarine sedimentation, Bristol Bay, Alaska. Quat. Res. 45: 59–72.

Keen, D. H., J. T. Andrews & R. S. Harmon, 1981. U-series and amino acid dates from Jersey. Nature 289: 162–164.

Keil, R. G., E. Tsamakis, J. I. Hedges, 2000. Preservation and diagenesis of amino acids and proteins. In Goodfriend, G. A., M. J. Collins, M. L. Fogel, S. A. Macko & J. F. Wehmiller (eds.) Perspectives in Amino Acid and Protein Geochemistry. Oxford University Press, New York: 69–82.

Kennedy, G. L., LaJoie, K. R. & J. F. Wehmiller, 1982. Aminostratigraphy and faunal correlations of Late Quaternary marine terraces, Pacific coast, USA. Nature 299: 545–547.

Kessels, H. J. & G. Dungworth, 1980. Necessity of reporting amino acid compositions of fossil bones where racemization analyses are used for geochronological applications: Inhomogeneities of *D/L* amino acids in fossil bones. In Hare, P. E., T. C. Hoering & K. King Jr. (eds.) Biogeochemistry of Amino Acids. Wiley, New York: 527–541.

Kimber, R. W. L. & C. V. Griffin, 1987. Further evidence of the complexity of the racemization process in fossil shells with implications for amino acid racemization dating. Geoch. Cosmoch. Acta 51: 839–846.

Kimber, R. W. L. & C. V. Griffin, 2000. Interpretation of D/L amino acid data from agricultural soils. In Goodfriend, G. A., M. J. Collins, M. L. Fogel, S. A. Macko & J. F. Wehmiller (eds.) Perspectives in Amino Acid and Protein Geochemistry. Oxford University Press, New York: 195–201.

Kimber, R. W. L. & P. E. Hare, 1992. Wide range of racemization of amino acids in peptides from human fossil bone and its implications for amino acid racemization dating. Geoch. Cosmoch. Acta 56: 739–743.

Kimber, R. W. L. & A. R. Milnes, 1984. The extent of racemization of amino acids in Holocene and Pleistocene marine mollusks in southern South Australia: Preliminary data on a time framework for calcrete formation. Austr. J. Earth Sci. 31: 279–286.

Kimber, R. W. L., C. V. Griffin & A. R. Milnes, 1986. Amino acid racemization dating: Evidence of apparent reversal in aspartic acid racemization with time in shells of *Ostrea*. Geoch. Cosmoch. Acta 50: 1159–1161.

Kimber, R. W. L., N. M. Kennedy & A. R. Milnes, 1994. Amino acid racemization dating of a 140,000 year old tephra-loess-palaeosol sequence on the Mamaku Plateau near Tororua New Zealand. Austr. J. Earth Sci. 41: 19–26.

King, K., Jr., 1980a. Applications of amino acid biogeochemistry for marine sediments. In Hare, P. E., T. C. Hoering & K. King Jr. (eds.) Biogeochemistry of Amino Acids. Wiley, New York: 377–391.

King, K., Jr, 1980b. γ-carboxyglutamic acid in fossil bone. In Hare, P. E., T. C. Hoering & K. King Jr. (eds.) Biogeochemistry of Amino Acids. Wiley, New York: 491–501.

King, K., Jr. & J. L. Bada, 1979. Effect of *in situ* leaching on amino acid racemisation rates in fossil bone. Nature 281: 135–137.

King, K., Jr. & P. E. Hare, 1972a. Amino acid composition of planktonic foraminifera: A paleobiochemical approach to evolution. Science 175: 1461–1463.

King, K., Jr. & P. E. Hare, 1972b. Amino acid composition of the test as a taxonomic character for living and fossil planktonic foraminifera. Micropaleontology 18: 285–293.

King, K., Jr. & C. Neville, 1977. Isoleucine epimerization for dating marine sediments: The importance of analyzing monospecific foraminiferal samples. Science 195: 1333–1335.

Knudsen, K. L. & H. P. Sejrup, 1988. Amino acid geochronology of selected interglacial sites in the North Sea area. Boreas 17: 347–354.

Knudsen, K. L & H. P. Sejrup, 1993. Pleistocene stratigraphy in the Devils Hole area, central North Sea foraminiferal and amino-acid evidence. J. Quat. Sci. 8: 1–14.

Kokis, J. E., 1988. Protein Diagensis Dating of *Struthio camelus* Eggshell: An Upper Pleistocene Dating Technique. Unpublished M.Sc. thesis, George Washington University, Washington, 154 pp.

Kowalewski, M., G. A. Goodfriend & K. W. Flessa, 1998. High resolution estimates of temporal mixing within shell beds: The evils and virtues of time-averaging. Paleobiology 24: 287–304.

Kriausakul, N. & R. M. Mitterer, 1980. Some factors affecting the epimerization of isoleucine in peptides and proteins. In Hare, P. E., T. C. Hoering & K. King Jr. (eds.) Biogeochemistry of Amino Acids. Wiley, New York: 283–296.

Kriausakul, N. & R. M. Mitterer, 1983. Epimerization of COOH-terminal isoleucine in fossil dipeptides. Geoch. Cosmoch. Acta 47: 963–966.

Krings, M., A. Stone, R. W. Schmitz, H. Krainitzki, M. Stoneking & S. Pääbo, 1997. Neanderthal DNA sequences and the origin of modern humans. Cell 90: 19–30.

Kvenvolden, K. A., 1975. Advances in the geochemistry of amino acids. Ann. Rev. Earth Planet. Sci. 3: 183–212.

Kvenvolden, K. A., 1980. Interlaboratory comparison of amino acid racemization in a Pleistocene mollusk, *Saxidomus giganteus*. In Hare, P. E., T. C. Hoering & K. King Jr. (eds.) Biogeochemistry of Amino Acids. Wiley, New York: 223–232.

Kvenvolden, K. & E. Peterson, 1973. Amino acids in late Pleistocene bone from Rancho LaBrea, California. Geol. Soc. Am., Abstr. 5: 704–705.

Kvenvolden, K. A., D. J. Blunt & H. E. Clifton, 1979a. Amino acid racemization in Quaternary shell deposits at Willpa Bay, Washington. Geoch. Cosmoch. Acta 43: 1505–1520.

Kvenvolden, K. A., D. J. Blunt, S. W. Robinson & G. Bacon, 1979b. Amino acid dating of an arcaheological site on Amaknak Island, Alaska. Geol. Soc. Am., Abstr. 11: A462.

Kvenvolden, K. A., D. J. Blunt & H. E. Clifton, 1981. Age estimations based on amino acid racemization: Reply. Geoch. Cosmoch. Acta 45: 265–267.

Kvenvolden, K. A., D. P. Glavin & J. L. Bada, 2000. Extra-terrestrial amino acids in the Murchison Meteorite: Re-evaluation after thirty years. In Goodfriend, G. A., M. J. Collins, M. L. Fogel, S. A. Macko & J. F. Wehmiller (eds.) Perspectives in Amino Acid and Protein Geochemistry. Oxford University Press, New York: 7–14.

Lafont, R., G. Perinet, F. Bazile & M. Icole, 1984. Racémisations d'acides aminés d'ossements fossiles du Paléolithique supérieure languedocien. C. r. Acad. Sci., Paris D299: 447–450.

LaJoie, K. R., J. F. Wehmiller & G. L. Kennedy, 1980a. Inter- and intrageneric trends in apparent racemization kinetics of amino acids in Quaternary mollusks. In Hare, P. E., T. C. Hoering & K. King Jr. (eds.) Biogeochemistry of Amino Acids. Wiley, New York: 305–340.

Lajoie, K. R., E. Peterson & B. A. Gerow, 1980b. Amino acid bone dating: a feasibility study, South San Francisco Bay Region, California. In Hare, P. E., T. C. Hoering & K. King Jr. (eds.) Biogeochemistry of Amino Acids. Wiley, New York: 477–489.

Lauritzen, S. E., J. E. Haugen, R. Lovlie & H. Gilje-Nielsen, 1994. Geochronological potential of isoleucine epimerization in calcite speleothems. Quat. Res. 41: 52–58.

Lee, C., J. L. Bada & E. Peterson, 1976. Amino acids in modern and fossil woods. Nature 259: 183–186.

Leonard, E. M. & J. F. Wehmiller, 1992. Low uplift rates and terrace reoccupation inferred from mollusk aminostratigraphy, Coquimbo Bay area, Chile. Quat. Res. 38: 246–259.

Liardon, R. & R. Jost, 1981. Racemization of free and protein-bound amino acids in strong mineral acid. Int. J. Pept. Protein Res. 18: 500–505.

Liverman, D. G. E., N. R. Catto & N. W. Rutter, 1989. Laurentide Glaciation in west central Alberta: A single (Late Wisconsinan) event. Can. J. Earth Sci. 26: 266–274.

Livingstone, D. A., 1980. Environmental changes in the Nile headwaters. In Williams, M. A. J. & H. Faure (eds.) The Sahara and the Nile. Balkema, Rotterdam: 339–360.

Lowenstam, H. A., 1980. Bioinorganic constituents of hard parts. In Hare, P. E., T. C. Hoering & K. King Jr. (eds.) Biogeochemistry of Amino Acids. Wiley, New York: 3–16.

Lowenstein, J. M., 1980. Immunospecificity of fossil collagens. In Hare, P. E., T. C. Hoering & K. King Jr. (eds.) Biogeochemistry of Amino Acids. Wiley, New York: 41–51.

Lycke, A. K., Mangerud, J. & H. P. Sejrup, 1992. Late Quaternary foraminiferal stratigraphy from Western Svalbard. Boreas 21: 271–288.

Macko, S. A. & A. E. Aksu, 1986. Amino acid epimerization in planktonic foraminifera suggests slow sedimentation rates for Alpha Ridge, Arctic Ocean. Nature 322: 730–732.

Macko, S. A., M. L. Fogel, P. E. Hare & T. C. Hoering, 1987. Isotopic fractionation of nitrogen and carbon in the synthesis of amino acids by microorganisms. Chem. Geol. 65: 79–92.

Macko, S. A., M. E. Uhle, M. H. Engel & V. Andrusevich, 1997. Stable nitrogen isotope analysis of amino acid enantiomers by gas chromatography combustion-isotope ratio mass spectrometry. Analyt. Chem. 69: 926–929.

Magee, J. W., J. M. Bowler, G. H. Miller & D. L. G. Williams, 1995. Stratigraphy, sedimentology, chronology, and paleohydrology of Quaternary lacustrine deposits at Madigan Gulf, Lake Eyre, South Australia. Palaeogeog. Palaeoclim. Palaeoecol. 113: 3–42.

Mahaney, W. C. & N. W. Rutter, 1989. Amino acid D/L ratio distribution in two Late Pleistocene soils in the Afro-alpine of Mt. Kenya, East Africa. Catena 16: 205–214.

Mahaney, W. C. & N. W. Rutter, 1992. Relative ages of the moraines of the Daijia Shan, northwestern China. Catena 19: 179–191.

Mahaney, W. C., M. G. Boyer & N. W. Rutter, 1986. Evaluation of amino acid composition as a geochronometer in buried soils on Mount Kenya, East Africa. Géog. phys. Quat. 40: 171–184.

Mahaney, W. C., R. G. V. Hancock, N. W. Rutter & K. Sanmugadas, 1991. Stratigraphy and geo-chemistry of a sequence of Quaternary paleosols in the Lower Teleki Valley, Mount Kenya, East Africa Implications for interregional correlations. J. Quat. Sci. 6: 245–256.

Manley, W. F., G. H. Miller & J. Gzywczynski, 2000. Kinetics of aspartic acid racemization in *Mya* and *Hiatella*: Modeling age and paleotemprature of high-latitude Quaternary mollusks. In Goodfriend, G. A., M. J. Collins, M. L. Fogel, S. A. Macko & J. F. Wehmiller (eds.) Perspectives in Amino Acid and Protein Geochemistry. Oxford University Press, New York: 202–218.

Mangerud, J., S. Gulliksen, E. Larsen, O. Longva, G. H. Miller, H. P. Sejrup & E. Sonstegaard, 1981. A Middle Weichselian ice-free period in western Norway: The Alesund Interstadial. Boreas 10: 447–462.

Martin, R. E., J. F. Wehmiller, M. S. Harris & W. D. Liddel, 1996. Comparative taphonomy of bivalues and formainifer from the Holocene tidal flat sdeiments, Bahia La Choya, Sonora, Mexico (Northern Gulf of Mexico): Taphonomic grades and temporal resolution. Paleobiology 22: 80–90.

Masters, P. M., 1982. Amino acid racemization in structural preoteins. In Raff, M. E. & M. D. Schneider (eds.) Biological Markers of Ageing. NIH, Washington: 120–137.

Masters, P. M., 1985. *In vivo* decompositon of phosphoserine and serine in noncollagenous protein from human dentine. Calc. Tissue Int. 37: 236–241.

Masters, P. M., 1986. Amino acid racemization dating. In Zimmerman, M. R. & J. L. Angel (eds.) Dating and Age Determination of Biological Materials. Cromm, Helm, Longwood, London: 39–58.

Masters, P. M., 1987. Preferential preservation of non-collagenous protein during bone diagenesis: Implications for chronometric and stable isotope measurements. Geochim. Cosmochim. Acta 51: 3209–3214.

Masters, P. M. & J. L. Bada, 1978a. Racemization of isoleucine in fossil molluscs from Indian middens and interglacial terraces in Southern California. Earth & Planet. Sci. Lett. 37: 173–183.

Masters, P. M. & J. L. Bada, 1978b. Amino acid racemization dating of bone and shell. Archaeological Chemistry 2 (American Chemical Society, Advances in Chemistry 171): 117–138.

Masters, P. M., J. L. Bada & J. S. Sigler Jr., 1977. Aspartic acid racemisation in the human lens during aging and in cataract formation. Nature 268: 71–73.

Matsu'ura, S. & N. Ueta, 1980. Fraction dependent variation of aspartic acid racemization age in fossil bone. Nature 286: 883–884.

McCartan, L., D. F. Belknap, B. W. Blackwelder, N. Kriausakul, R. M. Mitterer, J. R. Owens, B. J. Szabo & J. F. Wehmiller, 1982. Comparison of amino acid racemization geochronometry with lithostratigraphy, biostratigraphy, uranium-series coral dating, and magnetostratigraphy in the Atlantic coastal plain of the southeastern United States. Quat. Res. 18: 337–359.

McCoy, W. D., 1981. Quaternary Stratigraphy of the Bonneville and Lahontan Basins. Unpublished Ph.D. thesis, University of Colorado, Boulder.

McCoy, W. D., 1987a. Quaternary aminostratigraphy of the Bonneville Basin, Western United States. Geol. Soc. Am. Bull. 98: 99–112.

McCoy, W. D., 1987b. The precision of amino acid geochronology and paleothermometry. Quat. Sci. Rev. 6: 43–54.

McCoy, W. D., 1988. Amino acid racemization in fossil non-marine ostracod shells: A potential tool for the study of Quaternary stratigraphy, chronology, and palaeotemperature. In de Deckker, P., J. P. Colin & J. P. Peypouquet (eds.) Ostracods in the Earth Sciences. Monash University, Clayton: 219–229.

McCulloch, E. A. & G. G. Smith, 1976. Correction in the glacial-postglacial temperature difference computed from amino acid racemization. Science 191: 102–103.

McMenamin, M. A. S., D. J. Blunt, K. A. Kvenvolden, L. F. Marcus, S. E. Miller & R. R. Pardi, 1982. Amino acid geochemistry of fossil bones from the Rancho La Brea asphalt deposit, California. Quat. Res. 18: 174–183.

Meyer, V. R., 1991. Amino acid racemization: A tool for dating. American Chemical Society, Advances in Chemistry 471: 217–227.

Meyer, V. R., 1992. Amino acid racemization: A tool for fossil dating. Chemtech 22: 412–417.

Mienert, J., J. T. Andrews & J. D. Milliman, 1992. The east Greenland continental margin 65 °N since the last deglaciation: Changes in sea floor properties and ocean circulation. Mar. Geol. 106: 217–238.

Millard, A. R., 2000. A model for the effect of weaning on nitrogen isotope ratios in humans. In Goodfriend, G. A., M. J. Collins, M. L. Fogel, S. A. Macko & J. F. Wehmiller (eds.) Perspectives in Amino Acid and Protein Geochemistry. Oxford University Press, New York: 51–60.

Miller, B. B., N. K. Bleuer & W. D. McCoy, 1987. Stratigraphic potential of amino acid ratios in Pleistocene terrestrial gastropods: An example from west central Indiana, USA. Boreas 16: 133–138.

Miller, B. B., W. D. McCoy, W. J. Wayne & C. S. Brockman, 1992. Ages of the Whitewater and Fairhaven Tills in southwestern Ohio and southeastern Indiania. In Clark, P. U. & P. D. Lea (eds.) The Last Interglacial-Glacial Tranisition in North America. Geol. Soc. Am. Spec. Pap. 270: 89–98.

Miller, G. H., 1982. Quaternary depositional episodes, Western Spitsbergen, Norway: Aminostratigraphy and glacial history. Arct. Alp. Res. 14: 321–340.

Miller, G. H., 1985. Aminostratigraphy of Baffin Island shell-bearing deposits. In Andrews, J. T. (ed.) Quaternary Environments: Baffin Island, Baffin Bay, and West Greenland. Allen & Unwin, London: 394–427.

Miller, G. H. & P. B. Beaumont, 1989. Dating the Middle Stone Age at Border Cave, South Africa, by the epimerization of isoleucine in ostrich eggshell. Geol. Soc. Am., Abstr. 21: A235.

Miller, G. H. & J. K. Brigham-Grette, 1989. Amino acid geochronology: Resolution and precision in carbonate fossils. Quat. Int. 1: 111–128.

Miller, G. H. & P. E. Hare, 1975. Use of amino acid reactions in some arctic marine shells: Stratigraphic and geochronological indicators. Carnegie Inst. Washington Yrbk. 74: 612–617.

Miller, G. H. & P. E. Hare, 1980. Amino acid chronology: Integrity of the carbonate matrix and potential of molluscan fossils. In Hare, P. E., T. C. Hoering & K. King Jr. (eds.) Biogeochemistry of Amino Acids. Wiley, New York: 415–443.

Miller, G. H. & J. Mangerud, 1985. Aminostratigraphy of European marine interglacial deposits. Quat. Sci. Rev. 4: 215–278.

Miller, G. H. & A. Rosewater, 1995. The potential of isoleucine epimerization in fish otoliths to date Pleistocene archaeological sites. Geol. Soc. Am. Abstr. 27: A415.

Miller, G. H., J. T. Andrews & S. K. Short, 1977. The last interglacial-glacial cycle, Clyde foreland, Baffin Island, NWT: Stratigraphy, biostratigraphy, and chronology. Can. J. Earth Sci. 14: 2824–2857.

Miller, G. H., J. T. Hollin & J. T. Andrews, 1979. Aminostratigraphy of UK Pleistocene deposits. Nature 281: 539–543.

Miller, G. H., H. P. Sejrup, J. Mangerud & B. G. Andersen, 1983. Amino acid ratios in Quaternary molluscs and foraminifera from Western Norway: Correlation, geochronology, and paleotemperature estimates. Boreas 12: 107–124.

Miller, G. H., R. Paskoff & C. E. Stearns, 1986. Amino acid geochronology of Pleistocene littoral deposits in Tunisia. Zeitschrift für Geomorphologie, Supplementband 62: 197–207.

Miller, G. H., D. Q. Bowen, J. K. Brigham, A. J. T. Bull, T. Linick, J. Mangerud, H. P. Sejrup & D. Sutherland, 1987. Racemization-derived Late Devensian temperature reduction in Scotland. Nature 326: 593–595.

Miller, G. H., P. J. Hearty & J. A. Stravers, 1988. Ice sheet dynamics and glacial history of southeasternmost Baffin Island and outermost Hudson Strait. Quat. Res. 30: 116–136.

Miller, G. H., S. J. Lehman, S. L. Forman & H. P. Sejrup, 1989a. Glacial history and marine environmental change during the last interglacial-glacial cycle, Western Spitsbergen, Svalbard. Boreas 18: 273–296.

Miller, G. H., F. Wendorf, R. D. Ernst, R. Schild, A. E. Close, I. Friedman & H. P. Schwarcz, 1989b. Dating lacustrine episodes in the eastern Sahara by the epimerization of isoleucine in ostrich eggshells. Palaeogeog. Palaeoclim. Palaeoecol. 84: 175–189.

Miller, G. H., P. B. Beaumont, A. J. T. Jull & B. Johnson, 1992. Pleistocene geochronology and palaeothermometry from protein diagenesis in ostrich eggshells: Implications for the evolution of modern humans. Phil. Trans. R. Soc. London 337: 149–157.

Miller, G. H., J. W. Magee & A. J. T. Jull, 1997. Low-latitude glacial cooling in the southern hemisphere from amino acid racemization in emu eggshells. Nature 385: 241–244.

Miller, G. H., J. W. Magee, B. J. Johnson, M. L. Fogel, N. A. Spooner, M. T. McCulloch & L. K. Ayliffe, 1999. Pleistocene extinction of *Genyornis newtoni*: Human impact on Australian megafauna. Science 283: 205–208.

Miller, G. H., C. P. Hart, E. B. Roark & B. J. Johnson, 2000. Isoleucine epimerization in eggshells of the flightless Australian birds *Genyorins* and *Dromaius*. In Goodfriend, G. A., M. J. Collins, M. L. Fogel, S. A. Macko & J. F. Wehmiller (eds.) Perspectives in Amino Acid and Protein Geochemistry. Oxford University Press, New York: 161–181.

Miller, S. L. & J. L. Bada, 1988. Submarine hot springs and the origin of life. Nature 334: 609–611.

Milnes, A. R., R. W. Kimber & S. E. Phillips, 1987. Studies in calcareous aeolian landscapes of southern Australia. In Liu, T. (ed.) Aspects of Loess Research. China Ocean Press, Hong Kong: 130–139.

Mirecki, J. & B. B. Miller, 1994. Amino acid racermization dating of some southeastern coastal plain sites. Abstr. Am. Chem. Soc. 188: Geoc 19.

Mirecki, J. & J. F. Wehmiller, 1984. Aminostratigraphic correlation and geochronology of two Quaternary loess localities, Central Mississippi Valley. Quat. Res. 41: 289–297.

Mirecki, J., J. F. Wehmiller & A. F. Skinner, 1995. Geochronology of Quaternary coast plain deposits, southeastern Virginia, USA. J. Coastal Res. 11: 1135–1144.

Mitchell, L. & G. B. Curry, 1997. Diagenesis and survival of intracrystalline amino acids in fossil and recent mollusk shells. Palaeontology 40: 855–874.

Mitterer, R. M., 1974. Pleistocene stratigraphy in southern Florida based on amino acid diagenesis in fossil *Mercenaria*. Geology 2: 425–428.

Mitterer, R. M., 1975. Ages and diagenetic temperatures of Pleistocene deposits of Florida based on isoleucine epimerization in *Mercenaria*. Earth Planet. Sci. Lett. 28: 275–282.

Mitterer, R. M., 1993. The diagenesis of proteins and amino acids in fossil shells. In Engel, M. H. & S. A. Macko (eds.) Organic Geochemistry Principles and Applications. Plenum, New York: 739–753.

Mitterer, R. M. & N. Kriausakul, 1984. Comparison of rates and degrees of isoleucine epimerization in dipeptides and tripeptides. Org. Geoch. 7: 91–98.

Mitterer, R. M. & N. Kriausakul, 1989. Calculation of amino acid racemization ages based upon apparent parabolic kinetics. Quat. Sci. Rev. 8: 353–357.

Moir, M. E., 1985. Peptide Racemization Kinetics. Unpublished Ph.D. thesis, University of Alberta, Edmonton, 352 pp.

Moir, M. E. & R. J. Crawford, 1988. Model studies of competing hydrolysis and epimerization of some tetrapeptides of interest in amino acid racemization studies in geochronology. Can. J. Chem. 66: 2903–2913.

Muhs, D. R., H. M. Kelsey, G. H. Miller, G. L. Kennedy, J. F. Whelan & G. W. McInelly, 1990. Age estimates and uplift rates for Late Pleistocene marine terraces: Southern Oregon portion of the Cascadia Fore-Arc. J. Geophys. Res. 95: 6685–6698.

Murray-Wallace, C. V., 1991. Quaternary correlations using amino acid racemization: The Sydney Basin Province in a global context. Advances in the Study of the Sydney Basin: Proceedings of the Symposium 25: 170–177.

Murray-Wallace, C. V., 1993. A review of the application of the amino acid racemisation reaction to archaeological dating. The Artifact 16: 19–26.

Murray-Wallace, C. V., 2000. Quaternary coastal aminostratigraphy: Australian data in a global context. In Goodfriend, G. A., M. J. Collins, M. L. Fogel, S. A. Macko & J. F. Wehmiller (eds.) Perspectives in Amino Acid and Protein Geochemistry. Oxford University Press, New York: 279–300.

Murray-Wallace, C. V. & A. P. Belperio, 1994. Identification of remanie fossils using amino acid racemization. Alcheringa 18: 219–227.

Murray-Wallace, C. V. & R. P. Bourman, 1990. Direct radiocarbon calibration for amino acid racemization dating. Austr. J. Earth Sci. 37: 365–367.

Murray-Wallace, C. V. & A. Goede, 1995. Aminostratigraphy and electron spin resonance dating of Quaternary coastal neotectonism in Tasmania and the Bass Straight Islands. Austr. J. Earth Sci. 42: 51–67.

Murray-Wallace, C. V. & R. W. L. Kimber, 1987. Evaluation of the amino acid racemization reaction in studies of Quaternary marine sediments in South Australia. Austr. J. Earth Sci. 34: 279–292.

Murray-Wallace, C. V. & R. W. L. Kimber, 1989. Quaternary marine aminostratigraphy, Perth Basin, Western Australia. Austr. J. Earth Sci. 36: 553–568.

Murray-Wallace, C. V. & R. W. L. Kimber, 1993. Further evidence for apparent 'parabolic' kinetics in Quaternary molluscs. Austr. J. Earth Sci. 40: 313–317.

Murray-Wallace, C. V., A. P. Belperio, A. V. Gostin & J. H. Cann, 1993. Amino acid racemization and radiocarbon dating of interstadial marine strata (Oxygen Isotope Stage 3), Gulf of St. Vincent, South Australia. Mar. Geol. 110: 83–92.

Murray-Wallace, C. V., A. P. Belperio & R. W. L. Kimber, 1988a. Holocene paleotemperature studies using amino acid racemization reactions. Austr. J. Earth Sci. 35: 575–577.

Murray-Wallace, C. V., A. P. Belperio, R. W. L. Kimber & A. V. Gostin, 1988b. Aminostratigraphy of the Last Interglacial in Southern Australia. Search 19: 33–36.

Murray-Wallace, C. V., A. P. Belperio, K. Picker & R. W. L. Kimber, 1991. Coastal aminostratigraphy of the Last Interglaciation in Southern Australia. Quat. Res. 35: 63–71.

Murray-Wallace, C. V., M. A. Ferland, P. S. Roy & A. Sollar, 1996. Unraveling patterns of reworking in lowstand shelf deposits using amino acid racemization and radiocarbon dating. Quat. Sci. Rev. 15: 685–697.

Nakamiya, T., H. Mizuno, T. Meguro, H. Ryono & K. Takanamik, 1976. Antibacterial activity of lauryl ester of D, L-lysine. J. Fermen. Tech. 54: 369–373.

Nichol, S. L. & C. V. Murray-Wallace, 1992. A partially preserved last interglacial estuarine fill: Narrawallee Inlet, New South Wales. Austr. J. Earth Sci. 39: 545–553.

Nelson, A. R., 1982. Aminostratigraphy of Quaternary marine and glaciomarine sediments, Qivitu Peninsula, Baffin Island. Can. J. Earth Sci. 19: 945–961.

Nielsen, E., C. Causse, A. Morgan, A. V. Morgan, R. J. Mott & N. W. Rutter, 1986. Stratigraphy, paleoecology, and glacial history of the Gillam Area, Manitoba. Can. J. Earth Sci. 23: 1641–1661.

Oches, E. A. & W. D. McCoy, 1989. Amino acid paleotemperature estimates for the Last Glacial Maximum, Lower Mississippi Valley. Geol. Soc. Am. Abstr. 21: A210.

Oches, E. A. & W. D. McCoy, 1995. Aminostratigraphic evaluation of conflicting age estimates for the "Young Loess" of Hungary. Quat. Res. 44: 160–170.

Oches, E. A. & W. D. McCoy, 1996. Amino acid geochronology applied to the correlation and dating of central European loess deposits. Quat. Sci. Rev. 14: 767–782.

Oches, E. A., W. D. McCoy & D. Gnieser, 2000. Aminostratigraphic correlation of loess-paleosol sequences across Europe. In Goodfriend, G. A., M. J. Collins, M. L. Fogel, S. A. Macko &

J. F. Wehmiller (eds.) Perspectives in Amino Acid and Protein Geochemistry. Oxford University Press, New York: 331–348.

Ogino, T. & H. Ogino, 1988. Application to forensic odontology of aspartic acid racemization in unerupted and supernumerary teeth. J. Dent. Res. 67: 1319–1322.

Oldale, R. N., D. F. Belknap, B. W. Blackwelder, T. M. Cronin, E. C. Spiker, B. J. Szabo, P. C. Valentine & J. F. Wehmiller, 1982. Stratigraphy, structure, absolute age, and paleontology of the Upper Pleistocene deposits at Sankaty Head, Nantucket Island, Massachusetts. Geology 10: 246–252.

O'Neal, M. L., J. F. Wehmiller & W. L. Newell, 2000. Amino Acid geochronology of Quaternary coastal terraces on the northern margin of Delaware Bay, southern New Jersey, U.S.A. In Goodfriend, G. A., M. J. Collins, M. L. Fogel, S. A. Macko & J. F. Wehmiller (eds.) Perspectives in Amino Acid and Protein Geochemistry. Oxford University Press, New York: 301–319.

Osborn, G., R. Thomas, W. D. McCoy, B. B. Miller & A. Smith, 1991. Significance of a molluscan fauna to the physiographic history of the Calgary area, Alberta. Can. J. Earth Sci. 28: 1948–1955.

Ohtani, S. & K. Yamamoto, 1991. Age estimation using the racemization of amino acids in human dentin. J. Forensic Sci. 36: 792–800.

Ohtani, S. & K. Yamamoto, 1992. Estimation of age from a tooth by means of racemization of an amino acid, espeically aspartic acid: Comparison of enamel and dentin. J. Forensic Sci. 37: 1061–1067.

Ohtani, S., Y. Matsushima, H. Ohhira & A. Watanabe, 1995. Age-related changes in D-aspartic acid of rat teeth. Growth, Development & Aging 59: 55–61.

Oviatt, C. G., W. D. McCoy & R. G. Reider, 1987. Evidence for a shallow Early or Middle Wisconsin age lake in the Bonneville Basin, Utah. Quat. Res. 27: 248–262.

Perinet, G., R. Lafont & N. Petit-Marie, 1975. Prémiers résultats concernant les essais de fossilization d'un ossement en laboratoire. C. r. Acad. Sci., Paris 280D: 1531–1533.

Perinet, G., R. Lafont & J. Amic, 1977. Sur l'analyse thermique differentielle de la fraction organique de l'os. C. r. Acad. Sci., Paris 284D: 1927–1930.

Petit, M. G., 1974. Racemization rate constant for protein-bound aspartic acid in woodrat middens. Quat. Res. 4: 340–345.

Pickard, J., D. A. Adamson, R. K. Dell, D. M. Harwood, G. H. Miller & P. G. Quilty, 1986. Early Pliocene marine sediments in the Vestfold Hills, East Antarctica: Implications for coastline, ice sheet and climate. S. Afr. J. Sci. 82: 520–521.

Pickard, J., P. G. Quilty, G. H. Miller, D. A. Adamson, R. K. Dell & D. M. Harwood, 1988. Early Pliocene marine sediments, coastline, and climate of East Antarctica. Geology 16: 158–161.

Poinar, H. N., M. Höss, J. L. Bada & S. Pääbo, 1996. Amino acid racemization and the preservation of ancient DNA. Science 272: 864–866.

Pollard, A. M. & C. Heron, 1996. Amino acid stereochemistry and the first Americans. In Pollard, A. M. & C. Heron (eds.) Archaeological Chemistry. RCS Paperbacks, London: 271–301.

Pollock, G. E., Cheng, C. N. & Cronin, S. E., 1977. Determination of the D and L isomers of some protein amino acids present in soils. Analyt. Chem. 49: 2–7.

Powell, E. N., A. Logan, R. J. Stanton Jr., D. J. Davies & P. E. Hare, 1989. Estimating time since death from the free amino acid content of the mollusc shell: A measure of time-averaging in modern death assemblages? Description of the technique. Palaios 4: 16–31.

Prior, C. A., P. J. Ennis, E. A. Noltmann, P. E. Hare & R. E. Taylor, 1986. Variations in D/L aspartic acid ratios in bones of similar age and temperature history. In Olin, J. S. & M. J. Blackman (eds.) Proceedings of the 24th Archaeometry Symposium. Smithsonian Institution, Washington: 487–498.

Qian, Y. R., M. H. Engel, G. A. Goodfriend & S. A. Macko, 1995. Abundance and stable carbon isotope composition of amino acids in molecular weight fractions of fossil and artificially aged mollusk shells. Geoch. Cosmoch. Acta 59: 1113–1124.

Reasoner, M. A. & N. W. Rutter, 1988. Late Quaternary history of the Lake O'Hara region, British Columbia: An evaluation of sedimentation rates and bulk amino acid ratios in lacustrine records. Can. J. Earth Sci. 25: 1037–1048.

Ritz, S. & H. W. Schutz, 1993. Aspartic acid racemization in intervertebral disks as an aid to postmortem estimation of age at death. J. Forensic Sci. 38: 633–640.

Ritz, S., H. W. Schutz & C. Peper, 1993. Postmortem estimation of age at death in based on aspatric acid racemization in dentin: Its applicability for root dentin. Int. J. Legal Med. 105: 289–293.

Robbins, L. L., S. Andrews & P. H. Ostrom, 2000. Characterization of ultrastructural and biochemical characteristics of modern and fossil shells. In Goodfriend, G. A., M. J. Collins, M. L. Fogel, S. A. Macko & J. F. Wehmiller (eds.) Perspectives in Amino Acid and Protein Geochemistry. Oxford University Press, New York: 108–119.

Roof, S., 1997. Comparison of isoleucine epimerizationa dn leaching potential in the molluskan genera Astarte, Macoma, and Mya. Geoch. Cosmoch. Acta 61: 5325–5333.

Rosen, S. A. & G. A. Goodfriend, 1993. An early date for Gaza ware from the Northern Negev. Palestine Exploration Quaterly 125: 143–148.

Rutter, N. W. & B. A. Blackwell, 1995. Amino acid racemization dating. In Rutter, N. W. & N. R. Catto (eds.) Dating Methods for Quaternary Deposits. Geological Association of Canada, St. John's, GEOtext 2: 125–164.

Rutter, N. W. & R. J. Crawford, 1984. Utilizing wood in amino acid dating. In Mahaney, W. C. (ed.) Quaternary Dating Methods. Elsevier, Amsterdam: 195–209.

Rutter, N. W. & C. K. Vhalos, 1988. Amino acid racemization kinetics in wood: Applications to geochronology and geothermometry. In Easterbrook, D. J. (ed.) Dating Quaternary Sediments. Geol. Soc. Am., Spec. Pap. 227: 51–67.

Rutter, N. W., R. J. Crawford & R. D. Hamilton, 1979. Dating methods of Pleistocene deposits and their problems IV: Amino acid racemization dating. Geosci. Canada 6: 122–139.

Rutter, N. W., R. J. Crawford & R. D. Hamilton, 1980. Correlation and relative age dating of Quaternary strata in the continuous permafrost zone of northern Yukon with D/L ratios of aspartic acid of wood, freshwater molluscs and bone. In Hare, P. E., T. C. Hoering & K. King Jr. (eds.) Biogeochemistry of Amino Acids. Wiley, New York: 463–475.

Rutter, N. W., B. A. Blackwell, R. J. Crawford & Z. Karapissides, 1981. Relative dating of bones by D/L ratios of amino acids from la Caune de l'Arago á Tautavel, France. In de Lumley, H. & J. Labeyrie (eds.) Datations Absolus et les Analyses Isotopiques en Préhistoire: Méthodes et Limites. Prétirage. CNRS, Paris.

Rutter, N. W., E. J. Schnack, J. Del Rio, J. L. Fasano, F. I. Isla & U. Radtke, 1989. Correlation and dating of Quaternary littoral zones along the Patagonian coast, Argentina. Quat. Sci. Rev. 8: 213–234.

Rutter, N. W., U. Radtke & E. J. Schnack, 1990. Comparison of ESR and amino acid data in correlating and dating Quaternary shorelines along the Patagonia coast, Argentina. J. Coastal Res. 6: 391–411.

Rutter, N. W., F. W. Bachhuber & G. Lyons, 1993. The use of seeds in aminostratigraphy of a Wisconsin paleolimnological record from central New Mexico, USA. Sveriges Geologiska Undersóking 81: 307–312.

Saettem, J., D. A. R. Poole, K. L. Ellingsen & H. P. Sejrup, 1992. Glacial geology of outer Bjornoyrenna, southwestern Barents Sea. Mar. Geol. 103: 15–51.

Saettem, J., D. A. R. Poole, H. P. Sejrup & K. L. Ellingsen, 1991. Glacial geology of outer Bjornoyrenna, western Barents Sea: Preliminary results. Norsk Geologisk Tidsskrift 71: 173–177.

Sahel, N., D. Deutsch & E. Gil-Av, 1993. Racemization of aspartic acid in the extracellular matrix proteins of primary and secondary dentin. Calc. Tissue Int. 53: 103–110.

Saint-Martin, B., 1991. Etude des inluence geochemiques sure la vitess de racemization des acides amines dans les ossements fossiles. C. r. Acad. Sci., Paris 313: 655–660.

Saint-Martin, B. & A. Julg, 1991. Influence of interaction between asymmetry centers on the kinetics of racemization. J. Molec. Struct. 251: 375-383.

Salvigsen, O. & R. Nydel, 1981. The Weichselian Glaciation in Svalbard before 15,000 BP. Boreas 10: 433–446.

Schroeder, R. A. & J. L. Bada, 1973. Glacial-postglacial temperature difference deduced from aspartic acid racemization in fossil bones. Science 182: 479–482.

Schroeder, R. A. & J. L. Bada, 1975. Absence of β-alanine and γ-aminobutyric acid in cleaned foraminiferal shells: Implications for use as a chemical criterion to indicate removal of non-indigenous amino acid contaminants. Earth Planet. Sci. Lett. 25: 274–278.

Schroeder, R. A. & J. L. Bada, 1976. Review of the geochemical applications of the amino acid racemization reaction. Earth Sci. Rev. 12: 347–391.

Schroeder, R. A. & J. L. Bada, 1977. Kinetics and mechanism of decomposition of threonine in fossil foraminifera. Geoch. Cosmoch. Acta 41: 1087–1095.

Schroeder, R. A. & J. L. Bada, 1978. Aspartic acid racemization in late Wisconsinan Lake Ontario sediments. Quat. Res. 9: 193–204.

Scott, W. E., W. D. McCoy, M. Rubin & R. R. Shroba, 1983. Reinterpretation of the exposed record of the last 2 cycles of Lake Bonneville, western United States. Quat. Res. 20: 261–285.

Sejrup, H. P. & J. E. Haugen, 1992. Foraminiferal amino acid stratigraphy of the Nordic Seas: Geological data and pyrolysis experiments. Deep Sea Res. 39A: S603–S623.

Sejrup, H. P., G. H. Miller & K. Rokoengen, 1984a. Isoleucine epimerization in Quaternary benthonic foraminifera from the Norwegian Continental Shelf: A pilot study. Mar. Geol. 56: 227–239.

Sejrup, H. P., J. K. Brigham-Grette, D. M. Hopkins, R. Lovlie & G. H. Miller, 1984b. Amino acid epimerization implies rapid sedimentation rates in Arctic Ocean cores. Nature 310: 772–775.

Sejrup, H. P., I. Aarseth, K. L. Ellingsen, E. Reither, E. Jansen, R. Lovlie, A. Bent, J. K. Brigham-Grette, E. Larsen & M. Stoker, 1987. Quaternary stratigraphy of the Gladen area, central North Sea: A multidisciplinary study. J. Quat. Sci. 2: 35–58.

Sejrup, H. P., J. K. Nagy & J. K. Brigham-Grette, 1989. Foraminiferal stratigraphy and amino acid geochronology of Quaternary sediments in the Norwegian Channel, northern North Sea. Norsk Geologisk Tidsskrift 69: 111–124.

Sejrup, H. P., I. Aarseth & H. Haflidason, 1991. The Quaternary succession in the northern North Sea. Mar. Geol. 101: 103–111.

Serban, A., M. H. Engel & S. A. Macko, 1986. The distribution, stereochemistry and stable isotopic composition of amino acid constituents of fossil and modern mollusk shells. Org. Geoch. 13: 1123–1129.

Shimoyama, A. & K. Harada, 1984. An age determination of an ancient burial mound man by apparent racemization reaction of aspartic acid in tooth dentine. Chemical Letters 10: 1661–1664.

Shimoyama, A., S. Muraoka, G. Krampitz & K. Harada, 1989. Amino acids recovered from fossil egg shells of dinosaurs. Chemistry Letters 1989: 505–508.

Sigleo, A. C., P. E. Hare & G. R. Helz, 1983. The amino acid composition of estuarine colloidal material. Estu. Coastal & Shelf Sci. 17: 87–96.

Skelton, R. R., 1982. A test of the applicability of amino acid racemization dating for Northern California. Am. J. Phys. Anth. 57: 228–229.

Smith, G. G. & R. C. Evans, 1980. The effect of structure and conditions on the rate of racemization of free and bound amino acids. In Hare, P. E., T. C. Hoering & K. King Jr. (eds.) Biogeochemistry of Amino Acids. Wiley, New York: 257–282.

Smith, G. G. & G. S. Reddy, 1986. The effect of metal ions on the racemization of hydroxy amino acids. Abstr. Am. Chem. Soc. 192: Orgn 235.

Smith, G. G. & T. Sivakua, 1983. Mechanism of the racemization of amino acids: Kinetics of racemization of arylglycines. J. Org. Chem. 48: 627–634.

Smith, G. G., K. M. Williams & D. M. Wonacott, 1978. Factors affecting the racemization of amino acids and their significance to geochronology. J. Org. Chem. 43: 1–5.

Smith, G. G., A. Khatib & G. S. Reddy, 1982. The effect of nickel (II) ion on racemization of amino acids. Abstr. Am. Chem. Soc. 183: Orgn 201.

Soper, T. S., M. Manning, P. A. Marotte & C. T. Walsh, 1977. Inactivation of bacterial D-amino acid transaminase by β-chloro-D-alanine. J. Bioch. Res. 252: 1571–1575.

Stafford, T. W. Jr., P. E. Hare, L. Currie, A. J. T. Jull & D. J. Donahue, 1990. Accuracy of North American human skeleton ages. Quat. Res. 34: 111–120.

Stanley, D. J. & G. A. Goodfriend, 1997. Recent subsidence of the northern Suez Canal. Nature 388: 335–336.

Steinberg, S. M. & J. L. Bada, 1981. Diketopiperazine formation during investigations of amino acid racemization in dipeptides. Science 213: 544–545.

Steinberg, S. M. & J. L. Bada, 1983. The diagenetic production of alpha-ketoacids in heated and fossil *Chione* shells. Geoch. Cosmoch. Acta 47: 1481–1486.

Steinberg, S. M., J. L. Bada & P. M. Masters, 1984. The racemization of free and peptide-bound serine and aspartic acid at 100 °C as a function of pH: Implications for *in vivo* racemization. Bioorg. Chem. 12: 349–355.

Sykes, G. A., 1984. Comment on Andrews, J. T., G. H. Miller, and W. W. Shilts, "Multiple deglaciations of the Hudson Bay Lowlands, Canada, since deposition of the Missinaibi (Last Interglacial questionable) Formation". Quat. Res. 22: 247–252.

Sykes, G. A., 1988. Amino acids on ice. Chemistry in Britain 24: 235–240, 244.

Sykes, G. A., M. J. Collins & D. I. Walton, 1995. The significance of a geochemically isolated (intra-crystalline) organic fraction within biominerals. Org. Geoch. 23: 1059–1066.

Szabo, B. J., J. T. Andrews, G. H. Miller & M. Stuiver, 1981. Comparison of uranium-series, radiocarbon, and amino acid data from marine mollusks, Baffin Island, Arctic Canada. Geology 9: 451–457.

Szabo, B. J., J. T. Andrews, G. H. Miller & M. Stuiver, 1982. Comparison of uranium-series, radiocarbon, and amino acid data from marine mollusks, Baffin Island, Arctic Canada: Reply. Geology 10: 216.

Taylor, R. E., 1983. Nonconcordance of radiocarbon and amino acid racemization deduced age estimates on human bone. Radiocarbon 25: 647–654.

Taylor, R. E., 1987. Radiocarbon Dating: An Archaeological Perspective. Academic Press, New York.

Taylor, R. E., L. A. Payen, B. Gerow, D. J. Donahue, T. H. Zabel, A. J. T. Jull & P. E. Damon, 1983. Middle Holocene age of the Sunnyvale human skeleton. Science 220: 1271–1273.

Taylor, R. E., D. J. Donahue, T. H. Zabel, P. E. Damon & A. J. T. Jull, 1984. In Lambert, J. B. (ed.) Archaeological Chemistry 3 (American Chemical Society, Advances in Chemistry): 333–340.

Taylor, R. E., L. A. Payen, C. A. Prior, P. J. Slota Jr., R. Gillespie, J. A. J. Gowlett, R. E. B. Hedges, A. J. T. Jull, T. H. Zabel, D. J. Donahue & R. E. Berger, 1985a. Major revisions in the Pleistocene age assignments for North American human skeletons by C-14 accelerator mass spectrometry: None older than 11,000 C-14 years B.P. Am. Antiq. 50: 136–140.

Taylor, R. E., L. A. Payen & P. J. Slota Jr., 1985b. Impact of ^{14}C determinations on considerations of the anitquity of Homo sapiens in the Western Hemisphere. Nucl. Instr. Meth.

Taylor, R. E., P. J. Ennis, P. J. Slota Jr. & L. A. Payen, 1989. Non-age-related variations in aspartic acid racemization in bone from a radiocarbon-dated late Holocene archaeological site. Radiocarbon 31: 1048–1056.

Taylor, R. E., P. E. Hare, C. A. Prior, D. L. Kirner, L. J. Wan & R. R. Burky, 1995a. Geochemical criteria for thermal alteration of bone. J. Archaeol. Sci. 22: 115–119.

Taylor, R. E., P. E. Hare, C. A. Prior, D. L. Kirner, L. J. Wan & R. R. Burky, 1995b. Radiocarbon dating of biochemically characterized hair. Radiocarbon 37: 319–330.

Teece, M. A., N. Tuross, W. J. Kress, P. M. Peterson, G. Russell & M. L. Fogel, 2000. Preservation of amino acids in museum herbarium samples. In Goodfriend, G. A., M. J. Collins, M. L. Fogel, S. A. Macko & J. F. Wehmiller (eds.) Perspectives in Amino Acid and Protein Geochemistry. Oxford University Press, New York: 83–87.

Teller, J., N. W. Rutter & N. Lancaster, 1990. Sedimentology and history of Cenozoic lake deposits in the northern Namib Sand Sea, Namibia. Quat. Sci. Rev. 9: 343–364.

Torres, T., J. F. Llamas, L. Canoria, F. J. Coello, P. Garcia-Alonso & J. E. Ortiz, 2000b. Amino-stratigraphy of two Pleistocene marine sequences from the Mediterranean coast of Spain: Cabo de Huertas (Alicante) and Garrucha (Almería). In Goodfriend, G. A., M. J. Collins, M. L. Fogel, S. A. Macko & J. F. Wehmiller (eds.) Perspectives in Amino Acid and Protein Geochemistry. Oxford University Press, New York: 108–119.

Torres, T., J. F. Llamas, L. Canoira & P. Garcia-Alonso, 2000b. Aspartic acid racemization and protein preservation in the dentine of Pleistocene European bear teeth. In Goodfriend, G. A., M. J. Collins, M. L. Fogel, S. A. Macko & J. F. Wehmiller (eds.) Perspectives in Amino Acid and Protein Geochemistry. Oxford University Press, New York: 349–355.

Toscano, M. A. & L. L. York, 1992. Quaternary stratigraphy and and sea-level history of the U.S. middle Atlantic Coastal Plain. Quat. Sci. Rev. 11: 301–325.

Towe, K. M., 1980. Preserved organic ultrastructure: An unreliable indicator for Paleozoic amino acid geochemistry. In Hare, P. E., T. C. Hoering & K. King Jr. (eds.) Biogeochemistry of Amino Acids. Wiley, New York: 65–74.

Turban-Just, S. & S. Schramm, 1998. Stable carbon and nitrogen isotope ratios of individual amino acieds give new insights into bone collagen degradation. Bull. Soc. Géol. France 169: 109–114.

Turekian, K. K. & J. L. Bada, 1972. The dating of fossils bones. In Bishop, W. W. & J. K. Miller (eds.) Calibration of Hominid Evolution. University of Toronto Press, Toronto: 45–82.

Tuross, N., D. R. Eyre, M. E. Holtrop, M. J. Glimcher & P. E. Hare, 1980. Collaagen in fossil bones. In Hare, P. E., T. C. Hoering & K. King Jr. (eds.) Biogeochemistry of Amino Acids. Wiley, New York: 53–63.

Uhle, M. E., S. A. Macko, H. J. Spero, M. H. Engel & D. W. Lea, 1997. Sources of carbon and nitrogen in modern planktonic foraminifera: The role of algal symbionts as determined by bulk and compound-specific stable isotopic analyses. Org. Geoch. 27: 103–113.

van Duin, A. C. T. & M. J. Collins, 1998. The effects of conformational constraints on aspartic acid racemization. Org. Geoch. 29: 1227–1232.

Vander Borsch, C. C., J. L. Bada & D. L. Schwebel, 1980. Amino acid racemization dating of Late Quaternary strandline events of the coastal plain sequence near Robe, southeastern Australia. R. Soc. S. Austr. Trans. 104: 167–170.

Vauchskii, Y. P., N. A. Aksenova, M. G. Ryzhov & A. A. Velts, 1982. Polymeric catalysts of racemization of optically active amino acids. J. Appl. Chem. USSR 55: 1501–1507.

Vincent, J. S., S. Occhietti, N. W. Rutter, G. Lortie, J. P. Guilbaurt & B. de Boutray, 1983. Late Quaternary record of the Duck Hawk Bluffs, Banks Island, Canadian Arctic Archipelago. Can. J. Earth Sci. 20: 1694–1712.

Vlahos, C. K., 1985. Amino Acid Racemization Analyses in Wood. Unpublished M.Sc. thesis, University of Alberta, Edmonton.

Von Endt, D. W., 1979. Techniques of amino acid dating. In Humphrey, R. L. & D. Stanford (eds.) Pre-Llano Cultures of the Americas: Paradoxes and Possibilities. Anthropological Society of Washington, Washington: 71–100.

Von Endt, D. W., 1980. Protein hydrolysis and amino acid racemization in sized bone. In Hare, P. E., T. C. Hoering & K. King Jr. (eds.) Biogeochemistry of Amino Acids. Wiley, New York: 297–306.

Waite, E. R. & M. J. Collins, 2000. The interpretation of aspartic acid racemization of dentine proteins. In Goodfriend, G. A., M. J. Collins, M. L. Fogel, S. A. Macko & J. F. Wehmiller (eds.) Perspectives in Amino Acid and Protein Geochemistry. Oxford University Press, New York: 182–194.

Warnke, D. A., D. J. Blunt & G. E. Pollock, 1980. Enantiomeric ratios of amino acids in southern-ocean silicious oozes. In Hare, P. E., T. C. Hoering & K. King Jr. (eds.) Biogeochemistry of Amino Acids. Wiley, New York: 183–189.

Wehmiller, J. F., 1977. Amino acid studies of the Del Mar, California, midden site: Apparent rate constants, ground temperature models and chronological implications. Earth Planet. Sci. Lett. 37: 184–196.

Wehmiller, J. F., 1980. Intergenetic differences in apparent racemization kinetics in mollusks and foraminifera: Implications for models of diagenetic racemization. In Hare, P. E., T. C. Hoering & K. King Jr. (eds.) Biogeochemistry of Amino Acids. Wiley, New York: 341–355.

Wehmiller, J. F., 1981. Kinetic model options for interpretation of amino acid enantiomerization ratios in Quaternary mollusks: Comments. Geoch. Cosmoch. Acta 45: 261–264.

Wehmiller, J. F., 1982. A review of amino acid racemization studies in Quaternary mollusks: Strati-graphic and chronological applications in coastal and interglacial sites, Pacific and Atlantic coasts, United States, United Kingdom, Baffin Island, and tropical islands. Quat. Sci. Rev. 1: 83–120.

Wehmiller, J. F., 1984a. Relative and absolute dating of Quaternary mollusks with amino acid racemization: evaluation, applications, and questions. In Mahaney, W. C. (ed.) Quaternary Dating Methods. Elsevier, Amsterdam: 171–193.

Wehmiller, J. F., 1984b. Intertaboratory comparison of amino acid enantiomeric ratios in fossil Pleistocene mollusks: Quat. Res. 22: 109–120.

Wehmiller, J. F., 1986. Amino acid racemization geochronology. In Hurford, A. J., E. Jaeger & J. A. M. Ten Cate (eds.) Dating Young Sediments. University of Bern 16: 139–158.

Wehmiller, J. F., 1993. Applications of organic geochemistry for Quaternary research: Amino-stratigraphy and aminochronology. In Engel, M. H. & S. A. Macko (eds.) Organic Geochemistry Principles and Applications. Plenum, New York: 755–783.

Wehmiller, J. F. & D. F. Belknap, 1982. Amino acid age estimates, Quaternary Atlantic Coastal Plain: Comparison with U-series, biostratigraphy, and paleomagnetic control. Quat. Res. 18: 311–336.

Wehmiller, J. F. & P. E. Hare, 1970. Amino acid diagenesis in fossil carbonates. Geol. Soc. Am. Abstr. 2: 718.

Wehmiller, J. F. & P. E. Hare, 1971. Racemization of amino acids in marine sediments. Science 173: 907–911.

Wehmiller, J. F. & G. H. Miller, 1990. Amino acid racemization geochronology. In Edwards, T. W. D. (ed.) Examples and Critiques of Quaternary Dating Methods. AMQUA/CANQUA Short Course 3: 1–72.

Wehmiller, J. F. & G. H. Miller, 1998. Amino acid racemization. In Noller, J. S., J. M. Sowers & W. F. Lettis (eds.) Quaternary Geochronology: Applications to Quaternary Geology and Paleoseismology. US Nuclear Regulatory Commission, Washington, NUREG 5562: 307–361.

Wehmiller, J. F., K. R. Lajoie, K. A. Kvenvolden, E. Peterson, D. F. Belknap, G. L. Kennedy, W. O. Aldicott, J. G. Vedder & R. W. Wright, 1977. Correlation and Chronology of the Pacific Coast Marine Terraces of the Continental United States by Amino Acid Stereochemistry: Technique Evaluation, Relative Ages, Kinetic Models, Ages, and Geological Implications. US Geol. Survey, Open File Rep. 77–680, 196 pp.

Wehmiller, J. F., D. F. Belknap, B. S. Boulton, J. E. Mirecki, S. D. Rahaim & L. L. York, 1988. A review of the aminostratigraphy of Quarternary mollusks from the United States Atlantic Coastal Plain sites. In Easterbrook, D. J. (ed.) Dating Quaternary Sediments. Geol. Soc. Am. Spec. Pap. 227: 69–110.

Wehmiller, J. F., L. L. York, D. F. Belknap & S. W. Snyder, 1992. Theoretical correlations and lateral discontinuities in the Quaternary aminostratigraphic record of the U.S. Atlantic coastal plain. Quat. Res. 38: 275–291.

Wehmiller, J. F., L. L. York & M. L. Bart, 1995. Amino acid racemization geochronology of reworked Quaternary mollusks on US Atlantic coast beaches: Implications for chronostratigraphy, taphonomy, and coast sediment rarnsport. Mar. Geol. 124: 303–337.

Wehmiller, J. F., H. A. Strecher, III, L. L. York & I. Friedman, 2000. The thermal environment of fossils: Effective ground temperatures at aminostratigraphic sites on the U.S. Atlantic coastal plain. In Goodfriend, G. A., M. J. Collins, M. L. Fogel, S. A. Macko & J. F. Wehmiller (eds.) Perspectives in Amino Acid and Protein Geochemistry. Oxford University Press, New York: 219–250.

Weiner, S. & H. A. Lowenstam, 1980. Well-preserved fossil mollusk shells: Characterization of mild diagenetic processes. In Hare, P. E., T. C. Hoering & K. King Jr. (eds.) Biogeochemistry of Amino Acids. Wiley, New York: 95–114.

Weiner, S., Z. Kustanovich, E. Gil-Av & W. Traub, 1980. Dead Sea Scrolls parchments: Unfolding the collagen molecules and racemization of aspartic acid. Nature 287: 820–823.

Williams, K. M. & G. G. Smith, 1977. A critical evaluation of the application of amino acid racemization to geochronology and geothermometry. Origins of Life 8: 91–144.

Wendorf, F., A. E. Close, R. Schild, A. Gauthier, H. P. Schwarcz, G. H. Miller, K. Kowalski, H. Krolik, A. Bluszcz, D. Robins, R. Grün & C. McKinney, 1987. Chronology and stratigraphy of the Middle Paleolithic at Bir Tarfawi, Eqypt. In Clark, J. d. (ed.) Culutral Beginnings. Hablet, Bonn: 197–208.

Woodroffe, C. D., C. V. Murray-Wallace, E. A. Bryant, B. Brooke, H. Heijnis & D. M. Price, 1995. Late Quaternary sea-level highstands in the Tasman Sea: Evidence from Lord Howe Island. Mar. Geol. 125: 61–72.

Wyckoff, R. G. W., 1972. The Biochemistry of Animal Fossils. Scientechnica, Bristol.

Wyckoff, R. G. W. & F. D. Davidson, 1976. Pleistocene and dinosaur gelatins. Comp. Bioch. Physiol. 55B: 95–97.

Wyckoff, R. G. W., W. F. McCaughey & A. R. Doberenz, 1964. Amino acid composition of proteins from Pleistocene bones. Bioch. Biophys. Acta 93: 374–377.

Wyckoff, R. G. W., E. Wagner, P. Matter & A. R. Doberenz, 1963. Collagen in fossil bone. Proc. Natl. Acad. Sci. USA 50: 215–218.

York, L. L., T. M. Cronin, T. A. Ager & J. F. Wehmiller, 1989. Stetson Pit, Dare County, North Carolina: An integrated chronologic, faunal, and floral record of subsurface coastal Quaternary sediments. Palaeogeog. Palaeoclim. Palaeoecol. 72: 115–132.

York, L. L. & J. F. Wehmiller, 1992. Aminostretigraphic results from Cape Lookout, N.C., and their relation to the preserved Quaternary marine record of SE North Carolina. Sed. Geol. 80: 279–291.

Zhao, M. X. & J. L. Bada, 1989. Extraterrestrial amino acids in Cretaceous-Tertiary boundary sediments at Stevns Klint, Denmark. Nature 339: 463–465.

Zumberge, J. E., M. H. Engel & B. Nagy, 1980. Amino acids in Bristlecone pine: An evaluation of factors affecting racemization rates and paleothermometry. In Hare, P. E., T. C. Hoering & K. King Jr. (eds.) Biogeochemistry of Amino Acids. Wiley, New York: 503–525.

16. TEPHROCHRONOLOGY

C. S. M. TURNEY (c.turney@qub.ac.uk)
J. J. LOWE
Centre for Quaternary Research
Geography Department
Royal Holloway
University of London
Egham, Surrey
TW20 0EX, UK

Keywords: Tephra extraction; ash; rhyolitic; basaltic; andesitic; glass shards; microtephra; time-parallel marker horizons; geochronology; volcanic eruptions.

Introduction

Tephrochronology provides time-parallel marker horizons that allow precise correlation between environmental and climatic records of the recent geological past (Westgate & Gorton, 1981; Sarna-Wojcicki, 2000). Few (if any) geochronological techniques can provide the precision offered by tephrochronology. The virtually instantaneous atmospheric deposition of tephra following an eruption can often lead to clear tephra layers in a wide range of depositional environments, including, for example, lakes mires, peats, soils, loess, marine sediments and glacier ice masses. Here we focus on records in lake sediments and their unparalleled potential for testing hypotheses concerning the degree of synchroneity of environmental events reflected in stratigraphical records. The term tephra encompasses a wide range of airborne pyroclastic material ejected during a volcanic eruption, including blocks and bombs (>64 mm), lapilli (2–64 mm) and ash (<2 mm) (Thorarinsson, 1944). These, in turn, may comprise one or more of the following lithologies (Lowe & Hunt, 2001):

- rock pieces or 'lithics', including pumice and scoria;

- glassy material;

- crystals of felsic ('light' — e.g., plagioclase, K-feldspars) and mafic ('heavy' — e.g., ferromagnesian minerals, such as micas, amphiboles and pyroxenes) composition.

All of the above may be found in a single, discrete tephra horizon within limnological deposits. The distribution of active volcanoes throughout the world is widespread and associated with plate margins or intra-plate hotspot activity. Most of these have been active during the Quaternary (last 2.6 Myr), though the activity of many of these volcanic

451

centres can be traced over a much longer geological period. There has therefore been a long, complex history of tephra ejection and deposition and, as a result, many areas of the world have the potential for the application of tephrochronological studies. The first attempts to utilise tephra horizons for correlation and geochronological control were based in New Zealand (Berry, 1928; Oliver, 1931; Grange, 1931). These were soon followed by detailed tephrochronological studies in South America (Auer, 1932), Japan (Uragami et al., 1933) and Iceland (Bjarnason & Thorarinsson, 1940). Once its potential was established, tephrochronology was rapidly adopted as a key tool for the correlation and dating of stratigraphic horizons. Regional tephrochronological frameworks (the recognition of one or more 'marker' tephra layers of known provenance and/or unique geochemistry) have been developed for many parts of the world, including Africa (e.g., Chernet et al., 1998; Pyle, 1999), South America (e.g., Haberle & Lumley, 1998; Samaniego et al., 1998; Rose et al., 1999), North America (e.g., Begét, 1984; Hildreth & Fierstein, 1997; Child et al., 1998; Whitlock et al., 2000), Asia (e.g., Kamata & Kobayashi, 1997; Ponomareva et al., 1998; Machida, 1999), Antarctica (e.g., Smellie, 1999), Australasia (e.g., Froggatt & Lowe, 1990; Lowe et al., 1999; Shane, 2000) and Europe (e.g., Dugmore et al., 1995a; Juvigné et al., 1996; Pilcher & Hall, 1996; Pilcher et al., 1996; Vernet et al., 1998; Di Vito et al., 1999; Guest et al., 1999; Narcisi & Vezzoli, 1999; Haflidason et al., 2000; Turney et al., 2001).

Mechanisms of tephra deposition

The geographical distribution of particles deposited on the Earth's surface following a volcanic eruption depends upon (a) the size of the eruption, (b) the relative proportion of different types and sizes of ejecta, and (c) the direction and strength of the wind during the time the tephra remains in the atmosphere. For instance, following the eruption of the Laacher See Tephra (Eifel, Germany) at ~12,900 BP, three principle ash fans were generated (Fig. 1a), due to differences in wind direction with altitude (van den Bogaard & Schmincke, 1985). In contrast, a plume ejected from Mount St. Helens during the 1980 eruption was unidirectional, spreading as far as 1000 km from the volcanic centre in less than 10 hours (Fig. 1b; Sarna-Wojcicki et al., 1981). The most extensive tephra plumes and associated deposits are generally those associated with plinian and ignimbrite eruptions of large magnitude (Sparks et al., 1997).

Typically, tephra deposits show a decrease in horizon thickness as well as in mean grain size with increasing distance from eruptive sources (Pyle, 1989; Sparks et al., 1992), though this is not always the case, as, for example, is illustrated by the tephra deposited by the Toba volcanic eruption ca. 75,000 years ago (Rose & Chesner, 1987). The larger components of pyroclastic material (e.g., blocks, bombs and lapilli) tend to follow a ballistic trajectory, staying proximal to the eruption site and blanketing the immediate landscape. Conversely, ash particles are frequently erupted much higher into the atmosphere, especially during plinian eruptions, and have a much wider spread. They may account for a greater proportion of the overall tephra fall-out with increasing distance from the volcano, as is the case, for example, in tephra deposited after the Mount St. Helens eruption of 1980 (Carey & Sigurdsson, 1982). The deposition of coarse tephra will typically occur within hours to days following an eruption, largely through density settling (Sparks et al., 1997). In the case of finer ash particles, however, localised meteorological precipitation plays

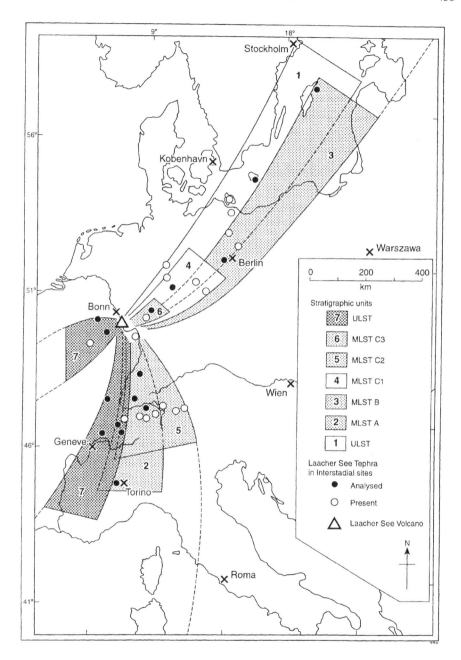

Figure 1a. Tephra fans associated with the Laacher See Tephra, ca. 12.9 ka (van den Bogaard & Schmincke, 1985). Note that the different volcanic eruption, which have different transport directions, are recorded as separate stratigraphic units.

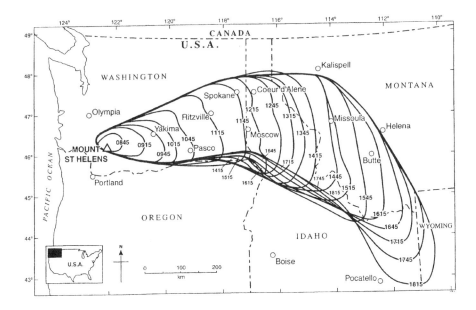

Figure 1b. Tephra fan associated with the Mount. St. Helens eruption, 1980 (Sarna-Wojcicki et al., 1981). Numbers refer to time of day at which fan reached location indication

an increasingly important role with increasing distance from the volcano (Dugmore et al., 1995a). Rain droplets remove light particles from the air column, and this leads to a much more 'patchy' pattern of deposition of ash, with greater concentrations found in areas experiencing heavier and/or more persistent rain. Regardless of the mechanisms involved, the period of time in which fine-grained tephra remains in the atmosphere is at least an order of magnitude lower than that in which fine sediments remain in suspension in most lakes. For all practical purposes, therefore, tephra horizons can be regarded as instantaneous time-parallel 'marker' horizons in palaeolimnological contexts.

Detection of tephra horizons

If located sufficiently proximal to an eruption site, limnological sediments may contain several distinct tephra horizons that can be observed by the naked eye (Fig. 2). A pre-liminary tephrochronological framework can therefore be developed quickly by measures or observations of the varying grain size, colour, mineralogy and thickness of different tephra layers. As an approximate 'rule of thumb', ash horizons comprising of largely basaltic material tend to be black, those of andesitic origin are grey, and rhyolitic material is predominantly white in colour. Colour of ash can, however, be misleading, and thus recourse should be made to geochemical analyses or other diagnostic laboratory tests to confirm the nature and composition of tephra layers (Westgate & Gorton, 1981). In the case of ash deposited more distally from the point of origin, the tephra particles may not be

Tuhua Tephra ——
(7000 BP)

Mamaku Tephra ——
(8050 BP)

Rotoma Tephra ——
(9500 BP)

Opepe Tephra (Unit-E) ——
(10200 BP)

Mangamate Tephra ——
(11750 BP)

Andesitic tephra (Tongariro) ——
(12900 BP)

Waiohau Tephra ——
(13800 BP)

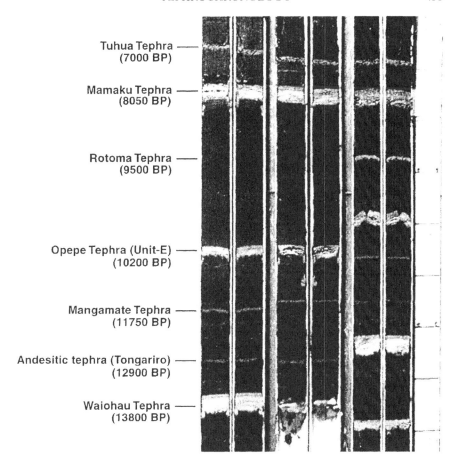

Figure 2. Late-Pleistocene tephra horizons in sediment sequences in Lake Rotomanuka, New Zealand (photo: D. Lowe).

visible to the naked eye and may be extremely low in concentration within lake sediments. More detailed laboratory techniques are therefore required to detect and characterise these 'micro-tephra' horizons (Dugmore et al., 1995a; Pilcher & Hall, 1996; Pilcher et al., 1996; Turney et al., 1997), for which some researchers have suggested the term *cryptotephra* to be more appropriate (Lowe & Hunt, 2001). Examples of glass 'ash-sized' shards from around the world, displaying a wide range of morphological types, can be seen in Figure 3. A more comprehensive description of volcanic ash types is given in Heiken & Wohletz (1985).

A large number of laboratory techniques are available for detecting the occurrence of micro-tephra particles in limnological sediments. Non-destructive detection methods (i.e., methods which leave the host sediment structure undisturbed) include analysis of magnetic properties of sediments (e.g., Oldfield et al., 1980; van den Bogaard et al., 1994; Child et al., 1998; Guerrero et al., 2000), X-ray analysis (e.g., Dugmore & Newton, 1992)

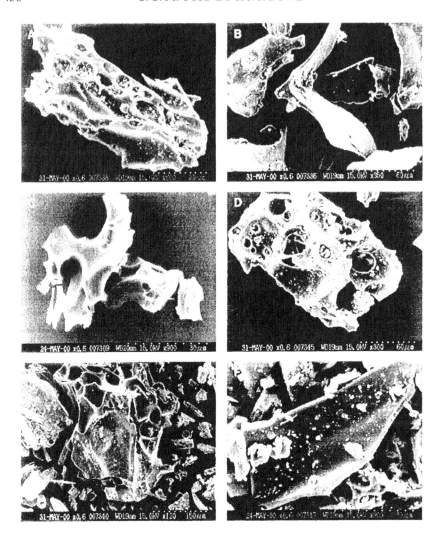

Figure 3. SEM photographs of ash shards from (a) Laacher See Tephra (ca. 12.9 ka, Germany; (b) Vedde Ash (ca. 12.0 ka), Iceland; (c) Hekla 4 (ca. 4.4 ka), Iceland; (d) La Solfatara (ca. 3.7 ka), Italy; (e) Taupo (ca. 1.8 ka), New Zealand; (f) Mount Hudson (1991), Chile (photos taken at the Electron Microscopy Unit, Royal Holloway, University of London).

and measurements of light reflectance of host sediments (e.g., Caseldine et al., 1999). Destructive detection methods (those which disturb the host sediments and/or remove, alter or even destroy the tephra particles) include the preparation of thin sections of the host sediments to reveal microscopically-thin tephra bands (e.g., Merkt et al., 1993), analysis of the bulk geochemistry of the sediments in which the tephra particles occur (e.g., Lowe & Turney, 1997), magnetic separation (Froggatt & Gosson, 1982; Mackie et al., 2001), ashing of organic sediments (Pilcher & Hall, 1992), organic silica digestion (Rose et al., 1996) and floatation of tephra shards (Turney, 1998).

Extraction of tephra shards

Although one or more of the above methods may reveal the presence of micro-tephra particles within deposits, the concentration of tephra shards is often significantly diluted by contemporaneous deposition of mineral sediments of non-volcanic origin, particularly within a limnological context. The detection of tephra in predominantly minerogenic sediments can therefore frequently be difficult to achieve.

One way to detect and quantify concentrations of glass shards in limnological sediments is to use ashing (combustion of the sediments, to remove any organic component — e.g., Pilcher & Hall, 1992) combined with floatation of the mineral residue in a heavy liquid medium, a procedure which has had notable success in the extraction of rhyolitic shards from minerogenic lake sediments (Turney, 1998). In this method, the aim is to use a liquid of known specific gravity, which will float off rhyolitic glass, leaving heavier mineral components to settle. The application of this method is summarised as follows. Contiguous samples of 5 cm vertical thickness are extracted from the sediment column and ashed in a muffle furnace at 550 °C for 1 hour (the use of 30% hydrogen peroxide may be equally effective in reducing or removing the organic components). The recovered material is soaked in 10% HCl overnight to disaggregate the material and dissolve any soluble inorganics (e.g., carbonates). If biogenic silica occurs within the sediments, this can be removed by treatment in 0.3 M NaOH at 80–90 °C for 3 hours (Rose et al., 1996). The surviving material is then sieved through meshes of 80 and 24 μm and soaked in a solution prepared to a specific gravity of 2.3–2.5 g cm^{-3}, the optimum density for recovering rhyolitic glass shards. A suitable, non-hazardous solution for this purpose can be made using sodium polytungstate, for example. The flotants are then examined for the presence of glass microshards using a polarising microscope, the shards being identified on the basis of their distinctive morphology and optical characteristics (including polarisation, vesicularity and the migration of the Becke line into the mounting medium). For those parts of the sediment column in which shards are detected, contiguous samples of 1 cm vertical interval or less can then be extracted (to a constant volume), and the procedures repeated, to pin-point maximal shard concentrations more precisely.

In order to isolate sufficient glass shards for successful geochemical analysis (see below), the shards are usually extracted from lake sediment samples using the acid digestion method described by Dugmore (1989). Although the ashing of sediment samples at furnace temperatures of 550°C should not melt glass shards (Swanson & Begét, 1994), it can result in changes in the alkali concentration (Dugmore et al., 1995a) as well as chemical reactions between glass surfaces and organic sediment components (Pilcher & Hall, 1992). In the case of micro-tephra (cryptotephra) horizons, the density separation ('floatation') method (Turney, 1998) is employed. While suitable for the extraction of rhyolitic shards, this method is less effective at separating out basaltic material. Basaltic tephra has a density (>2.7 g cm^{-3}) which is too close to the density values of many other minerals, but especially to that of quartz, which tends to be the most abundant mineral in minerogenic lake sediments (Turney, 1998; Wastegård et al., 1998). In order to extract basaltic shards, therefore, some alternative means has to be employed, which exploits the physical and/or chemical characteristics of this material. Within minerogenic sediments, the most successful approach has been to utilise flotation (Turney, 1998) followed by magnetic separation (Mackie et al., 2002). Ashing or the application of hydrogen peroxide can produce workable concentrations where the organic content of the sediment samples exceeds ca. 80% (Pilcher & Hall, 1992).

Correlation and dating of tephras: building a tephrochronological framework

The use of tephra layers to define 'time-lines' or 'marker' horizons, traceable between different stratigraphic successions, is only possible if each horizon has unique characteristics which not only aid detection and identification, but also reduce the possibility of invalid correlations between tephras of different ages but with similar physical or chemical properties. Some of the early research in tephrochronology relied upon visible properties of tephra deposits where, for example, colour and characteristic shard morphology were assumed to be sufficiently diagnostic. Examples of tephras with distinctive morphological characteristics (both late-glacial in age and distributed quite widely in north-west Europe) are the Laacher See Tephra, which was extruded from an eruption centre in the Eifel area in Germany (van den Bogaard & Schmincke, 1985), and which has highly vesicular shards (Fig. 3a), and the Vedde Ash, of Icelandic origin, and within which 'butterfly'-shaped (three-winged) shards (Fig. 3b) are frequently encountered (Mangerud et al., 1984; Wohlfarth et al., 1993). However, shard shape and colour are, by themselves, rarely sufficient to distinguish individual eruption events, since these characteristics may be common to tephras derived from the same eruption centre, or from different eruption centres which have similar lithological compositions. It should also be noted that certain minerals of non-volcanic origin, for example opaline silica, can assume the characteristic shapes and some of the optical properties of tephra (Fig. 4). Care has to be exercised, therefore, to distinguish between these materials. For these reasons, recourse is now usually made to a study of the mineralogical and chemical properties of glass shards, as a means of confirming that the shards are of volcanic origin, and for 'finger-printing' individual tephra horizons.

The characterisation of the mineral assemblage (e.g., relative abundance of ferromagnesian minerals) of a tephra horizon proximal to the eruption centre can be a powerful correlative tool (Westgate & Gorton, 1981; Shane, 2000), though dissolution of mafic minerals in acidic environments (Hodder et al., 1991) and preferential settling of tephra with distance from the centre (Sparks et al., 1997) can lead to significant variations in mineralogical composition over relatively short distances. Such mineralogical assessments also have other limitations, because tephras derived from the same centre, or from several centres with similar geological characteristics and histories, will tend to have the same mineralogical composition. In the search for a more diagnostic basis for distinguishing between tephras of similar mineralogy, therefore, scientists have turned to geochemical techniques, an approach made increasingly more precise and accessible by the development of high-precision analytical machines within the last two decades or so. A wide range of geochemical methods have been utilised by the tephrochronologist for the geochemical 'finger-printing' of tephra shards, including, for example, X-Ray Fluorescence (XRF — e.g., Norddahl & Haflidason, 1992), Instrumental Neutron Activation Analysis (INAA — e.g., Westgate & Gorton, 1981) and Inductively Coupled Plasma Mass Spectrometry (ICP-MS — e.g., Eastwood et al., 1999). These particular methods also have their limitation, however, because they generally require relatively large sediment samples, and as a result

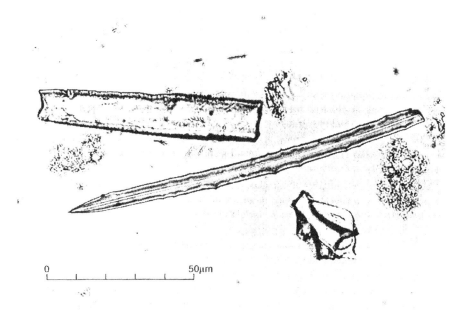

Figure 4. Opaline silica obtained from Finglas River, SW Ireland. The diatom (?) and sponge spicule (upper left and middle) are clearly distinguishable from tephra. The feature at bottom right is much less distinguishable. However, note the high relief and 'mottled' surface that characterises this material type and helps to differentiate it from volcanic glass shards.

are susceptible to contamination effects (Shane, 2000). In addition, analytical targets must be of sufficient size to be representative of the original magmatic components (Shane, 2000), which is particularly difficult to achieve in the case of micro-tephra layers, where the particles are extremely small, and often in very low concentrations. Other methods, such as electron microprobe analysis and laser ablation ICP-MS, offer a much higher analytical precision, which makes it possible to obtain detailed geochemistry from individual tephra shards. These developments have proved extremely useful for establishing the spectrum of geochemical variation within a single tephra horizon.

The electron microprobe uses Energy Dispersive Spectrometry (EDS) or Wavelength Dispersive Spectrometry (WDS) to measure variations in the concentration of the oxides of the major elements within individual tephra shards, WDS being the preferred option because it enables the relative importance of each major oxide to be monitored during analysis. This is particularly important with respect to the oxides of alkali metals, which are vulnerable to mobilisation through the tephra shards during analysis. For all results obtained using electron microprobes, the values are reported as percentages of sample weight. It is rare for the total analysis to reach 100%, because of small inaccuracies introduced during sample preparation, variations in shard characteristics (e.g., thickness and structure of shards walls, which alter susceptibility to impurities, including the mounting medium) and the water content of the shards (Hunt & Hill, 1993). To facilitate comparisons between different sets of analyses produced by different laboratories, therefore, it has been recommended that all geochemical data be normalised to 100% (Froggatt, 1992). This procedure has been criticised by Hunt & Hill (1993), however, who argue that no adjustment of the original measures be undertaken but that all samples yielding oxide measure totals of less than 95% be rejected. This is because the variance of the values obtained for individual oxide measures remains insufficiently precise until more than 95% of the total is accounted for. Comparisons of electron microprobe data obtained from different ash samples are typically based on simple bivariate or ternary plots (see e.g., Hunt et al., 1995; Turney et al., 2001) or on the numerical analysis of the data using either similarity coefficients (e.g., Borchardt et al., 1972) or discriminant function analysis (Stokes & Lowe, 1988; Charman & Grattan, 1999).

Although electron microprobe analysis of individual glass shards is a relatively sophisticated and rigorous technique, there are problems associated with this approach. Inter-laboratory comparisons have so far shown that more rigorous standardisation procedures need to be undertaken to facilitate and improve the accuracy of comparative data (Hunt & Hill, 1996), because some variation in results is introduced through differences in laboratory procedures and standards, in machine precision and operator error. Furthermore, post-depositional alteration of glass shards can occur, through hydration and alkali exchange (Shane, 2000), processes which are dependent upon the duration of any subaerial exposure and differences in depositional environment (Dugmore et al., 1992). Finally, even the detailed examination of major oxide content by electron microprobe may fail to produce confident distinctions between successive tephra layers which have emanated from the same volcanic source, despite the application of powerful discriminant statistical methods (e.g., Stokes et al., 1992). In recent years, therefore, attention has switched from the analysis of major element concentration values to that of variations in the relative proportions or concentrations of minor, trace and rare earth elements. This has become possible through the development of more sophisticated element analysers, such as laser ablation ICP-MS, which can derive a comprehensive suite of geochemical data from extremely small surface areas on individual glass shards (e.g., Westgate et al., 1994; Eastwood et al., 1998; Pearce et al., 1999). Measures of these indices may prove to be more discriminating than measures of major element content, and thus lead to a more reliable 'finger-printing' methodology.

Tephras can be dated directly, using a variety of methods, and where this is possible then tephrochronology not only provides a basis for correlation but also for dating events as well. Because tephra horizons are time-parallel features, then a reliable age need only be established for one locality: the calculated age becomes immediately valid for all other occurrences, providing that the tephra at any one locality has not been subject to redeposition. Techniques that have been employed to date tephras directly include fission track analysis (e.g., Seward, 1974; Gleadow, 1980; Black et al., 1996), thermoluminescence (TL) dating (e.g., Berger, 1987; Berger, 1992; Haberle, 1998; Lian & Huntley, this volume) and $^{40}K/^{40}Ar$ or $^{40}Ar/^{39}Ar$ dating (e.g., McDougall et al., 1980; McDougall, 1981; van den Bogaard, 1995; Lanphere, 2000). Tephra layers can also be dated indirectly, by the dating of the host sediments in which the tephras occur, or materials adjacent to horizon. Examples are radiocarbon dating of organic materials closely associated with the ash layers (Mack et al., 1979; Foit et al., 1993; Dugmore et al., 1995b; Pilcher et al., 1995; Birks et al., 1996; Turney et al., 1997; Björck & Wohlfarth, this volume) and, for more recent events, ^{137}Cs and ^{210}Pb dating of associated sediments (e.g., Stihler et al., 1992). Tephras have also been detected in ice cores, and their ages can therefore be determined by ice-layer counting (e.g., Grönvold et al., 1995) while varve counting can establish the age of ashes located within varved lakes sediments (e.g., Wohlfarth et al., 1993; Lamoureux, this volume). However, all dating methods are subject to a range of potential errors, and all have limits to analytical precision. Correlation of tephra layers should not be based on geological dating methods alone, but should always be confirmed by the geochemical 'finger-printing' methods referred to above.

Greater confidence can be attached to correlations and the chronology of events where several tephra layers have been geochemically 'finger-printed' and dated by direct or indirect means. Where, for example, separate stratigraphic successions show the same sequence of tephra layers over the same time interval, and each sequence suggests (independently) the same geochemical variation between the various ash layers, as well as the same geological ages, then greater confidence can be attached to the overall scheme and to the dating of any specific event within that scheme. By building up such a comprehensive tephrochronolog-ical framework for a region, incidences of hiatuses, reworked tephras (through recycling from disturbed, older deposits — see Boygle, 1999) and anomalous geochemical results can be more readily identified.

Applications of tephrochronology

Well-developed regional tephrochronological frameworks can contribute to an understand-ing of a wide range of environmental issues, too diverse to address adequately in this short chapter. A few examples will suffice which illustrate the versatility and increasing importance of tephrochronology in palaeoenvironmental research. The main advantage that tephrochronology brings to inter-disciplinary palaeoenvironmental research is the improved precision and confidence with which sequences can be correlated and dated, but tephra deposits also record the incidence of airborne volcanic ejecta, which many consider to be important agents of environmental change. Within archaeological con-texts, tephrochronology provides the potential for testing hypotheses of the timing and pattern of human evolution by the identification of remains/artefacts (e.g., Steen-McIntyre, 1981; Brown et al., 1985) and environmental indicators of human activity (e.g., Newn-ham et al., 1998; Lowe et al., 2000) within a tephrochronological framework. In ad-dition, cataclysmic volcanic events, or the tsunami that they initiate, have long been considered likely causes of certain catastrophic events in human history, such as the destruction of the Minoan culture around 1600 B.C. (Burgess, 1989) and influencing the development of civilisation in central Mexico in the first century A.D. (Plunket & Uruñela, 1998).

The full extent of the ways by which tephra fall-out affects surface processes is still imperfectly understood. While there is little doubt that they can affect vegetation succes-sion, lake systems and soils proximal to the eruptive source (e.g., Lotter & Birks, 1993; Wilmshurst et al., 1996; Giles et al., 1999), the effects of ash-falls more distal from source are difficult to assess. While some have concluded that there is evidence to suggest that a moderate fall-out of fine-grained (micro-) tephra may have disrupted ecosystems on a local scale (e.g., Blackford et al., 1992; Baker et al., 1995), others have argued that the evidence for this is far from conclusive (e.g., Birks, 1994; Hall et al., 1994; Caseldine et al., 1998).

There is growing evidence to suggest a causal link between volcanic activity and short-lived deteriorations in global climate (e.g., Rampino & Self, 1982; 1992; Kelly & Sear, 1984; Sear et al., 1987; Baillie & Munro, 1988; Kelly et al., 1996; Grattan & Charman, 1994; Dawson et al., 1997; Briffa et al., 1998; Jacoby et al., 1999; Lowe & de Lange, 2000; Zielinski, 2000) which can be investigated by tephrochronology. Rampino et al. (1988) refer to the consequences of significant tephra 'veils' in the atmosphere as 'volcanic winters', since they appear to lead to higher reflection of incoming solar radiation, and thus reduced global temperatures. The worldwide effects of these events on human populations and ecosystems, and the extent to which they can be detected in palaeolimnological records, have still to be established.

A further application of tephrochronological studies is in establishing the recurrence intervals of major episodes of volcanic activity as well as of those phenomena that they initiate. Detailed stratigraphical records may reveal temporal sequences of volcanic activity that may provide a basis for predicting the likely frequency and magnitude of volcanic activity in the future (e.g., Kamata & Kobayashi, 1997; Newnham et al., 1999; Palumbo, 1999). Records of volcanic eruptions during the late Quaternary in Indonesia, the Mediterranean and Japan have all suggested a close relationship between Milankovitch climatic periodicities and eruption recurrence intervals (Rampino and Self, 1992). Other studies have linked the history of eruptions to changes in the volume of glacier ice or of variations in sea-surface altitude (e.g., Sigvaldson et al., 1992; Nakada & Yokose, 1992).

Tephrochronology also provides constant-age marker horizons for calibrating radiocarbon ages, an application that is likely to gain higher prominence in future years, as attempts continue to be made to obtain radiocarbon dates of higher precision. One major difficulty at the present time concerns the precise comparison of terrestrial and marine sequences dated by radiocarbon. Marine sequences are liable to acquire an 'ageing' error, because seawater tends to have a significantly older age than contemporaneous air or terrestrial organisms, due to slow cycling within the ocean column (the marine 'reservoir effect'). The magnitude of this error varies spatially and over time, and so there is no universal correction factor that can be applied. One way to establish the magnitude of the correction factor that should be applied in one sector of the oceans for a particular time interval is to establish the difference in radiocarbon age between marine and terrestrial samples that can be shown to be contemporaneous by tephrochronology (see e.g., Bard et al., 1994; Austin et al., 1995; Voelker et al., 1998; Sikes et al., 2000). Tephra layers are also becoming increasingly influential in the dating and correlation of late-Quaternary lake sequences of pre-Holocene age, where they are used to test the veracity of high-precision correlations between sequences. Formerly such correlations were attempted using radiocarbon dating alone, but the radiocarbon timescale for the period 10,000 to 50,000 B.P. is subject to a number of distortions, which cannot (yet) be corrected for by dendrochronological calibrations in the manner employed for Holocene investigations. Chief among these are the general divergence between the radiocarbon and calendar timescales, and a series of episodes of near-constant atmospheric activity levels — the so-called 'radiocarbon plateaux' (Fig. 5). Fortunately, a number of tephra layers have been discovered that are widespread over the North Atlantic region (Turney et al., 2001) and some of which are represented in the Greenland ice-core records (Grönvold et al., 1995; Zielinski et al., 1996), and these provide the potential for correlating sequences independently, as well as for testing the validity of

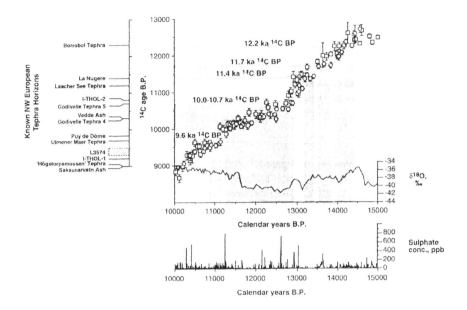

Figure 5. Known tephra horizons for NW Europe (Turney et al., 2001), the GISP2 $\delta^{18}O$ record (Stuiver & Grootes, 2000), sulphate peaks attributed to volcanic events (Zielinski et al., 1996) and radiocarbon calibration curve for the last glacial-interglacial transition (Hughen et al., 1998; Kitagawa & van der Plicht, 1998). The grey-boxed areas mark periods of known constant radiocarbon age.

calibration models for the pre-Holocene interval over which radiocarbon dating can apply (Kitagawa & van der Plicht, 1998).

Perhaps the most exciting development at present is the use of tephrochronology to effect precise correlations between marine, ice-core and terrestrial records (e.g., Ruddiman & Glover, 1972; Ram & Gayley, 1991; Palais et al., 1991; Björck et al., 1992; Palais et al., 1992; Grönvold et al., 1995; Zielinski et al., 1996; Hammer et al., 1997; Zielinski et al., 1997; Kirby, 1998; Kroon et al., 2000). This holds the key to testing current theories about the degree of synchroneity (or otherwise) of climate changes, at regional, continental and global scales. While many researchers have suggested that the rapid climate fluctuations of the last glacial-interglacial transition (15–10 ka) were more-or-less synchronous on a global scale, Coope et al. (1998) and Witte et al. (1998) have presented evidence that suggests that some of these changes were probably time-transgressive over north-west Europe. This is a hypothesis that is difficult to test using conventional radiocarbon dating approaches, but which could be tested by the application of tephrochronology. The development of methods for detecting micro-tephra in deposits spanning this time interval has greatly enlarged the areas over which such correlations can be effected. Thus, for example, micro-tephra horizons originating from Iceland and dated to between ca. 14.5 and 11 ka years ago (calibrated ^{14}C) have been discovered in lake records from sites in the British Isles,

Sweden, Norway, Iceland and Russia (Turney et al., 1997; 1998; Turney, 1999; Wastegård et al., 2000; Wastegård et al., 1998; Björck & Wastegård, 1999; Wastegård et al., 2000), from sea-floor sediment cores in the North Atlantic and adjacent seas (Kvamme et al., 1989; Haflidason et al., 1995; Austin et al., 1995; Kroon et al., 2000; Eiríkísson et al., 2000) and in the Greenland GRIP ice core (Grönvold et al., 1995; Zielinski et al. 1997). Tephrochronology will play a key role in developing a master chronology and correlation scheme for all of these records, and thereby help to test prevalent theories about the pattern and timing of global climate changes as well as the environmental responses that they foster.

Summary

Tephrochronology provides time-parallel marker horizons that allow a level of precision few (if any) other geochronological techniques can offer. The wide range of extraction techniques and increasingly sophisticated methods of 'finger-printing' tephra are helping to extend the provenance of known horizons and the identification of previously unrecorded events, making tephrochronology a powerful geochronological tool when applied to lake sediment sequences. Consequently, tephra layers are now routinely detected and identified in both visible and micro-tephra (or cryptotephra) forms, and are used to test important scientific hypotheses, in the fields of archaeology, climate research and environmental reconstruction.

Acknowledgements

Enormous thanks to Valerie Hall for all her inspiration and to Peter Hill, for his tireless explanations and support in microprobe analysis. Also, many thanks to the Electron Microscopy Unit, Royal Holloway, University of London, who have aided identification of many tephras at initial stages of discovery and helped with photography. Finally, we would like to thank our tephrochronological colleagues and friends who kindly provided references and help, particularly David Lowe. We are grateful to Justin Jacyno who produced the figures.

References

Auer, V., 1932. Palaeogeographische Untersuchungen in Feuerland und Patagonien. Sitz. Berichte Acad. Sci. Fennicae: 1–17.

Austin, W. E. N., E. Bard, J. B. Hunt, D. Kroon & J. D. Peacock, 1995. The ^{14}C age of the Icelandic Vedde Ash: implications for Younger Dryas marine reservoir age corrections. Radiocarbon 37: 53–62.

Baillie, M. G. L. & M. A. R. Munro, 1988. Irish tree rings, Santorini and volcanic dust veils. Nature 332: 344–346.

Baker, A., P. L. Smart, W. L. Barnes, R. L. Edwards & A. Farrant, 1995. The Hekla 3 volcanic eruption recorded in a Scottish speleothem. The Holocene 5: 336–342.

Bard, E., M. Arnold, J. Mangerud, M. Paterne, L. Labeyrie, J. Duprat, M.-A. Méliéres, E. Sønstegaard & J. C. Duplessy, 1994. The North Atlantic atmosphere-sea surface ^{14}C gradient during the Younger Dryas climatic event. Earth Plan. 126: 275–287.

Begét, J. E., 1984. Tephrochronology of Late Wisconsin deglaciation and Holocene glacier fluctuations near Glacier Peak, North Cascade Range, Washington. Quat. Res. 21: 304–316.

Berger, G. W., 1987. Thermoluminescence dating of the Pleistocene Old Crow tephra and adjacent loess, near Fairbanks, Alaska. Can. J. Earth Sci. 24: 1975–1984.

Berger, G. W., 1992. Dating volcanic ash by use of thermoluminescence. Geology 20: 11–14.

Berry, A. J., 1928. The volcanic ash deposits of Scinde Island with special reference to the pumice bodies called chalazoidites. Trans. New Zeal. Inst. 59: 571–608.

Birks, H. H., S. Gulliksen, H. Haflidason, J. Mangerud & G. Possnert, 1996. New radiocarbon dates for the Vedde Ash and the Saksunarvatn Ash from Western Norway. Quat. Res. 45: 119–127.

Birks, H. J. B., 1994. Did Icelandic volcanic eruptions influence the post-glacial vegetational history of the British Isles? Trends Ecol. Evol. 9: 312–314.

Bjarnason, H. & S. Thorarinsson, 1940. Datering av vulkaniska asklager i isländsk jordmån. Geogr. Tidsskr. 43: 5–30.

Björck, J. & S. Wastegård, 1999. Climate oscillations and tephrochronology in eastern middle Sweden during the last glacial-interglacial transition. J. Quat. Sci. 14: 399–410.

Björck, S., O. Ingólfsson, H. Haflidason, M. Hallsdóttir & N. J. Anderson, 1992. Lake Torfadals-vatn: a high resolution record of the North Atlantic ash zone I and the last glacial-interglacial environmental changes in Iceland. Boreas 21: 15–22.

Black, T. M., P. A. R. Shane, J. A. Westgate & P. C. Froggatt, 1996. Chronological and palaeomagnetic constraints on widespread welded ignimbrites of the Taupo volcanic zone, New Zealand. Bull. Volcan. 58: 226–238.

Blackford, J. J., K. J. Edwards, A. J. Dugmore, G. T. Cook & P. C. Buckland, 1992. Icelandic volcanic ash and the mid-Holocene Scots pine (*Pinus sylvestris*) pollen decline in northern Scotland. The Holocene 2: 260–265.

Borchardt, G. A., P. J. Aruscavage & H. T. Millard Jr., 1972. Correlation of the Bishop Ash, a Pleistocene marker bed, using instrumental neutron activation analysis. J. Sed. Petrol. 42: 301–306.

Boygle, J., 1999. Variability of tephra in lake and catchment sediments, Svínavatn, Iceland. Global Planet. Change 21: 129–149.

Briffa, K. R., P. D. Jones, F. H. Schweingruber & T. J. Osborn, 1998. Influence of volcanic eruptions on Northern Hemisphere summer temperature over the past 600 years. Nature 393: 450–455.

Brown, F., J. Harris, R. Leakey & A. Walker, 1985. Early *Homo erectus* skeleton from west Lake Turkana, Kenya. Nature 316: 788–792.

Burgess, C., 1989. Volcanoes, catastrophe and the global crisis of the late second millenium BC. Curr. Anthro. 117: 325–329.

Carey, S. N. & H. Sigurdsson, 1982. Influence of particle aggregation on deposition of distal tephra from the May 18, 1980, eruption of Mount St. Helens volcano. J. Geophys. Res. 87: 7061–7072.

Caseldine, C., A. Baker & W. L. Barnes, 1999. A rapid, non-destructive scanning method for detecting distal tephra layers in peats. The Holocene 9: 635–638.

Caseldine, C., J. Hatton, U. Huber, R. Chiverrell & N. Woolley, 1998. Assessing the impact of volcanic activity on mid-Holocene climate in Ireland: the need for replicate data. The Holocene 8: 105–111.

Charman, D. J. & J. Grattan, 1999. An assessment of discriminant function analysis in the ientification and correlation of distal Icelandic tephras in the British Isles. In Firth, C. R. & W. J. McGuire (eds.) Volcanoes in the Quaternary. Geol. Soc., London, Special Publications 161: 147–160.

Chernet, T., W. K. Hart, J. L. Aronson & R. C. Walter, 1998. New age constraints on the timing of volcanism and tectonism in the northern Main Ethiopian Rift-southern Afar transition zone (Ethiopia). J. Volc. Geo. Res. 80: 267–280.

Child, J. K., J. E. Begét & A. Werner, 1998. Three Holocene tephras identified in lacustrine sediment cores from the Wonder Lake area, Denali National Park and Preserve, Alaska, USA. Arct. Alp. Res. 30: 89–95.

Coope, G. R., G. Lemdahl, J. J. Lowe & A. Walkling, 1998. Temperature gradients in northern Europe during the last glacial-Holocene transition (14–9 ^{14}C kyr BP) interpreted from coleopteran assemblages. J. Quat. Sci. 13: 419–433.

Dawson, A. G., K. Hickey, J. McKenna & I. D. L. Foster, 1997. A 200-year record of gale frequency, Edinburgh, Scotland: possible link with high-magnitude volcanic eruptions. The Holocene 7: 337–341.

Di Vito, M. A., R. Isaia, G. Orsi, J. Southon, S. de Vita, M. D'Antonio, L. Pappalardo & M. Piochi, 1999. Volcanism and deformation since 12,000 years at the Campi Flegrei caldera (Italy). J. Volc. Geo. Res. 91: 221–246.

Dugmore, A., 1989. Icelandic volcanic ash in Scotland. Scot. Geog. M. 105: 168–172.

Dugmore, A. J. & A. J. Newton, 1992. Thin tephra layers in peat revealed by X-Radiography. J. Arch. Sci. 19: 163–170.

Dugmore, A., G. Larsen & A. J. Newton, 1995a. Seven tephra isochrones in Scotland. The Holocene 5: 257–266.

Dugmore, A. J., A. J. Newton, D. E. Sugden & G. Larsen, 1992. Geochemical stability of fine-grained silicic Holocene tephra in Iceland and Scotland. J. Quat. Sci. 7: 173–183.

Dugmore, A. J., G. T. Cook, J. S. Shore, A. J. Newton, K. J. Edwards & G. Larsen, 1995b. Radiocarbon dating tephra layers in Britain and Iceland. Radiocarbon 37: 379–388.

Eastwood, W. J., N. J. G. Pearce, J. A. Westgate & W. T. Perkins, 1998. Recognition of Santorini (Minoan) Tephra in lake sediments from Gölhisar Gölü, southwest Turkey by laser ablation ICP-MS. J. Arch. Sci. 25: 677–687.

Eastwood, W. J., N. J. G. Pearce, J. A. Westgate, W. T. Perkins, H. F. Lamb & N. Roberts, 1999. Geochemistry of Santorini tephra in lake sediments from southwest Turkey. Global Planet. Change 21: 17–29.

Eiríksson, J., K. L. Knudsen, H. Haflidason & P. Henriksen, 2000. Late-glacial and Holocene palaeoceanography of the North Icelandic shelf. J. Quat. Sci. 15: 23–42.

Foit, F. F., Jr., P. J. Mehringer Jr. & J. C. Sheppard, 1993. Age, distribution, and stratigraphy of Glacier Peak tephra in eastern washinton and western Montana, United States. Can. J. Earth Sci. 30: 535–552.

Froggatt, P. C., 1992. Standardization of the chemical analysis of tephra deposits. Report of the ICCT Working Group. Quat. Int. 13/14: 93–96.

Froggatt, P. C. & G. J. Gosson, 1982. Techniques for the Preparation of Tephra Samples for Mineral and Chemical Analysis and Radiometric Dating. Department of Geology Publication No. 23, Victoria University of Wellington, 12 pp.

Froggatt, P. C. & D. J. Lowe, 1990. A review of late Quaternary silicic and some other tephra formations from New Zealand: their stratigraphy, nomenclature, distribution, volume and age. New Zeal. J. Geol. Geophys. 33: 89–109.

Giles, T. M., R. M. Newnham, D. J. Lowe & A. J. Munro, 1999. Impact of tephra fall and environmental change: a 1000 year record from Matakana Island, Bay of Plenty, North Island, New Zealand. In Firth, C. R. & W. J. McGuire (eds.) Volcanoes in the Quaternary. Geol. Soc., London , Special Publications 161: 11–26.

Gleadow, A. J. D., 1980. Fission track age of the KBS Tuff and associated hominid remains in northern Kenya. Nature 284: 225–230.

Grange, L. I., 1931. Volcanic ash showers: a geological reconnaissance of volcanic ash showers of the central part of the North Island. New Zeal. J. Sci. Tech. 12: 228–240.

Grattan, J. & D. J. Charman, 1994. Non-climatic factors and the environmental impact of volcanic volatiles: implications of the Laki fissure eruption of AD 1783. The Holocene 4: 101–106.

Grönvold, K., N. Oskarsson, S. J. Johnsen, H. B. Clausen, C. U. Hammer, G. Bond & E. Bard, 1995. Ash layers from Iceland in the Greenland GRIP ice core correlated with oceanic and land sediments. Earth Planet. 35: 149–155.

Guerrero, B. O., R. Thompson & J. U. Fucugauchi, 2000. Magnetic properties of lake sediments from Lake Chalco, central Mexico, and their palaeoenvironmental implications. J. Quat. Sci. 15: 127–140.

Guest, J. E., J. L. Gaspar, P. D. Cole, G. Queiroz, A. M. Duncan, N. Wallenstein, T. Ferreira & J.-M. Pacheco, 1999. Volcanic geology of Furnas Volcano, São Miguel, Azores. J. Volc. Geo. Res. 92: 1–29.

Haberle, S. G., 1998. Late Quaternary vegetation change in the Tari Basin, Papua New Guinea. Palaeogeogr. Palaeoclim. Palaeoecol. 137: 1–24.

Haberle, S. G. & S. H. Lumley, 1998. Age and origin of tephras recorded in postglacial lake sediments to the west of the southern Andes, 44 °S to 47 °S. J. Volc. Geo. Res. 84: 239–256.

Haflidason, H., J. Eiriksson & S. van Krefeld, 2000. The tephrochronology of Iceland and the North Atlantic region during the Middle and Late Quaternary: a review. J. Quat. Sci. 15: 3–22.

Haflidason, H., H. P. Sejrup, D. K. Kristensen & S. Johnsen, 1995. Coupled response of the late glacial climatic shifts of northwest Europe reflected in Greenland ice cores: evidence from the northern North Sea. Geology 23: 1059–1062.

Hall, V. A., J. R. Pilcher & F. G. McCormac, 1994. Icelandic volcanic ash and the mid-Holocene Scots pine (*Pinus sylvestris*) decline in the north of Ireland: no correlation. The Holocene 4: 79–83.

Hammer, C. U., H. B. Clausen & C. C. Langway Jr., 1997. 50,000 years of recorded global volcanism. Clim. Chan. 35: 1–15.

Heiken, G. & K. Wohletz, 1985. Volcanic Ash. University of California Press, Berkeley, 246 pp.

Hildreth, W. & J. Fierstein, 1997. Recent eruptions of Mount Adams, Washington Cascades, USA. Bull. Volcanol. 58: 472–490.

Hodder, A. P. W., P. J. de Lange & D. J. Lowe, 1991. Dissolution and depletion of ferromagnesian minerals from Holocene tephra layers in an acid bog, New Zealand, and implications for tephra correlation. The Holocene 6: 195–208.

Hughen, K. A., J. T. Overpeck, S. J. Lehman, M. Kashgarian, J. Southon, L. C. Peterson, R. Alley & D. M. Sigman, 1998. Deglacial changes in ocean circulation from an extended radiocarbon calibration. Nature 391: 65–68.

Hunt, J. B. & P. G. Hill, 1993. Tephra geochemistry: a discussion of some persistent analytical problems. The Holocene 3: 271–278.

Hunt, J. B. & P. G. Hill, 1996. An inter-laboratory comparison of the electron probe microanalysis of glass geochemistry. Quat. Int. 34–36: 229–241.

Hunt, J. B., N. G. T. Fannin, P. G. Hill & J. D. Peacock, 1995. The tephrochronology and radiocarbon dating of North Atlantic, late-Quaternary sediments: an example from the St. Kilda Basin. In Scrutton, R. A., M. S. Stoker, G. B. Shimmield & A. W. Tudhope (eds.) The Tectonics, Sedimentation and Palaeoceanography of the North Atlantic Region. Geol. Soc., London, Special Publications 90: 227–248.

Jacoby, G. C., K. W. Workman & R. D. D. D'Arrigo, 1999: Laki eruption of 1783, tree rings, and disaster for northwest Alaska Inuit. Quat. Sci. Rev. 18: 1365–1371.

Juvigné, E., B. Bastin, G. Delibrias, J. Evin, M. Gewelt, E. Gilot & M. Streel, 1996. A comprehensive pollen- and tephra-based chronostratigraphic model for the Late Glacial and Holocene period in the French Massif Central. Quat. Int. 34–36: 113–120.

Kamata, H. & T. Kobayashi, 1997. The eruptive rate and history of Kuju volcano in Japan during the past 15,000 years. J. Volc. Geo. Res. 76: 163–171.

Kelly, P. M. & C. B. Sear, 1984. Climate impact of explosive volcanic eruptions. Nature 311: 740–742.

Kelly, P. M., P. D. Jones & J. Pengqun, 1996. The spatial response of the climate system to explosive volcanic eruptions. Int. J. Clim. 16: 537–550.

Kirby, M. E., 1998. Heinrich event-0 (DC-0) in sediment cores from the northwest Labrador Sea: recording events in Cumberland Sound. Can. J. Earth Sci. 35: 510–519.

Kitagawa, H. & J. van der Plicht, 1998. Atmospheric radiocarbon calibration to 45,000 yr B.P.: Late Glacial fluctuations and cosmogenic isotopic production. Science 279: 1187–1190.

Kroon, D., G. Shimmield, W. E. N. Austin, S. Derrick, P. Knutz & T. Shimmield, 2000. Century to millenial-scale sedimentological-geochemical records of glacial-Holocene sediment variations from the Barra Fan (NE Atlantic). J. Geol. Soc. 157: 643–653.

Kvamme, T., J. Mangerud, H. Furnes & W. F. Ruddiman, 1989. Geochemistry of Pleistocene ash zones in cores from the North Atlantic. Nor. Geol. Tidsskr. 69: 251–272.

Lanphere, M. A., 2000. Comparison of conventional K-Ar and ^{40}Ar/^{39}Ar dating of young mafic volcanic rocks. Quat. Res. 53: 294–301.

Lotter, A. F. & H. J. B. Birks, 1993. The impact of the Laacher See Tephra on terrestrial and aquatic ecosystems in the Black Forest, southern Germany. J. Quat. Sci. 8: 263–276.

Lowe, D. J. & W. P. de Lange, 2000. Volcano-meteorological tsunamis, the c. AD 200 Taupo eruption (New Zealand) and the possibility of a global tsunami. The Holocene 10: 401–407.

Lowe, D. J. & J. B. Hunt, 2001. A summary of terminology used in tephra-related studies. In Juvigne, E. & J.-P. Raynal (eds.) Tephras: Chronology and Archeology. Les dossiers de l'Archéo-Logis No. 1, Goudet, 18–22.

Lowe, D. J., R. M. Newnham & C. M. Ward, 1999. Stratigraphy and chronology of a 15 ka sequence of multi-sourced silcic tephras in a montane peat bog, eastern North Island, New Zealand. New Zeal. J. Geol. Geophy. 42: 565–579.

Lowe, D. J., R. M. Newnham, B. C. McFadgen & T. F. G. Higham, 2000. Tephras and New Zealand archaeology. J. Archaeol Sci. 27: 859–870.

Lowe, J. J. & C. S. M. Turney, 1997. Vedde Ash layer discovered in small lake basin on Scottish mainland. J. Geol. Soc. 154: 605–612.

Machida, H., 1999. The stratigraphy, chronology and distribution of distal marker-tephras in and around Japan. Glob. Plan. Chan. 21: 71–94.

Mack, R. N., R. Okazaki & S. Valastro, 1979. Bracketing dates for two ash falls from Mount Mazama. Nature 279: 228–229.

Mackie, E.A.V., S.M. Davies, C.S.M. Turney, K. Dobbyn, J.J. Lowe & P.J. Hill, 2002. The use of magnetic separation techniques to detect basaltic microtephra in last glacial-interglacial transition (L.G.I.T.; 15-10 ka cal. B.P.) sediment sequences in Scotland. Scot. J. Geol. 38: 21-30

Mangerud, J., S. E. Lie, H. Furnes, I. L. Kristiansen & L. Lømo, 1984. A Younger Dryas ash bed in Western Norway, and its possible correlations with tephra in cores from the Norwegian Sea and the North Atlantic. Quat. Res. 21: 85–104.

McDougall, I., 1981. ^{40}Ar/^{39}Ar age spectra from the KBS Tuff, Koobi Fora Formation. Nature 294: 120–124.

McDougall, I., R. Maier & P. Sutherland-Hawkes, 1980. K-Ar age estimate for the KBS Tuff, East Turkana, Kenya. Nature 284: 230–234.

Merkt, J., H. Müller, W. Knabe, P. Müller & T. Weiser, 1993. The early Holocene Saksunarvatn tephra found in lake sediments in NW Germany. Boreas 22: 93–100.

Nakada, M. & H. Yokose, 1992. Ice age as a trigger of active Quaternary volcanism and tectonism. Tectonophysics 212: 321–329.

Narcisi, B. & L. Vezzoli, 1999. Quaternary stratigraphy of distal tephra layers in the Mediterranean-an overview. Glob. Planet. Ch. 21: 31–50.

Newnham, R. M., D. J. Lowe & B. V. Alloway, 1999. Volcanic hazards in Auckland, New Zealand: a preliminary assessment of the threat posed by central North Island silic volcanism based on the Quaternary tephrostratigraphical record. In Firth, C. R. & W. J. McGuire (eds.) Volcanoes in the Quaternary. Geol. Soc., London, Special Publications 161: 27–45.

Newnham, R. M., D. J. Lowe, M. S. McGlone, J. M. Wilmshurst & T. F. G. Higham, 1998. The Kaharoa Tephra as a critical datum for earliest human impact in northern New Zealand. J. Arch. Sci. 25: 533–544.

Norddahl, H. & H. Haflidason, 1992. The Skogar Tephra, a Younger Dryas marker in North Iceland. Boreas 21: 23–41.

Oldfield, F., P. G. Appleby & R. Thompson, 1980. Palaeoecological studies of lakes in the Highlands of Papua New Guinea. J. Ecol. 68: 457–477.

Oliver, W. R. B., 1931. An ancient maori oven on Mount Egmont. J. Poly. Soc. 40: 73–80.

Palais, J. M., M. S. Germani & G. A. Zielinski, 1992. Inter-hemispheric transport of volcanic ash from a 1259 A.D. volcanic eruption to the Greenland and Antarctic ice sheets. Geo. Res. Lett. 19: 801–804.

Palais, J. M., K. Taylor, P. A. Mayewski & P. Grootes, 1991. Volcanic ash from the 1362 A.D. Oraefajokull eruption (Iceland) in the Greenland ice sheet. Geo. Res. Lett.18: 1241–1244.

Palumbo, A., 1999. The activity of Vesuvius in the next millenium. J. Volc. Geo. Res. 88: 125–129.

Pearce, N. J. G., J. A. Westgate, W. T. Perkins, W. J. Eastwood & P. Shane, 1999. The application of laser ablation ICP-MS to the analysis of volcanic glass shards from tephra deposits: bulk glass and single shard analysis. Glob. Planet. Ch. 21: 151–171.

Pilcher, J. R. & V. A. Hall, 1992. Towards a tephrochronolgy for the Holocene of the north of Ireland. The Holocene 2: 255–259.

Pilcher, J. R. & V. A. Hall, 1996. Tephrochronological studies in northern England. The Holocene 6: 100–105.

Pilcher, J. R., V. A. Hall & F. G. McCormac, 1995. Dates of Holocene Icelandic volcanic eruptions from tephra layers in Irish peats. The Holocene 5: 103–110.

Pilcher, J. R., V. A. Hall & F. G. McCormac, 1996. An outline tephrochronology for the Holocene of the north of Ireland. J. Quat. Sci. 11: 485–494.

Plunket, P. & G. Uruñuela, 1998. The impact of the Popocatepetl Volcano on preclassic settlement in Central Mexico. Quaternaire 9: 53–59.

Ponomareva, V. V., M. M. Pevzner & I. V. Melekestsev, 1998. Large debris avalanches and associated eruptions in the Holocene eruptive history of Shiveluch Volcano, Kamchatka, Russia. Bull. Volcanol. 59: 490–505.

Pyle, D. M., 1989. The thickness, volume and grainsize of tephra fall deposits. Bull. Volcanol. 51: 1–15.

Pyle, D. M., 1999. Widely dispersed Quaternary tephra in Africa. Glob. Plan. Ch. 21: 95–112.

Ram, M. & R. I. Gayley, 1991. Long-range transport of volcanic ash to the Greenland ice sheet. Nature 349: 401–404.

Rampino, M. R. & S. Self, 1982. Historic eruptions of Tambora (1815), Krakatau (1883), and Agung (1963), their stratospheric aerosols, and climatic impact. Quat. Res. 18: 127–143.

Rampino M. R. & S. Self, 1992. Volcanic winter and accelerated glaciation following the Toba super-glaciation. Nature 359: 50–52.

Rampino, M. R., S. Self & R. B. Stothers, 1988. Volcanic Winters. Ann. Rev. Earth Planet. Sci. 16: 73–99.

Rose, N. L., P. N. E. Golding & R. W. Battarbee, 1996. Selective concentration and enumeration of tephra shards from lake sediment cores. The Holocene 6: 243–246.

Rose, W. I. & C. A. Chesner, 1987. Dispersal of ash in the great Toba eruption, 75 ka. Geology 15: 913–917.

Rose, W. I., F. M. Conway, C. R. Pullinger, A. Deino & W. C. McIntosh, 1999. An improved age framework for late Quaternary silicic eruptions in northern Central America. Bull. Volcanol. 61: 106–120.

Ruddiman, W. F. & L. K. Glover, 1972. Vertical mixing of ice-rafted volcanic ash in north Atlantic sediments. Geol. Soc. Am. Bull. 83: 2817–2836.

Samaniego, P., M. Monzier, C. Robin & M. L. Hall, 1998. Late Holocene eruptive activity at Nevado Cayambe Volcano, Ecuador. Bull. Volcanol. 59: 451–459.

Sarna-Wojcicki, A. M., 2000. Tephrochronology. In Noller, J. S., J. M. Sowers & W. R. Lettis (eds.) Quaternary Geochronology: Methods and Applications. American Geophysical Union Reference Shelf 4: 357–377.

Sarna-Wojcicki, A. M., S. Shipley, R. Waitt, D. Dzurisin & S. Woods, 1981. Areal distribution, thickness, mass, volume, and grain size of airfall ash from the six major eruptions of 1980. US Geol. Sur. Prof. Paper 1250: 577–600.

Sear, C. B., P. M. Kelly, P. D. Jones & C. M. Goodess, 1987. Global surface-temperature responses to major volcanic eruptions. Nature 330: 365–367.

Seward. D., 1974. Age of New Zealand Pleistocene substages by fission-track dating of glass shards from tephra horizons. Earth Plan. 24: 242–248.

Shane, P., 2000. Tephrochronology: a New Zealand case study. Earth Sci. Rev. 49: 223–259.

Sigvaldason, G. E., K. Annertz & M. Nilsson, 1992. Effect of glacier loading/deloading on volcanism: postglacial volcanic production rate of the Dyngjufjoll area, central Iceland. Bull. Volcan. 54: 385–392.

Sikes, E. L., C. R. Samson, T. P. Guilderson & W. R. Howard, 2000. Old radiocarbon ages in the southwest Pacific Ocean during the last glacial period and deglaciation. Nature 405: 555–559.

Smellie, J. L., 1999. The upper Cenozoic tephra record in the south polar region: a review. Glob. Plan. Ch. 21: 51–70.

Sparks, R. S. J., M. I. Bursik, G. J. Ablay, R. M. E. Thomas & S. N. Carey, 1992. Sedimentation of tephra by volcanic plumes. Part 2: controls on thickness and grain-size variations of tephra fall deposits. Bull. Volcanol. 54: 685–695.

Sparks, R. S. J., M. I. Bursik, S. N. Carey, J. S. Gilbert, L. S. Glaze, H. Sigurdsson & A. W. Woods, 1997. Volcanic Plumes. John Wiley and Sons, Chichester, 574 pp.

Steen-McIntyre, V., 1981. Tephrochronology and its applications to problems in New-World archaeology. In Self, S. & R. S. J. Sparks (eds.) Tephra Studies. Dordrecht, Reidel: 355–372.

Stihler, S. D., D. B. Stone & J. E. Begét, 1992. "Varve" counting vs. tephrochronology and ^{137}Cs and ^{210}Pb dating: A comparative test at Skilak Lake, Alaska. Geology 20: 1019–1022.

Stokes, S. & D. J. Lowe, 1988. Discriminant function analysis of late Quaternary tephras from five volcanoes in New Zealand using glass shard major element chemistry. Quat. Res. 30: 270–283.

Stokes, S., D. J. Lowe & P. C. Froggatt, 1992. Discriminant function analysis and correlation of late Quaternary rhyolitic tephra deposits from Taupo and Okataina volcanoes, New Zealand, using glass shard major element composition. Quat. Int. 13/14: 103–117.

Stuiver, M. & P. M. Grootes, 2000. GISP2 oxygen isotope ratios. Quat. Res. 53: 277–284.

Thorarinsson, S., 1944. Tefrokronologiska studier på Island. Geog. Ann. 26: 1–217.

Turney, C. S. M., 1998. Extraction of rhyolitic component of Vedde microtephra from minerogenic lake sediments. J. Paleolim. 19: 199–206.

Turney, C. S. M., 1999. Lacustrine bulk organic δ^{13}C in the British Isles during the Last Glacial-Holocene transition (14–9 ka ^{14}C BP). Arct. Ant. Alp. Res. 31: 71–81.

Turney, C. S. M., D. D. Harkness & J. J. Lowe, 1997. The use of micro-tephra horizons to correlate Lateglacial lake sediment successions in Scotland. J. Quat. Sci. 12: 525–531.

Turney, C. S. M., D. D. Harkness & J. J. Lowe, 1998. Carbon isotope variations and chronology of the last glacial-interglacial transition (14–9 ka ^{14}C BP). Radiocarbon 40: 873–881.

Turney, C. S. M., J. J. Lowe, S. Wastegård, R. Cooper & S. J. Roberts, 2001. The development of a tephrochronological framework for the last glacial-Holocene transition in NW Europe. In Juvigne, E. & J.-P. Raynal (eds.) Tephras: Chronology and Archeology. Les dossiers de l'Archéo-Logis No. 1, Goudet, 101–109.

Uragami, K., S. Yamada & Y. Naganuma, 1933. Studies on the volcanic ashes in Hokkaido. Bull. Vol. Soc. Jap. 1: 44–60.

van den Bogaard, C., W. Dorfler, P. Sandgren & H.-U. Schmincke, 1994. Correlating the Holocene Records: Icelandic Tephra Found in Schleswig-Holstein (Northern Germany). Naturwissenschaften 81: 554–556.

van den Bogaard, P., 1995. ^{40}Ar/^{39}Ar ages of sanidine phenocrysts from Laacher See Tephra (12,900 yr BP): chronostratigraphic and petrological significance. Earth Plan. 133: 163–174.

van den Bogaard, P. & H. Schmincke, 1985. Laacher See Tephra: a widespread isochronous late Quaternary tephra layer in central and northern Europe. Geol. Soc. Am. Bull. 96: 1554–1571.

Vernet, G., J. P. Raynal, J. Fain, D. Miallier, M. Montret, T. Pilleyre & S. Sanzelle, 1998. Tephrostratigraphy of the last 160 ka in western Limagne (France). Quat. Int. 47/48: 139–146.

Voelker, A. H. L., M. Sarnthein, P. M. Grootes, H. Erlenkeuser, C. Laj, A. Mazaud, M.-J. Nadeau & M. Schleicher, 1998. Correlation of marine ^{14}C ages from the Nordic Seas with the GISP2 isotope record: implications for ^{14}C calibration beyond 25 ka BP. Radiocarbon 40: 517–534.

Wastegård, S., S. Björck, G. Possnert & B. Wohlfarth, 1998. Evidence for the occurence of Vedde Ash in Sweden: radiocarbon and calendar age estimates. J. Quat. Sci. 13: 271–274.

Wastegård, S., C. S. M. Turney, J. J. Lowe & S. J. Roberts, 2000. The Vedde Ash in NW Europe: distribution and geochemistry. Boreas 29: 72–78.

Wastegård, S., B. Wohlfarth, D. A. Subetto & T. V. Sapelko, 2000. Extending the known distribution of the Younger Dryas Vedde Ash into north-western Russia. J. Quat. Sci. 15: 581–586.

Westgate, J. A. & M. P. Gorton, 1981. Correlation techniques in tephra studies. In Self, S. & R. S. J. Sparks (eds.) Tephra Studies. Dordrecht, Reidel: 73–94.

Westgate, J. A., W. T. Perkins, R. Fuge, N. J. G. Pearce & A. G. Wintle, 1994. Trace-element analysis of volcanic glass shards by laser ablation inductively coupled plasma mass spectrometry: application to tephrochronogical studies. Appl. Geochem. 9: 323–335.

Whitlock, C., A. M. Sarna-Wojcicki, P. J. Bartlein & R. J. Nickmann, 2000. Environmental history and tephrostratigraphy at Carp Lake, southwestern Columbia Basin, Washinton, USA. Palaeogeogr. Palaeoclim. Palaeoecol. 155: 7–29.

Wilmshurst, J. M. & M. S. McGlone, 1996. Forest disturbance in the central North Island, New Zealand, following the 1850 BP Taupo eruption. The Holocene 6: 399–411.

Witte, H. J. L., G. R. Coope, G. Lemdahl & J. J. Lowe, 1998. Regression coefficients of thermal gradients in northwestern Europe during the last glacial-Holocene transition using beetle MCR data. J. Quat. Sci. 13: 435–445.

Wohlfarth, B., S. Björck, G. Possnert, G. Lemdahl, L. Brunnberg, J. Ising, S. Olsson & N. Svensson, 1993. AMS dating Swedish varved clays of the last glacial/interglacial transition Bjand the potential difficulties of calibrating Late Weichselian 'absolute' chronologies. Boreas 22: 113–128.

Zielinski, G. A., 2000. Use of paleo-records in determining variability within the volcanism-climate system. Quat. Sci. Rev. 19: 417–438.

Zielinski, G. A., P. A. Mayewski, L. D. Meeker, S. Whitlow & M. S. Twickler, 1996. A 110,000-yr record of explosive volcanism from the GISP2 (Greenland) ice core. Quat. Res. 45: 109–118.

Zielinski, G. A., P. A. Mayewski, L. D. Meeker, K. Grönvold, M. S. Germani, S. Whitlow, M. S. Twickler & K. Taylor, 1997. Volcanic aerosol records and tephrochronology of the Summit, Greenland, ice cores. J. Geophy. Res. 102: 26625–26640.

Glossary, acronyms and abbreviations

α: Catchment/lake area ratio (A_C/A_L).

α-amino butyric acid: An organic compound with the formula $C_2H_5CH(NH_2)CO_2H$ formed from glutamic acid degradation, also called 2-aminobutyric acid.

α Efficiency factor (k_α): A factor used in ESR and luminescent dating that relates the effectiveness of α radiation to that of γ radiation in producing trapped charges.

β-alanine: An amino acid with the formula $HN_2CH_2CO_2H$, also called (β-amino-proprionic acid, produced by the degradation of other amino acids and organic compounds.

δ: The conventional notation for expressing the ratio between two stable isotopes of the same element. The δ value for isotopes x and y of element A (where y is the heavier isotope) is given by the expression:

$$\delta^y A = \left[\frac{(^y A/^x A)_{\text{sample}}}{(^y A/^x A)_{\text{standard}}} - 1 \right] \times 1000.$$

$\delta\,^{14}C$: The difference between the ^{14}C content of the pre-industrial atmosphere as compared to the atmospheric ^{14}C content at a certain time in the past.

ε_0: Permittivity of free space.

γ Spectroscope: A devise emplaced within the sediment at a site to measure the current external dose rate, from sedimentary γ and cosmic sources over 0.5–2.0 hours.

κ: Magnetic volume susceptibility.

κ^*: Complex dielectric constant.

κ': Dielectric constant.

κ'': Loss factor.

λ: 1) Radioactive decay constant. 2) Geographic latitude of a sampling site on the Earth's surface. 3) Wavelength.

μ: 1) Mass attenuation coefficient. 2) X-ray absorption coefficient.

μ_0: Magnetic permeability of free space.

η: Fluid viscosity.

η, (η_{Pb}): Catchment-lake fallout (^{210}Pb) transport parameter.

ω: 1) Angular velocity. 2) Frequency in radians.

ϕ: 1) Fractional porosity or liquid content in completely saturated samples; sediment porosity. 2) See phi grade scale.

φ: φ^2: A measure of sediment tortuosity (dimensionless).

Φ: The fraction of the total atmospheric solute concentration on the particle phase (dimensionless).

$\Phi(t)$: Atmospheric flux of a fallout radionuclide.

$\Psi_C(t)$: Rate of transport of a fallout radionuclide deposited on the catchment from the catchment to the lake.

ρ: Density.

ρ_B: Bulk density.

P_b: Sediment bulk density.

σ: 1) Standard error or standard deviation. 2) Mineral density.

τ: 1) Intrinsic resolution. 2) ESR or luminescent signal lifetime. 3) Residence time.

A: Long axis diameter.

A: Unsupported ^{210}Pb inventory below a specified depth in the sediment record.

$A(0)$: Unsupported ^{210}Pb inventory in the whole sediment core.

$\hat{A}$: Unsupported ^{210}Pb inventory above a specified depth in the sediment record.

Å: The Ångström, a unit of length principally used to express the wavelength of electromagnetic radiation. 1 Ångström is equal to 0.1 nanometre (10^{-10} m).

$\mathscr{A}_L$, $\mathscr{A}_C$: Areas of a lake and its catchment.

A_Σ: The symbol for the accumulated dose (used in ESR and luminescent dating methods).

A_{ext}: The symbol for the external component in the accumulated dose (used in ESR and luminescent dating methods).

A_{int}: The symbol for the internal component in the accumulated dose (used in ESR and luminescent dating methods).

AAAA: Automatic amino acid analyzer.

AAR: Amino acid racemization.

AAR dating: Amino acid racemization dating.

Accelerator mass spectrometry (AMS): In an accelerator particles are subjected to large voltage differences so that they travel at very high speed. This allows detecting atoms of specific elements based on differences in atomic weight.

Accumulated dose (A_Σ, AD): The ionizing radiation dose equivalent to the ESR or luminescent signal intensity seen in a sample, also known as the equivalent or archaeological dose.

Accumulation rate: The rate of accumulation of particles or elements on the lakebed. Normally expressed as mass per unit area per unit time (e.g., mg m^{-2} yr^{-1}).

Accuracy: 1) The degree of agreement between an analysis and a standard analysis (also known as quality assurance). 2) The degree of agreement between an analysis and the true value (see also precision).

AD: 1) *Anno Domino.* 2) The accumulated dose (used in ESR and luminescent dating methods).

Added doses: The artificial dose added to a sample during the additive dose method to assess the accumulated dose (used in ESR and the luminescent dating methods).

Additive-dose method: 1) A method of determining the equivalent dose. Aliquots of the sample are given a variety of laboratory radiation doses, and then usually preheated. The thermoluminescence or the optically stimulated luminescence is then measured. The luminescence intensity vs. dose response is extrapolated to the dose axis and the equivalent dose taken to be the dose intercept. 2) A protocol for determining the accumulated dose for ESR, TL, OSL, and RL dating methods, using 10–15 aliquots of homogenized sample. Each aliquot is irradiated to a different, but precisely known artificial radiation dose to produce a curve that compares signal intensity versus added dose.

Aeolian: Refers to an allogenic particle transported by wind or deposits associated with wind transport.

AES: Atomic emission spectroscopy.

AF: Alternating field.

AGC: Automatic gain control accelerator mass spectrometry (AMS): In an accelerator particles are subjected to large voltage differences so that they travel at very high speed. This allows detecting atoms of specific elements based on differences in atomic weight.

Age model: An age model is often displayed as a curve where the data points are ages (x-axis) and sediment depths (y-axis) with interpolations between each age/depth point, which in theory means that each sediment level obtains an age. The shape of such a sedimentation curve should not only depend on the age of each dated level, but also on changes in lithology, i.e., possible changes in sedimentation rate. Attempts to find a mathematic function, fitting to the curve shape, can sometimes result in mathematically based age models.

476

Alle: Alloisoleucine (an amino acid).

Ala: Alanine (an amino acid).

Alanine (Ala): A uniasymmetric amino acid occurring abundantly in natural proteins with the formula $CH_3CH(NH_2)CO_2H$, also called aminoproprionic acid and α-alanine.

Aliquot: A geochemically identical subsample from a larger solution or other sample (see also subsample).

Alkali-insoluble (INS): Non-dissolvable fraction of a sediment after alkali (NaOH) treatment (see also humin-insoluble).

Alkali-soluble (SOL): Term to denote the soluble, humic fraction of sediment. It is obtained by subjecting a sample to NaOH (alkali) treatment, which is followed by acidification (pH value of $\sim$3) to precipitate the alkali-soluble, humic acids.

Allochthonous: 1) Minerogenic and/or organic material transported from its place of origin to form a sediment (or part of a sediment) at another locality. 2) A material (e.g., rock, mineral, fossil) formed elsewhere than its present resting place (cf. autochthonous).

Allogenic: Sediment component(s) formed outside the lake; specifically refers to rock, mineral, and organic particles that were derived form pre-existing materials, eroded and transported to the depositional site.

Alloisoleucine: A multiasymmetric isoleucine epimer produced by isoleucine epimerization, with the formula $CH_3CH_2CH(CH_3)CH(NH_2)CO_2H$. It only occurs naturally in the D form (2R, 3S-isoleucine).

Aluminum (Al ESR) signal: A complex ESR signal arising from an $(AlO_4)^0$ defect, seen mainly in quartz and related minerals, that is measured at $70\,°K$.

Alpha (α) particle: A particle emitted during the decay of some nuclei; it consists of 2 protons and 2 neutrons, and thus is the same as the nucleus of a ^4He atom.

Alpha spectrometry: Measurement of alpha particles emitted by radioactive substances.

^{241}**Am:** Artificial radionuclide with a 432 year half-life produced by the radioactive decay of ^{241}Pu.

Ambient temperatures: Current average temperatures.

Amino acid: Any organic compound, containing both an amine (-NH_2) and carboxyl (-COOH) groups, having the general configuration $(HO_2C)R^1CR^2(NH_2)$. The major constituent in proteins, amino acids are essential to life. All but glycine are optically active. Most organisms produce only L-amino acids.

Amino acid racemization (AAR) dating: A chemically based geochronological method that relies on the natural conversion of L-amino acids to a mixture of D and L-amino acids during protein diagenesis.

Amino acid racemization (AAR) rate (k_i): The speed at which the amino acid racemization reaction occurs. In all materials, the rate depends on temperature, but in some materials such as bone, it also depends on other physical, geochemical, and biological factors.

Aminostratigraphic analysis: A process that involves defining aminozones, which can be correlated over large areas.

Aminostratigraphy: An aminozone sequence usually used for regional correlations.

Aminozones: Stratigraphic units defined by amino acid racemization (AAR) ratios from two or more amino acids determined for *in situ* molluscs, foraminifera, or other fossils, that are widely used to correlate marine, lacustrine, and loess sequences.

Amphibole: Ferromagnesian silicate mineral.

AMS: 1) Accelerator mass spectrometry. 2) Anisotropy of magnetic susceptibility.

AMS ^{14}C (dating): ^{14}C dating performed by accelerator mass spectrometry. (See ^{14}C dating).

Andesite: A volcanic rock with 52 to 63% SiO_2.

Anhydrite: An orthorhombic mineral with the formula $CaSO_4$; also called cube spar.

Anhysteretic: Free of hysteresis, does not display hysteresis properties.

Anhysteretic remanent magnetization (ARM): The magnetization that remains in a sample after exposure to a smooth decaying strong alternating field in the presence of a weak direct field.

Anisotropic: 1) Giving incomplete extinction when rotated under cross-polarised light. 2) Opposite of isotropic. That is, a medium in which certain physical properties are different in different directions. Such properties may relate to physical strength, light transmission electrical resistivity, etc.

Annealing: Any mild, short-lived heating process used to remove ESR signals with short lifetimes and low thermal stabilities, thereby improving the ESR signal resolution for the signals of interest.

Annually laminated sediments: Sediments with annual layers (usually couplets) produced by the seasonal delivery of minerogenic material, by seasonal precipitation and/or biological activity. Formed in deep lakes with oxygen-deficient bottom waters and little or no bioturbation; one yearly package is often called a varve.

Anomalous fading: An electron in a deep trap is expected to stay there for at least 10^5 years at room temperature. In some minerals, most notably feldspars, electrons in deep traps are observed to leave some of the traps on a laboratory time scale; consequently the intensity of the thermoluminescence or optically-stimulated luminescence decreases with increasing length of time between laboratory irradiation and measurement. When

478

first discovered, this decrease was called anomalous fading as it was not understood. The name persists even though it is now widely thought that quantum-mechanical tunneling is the cause.

Anoxic: Strictly speaking, the absence of oxygen. As dissolved oxygen concentrations cannot be measured in ancient sediments, an absence of all evidence for higher life forms (metazoans) has commonly been taken as a strong indication of anoxic conditions. Today such organisms are typically excluded when dissolved oxygen levels fall below ca. 0.01 ml l^{-1}, although some specialized organisms are capable of making "visits" to environments with lower O_2 concentrations.

Anthropogenic: Sediment component(s) derived from human sources.

Antler: 1) Histology: A calcified deciduous tissue chemically and histologically similar to bone, but lacking a cortical bone sheath (see bone). 2) Anatomy: A large deciduous tissue growth, grown annually by adult male Cervidae and both reindeer sexes *(Rangifer)*.

AOTF: Acousto-Optical Tuneable Filter

AP: Apatite phosphorus.

Apatite phosphorus: That fraction of the total phosphorus concentration that is in the form of apatite.

^{39}Ar/^{40}Ar (dating): A radiometric geochronological method often used to date volcanic and igneous rocks. A type of K/Ar dating, it uses the ratio between the argon isotope ^{39}Ar produced by irradiation of the naturally occurring isotope ^{39}K with high energy neutrons, and the ^{40}Ar produced by radioactive ^{40}K decay to determine the age.

Aragonite: 1) *Sensu stricto*: An orthorhombic carbonate with the formula $CaCO_3$, which is trimorphous with calcite and vaterite. Also known as Aragon spar; it is a common matrix mineral in invertebrate shells. 2) *Sensu lato*: A group of orthorhombic carbonate minerals, including aragonite, strontianite, cerussite, alstonite, and witherite.

Archaeological dose (A_{Σ}, AD): The ionizing radiation dose equivalent to the ESR or luminescent signal intensity seen in a sample, more commonly known as the accumulated or equivalent dose.

Arg: Arginine (an amino acid).

Arginine (Arg): An amino acid occurring in natural proteins with the formula $(H_2N)C(NH_2)NH(CH_2)_3CH(NH_2)(CO_2H)$; properly called 2-amino-5-guanido-pentanoic acid.

ARM: Anhysteretic remanent magnetization; an artificial magnetization acquired in a strong decaying alternating magnetic field superimposed on a weak static field.

Arrhenius equation: An equation often used in kinetic experiments to determine the rate constant for a reaction:

$$k_i = Ae^{-E_a/(RT_i)},$$

where A = the frequency factor, a constant, E_a = the activation energy at temperature, T_i, R = Boltzman's constant, 1.9872 cal/mole-°K, and T_i = temperature (°K).

Arrhenius plot: A graph of the natural logarithm of the reaction rate constant versus the inverse of absolute temperature, T^{-1}, used to determine the parameters in the Arrhenius equation.

Asp: Aspartic acid (an amino acid).

Aspartic acid (Asp): A uniasymmetric amino acid occurring abundantly in natural proteins with the formula $HO_2CCH_2CH(NH_2)CO_2H$; also called aminosuccinic acid and asparginic acid.

Asymmetric carbon atoms: Organic molecules in which the central carbon atom is bonded to four different atoms or ligands. These molecules show optically activity in plane-polarized light.

Attenuation: 1) Radiation: A reduction in incident radiation energy. Water can significantly attenuate most terrestrially produced radiation. 2) General: A reduction in a signal's amplitude or energy.

Attenuation coefficient: The proportion of the energy in a pulse that is converted into other types of energy (e.g., heat) thus reducing the strength of the signal.

Australian slide method: A method of determining the equivalent dose. Two similar sets of aliquots of the sample are prepared. Aliquots of the first set are given a variety of radiation doses. Aliquots of the second set are given a light exposure to empty the traps and then given a variety of radiation doses. Both sets are then preheated, measured and the luminescence intensity plotted as a function of the laboratory dose. The luminescence intensities from the first set are then matched to those of the second set by a shift along the dose axis; this shift is taken to be the equivalent dose, after some minor corrections. It is necessary to ensure that a satisfactory match is obtained with the shift; if not, no equivalent dose can be determined.

Authigenic: 1) Sediment component(s) formed within the sediment after deposition of the original material. 2) Sometimes used synonymously with endogenic.

Authigenic cement: Cement generated locally within the stratigraphic sequence; often a carbonate or silica cement.

Authigenic mineral: Any mineral precipitated within the sediment or rock from geochemical compounds found locally in the rock or sediment.

Autochthonous: 1) Rocks, sediments or deposits formed and deposited at the same place as where they are found today; fossils deposited at the place where the animal lived or the plant grew. 2) A material (e.g., rock, mineral, fossil) formed in the basin or water body where it is found (cf. allochthonous).

Automatic amino acid analyzer (AAAA): A specially modified HPLC able to detect amino acids at femtomole concentrations.

Automatic gain control: A system by which the gain applied along a trace is varied according to the original amplitude of returns within a each section of the trace.

Background amino acid racemization: The racemization caused by the amino acid racemization analysis process itself, also called the induced racemization.

Backscattering: The reflection and dispersion of ionizing radiation off a mineralogical boundary or surface that reduces the effective radiation transmission into the surface.

Barite: A sulfate mineral with the formula $BaSO_4$; also known as barytes, heavy spar.

Barrier: A wave- and current-constructed depositional landform that acts as a barricade protecting the land from direct attack of open-water waves. Barriers are separated from the mainland by a lagoon. Common types of lacustrine barriers include bayhead, baymouth, and cuspate.

Basal sedimentation rates: Estimated sedimentation rates at the base of the unsupported ^{210}Pb record used in calculating corrections to CRS model ^{210}Pb dates.

Basalt: A volcanic rock with 44 to 52% SiO_2.

Bathymetric map: Map indicating water depth.

BC: Before Christ.

BDP (Baikal Drilling Project): A project to reconstruct the paleoclimate and tectonic history of central Asia for the late Neogene.

Beach: A primarily wave- and current-constructed coastal depositional landform attached to the mainland along its entire planimetric length, with a length to width ratio greater than 1.0.

Beach ridges: Approximately parallel, alternating ridges and swales that comprise many beaches, forelands, or barrier beaches.

Belemnites: A fossil octopus group living between 370–65 million years ago; an early branch of the modern octopus.

Benthic: Bottom-living.

Beta particle (β): A particle emitted during the decay of some nuclei; it is a high-energy electron.

BIK: Botanisches Institut der Universität, Kiel, Germany.

Bioturbation: 1) Mixing of sediment by biological activity (e.g., burrowing). Occurs in oxygenated bottom waters, but is typically absent from anoxic bottom waters. Original sedimentary structures are usually destroyed and the stratigraphy is altered in bioturbated sediments. 2) Disturbance of sediment structures, layers or horizons by the activity of borrowing and/or grazing benthic organisms.

Bleach (a signal): See zeroing.

Bølling: Term derived from Lake Bølling Sø on Jylland in Denmark, where the first pollen evidence for the first warming after the last deglaciation was found. It was originally the name of this birch dominated pollen zone in NW Europe and later also designated the chronozone between 13,000–12,000 ^{14}C years BP. Now it usually denotes the first warm period during the last deglacial phase in the North Atlantic region, and is commonly dated to 12,700–12,100 ^{14}C years BP or to ca. 14,700–14,050 GRIP ice core years.

Bone: 1) Tissue: A biologically produced biomineralic compound containing an organic matrix high in collagen and a hydroxyapatite mineral matrix, which forms the skeleton of most vertebrates. Fossil bone can contain other apatite group minerals, including dahllite, francolite, and vivianite, and may lack most of or all its organic matrix. 2) Anatomy: Any individual piece from the bony vertebrate skeleton.

Boomer: A marine seismic source that uses capacitors to discharge a high voltage through a transducer in the water. Transducer is a flat coil with a spring-loaded aluminum plate. Generates a seismic signal with a bandwidth of $\sim$500–3000 Hz.

Borehole velocity logs: Records of P-wave velocity from a borehole.

Bow wave effect: The pressure wave set up in the water ahead of an advancing body. From a sampling point of view, this condition is important as material (sediment) may be deflected away from the sampler if a bow wave is present.

BP: Literally, before present, but actually before 1950. Defined to be 1950 when ^{14}C dates first began to be published, "present" is used as a calibration point only for ^{14}C dates or ages (see also ky BP).

Bq: SI unit of radioactivity, measured as 1 disintegration per second.

Breccia: 1) Geology: A coarsely grained clastic sediment or rock containing angular, and often broken, rock fragments lithified by mineral cements or supported in a more finely grained matrix. 2) Archaeology *sensu stricto*: A well-cemented coarsely grained sedimentary unit, often found in a cave. 3) Archaeology *sensu lato*: Any well-cemented sedimentary unit.

BSE: Backscattered scanning electron microscopy. Backscattered electrons are the result of elastic collisions between energetic beam electrons and atoms within the target, for example a thin-section.

Buffer: 1) Any pair of solutes that can maintain a solution's pH at an almost constant level despite the addition of moderate quantities of acids or bases. 2) A solution containing such solute pairs.

Buffered solutions: see buffer (2).

Bulk density: Specific weight or density of sediments including their water content; weight per unit volume of a sediment.

Bulk geochemical analysis: Any analytical protocol used for ESR or luminescent dating that involves determining the concentrations for the important radioactive elements using geochemical techniques, such as NAA, XRF, etc.

Bulk sediment: Sediment sample that has not been sub-divided into different fractions

^{14}**C (dating):** A radiometric geochronological method often used to date Late Quaternary fossils and charcoal <30–45 ka. It uses the quantity of ^{14}C occurring in the sample to determine the age. Most ^{14}C dates are now calibrated against local dendrochronological curves and corrected changes in the cosmic radiation flux (also known by the outdated term radiocarbon dating).

C: Mass specific radionuclide concentrations in sediments.

C_{tot}, C_{sup}, C_{uns}: Total, supported and unsupported ^{210}Pb concentrations.

$\dot{C}$ **or** $\dot{}C$ **(ESR) signal**: An ESR signal associated with carbon radicals, often seen associated with heated or burnt organic molecules.

C-**terminus (terminal position or end):** The end of an amino acid, peptide, or protein containing the carboxyl (-COOH) group.

Cable-operated devices: Those devices that are suspended, operated and recovered by a cable, wire line or other type of line.

Cainozoic: see Cenozoic.

Calcrete: A conglomeratic crust on or in surficial sediment cemented by secondary carbonate minerals, often formed by carbonate precipitation from evaporating groundwater (see also caliche, silcrete, gypcrete).

Calendar years: Absolute or sideral years; can be obtained from e.g., historical documents, a tree-ring chronology (dendrochronology) or counting of annually laminated sediments from present back in time.

Calibrated date: Transformation of a ^{14}C date into absolute ages by comparing it to the radiocarbon calibration curve. Different programs exist to calibrate a radiocarbon measurement.

483

Calibration curve: Denotes here the radiocarbon calibration curve (INTCAL98). This curve is composed of a ^{14}C dated, continuous tree ring chronology (from present back to ca. 12,000 dendro years), a ^{14}C-U/Th dated coral record (extending back to ca. 24,000 U/Th years) and ^{14}C dated, annually laminated marine sediments (extending back to ca. 14,500 varve years).

Calibration tests: Geochronology: Tests used to ensure the accuracy of a geochronological method in which multiple dating methods are used to ascertain the age for particular samples. Usually, several different tests are necessary covering different time ranges and environmental conditions.

Caliche: 1) *Sensu lato*: A secondary sedimentary accumulation cemented predominantly by secondary calcitic cements that can include dolomite, aragonite, gypsum, and silica. Formed by groundwater capillary action in regions with high evaporation rates, it is also called hardpan, duricrust, calcrete, kankar, calcareous crust, nari, sabach, and tepetate. 2) *Sensu stricto*: Discrete calcrete nodules that do not form a continuous crust (see also calcrete, silcrete, gypcrete).

Cambrian: The oldest geologic period between 570–510 million years ago with abundant animal fossils and the first period of the Paleozoic Era (570–245 million years ago); also denotes the system of strata deposited during this period.

Casing (cored holes): The lining of bored or cored holes to support the surrounding sediments and provide a mechanical guide to the sampler that passes through the casing. In lake coring, the casing is often made up of tubular or pipe sections that support the coring device in the water column as well as in the sediment.

Catalysis: Any process or chemical reaction(s) which increase(s) the reaction rate for a chemical reaction caused by the presence of a small amount of a substance or substances that are not permanently affected by the reaction.

Catchment: Used synonymously with watershed.

Cation exchange column: A glass column packed with a compound able to adsorb cations to and then desorb cations from its surface. These are often used to desalt solutions before geochemical analysis.

Cementation: The diagenetic process whereby clastic sedimentary grains are lithified or consolidated into sedimentary rocks, often by mineral deposition or precipitation on the grain surfaces or in the interstitial spaces.

Cementum (dental): A biologically produced biomineralic compound containing an organic matrix high in collagen and an hydroxyapatite mineral matrix, which forms the thin interface between the dentine and the bone in most vertebrate teeth, but can also occur on the occlussal or other surfaces of some ungulates' and rodents' teeth. Cementum, also called cement, has less mineral matrix than enamel, and a histology very similar to bone. Fossil cementum can contain other apatite group minerals, including dahllite, francolite, and vivianite, and may lack most or all of its organic matrix.

Cenozoic (in American English), **Cainozoic** (in British English): The most recent geological era that began approximately 66.5 Ma (also known as Cainozoic, Kainozoic, the Mammal Age).

CEREGE: Mixed center of research and teaching in geosciences of the environment in France.

Chert: A hard dense micro- or cryptocrystalline sedimentary rock with interlocking quartz crystals that displays concoidal fracture, prized by hominids as a material for making tools. Occurring as nodules in limestones or dolomites, it is often considered synonymous with flint. Chert is sometimes considered only to be the varieties with visible colors, the cryptocrystalline types, or varieties lacking visible fossil inclusions (see flint).

Chiral phase: A coating on a chromatographic column that allows isomers to be absorbed and desorbed at different rates, permitting their separation by the column.

Chirp: An acoustic measurement system that utilizes a broad-band, swept FM pulse as a signal, rather than an impulsive or percussive pulse.

Chitin: A resistant organic compound similar in structure to cellulose containing repeating N-acetylglucosamine units, which forms the hard skeleton in many invertebrate and foraminiferal inner tests.

Chronology: 1) A data series or framework referenced with respect to time (i.e., a time series). For lake sediment records, often the chronology describes the sediment depth-age relationship. 2) Time scale, which may be obtained through radiometric dating (e.g., ^{14}C, U/Th) or counting of annual layers.

Chronometer: Any of several radiometric or geochemical geochronological methods that can calculate absolute ages for geological or paleontological samples.

Chronostratigraphy: 1) Definition and subdivision of the stratigraphic record in terms of time. 2) Nordic: The Nordic chronostratigraphy-chronostratigraphic scheme is based on ^{14}C dated Late Weichselian and Holocene pollen zone boundaries in Scandinavia. See also chronozone.

Chronozone: A formal time-stratigraphic unit used to define and correlate strata of equivalent time span. The time span of this basic chronostratigraphic unit is often defined in terms of the time span of a previously designated stratigraphic unit, such as a formation or a member or a biozone.

CIC: Model for calculating ^{210}Pb dates in lake sediments assuming a Constant Initial ^{210}Pb Concentration for each sediment layer.

Clay: 1) Textural: The term applied to sediment particles finer than silt; particles having a diameter of smaller than 3.91 microns or sometimes 1.95 microns. 2) Mineralogical: A poorly defined group of aluminum silicate minerals.

CLEU: A coupled U uptake model (ESR and luminescent dating) in which the enamel is assumed to follow linear uptake and all other tissues early uptake.

Clinoform: A stratigraphic geometry associated with a sloping sea or lake surface and progradation of strata.

CMP: Common midpoint.

Coastal bar: A submerged ridge of sediment deposited by waves and currents along and approximately parallel to a portion of the coastline.

Coastal geomorphology: Study of the landforms and landforming processes found along marine and lacustrine coasts.

Collagen: Any of several similar complexly organized organic proteinaceous compounds of molecular weight approximately 300,000 u, with 33 mol% glycine, as well as significant alanine, proline, hydroxyproline, aspartic and glutamic acids. A major constituent in the organic matrix in bone, dentine, dental cementum, and antler, it contains five trimers coiled into microfibrils that are staggered to form fibril bundles averaging $0.3-0.5\ \mu m$ in diameter and $4-20\ \mu m$ long with a characteristic 6.4 nm banding. Each trimer contains three coiled crosslinked monomers of 1055 amino acids.

Common midpoint method (CMP): A reflection profiling technique that sums seismic traces that share the same midpoint between source and receiver. Also called common reflection point or common depth point method.

Common midpoint velocity survey: A GPR survey in which the antennas are progressively moved further apart while always remaining centered on the same point.

Component: A general term used to denote the separate materials which, when mixed together, make up soils and sediments.

Compressional wave transducers: Electronic components that change electrical signals into ultrasonic compressional (P) waves and back again.

Compton attenuation coefficient: A material specific constant.

Compton scattering: Scattering of incident photons by electrons resulting in an energy loss of gamma rays passing through sediments. This is the dominant factor controlling gamma ray attenuation.

Conduction band: The energy levels at which an electron can move freely about within a given mineral phase. In many materials, this is an excited state (see also valence band, ground state, excited state).

Conventional radiocarbon dating technique: This technique is based on the detection and counting of β emissions from ^{14}C atoms over a period of time in order to determine the rate of emissions and hence the activity of the samples. Two methods are employed: gas proportional counting and liquid scintillation counting.

Convolution: Linear filtering; a mathematical operation between two functions (e.g., a physical property distribution along a sediment core axis and a response function of the measuring sensor) generally resulting in a low-pass filtering of the real signal.

Core aspect ratio: A ratio of core diameter to its length. Expressed as length/diameter.

Core catcher/retainer: A device, of which there are many designs, installed inside the core tube or around the core head that allows sediment to enter the core barrel during penetration of the corer, but which blocks sediment from falling out during raising of the core barrel.

Core compaction: See core shortening.

Core compression: See core shortening.

Core imaging system: Acquisition, storing, management, and manipulation of digital sediment core images obtained by frame cameras, line-scan cameras, spectral imaging systems and by digital X-ray systems.

Core liner: A thin plastic pipe, commonly transparent, for easy removal of the sediment from core barrel or for sediment storage. It is inserted inside the core barrel before coring and is withdrawn from the barrel when filled with sediment. A new liner is necessary for each drive.

Core shortening: The degree to which the recovered length of the core sample is less then the penetration distance to which the core tube is driven. Core shortening is also sometimes referred to as core compaction or core compression.

Corer head: The leading edge of the core barrel commonly sharpened or beveled and containing the core catcher, if used.

Coring drive: Advancement of the coring tool until the core barrel is full.

CORPAC (Correlation Package): A software package that uses an inverse statistical approach to correlate stratigraphic data sets.

CORTEX: XRF core-scanner Texel.

Cosmic dose or dose rate: The ionizing radiation dose or dose rate generated by highly energetic particles impacting the Earth from space.

Cosmogenic isotope dating: A method using the amount of a cosmogenic isotope accumulated in a rock to determine the length of time the rock has been exposed to the atmosphere.

Coupled ESR-^{230}Th/^{234}U dating: A geochronological method that combines ^{230}Th/^{234}U dentine dates and ESR enamel dates to derive a unique age and the most applicable U uptake model, generally limited to teeth <250–400 ka.

Coupled (U) uptake (CU): 1) An U uptake model used for calculating ESR ages that assumes that the tooth enamel, dentine, cementum, and any attached bone absorbs U by a combination of uptake models. The most common assumes LU for the enamel and EU for the remaining the tissues, producing ages slightly younger than models that assume LU for both dentine and enamel, but significantly older than those that use EU model for both. 2) The equivalent U uptake model used calculating external radiation dose rates in sediment with multiple components that absorb secondary U.

Cromerian: 1) Time: A Middle Pleistocene interglacial phase recognized in Northern Europe and Britain, occurring after the Menapian (Northern Europe) and Beestonian (Britain), but before the Elsterian (Northern Europe) and Anglian (Britain). Thought to be equivalent to Günz-Mindel or Mindel-Riss interglacial in the Alpine sequence, age estimates range from approximately 300–600 ka. Largely abandoned in favor of Oxygen Isotope Stages, the term may encompass two or more interglacial phases from Oxygen Isotope Stages 9, 11, 13 and/or 15. 2) Rocks: The sediment and rock formed during the Cromerian interglacial phase.

Cross dating: The process of comparing and matching of two or more time series to identify the ages of the series relative to each other. Can also be used to identify where missing or extra elements occur in one or more of the series.

CRS: Model for calculating ^{210}Pb dates in lake sediments assuming a Constant Rate of Supply of fallout ^{210}Pb.

^{137}Cs: Radioactive isotope of cesium with a 30.2 year half-life produced as a fission product of nuclear weapons explosions or in nuclear reactors.

CU: Coupled U uptake model (ESR and luminescent dating).

Cumulative unsupported ^{210}Pb inventory: Total quantity of unsupported ^{210}Pb above each sediment layer.

Cystine: A multiasymmetric amino acid occurring abundantly in natural proteins with the formula $[HO_2CCH(NH_2)CH_2S\text{-}]_2$, also called dicystine.

D: Diffusivity of ^{222}Rn in soils.

D: 1) Distance. 2) Depth. 3) Declination.

D_e, D_{eq}: Equivalent dose.

$D_\Sigma(t)$: The symbol often used for the total dose rate (ESR and luminescent dating methods).

$D_{ext}(t)$: The symbol often used for the external dose rate (ESR and luminescent dating methods).

$D_{ext\beta}(t)$: The symbol often used for the external dose rate from β radiation (ESR and luminescent dating methods).

$D_{ext\gamma}(t)$: The symbol often used for the external dose rate from γ radiation (ESR and luminescent dating methods).

$D_{int}(t)$: The symbol often used for the internal dose rate (ESR and luminescent dating methods).

D amino acids: Optically active amino acids not normally produced by most living organisms. These are also produced by the geochemical or biochemical degradation of L-amino acids (see also L-amino acids).

Daughter isotope: Any radiogenic isotope.

Dalton: One atomic mass unit, u, also called an atomic mass unit (amu).

DC: Direct current.

DC conductivity loss: The conversion of EM energy into electrical energy as an EM signal propagates through a material. The amount EM energy loss (or energy conversion) is a function of the number of free electrons within the material.

DEC: Shorthand notation for dentine-enamel-cementum configuration used in describing tooth enamel samples for ESR dating.

Decay counting: See radioactive decay.

Deciduous teeth (dentition): The teeth initially erupted in juvenile mammals. Also known as milk teeth, they have poorly crystallized hydroxyapatite with higher carbonate apatite concentrations than the mature dentition that erupts in late adolescence.

Declination: 1) The deviation in degrees of a compass needle or the Earth's magnetic field from geographical or true north in the horizontal plane. 2) A protocol for measuring spectral peak heights or areas in complex peaks. A complex peak is separated mathematically into several theoretical peaks whose peaks are then measured.

Deconvolution: Inverse filtering, mathematical operation, in practice only an approximation, that compensates the (low-pass filtering) effect of a convolution.

DED: Shorthand notation for dentine-enamel-dentine configuration used in describing tooth enamel samples for ESR dating.

DEDC: Shorthand notation for dentine-enamel-dentine-cementum configuration used in describing tooth enamel samples for ESR dating.

Deep trap: Traps are defects in a crystal to which free electrons (those given enough energy by irradiation to leave their normal atomic sites) are attracted and may become trapped; a trapped electron is in a metastable state. An electron can be evicted from a trap by giving it sufficient energy; if the required energy is relatively small, the trap is said to be a shallow trap; if the energy is relatively large, it is said to be a deep trap. In the present context, "small" would be energies <1 electron volt (eV) and "large" would be energies >1.5 eV.

Deflation: A form of wind erosion that removes dry finely grained clays and silts from surficial sediment.

Degassing: Exsolution (change from a solution to a gas state) and/or expansion of gas, commonly methane caused by decreasing pressure and increasing temperature as a sediment core is retrieved. Cases are known when core liners have actually exploded.

Deglaciation: Exposure of a land surface due to melting and retreat of a glacier or an ice sheet.

Delayed neutron counting (DNC): A form of neutron activation analysis (NAA).

Denatured proteins or peptides: Proteins or peptides that have been partially decomposed, often through interfibril or intermonomer bond breakage. They are often partially or totally uncoiled and may be highly reactive.

Dendrochronology: Tree-ring time scale, which is based on the successive matching of overlapping tree ring thickness curves to obtain an annual chronology from present back in time.

Dendro years: Tree-ring years.

Dentine: Any of four biologically produced biomineralic compounds containing an organic matrix high in collagen and a hydroxyapatite mineral matrix, which forms part of the teeth in most vertebrates. Dentine has less mineral matrix and is softer than enamel, and a different histology from dental cementum. Fossil dentine can contain other apatite group minerals, including dahllite, francolite, and vivianite, and may lack most or all its organic matrix.

Depositional sequence: A stratigraphic unit bounded top and bottom by unconformities or correlative conformities.

DES: Shorthand notation for dentine-enamel-sediment configuration used in describing tooth enamel samples for ESR dating.

Desalt: To remove the dissolved mineral ions from a solution.

Detrital: Pertaining to material derived from detritus (i.e., loose rock, sediment, or other materials that are removed by mechanical means and transported to the depositional site).

Dextral (d) rotation (of light): The rightward rotation of plane-polarized light when viewed toward the oncoming light.

Dextrorotary: Any substance that can cause plane-polarized light to rotate dextrally.

DGPS: Differential global positioning systems.

DGT: Diffusive gradient in thin films.

Diagenesis: 1) Sedimentary: All the geochemical, physical, and biological processes that affect sediment after their initial deposition in the sediment, and during and after lithification, except surficial alteration (weathering) and metamorphosis. It can include compaction, cementation, leaching, reworking, authigenesis, replacement, crystallization, recrystallization, biological alteration, concretion formation, and any process at temperatures < 100–300 °C and pressures < 1 kb. Some restrict the term to the period before lithification. 2) Fossil: All the geochemical, physical, and biological processes that affect fossils, during weathering, after their initial deposition in the sediment, and during and after lithification, except metamorphosis.

Diagenetic: Sediment component(s) derived through modification of material after burial of the sediment.

Diagenetic alteration: Any geochemical changes arising from diagenesis.

Diagenetic temperature (T_d): An effective paleotemperature equivalent to the modern necessary to induce the amino acid racemization seen in a fossil.

Diamict: Term used to denote non-sorted sediments and rocks containing a wide range of particle sizes, regardless of genesis.

Diamicton: Poorly sorted unconsolidated sediment (e.g., till).

Diatom: A microscopic single-celled plant from the class Bacillariophyceae that lives in fresh and marine water and precipitate siliceous tests (frustules).

DIC: See dissolved inorganic carbon.

Dielectric constant: A measure of the amount of electric charge a substance can withstand at a given electric field strength.

Dielectric loss factor: The conversion of EM energy into another form (e.g., heat) as an EM signal propagates through a material. The amount EM energy loss (or energy conversion) is a function of the dielectric properties of the material.

Differential global positioning system: A two receiver GPS system that enables the subtraction of errors that are introduced into the GPS signal along its route.

Diffusional U uptake, leaching, or redistribution: Any nonrandom secondary U uptake, leaching, or redistribution event in which the change in U concentrations is not uniformly or proportionally distributed across the subsamples. (Also see equilibrative diffusional leaching, uptake, and redistribution, hot-atom diffusional uptake and leaching, hole filling diffusional uptake and redistribution.)

Dipole: A magnet that has a concentration of positive charge at one end and negative charge at the other end. A bar magnet is a good example of a dipole.

Dispersive diffusional U uptake, leaching, or redistribution: A secondary diffusional U uptake, leaching, or redistribution event in which the subsamples show more variation in their U concentrations after than before the secondary event, also known as disequilibrative diffusional uptake. (Also see diffusional uptake, leaching, and redistribution, hot-atom diffusional uptake and leaching, hole filling diffusional uptake and redistribution.)

Dissolved inorganic carbon (DIC): 1) Carbon contained in carbon dioxide, and bicarbonate and carbonate ions dissolved in water. 2) The sum of all dissolved inorganic carbon species ($CO_2 + HCO_3^- + CO_3^{2-} + H_2CO_3$) in an aqueous solution.

Dissolved organic carbon (DOC): 1) Carbon contained in colloidal ($< 10\,\mu$m) organic particles and in soluble organic compounds in water. 2) Organic molecules representing decay products of vegetation that pass through a $0.45\,\mu$m filter.

DNC: Delayed neutron counting (see neutron activation analysis).

DOC: See dissolved organic carbon.

Dolomite: 1) Mineralogy: A common, well-ordered, carbonate mineral with the formula $CaMg(CO_3)_2$; also called magnesian spar. 2) Sedimentology *sensu stricto*: A carbonate rock with more than 90% dolomite mineral, also known as dolostone. 3) Sedimentology *sensu lato*: A carbonate rock with more than 50% dolomite mineral, also known as magnesian limestone.

DOSECC: Drilling, Observation and Sampling of the Earth's Continental Crust, Inc.

(total) Dose rate ($D_\Sigma(t)$)**:** The total radiation dose rate that produces an ESR or luminescent dating signal in a sample (see internal and external dose rate).

Dosimeter (dosemeter): A device that can record or measure a radiation dose or dose rate.

Dosimetry: Techniques used for determining a radiation dose or dose rate.

Downlap: Initially inclined strata terminate downdip progressively against an initially horizontal or inclined surface.

Drift correction: Correction of a measured data set by subtracting the instrument drift, normally determined by regularly measuring the background signal, i.e., without sample/material.

Drill string: The set of pipes (or rods) often inside a casing to which the drill bits or coring tools are attached. The bits or tools can be either fixed to the end of the drill string and then lowered, or they may be lowered down the center of the drill string on a wire and latched or unlatched from the end of the drill string while it is in the hole.

Drilling muds: Fluids pumped through the annular space between the casing and the drill string in order to remove cuttings from the drill bit to the surface. In petroleum applications they are also used to resist pressure from compressed oil or gas. The mud is made by mixing water and clay, usually bentonite, in varying proportions.

Dry bulk density: Dry mass of sediment per unit *in situ* volume.

Dry density: Specific weight or density of all solid sediment components excluding water.

E' (ESR) signal: An ESR signal at $g = 2.0001$, which is easily measured at room temperature and used to date quartz, flint, and fault gouge.

EAAS: Electrothermal atomic absorption coefficient.

Early (U) uptake (EU): 1) An U uptake model used for calculating ESR ages for tooth enamel, mollusc shells, and other fossils tissues that can absorb U secondarily. It assumes that the sample absorbed all its U soon after burial, providing the youngest age given the accumulated dose, A_Σ, and external dose rate, $D_{ext}(t)$. 2) The equivalent U uptake model used calculating external radiation dose rates in sediment with components that absorb secondary U.

Echo sounding: A process of determining water depth through timing of sonic reflections from the water bottom.

ED: The equivalent dose, more commonly called the accumulated dose (used in ESR and luminescent dating methods).

EDS: Energy dispersive X-ray spectroscopy. X-ray detector typically attached to a scanning electron microscope (SEM) or electron microprobe.

Electrical conductivity: The ease with which electrons flow through a material.

Electric permittivity: See dielectric constant.

Electrical resistivity: Logging method based on the resistivity of sediments to the flow of electric currents.

Electromagnetic energy (EM): A form of energy, propagated through space or through a media in the form of an advancing disturbance in electric and magnetic fields. Examples of electromagnetic energy are light and radio signals.

Electromagnetic energy (EMR): See electromagnetic energy.

Electron microprobe analysis: Analytical technique where samples are exposed to electron beam. Characteristic secondary X-rays are generated and can be converted to concentration.

Electron paramagnetic resonance (EPR): A synonym for electron spin resonance.

Electron spin resonance (ESR) dating: A radiochemically based geochronological method in which the electrons trapped in crystal defects formed by ionizing radiation are measured by their resonant response to a strong magnetic field (see also luminescent dating techniques).

Electron spin resonance (ESR) isochron dating analysis: A variant on the standard ESR geochronological method used for tooth enamel, stalagmites, and ash, in which several subsamples are analyzed by the standard ESR method. If the sample has not suffered nonuniform U uptake, the method will yield either the age and external dose, A_{ext}, simultaneously, or the U uptake model for teeth, if the age and external dose rate are known independently.

Electron spin resonance microscope (ESRM): An ESR spectrometer that has been modified to scan solid mineral surfaces to determine the spin concentrations for a preset signal at various points, capable of producing 2D, 3D, and 4D images, and combination with other scanning systems, such as NMR and CT.

Electron spin resonance (ESR) spectrometer: A machine that uses a microwave signal to create resonance between the unpaired electrons in minerals or liquids and an externally applied strong magnetic field, thereby allowing their identification and measurement.

Electron volt (eV): A unit of energy; $1\,eV = 1.602 \times 10^{-19}\,J$.

EM: Electromagnetic energy.

EMR: Electromagnetic radiation.

Enamel: The hardest biologically produced biomineralic compound containing a hydroxyapatite mineral matrix with a diffuse poorly characterized organic matrix, which forms part of the teeth in most vertebrates. Enamel has 98 wt% mineral matrix, more than other organic tissue. Fossil enamel can contain other apatite group minerals, including dahllite, francolite, and vivianite, and may lack all its organic matrix.

Enantiomers: Stereoisomers whose molecule structures mirror each other at one or more asymmetrical carbon atoms (also called isomers, stereoisomers; see also racimers, epimers).

Enantiomerization: 1) Any process by which an asymmetric carbon atom in optically active organic molecule changes its optical configuration. For amino acids in living tissues, the process usually involves converting L-amino acids to a mixture of D- and L-amino acids (see also epimerization, racemization). 2) Any process by which an optically active stereoisomer is converted into a mixture of two isomers having no optical activity.

Endogenic: 1) Sediment component(s) formed within the basin, specifically from within the water column. 2) Sometimes used synonymously with authigenic.

Endogenic geomorphic processes: Processes originating within the Earth that create relief on the Earth's surface. Tectonism and volcanism are the principal endogenic processes.

Energy dispersive x-ray spectroscopy (EDS): A method for elemental analysis based on using a focussed electron beam (in an electron microscope) to cause the fluorescence of secondary x-rays from atoms in a specimen.

Epimers: Stereoisomers whose molecule structures mirror each other at only one of two or more asymmetrical carbon atoms (also called isomers, stereoisomers; see also racimers, enantiomers).

EPR: Electron paramagnetic resonance, also called electron spin resonance (ESR).

Equilibrative diffusional U uptake, leaching, or redistribution: A secondary diffusional U uptake, leaching, or redistribution event in which the subsamples show less variation in their U concentrations after than before the secondary event. Also see diffusional uptake, leaching, and redistribution, dispersive diffusional leaching, uptake, and redistribution, hot-atom diffusional uptake and leaching, hole filling diffusional uptake and redistribution.

Equivalent dose (A_Σ, ED): 1) The laboratory beta or gamma radiation dose that results in the same measured thermoluminescence or optically-stimulated luminescence as does the natural radiation dose. 2) The ionizing radiation dose equivalent to the ESR or luminescent signal intensity seen in a sample, usually known as the accumulated dose and occasionally as the archaeological dose.

Erosion: The general process by which Earth materials are loosened and moved from on place to another.

Erosional truncation: A type of stratal geometry typified by an angular unconformity; dipping beds terminating against upper boundary, as a consequence of erosion.

ESR: Electron spin resonance, also called electron paramagnetic resonance (EPR).

ESR signal: A short-lived ESR signal at or near $g = 2.0001$ with $\tau = 40\,$y, that often interferes with the E' signal.

Ester: An organic molecule containing an ester group (-C(O)OC-).

Esterify: To form an ester from an organic compound.

EU: Early U uptake model (ESR and luminescent dating).

Eustasy: Fluctuations in sea level attributed to changes in the absolute quantity of seawater in the oceans.

Excited state: Any energy level for an electron that is above its ground state. In many materials, the conduction band is an excited state (see also conduction band, ground state, valence band).

Extension rods: Extensions added to the top of the drill string or core barrel to lower it progressively to greater depths.

External (accumulated) dose (A_{ext}): The component in the accumulated dose arising from sources external to the sample, usually from the sediment and cosmic radiation sources (see accumulated dose).

External dose rate, $D_{ext}(t)$: The radiation dose rate produced by radioactive elements outside the dating sample. For teeth, this excludes the enamel, dentine, cementum, and any attached bone, which are all considered to produce the internal dose rate. The external dose does include dose arising from the adjacent sediment ($D_{sed}(t)$) and cosmic sources ($D_{cos}(t)$).

Extractable: That part of the total element concentration that can be extracted using a specified extractant.

Extrusion: Removal of sediment from core barrel by pressure. A technique used when there is no liner.

F: Geomagnetic intensity.

$\mathscr{F}$, ($\mathscr{F}_{Pb}$): Fraction of input of fallout (^{210}Pb) to the lake delivered to the sediment record.

f_D: Fraction of radionuclides in water column attached to suspended particles.

Fault gouge: Soft finely grained sediment formed by comminuting rock or sediment between fault planes. Gouge minerals may be recemented.

Feldspar: A mineral group comprising silicates of the general formula: XZ_4O_8, where X = Ba, Ca, K, Na, NH_4, Sr; Z = Al, B. Si.

Felsic: Light-colored rocks, rich in feldspar and quartz.

Ferricrete: A conglomeratic surficial sedimentary crust cemented by iron and aluminum oxides, often formed under strongly evaporative conditions or in deforested tropical soils (see also laterite, caliche, calcrete, gypcrete, ferricrete).

Fickian diffusion: Theory of diffusion proposed by Fick in the mid-nineteenth century in which the rate of transfer of the diffusing substance is proportional to the concentration gradient.

First order reversible linear kinetics: A chemical reaction that can proceed in either the forward or backward reaction depending on geochemical conditions and does produce a linear relationship between the reaction rate constant, k_i, and the inverse of temperature, T^{-1}.

Fission track analysis: Geochronological tool for dating volcanic rocks, utilizing counts in crystals resulting from radioactive decay.

Flint: A hard dense micro- or cryptocrystalline sedimentary rock with interlocking quartz crystals that displays concoidal fracture, prized by hominids as a material for making tools. Often occurring as nodules in limestones or dolomites, it is considered by some to be synonymous with chert. Flint is sometimes considered only to be gray or black varieties, the varieties with visible fossil inclusions, those types with more visible crystals or a sugary texture (see chert).

Fluvial: Pertaining to streams.

Food irradiation: Any treatment used to sterilize or preserve freshness in food involving dosage with ionizing radiation. γ radiation is most frequently used.

Foraminifera: Microscopic protozoans belonging to the class Foraminifera, characterized by tests containing one or more chambers, made of calcite, silica, aragonite, or agglutinated particles. Also called foram, foraminiferid.

Foreland: A wave- and current-constructed coastal depositional landform attached to the mainland along its entire planimetric length, and built out from the mainland so that its length to width ratio is less than 1.0.

Fossil: 1) *Sensu stricto*: Any evidence for past lifeforms in which the original minerals or tissues have been completely replaced by new minerals. 2) *Sensu lato*: Any evidence for past lifeforms in which the original minerals or tissues have experienced some secondary cementation or have suffered some diagenetic alteration. 3) *Sensu lato extremis*: Any dead organism or other evidence for past life.

Fractionation: 1) Selective separation of chemical elements or isotopes through physical, chemical or biochemical processes. For example, the fractionation of carbon isotopes: of the naturally occurring carbon isotopes ~98.9% is ^{12}C, 1.1% is ^{13}C and only 1 part in 10^{10}% is ^{14}C. In nature, however, a fractionation of this ratio occurs (e.g., photosynthesis results in an enrichment of ^{12}C relative to the other isotopes in most plant tissues). Based on thermodynamical laws, which show that the heavier isotope ^{14}C is twice as enriched as ^{13}C, radiocarbon laboratories correct for the probable effects of fractionation. ^{13}C can be measured in a sub-sample of the material to be dated and the ^{13}C : ^{12}C ratio is then compared with a standard (PDB) and published as deviation from this standard. See also normalization. 2) In physical and chemical processes involving the isotopes of a particular element, the relative abundance of the isotopes may change between the initial substance (the "substrate") and its product. In isotope geochemistry, this change is commonly referred to as fractionation. (See also: fractionation factor; isotope effects).

***g* value (Lande's factor):** A dimensionless number that uniquely describes the characteristics for an ESR peak. Most geologically or archaeologically interesting ESR signals fall within $3 < g < 1.9$.

GAD: Geocentric axial dipole.

Gain function: A formula by which the amplitude of recorded energy is increased. Gain functions are applied to emphasize weaker reflections.

Gamma ray (γ): A high-energy photon. Photons emitted during nuclear decay are referred to as gamma rays.

Gamma rays (natural): Energy emitted by natural radioactive substances corresponding to X-rays and visible light but with much shorter wavelengths.

Gamma ray attenuation: Absorption of gamma rays passing through the sediment is caused by Compton attenuation (scattering), pair production, and by photoelectric absorption. The amount of gamma ray attenuation is a measure of bulk density, from which water content and porosity can be calculated.

Gamma ray attenuation porosity evaluator: Instrument that measures gamma ray attenuation through unsplit cores and provides an estimate for bulk density, water content, and porosity.

Gamma spectrometry: Measurement of gamma particles emitted by radioactive substances.

Gar (fish, pike): Any ganoid fish of the genus *Lepisosteus* or several closely related extinct genera, with elongated bodies, long snouts, and enameloid scales, found in fresh and brackish waters in North America.

Gas chromatography (GC): A process for separating and analyzing mixed gases or vapors in which a carrier gas (e.g., helium, argon, nitrogen, hydrogen) carrying the dissolved gaseous or vaporous mixture is passed over a nonvolatile liquid (gas-liquid chromatography) or solid (gas-solid chromatography) coating an inert porous solid. The analytes are resolved by the differential mobility rates of the constituents as they repeatedly adsorb to, and then desorb from, the inert solid surfaces.

GC: Gas chromatograph or gas chromatography.

Geocentric axial dipole (GAD): The Earth's geomagnetic field, when averaged over several thousand years, approximates a dipole or bar magnet located at the center of the Earth with its long axis along the Earth's axis of rotation. The Earth's field can therefore be described as a geocentric axial dipole.

Geomagnetic excursion: Brief (10^3 years) but significant departure of the Earth's field from the geocentric axial dipole configuration where the field can approach a polarity transition in magnitude, but then returns to its pre-existing polarity.

Geomagnetic field: The magnetic field produced by the motion of the inner and outer cores of Earth.

Geomagnetic intensity (F): Total intensity of the Earth's magnetic field, in units of Gauss (cgs) or wbm^{-2} (SI). Intensity can be viewed as the magnitude of a vector (direction of the field).

Geomagnetic Secular Variation (SV): The typical or normal low amplitude, high frequency temporal variation of the Earth's magnetic field between polarity transitions.

Geomorphology: The scientific study of landforms and landforming processes; geomorphology is a subfield of both physical geography and geology.

Germanium (Ge ESR) signal: A complex ESR signal, arising from overlapping $(GeO_4/Li^+)^0$ and $(GeO_4/Na^+)^0$ defects. More easily bleached than most others in many quartz or related minerals, it is measured at $70\,°K$.

GFZ: GeoForschungsZentrum, Potsdam in Germany.

GISP: Greenland Ice Sheet Project led by American scientists.

GLAD800: Global lake drilling rig designed to penetrate up to 800 m of water plus sediment.

Global positioning system (GPS): 1) A series of satellites with well-known orbits that transmit a coded signal to receivers which can be used to determine the position of the receiver. 2) Determines x, y, and z locational coordinates of a point on the Earth by evaluating satellite position information sent from multiple satellites to a receiver at the point of interest on Earth.

Glu: Glutamic acid (an amino acid).

Glutamic acid (Glu): An uniasymmetric amino acid occurring abundantly in natural proteins with the formula $HO_2CCH_2CH_2CH(NH_2)CO_2H$, also called α-aminoglutaric acid and 2-aminopentanedioic acid.

Gly: Glycine (an amino acid).

Glycine (Gly): The only symmetric amino acid occurring abundantly in natural proteins with the formula $H_2NCH_2CO_2H$, also called aminoacetic acid.

GPR: Ground-penetrating radar.

GPS: Global Positioning System.

Grade scale: A systematic, arbitrary division of the essentially continuous range of particle sizes of a sediment or rock into a series of classes (grades) for the purpose of terminology and statistical analysis.

Grain size: Size of particles that make up a rock or sediment.

GRAPE: Gamma Ray Attenuation Porosity Evaluator.

Graphite: An opaque, lustrous black to steel gray hexagonal mineral composed of C.

Gravity corer: Any type of corer that uses its own submerged weight to drive the sample tube into the sediment.

Gravity system: A drilling system using only the force of gravity to penetrate the sediment, for example the Kullenberg corer.

Gray (Gy): 1) The S.I. unit of radiation dose. 1 Gy = 1 J of radiation energy absorbed per kg of matter. 2) A unit of ionizing radiation or absorbed dose, equivalent to 0.1 rads: 1 Gray = 1 J/kg = 1 m^2/s^2.

GRIP ice core: Acronym for Greenland Ice Core Project, a European effort, which led to obtaining a ca. 350,000 year long ice core from the Greenland summit.

Ground-penetrating radar: A geophysical technique used to image the subsurface. Pulses of electromagnetic energy are radiated into the subsurface and return time and strength of reflections from interfaces between differing materials is recorded.

Ground state: The lowest energy level at which a given electron can exist in an atom. This includes the valance band (see also conduction band, excited state, valence band).

Gypcrete: A conglomeratic surficial sedimentary crust cemented by secondary gypsum or anhydrite. Often formed under strongly evaporative conditions, it is common in playa lake beachrock (see also caliche, calcrete, silcrete, ferricrete).

Gypsum: A common hydrated mineral with the formula $CaSO_4 \cdot 2H_2O$, frequently associated with anhydrite and halite. Also called gypsite, plaster of Paris, gyp.

Gyttja: 1) Low density organic mud found at the bottom of lakes. 2) A Swedish term, introduced by H. von Post, denoting organic lake sediments, which are formed mainly under anaerobic conditions and have an organic content of $>30\%$. It consists of a mixture of allochthonous and autochthonous material. Coarse detritus gyttja is rich in macro- and microfossils and occurs in shallow water, while homogenous fine detritus gyttja is deposited in deeper water. Drift gyttja with larger, rounded plant fragments and sand is found in the littoral zone. Algae gyttja is mainly composed of autochthonous algae detritus and to a minor extent of minerogenic particles and often forms in shallow, high-productive lakes. Calcareous gyttja, which is usually rich in fragments from carbonate algae and shells, has a $CaCO_3$ content of 20–80% and forms in shallow water. Shell-gyttja is mainly composed of shell fragments, but may have a matrix of algae or fine detritus. Gyttja clay has an organic carbon content of 3–6% and clay gyttja of 6–30%. 3) A dark anaerobic mud, rich in organic matter, formed primarily in well-oxygenated lakes and marshes rich in nutrients.

H: 1) Magnetic field, measured in Am^{-1}. 2) Henry; SI unit of magnetic inductance.

Half-life: Period of time required to reduce a given quantity of the parent nuclide to one half.

Halite: A common evaporitic mineral having the chemical formula NaCl; also known as rock salt.

HAP: Hydroxyapatite.

Hard water lakes: Lakes with a fairly high pH (>7.5) and alkalinity. They are often situated in areas with carbonate-rich bedrock or till where the lake and groundwater is enriched in carbonate ions.

Hard-water effect: Sub-aquatic photosynthesis, water uptake from carbonate-rich groundwater, and carbonate secretion by freshwater or offshore organisms lead to an aging effect of bulk sediments and aquatic organisms due to the often high age of the DIC.

Heave: Vertical movement of the surface of a water body, and therefore of the vessel.

Heinrich Events: Marine sediment layers in the North Atlantic consisting of ice-rafted, detrital material that was released from melting icebergs, mainly related to icebergs coming from the Laurentide (North American) ice sheet, but also from the Scandinavian ice sheet.

Hemiarid basin: A closed drainage basin that has humid-climate highlands and arid lowlands so that runoff produced mostly in the highlands is primarily collected and evaporated in the lowlands.

High performance liquid chromatography: Technique for separating chemicals in solution by passing an inert solvent over a sorbent.

High precision dating: High-precision radiocarbon dating; dense set of radiocarbon dates with small confidence intervals.

"hole-filling" diffusional uptake: A secondary U uptake event that affects those subsamples with lower initial U concentrations gain, while those higher initial concentrations may gain little or no U.

"hole-filling" diffusional redistribution: A penecontemporaneous secondary U uptake and leaching event in which those subsamples with initially higher U concentrations lose U while the less uraniferous subsamples gain U.

Holocene (Recent): 1) Time: The last epoch within the Quaternary Period and the Cenozoic Era, equivalent to Oxygen Isotope Stage 2. Traditionally defined to have begun at 10 ka, some now place the boundary at approximately 10.8 ka, at the end of the Younger Dryas event. 2) Rocks: The rocks or sediment formed during the Holocene Epoch, also known as the Flandrian in Britain.

Hominid: Any member of the taxonomic group, Hominidae, now considered to include the genera, *Homo, Australopithecus, Paranthropus*, and sometimes, *Ardipithecus*.

"Hot-atom" diffusional U leaching or uptake: A secondary U uptake or leaching event that affects those subsamples with higher initial U concentrations more strongly than the less uraniferous subsamples (i.e., more uraniferous subsamples gain or lose more U secondarily), due to the hot-atom effect.

Hot spring: A spring whose temperature is above $37\,°C$.

hp: Horse power, unit of power corresponding to 75 kg per sec., i.e., 736 watts.

HPLC: High performance (or precision) liquid chromatography.

Humic-acids: 1) Organic material originating from almost any thermodynamically less stable organic species, which is potentially highly mobile in sediments. 2) A complex mixture of organic molecules extracted from soil, low-rank coal, and decayed plant matter by alkalic solutions.

Humic acid (ESR) signal: An ESR signal at $g = 2.0040$; seen in some travertine and tufa.

Humic-soluble: See alkali-soluble.

Humin-insoluble: Insoluble, but humic-like material, which may grade chemically into graphite; can be oxidized by treatment with sodium hypochloride to remove lignins from cellulose; see also alkali-insoluble.

Hydrograph: Graph depicting the change in size of a water body over time; most paleolake hydrographs plot water level elevation versus time, but lake surface area or volume versus time may also be used.

Hydrography: The spatial configuration of a water body.

Hydrophone: A detector sensitive to pressure variations, such as the passage of an acoustic pulse in the water column.

Hydrostatic pressure: The pressure or force imparted on an object by its immersion in a liquid. If the fluid is at rest, the unit pressure is the same in all directions. This principle is known as Pascal's Law.

Hydroxyapatite (HAP): An apatite group mineral with the formula $[Ca_5(PO_4)_3(OH)]_2$, also called hydroxylapatite. It forms the major mineral phase in vertebrate bone, enamel, dentine, dental cementum, antler, and some fish scales, as well as some other connective tissues.

Hydroxyapatite (HAP ESR) signal: A radiation-sensitive ESR signal at $g = 2.0018$ arising from a defect related to a carbonate substituting for phosphate, with a signal lifetime, $(\tau \sim 10^{19}$ y. Observed in tooth enamel, bone, dentine, dental cementum, and some fish scales, it is used extensively for ESR dating tooth enamel.

Hydroxylysine: A rare multiasymmetric amino acid occurring in trace amounts in some natural proteins with the formula $H_2N(HO)CH(CH_2)_3CH(NH_2)CO_2H$. A minor constituent in collagen, it only occurs naturally in the L form (2S, 3S-hydroxylysine) and epimerizes into D-allohydroxylysine.

Hydroxyproline: A rare uniasymmetric amino acid occurring in trace amounts in some natural proteins with the formula $HNCH_2CH(OH)CH_2CHCO_2H$, also known as (α-4-hydroxyproline. A major constituent in collagen, it only occurs naturally in the L form (2S,3S-hydroxyproline) and epimerizes into D-allohydroproline.

Hypersaline: 1) *Sensu lato*: Having a total dissolved solid content of greater than about 40 ppt. 2) *Sensu stricto*: Having salinity above the lowest salinity needed to precipitate halite.

Hypertectonism: High rates of movement or deformation of the lithosphere (crust and upper mantle) caused by processes originating within the Earth.

Hypsograph: Graph showing how lake surface area changes with water level.

Hz: Hertz, a unit for frequency, $1\,Hz = 1\,s^{-1}$.

I: Inclination; the dip angle of the magnetic vector below the horizontal plane.

I_0: Gamma ray beam intensity at its origin.

I_{max}: The symbol commonly used for maximum intensity (ESR or luminescent dating).

*i*ø: Inner diameter of a core tube or barrel.

ICDP: International Continental Drilling Program.

ICP-MS: Inductively coupled plasma-mass spectrometer.

IGBP: International Geosphere-Biosphere Programme.

Ignimbrite: Pyroclastic material fused following volcanic eruption.

Ile: Isoleucine (an amino acid).

Incandescence: The light emitted by anything that is hot, such as that emitted by an ordinary light bulb or a red-hot stove.

Inclination (*I*): The angle of deviation in degrees of a compass needle or the Earth's magnetic field below the horizontal plane. Also called the dip angle.

Incomplete zeroing: Any exposure to intense heat, light, or pressure that does not result in an ESR signal with zero intensity (see zeroing, residual signal).

Induced amino acid racemization (AAR): The racemization caused by the amino acid racemization analysis process itself; also called the background racemization.

Inductively coupled plasma: An emission spectrometric method where excitation is achieved using a plasma. Detection can be photometric or mass spectrophotometric.

Inductively coupled plasma mass spectrometry (ICP-MS): Analytical technique where samples are injected into a plasma as solution. Ion are generated and focused into a mass spectrometer for detection.

Indurated sediment: Sediment that has become hardened.

Infauna: 1) *Sensu stricto*: Aquatic animals living within, rather than on, the sediment. 2) *Sensu lato*: Any animal that lives within the sediment or within another organism or fossil.

Infaunal: An adjective meaning "living in the soft sediment near the sediment-water interface". Can refer to any substrate; not necessarily restricted to sediment.

Infestation: A colonization or invasion by large numbers of organisms.

Inflection points: A point on a graphical line or surface where the slope changes in a non-linear fashion. In amino acid racemization dating, it indicates a change in racemization rate constant, while in ESR or luminescent technique additive dose curves it indicates the peak under consideration is a complex peak.

Insoluble (INS): See alkali-insoluble.

Insoluble old carbon: Bedrock carbon particles, such as e.g., coal, which cannot be dissolved using the conventional radiocarbon acid-alkali-acid pre-treatment methods.

Instrumental Neutron Activation Analysis (INAA): Analytical technique where samples are irradiated, generating new, short-lived radioactive isotopes, which emit characteristic gamma radiations.

Insufficient sunlight exposure: An assumption in optical dating is that the sunlight exposure just before burial emptied all of the relevant electron traps; if the sunlight exposure did not do this, it was "insufficient".

Intensity of magnetization (M): The magnitude of magnetic remanence.

(amino) Interacid ratios: Concentration ratios between two or more amino acids.

(signal) Interference: 1) The inhibition or prevention of clear signal reception for any electronic signal. 2) The distorted portion in an electronic signal.

Interlaboratory calibration (tests): Any series of standardized analyses using a single or a series of unknown samples used to determine the collective reliability for an analytical method used by many laboratories.

Internal (accumulated) dose (A_{int}): The component in the accumulated dose arising from sources within the sample (see accumulated dose).

Internal dose rate ($D_{int}(t)$): The radiation dose rate produced by radioactive elements within the dating sample. For teeth, this includes the enamel ($D_{en}(t)$), dentine ($D_{den}(t)$), cementum ($D_{cem}(t)$), and any attached bone ($D_{bone}(t)$), regardless of the tissue being dated (see external dose rate).

Intrinsic resolution: Temporal resolution within a stratigraphy, i.e., the amount of time represented in a stratigraphy that is obscured by mixing processes.

INV_w: Inventory of total suspended sediment in the water column on an areal basis (mg cm^{-2}).

Inventory: Total quantity of a substance in a defined volume.

Ion chromatograph: Analytical device that determines the concentration of both cations and anions in solution.

Ionizing radiation: Any electromagnetic or particulate radiation that displaces electrons from atoms. It includes α, β, γ, χ, and cosmic radiation.

IR: Infrared or InfraRed.

IRSL: Infrared stimulated luminescence, usually referring to dating unless otherwise specified. A form of optically stimulated luminescence.

Isochromous: Equivalent in age or time.

Isochron: Any line or surface on a map or graph that links data points having the same chronological age.

Isoleucine (Ile): A multiasymmetric amino acid occurring abundantly in natural proteins with the formula $CH_3CH_2CH(CH_3)C(NH_2)HCO_2H$. It only occurs naturally in the L form (2S,3S-isoleucine) and epimerizes into D-alloisoleucine.

Isomers: Chemical compounds having the same formula, but different three-dimensional structures.

Isopleth: 1) Geochemistry *sensu stricto*: A line or surface on a map or chart representing a constant physical or geochemical quantity, such as concentration, temperature, etc. 2) Geochemistry *sensu lato*: An "isocompositional" line or surface, i.e., a line depicting a constant concentration, or concentration ratio. 3) General *sensu lato:* Any line or surface representing a constant physical or geochemical quantity, including isograds, isolats, isohyets, isopachs, and isorithms.

Isostatic deflection: The amount of isostatic depression or rebound achieved at a given point on the Earth's surface. Varies spatially and temporally over a load-impacted area of uniform geophysical characteristics as a consequence of the load distribution and the loading/unloading history.

Isostatic rebound: 1) The unloading effect of melting and retreating ice sheets on the Earth's mantle and crust, resulting in local/regional land uplift. 2) A regional increase in Earth surface elevation due to removal of a large load of ice, rock, sediments, or ocean or lake water.

Isotropic: 1) Giving complete extinction when rotated under cross-polarised light. 2) Denoting a medium in which a particular physical property is independent of direction.

J: Joule, the S.I. unit of energy.

Jurassic: The geologic period between 208–146 million years ago, which is the middle of three periods of the Mesozoic Era (245–65 million years ago); also denotes the system of strata deposited during this period.

K: Absolute temperature in degrees Kelvin.

K_1: The symbol usually associated with the forward rate constant for a chemical reaction.

K_2: The symbol usually associated with the backward rate constant for a chemical reaction.

K': The symbol usually associated with the total (forward + backward) rate constant for a chemical reaction.

k_α: The symbol used for the α efficiency factor.

k_i: The symbol often used for the racemization rate constant at a temperature, T_l .

k_0: Propagation constant of electromagnetic energy in free space.

ka: Literally, thousand years ago. This time unit is used to delineate dates or ages, and is calibrated (or linked) to the present (i.e., "10 ka" implies "10 thousand years ago"; see also ky, ky BP). It is used for dates for most radiometric and many chemical methods, except for ^{14}C ages.

K/Ar (dating): A radiometric geochronological method often used to date volcanic and igneous rocks. It uses the quantity of ^{40}Ar produced by radioactive ^{40}K to determine the age.

Kappabridge: Instrument for measuring magnetic susceptibility.

KeV: Kilo electron volts.

kHZ: Kilo Hertz.

kN: Kilonewton, a unit of force (see Newton).

kW: Kilowatt, a unit of power.

ky: Literally, thousand years. This time unit is used to delineate time spans, and is not calibrated to any particular time (i.e.,"10 ky" implies "10 thousand years"; see also ka, ky BP).

ky BP: Literally, thousand years before present. This time unit is used to delineate ^{14}C dates or ages, and is calibrated (or linked) to 1950 (i.e., "10 ky BP" implies "10 thousand years before 1950"; see also ka, ky).

L-amino acids: Optically active amino acids normally produced by most living organisms. They are also produced by the geochemical or biochemical degradation of D-amino acids (see also D-amino acids).

L-band ESR: ESR spectrometry using microwave frequencies near 1.5 GHz under a 100 kHz field modulation, with a field center near 53.5 mT for $g\sim2$.

Lake isolation: The emergence of a basin threshold from the sea, e.g., due to isostatic rebound, resulting in that a former marine bay or lagoon becomes isolated from the sea and becomes a freshwater lake.

Laminae: Sedimentary beds or layers that are less than 10 mm in thickness.

Lande's factor (_g_ value): A dimensionless number that uniquely describes the ESR characteristics for a peak.

Lapilli: Pyroclastic material 2–64 mm.

Last Glacial Maximum: The last really cold phase of the last glacial cycle at ca. 20,000 years ago, when most glaciers and ice sheets reached their maximum extension.

Last Interglacial: The warmer time period between the penultimate and last (Wisconsinan, Würm, Weichselian) glacial advance. Although some controversy surrounds the exact period covered by this interval, depending on where one places the start of the most recent glacial advance, the time period is often cited as approximately 128–74 ka (Oxygen Isotope Stages 5e–5a) for those who opt for a long chronology, or alternatively, 128–115 ka (only Oxygen Isotope Stages 5e) for the short chronology.

Laterite: A well-weathered tropical soil rich in iron and aluminum oxides, with quartz and kaolinite, often cemented by secondary iron-aluminum oxide and silica cements into hardgrounds.

Late Weichselian: Term used to denote the youngest part of the last glacial cycle (ca. 30,000–11,500 years ago) in northern Europe; often referred to as Late Glacial time. It corresponds to the Late Devensian on the British Isle, to the Late Würmian in central Europe, to the Late Valdaj in Russia and to the Late Wisconsinan in North America. See also Last Glacial Maximum.

Late Wisconsinan: The time period or physical events associated with the last glacial advance within North America, whose maximum extent occurred at about 18 ka. The time range is currently estimated at approximately 25–10.8 ka, and is considered equivalent to Oxygen Isotope Stage 2, the Late Würm, or Late Weichselian.

Law of Superposition: The law or principle of superposition. The law of statigraphic sequence in which older beds are successively covered by younger and younger layers. This law was first most clearly stated by James Hutton with respect to geologic strata.

LBHP: Laboratoire de Botanique Historique et Palynologie in France.

Leaching: 1) The separation, selective removal, or dissolution of soluble constituents from a rock or fossil by percolating groundwater or hydrothermal solutions. 2) The removal in solution of nutritive or harmful constituents, including mineral salts and organic matter, from an upper to a lower soil horizon or sedimentary layer by naturally (e.g., rainwater, percolating groundwater) or artificially (e.g., irrigation).

Leu: Leucine (an amino acid).

Leucine (Leu): A uniasymmetric amino acid occurring abundantly in natural proteins with the formula $(CH_3)_2CHCH_2CH(NH_2)CO_2H$, also known as α-aminoisocaproamic acid.

Levral (l) rotation (of light): The leftward rotation of plane-polarized light when viewed toward the oncoming light.

Levrorotary: Any substance that can cause plane-polarized light to rotate levrally.

LGQ: Laboratoire de Géologie du Quaternaire in France.

Life or living position: The normal orientation and stance for an organism while alive.

Light sensitivity: The propensity for an ESR or luminescent signal to partially zero when exposed to the sun or other light.

Lignite: Coal of relative recent origin, intermediate between peat and bituminous coal, often contains patterns from the wood from which it formed.

Limnic: Lacustrine; often implying freshwater.

Linear (U) uptake (EU): 1) An U uptake model used for calculating ESR ages for tooth enamel, mollusk shells, and other fossils tissues that can absorb U secondarily. It assumes that the sample absorbs U at a constant rate throughout its burial history, giving a median age given the accumulated dose, A_Σ, and external dose rate, $D_{ext}(t)$. 2) The equivalent U uptake model used calculating external radiation dose rates in sediment with components that absorb secondary U.

Liquid chromatography (LC): A process for separating and analyzing ions or other solutes in which a liquid solvent (e.g., water, benzene) carrying the dissolved solute mixture passes over a nonvolatile liquid (liquid-liquid chromatography) or solid (liquid-solid chromatography) coating an inert porous solid. The analyte solutes are resolved by the differential mobility rates of the constituents as they repeatedly adsorb to, and then desorb from, the inert solid surfaces. Also called column chromatography.

Liquefaction: The temporary transformation of a stable granular material into a fluid by application of shock forces or vibration in association with a liquid.

Lithology: Composition of rock or sediment.

Lithostratigraphy: The description and subdivision of stratified rock and sediment units based on a unit's lithologic characteristics, such as mineralogical composition, grain size, sorting, fossil content, and sedimentary structures.

Lithostatic pressure: The pressure imposed by the weight of lithologic material deposited on top of the sample, sometimes referred to as overburden pressure.

Loess: Aeolian sediment, usually predominantly silt-sized grains, but possibly including clays and fine sands, produced by glacial activity, deposited up to thousands of kilometres away from the glacier. A common sediment in Quaternary and late Pliocene stratigraphic sequences in China, Eastern Europe, and the US, it may be interstratified with paleosols, tufa travertines, glacial and periglacial sediment packages.

Logging: 1) Systematic (generally equidistant) measurement of a physical property along the axis of a core or a borehole. 2) Non-destructive measurements of physical sediment properties down a borehole or along a sediment core at frequent intervals or on a continuous basis.

LRC: Limnological Research Center at the University of Minnesota in the United States.

LU: Linear U uptake model (ESR and luminescent dating).

Luminescence: The light emitted by something in response to a stimulus. Specific names are used for different stimuli, e.g., thermoluminescence, optically-stimulated luminescence, photoluminescence, triboluminescence etc.

Luminescence age: An age obtained by either thermoluminescence dating or optical dating.

Luminescent (TL) dating techniques: A family of radiochemically based trapped-charge geochronological techniques that measures the light emitted by electrons trapped in crystal defects formed by ionizing radiation upon their evacuation from their traps when stimulated by various energy sources (see thermoluminescence dating, optically stimulated luminescence dating, radioluminescence dating).

Lumpy sites or units: Sedimentary units or sites in which the sediment show inhomogeneity in grain size and/or composition, and therefore will likely have significant inhomogeneity in dose production rates over short distances (see also smooth unit).

Lys: Lysine (an amino acid).

Lysine (Lys): A uniasymmetric amino acid occurring in natural proteins with the formula $H_2N(CH_2)_4CH(NH_2)CO_2H$, also known as α,ε-diaminocaproic acid.

m_{Mg}: Moles of magnesium per kilogram of water.

$M(m)$: Depth in sediment core measured as cumulative dry mass.

M: Magnetization, either induced and/or permanent, measured in Am^{-1}.

Ma: One million years ago.

Macrofossils: Remains of plants or animals that can be seen by eye or under a stereo microscope in a sediment.

Mafic: Dark-colored, iron and magnesian-rich minerals found in basic, igneous rocks.

Magnetic flux induction: The magnetic field strength generated within a material resulting from the material being placed within a magnetic field.

Magnetic permeability: The property of a material to modify the magnetic induction resulting when the material is subjected to a magnetic field or magnetizing force.

Magnetic susceptibility: 1) The proportionality factor between an induced magnetization and the inducing magnetic field. 2) The property of a material that determines the size of an applied magnetic field required to generate a certain level of magnetism in the material.

Magnetite: An iron oxide (Fe_3O_4) that is most important magnetic mineral. It is strongly ferrimagnetic, and fine-grained magnetite ($<2\,\mu$m) forms a very stable magnetic signal. Can undergo diagenesis under anoxic conditions.

Magnetometer: Instrument that measures magnetic intensities.

Magnetostratigraphy: A stratigraphic classification and dating system that organizes rocks and sediments according to the sampling site-dependent polarity acquired at the time of emplacement. The time control is provided by radiometric dating, seafloor spreading rates, biostratigraphic constraints, and orbital tuning.

Mandrel: A mechanical device consisting of a bar or tubular section for holding or positioning a part for forming, machining or extruding. Some tubes are formed by drawing them over a fixed mandrel (known as D.O.M. stock). With extrusion of cores, the mandrel is used to support the core tube and expel the core.

Marine reservoir effect: ^{14}C is transferred as $^{14}CO_2$ from the atmosphere to the oceans only across the ocean surface. However, the mixing rate of surface and deep ocean waters is very slow and ^{14}C decays in deep ocean waters without replenishment. Hence, seawater has a reservoir effect or an apparent age (e.g., ~400 years in North Atlantic surface waters, ~580 years in parts of the equatorial East Pacific, ~2000 years in the deep ocean). ^{14}C dates on foraminifera from deep ocean sediments and on fossils from coastal regions, where upwelling of deep water occurs, have, therefore, to be corrected for the age of the seawater.

Marker bed: A distinctive sedimentary unit that is recognizable throughout the depositional basin.

Mass spectrometer (MS): 1) An instrument for measuring the relative abundance of isotopes. 2) An apparatus used to separate a stream of ions according to their mass to charge ratio.

Mass susceptibility: Magnetic susceptibility related to the mass of a test sample, measured in $m^3\ kg^{-1}$.

Mastodon(t): Any member of the extinct elephantid group, including *Mammuthus*, having low-crowned teeth with closed roots.

Maximum intensity (I_{max}): The largest intensity attained by an ESR or luminescent signal, usually at saturation or steady state.

Mazier corer: A semi-automatic triple-walled corer (two metallic ones for the barrel and one for the liner) with a core catcher. It compensates automatically for the difference in sediment hardness. It has been invented by a French engineer from whom it derives its name.

Messenger: A small weight, often used in conjunction with cable-operated devices, that is clipped or otherwise attached to the cable in such a way that it can be released from the surface and slide down the cable to the device in order to trigger some sort of remote operation.

MHz: Megahertz.

Mica: A mineral group with sheet-like structure, comprising polymerised sheets of silicate terahedra having the general formula: $(K, Na, Ca)(Mg, Fe, Li, Al)_2 (Al, Si)_4O_{10} (OH, F)_2$.

Microenvironment: Any very restricted locus within a molecule, sediment, fossil, or geographic area with its own geochemical environment that can be characterized as different from other similar adjacent areas using geochemical, physical, or biological parameters.

Migration algorithm: A step-by-step numerical procedure for repositioning reflections on a profile into their geometrically correct position.

MilliTesla (mT): See Tesla.

Mineral matrix: 1) A general term for the inorganic material in a sedimentary deposit. 2) The interlinked mineral structure that forms the hard structure in complex biologically produced hard parts, including hydroxyapatite in bones and teeth, calcite or aragonite in shells and corals, apatite minerals in chitin, etc.

Mineralisation: As used in organic geochemistry, and studies of carbon and nutrient cycling, the term implies breakdown of organic matter to simple dissolved carbon, nitrogen, etc. species (e.g., CO_2, NO_3^{2-}, NH_3, etc.).

Modular raft or barge: It is system whereby a raft or barge is built with a series of independent floating devices providing safety in case one of the units fills in with water. The best units are either aluminum or heavy-duty plastic, light and strong, easily assembled and very stable and buoyant.

Modulation amplitude: The size of the sinusoidal 100 kHz magnetic field used in ESR spectrometry that can be varied, unlike the modulation frequency, to detect small changes in magnetic absorption. Theoretically, the modulation amplitude should not exceed or even approach the linewidth for the ESR absorption (see overmodulation).

Molar: A tooth with a broad occlussal surface and multiple roots located posteriorly to the other teeth in mammalian jaws.

Monohydrocalcite: A hydrated carbonate mineral with the formula $CaCO_3 \cdot H_2O$, also known as hydrocalcite.

Montmorillonite: 1) A group of expandable clay minerals with the general formula $R_{0.33}Al_2Si_4O_{10}(OH)_2 \cdot nH_2O$, where R can be Na^+, K^+, Mg^{2+}, Ca^{2+}, among others. Also known as smectite. May also have Mg or Fe substituting for Al and Al for Si. 2) The clay mineral with the formula $Na_{0.33}Al_{1.67}Mg_{0.33}Si_4O_{10}(OH)_2 \cdot nH_2O$; also known as beidellite. 3) Any member of the montmorillonite group of expandable clay minerals.

Moon pool: A cylinder in the center of a boat, raft or barge, for which the bottom hole is below water level and the top hole is above water level. It is used to lower the core barrel from the center of the craft.

MS: 1) Mass spectrometer. 2) Magnetic susceptibility.

MSC: Multiplicative signal correction.

Multiasymmetric amino acids: Amino acids having more than one asymmetric carbon atom, including isoleucine, threonine, hydroxylysine, and cystine.

Multichannel seismic reflection profiling: Seismic reflection method that uses multiple receivers (hydrophones or geophones) to recover seismic signals, generally over a range of angles of incidence.

Multiple-aliquot technique: A method of determining an equivalent dose using many aliquots, typically in the range 20 to 60, of grains separated from the sample; this is to be contrasted with a single-aliquot technique in which only one aliquot is used.

Multiple entry systems: A drilling system wherein it is possible to return to the same hole and sample deeper.

Multiplicative signal correction (MSC): A mathematical treatment of spectra used prior to regression to correct for baseline and offset shifts caused by e.g., difference in particle size or in colour between samples. Based on a standard spectrum (usually a mean of several good spectra) every sample is corrected for a slope and an offset calculated by least squares regression on the standard spectrum.

Mylonite: 1) A compact streaked or banded chert-like rock produced by extreme comminution and shearing of rocks in faults or metamorphic zones. 2) A microbreccia with flow texture.

N: A Newton, a unit of force that communicates to a 1 kg-body an acceleration of 1 m s^{-1} per second

N-terminus (terminal position or end): The end of an amino acid, peptide, or protein containing the amino ($-NH_2$) group.

NAA: Neutron activation analysis.

NaOH-soluble fraction: See alkali-soluble.

Natural Remanent Magnetization (NRM): The magnetization of a sample that has not been subject an external magnetic field.

NEDRA (FGUP NPC NEDRA): The Federal State Unitary Enterprise "Scientific-Industrial Center for Superdeep Drilling and Comprehensive Studies of the Earth's Interior", located in Russia.

Neogene: The last ~23 million years of the Geologic Time Scale.

Neotectonism: Geologically recent movement or deformation of the lithosphere caused by processes originating within the Earth.

512

Neutron activation: A radioactivity logging method mainly used in boreholes. A neutron source provides neutrons that enter the sediment and induce additional gamma radiation that is recorded. Induced gamma radiation is related to the hydrogen content.

Neutron activation analysis (NAA): A geochemical analytical tool for measuring trace and minor element concentrations in which samples are exposed to highly energetic neutron fluxes in a reactor. Analyte concentrations are determined by α, β, γ or neutron counting the isotopes created when the neutron interacts with the analyte element.

NMR: Nuclear magnetic resonance.

Nodular: Having the shape of a nodule, or occurring in the form of nodules.

Non-linear kinetics: A chemical reaction that does not produce a linear relationship between the reaction rate constant, k_i, and the inverse of absolute temperature, T^{-1}.

NRM: Natural remanent magnetization.

NSF: National Science Foundation of the United States of America.

Nutation: A nodding motion or oscillation about a fixed or rotational axis. Usually caused by an imbalance of forces acting to stabilize the motion of an object. Motion due to nutation is sometimes referred to as precession.

Ø: Diameter.

iØ: Inner diameter of a core tube or barrel.

oØ: Outer diameter of a core tube or barrel.

ODP: Ocean Drilling Program

Offline reflections: Reflections recorded from objects or interfaces not directly below the antennas, but off to the side. This is a function of how well focused the energy is that is propagated into the ground.

Older Dryas: Short cold interval, found in many North Atlantic records, during the last deglaciation, which is dated to 12,100–12,000 ^{14}C years BP or to 14,050–13,900 GRIP years BP. See also Late Weichselian.

OM: Organic matter.

Onlap: The upper termination of relatively flat-lying strata onto an inclined surface.

Open-barrel gravity corer: A sediment coring device, the penetration of which is accomplished by the mass of the device that is used to drive a tube, open at both ends, into the sediment.

Optical dating: A method of determining the time elapsed since a mineral sediment (usually quartz or feldspar) was last exposed to light (usually sunlight). The technique involves shining a beam of laboratory photons of one energy, or energy range, onto the sample and measuring the intensity of photons emitted at a higher energy; the higher the measured intensity, the larger the past radiation dose and hence the longer the time elapsed since the natural light exposure that is being dated.

Optical isomers: See Enantiomer.

Optically-stimulated luminescence (OSL): 1) Luminescence is the emission of light from non-conducting solids in addition to their black-body radiation. It is caused by the stimulation of trapped electrons from metastable energy levels, which are related to crystal defects and/or impurities, and their subsequent recombination under photon emission. Luminescence dating exploits the fact that ionizing radiation from natural radioactivity and cosmic rays produces free charge carriers, which are partly stored in the crystal lattice. Since these charge carriers accumulate with time, their amount and thus the intensity of the luminescence signal can be used for dating. For optically stimulated luminescence the phenomenon of luminescence is stimulated by visible light. 2) A dating method in which visible or near infrared radiation is used to release luminescence in Quaternary sediment samples. The nature of the resulting luminescence signal is used to determine elapsed time since burial of the sediment sample.

Ordovician: The geologic period between 510–439 million years ago, and the second period of the Paleozoic Era (see Cambrian). It also denotes the system of strata deposited during this period.

Organic matrix: The interlinked organic molecular structure that forms the "soft" structure in complex biologically produced hard parts, including the collagenous and non-collagenous matrices in bones and dentine.

Organic matter (OM): Complex substances consisting mainly of C, H and O, with subordinate quantities of N, S, and other elements, produced by the growth of organisms. OM is present in trace amounts in virtually all sediments, is particularly abundant in some muds and silts, and is the dominant component of gyttja, peat and coal.

Oriented cores: sediment cores for which the original orientation of the sediment is known, either because the sediment has not turned inside the core barrel or because the sediment rotation path can be traced. The orientation is an essential parameter for palaeomagnetic studies.

Orn: Ornithine (an amino acid).

Ornithine (Orn): A uniasymmetric amino acid occurring abundantly in natural proteins with the formula $NH_2(CH_2)_3CH(NH_2)CO_2H$, also known as 2.5-diaminopentanoic acid.

Orographic effect: A mountain barrier causing air mass lifting, adiabatic cooling, and therefore higher precipitation values than would have occurred in the region if the mountain had not been there.

OSL: Optically-stimulated luminescence; an experimental dating technique.

Ostracode (also Ostracod): A millimeter-sized crustacean with the soft body enclosed in a carapace comprised of a pair of kidney-shaped, spherical or lenticular shells made of chitin and low-Mg calcite.

Outrigger cable (trigger cable): A weighted cable suspended alongside or below a sampling device in such an arrangement that allows it to strike the sediment in advance of the main sampler and trigger some action. Using the Kullenberg corer as an example, the outrigger is used to release the driving weight.

Overmodulation: Using a modulation amplitude greater than the theoretical value in order to obtain a smoother ESR spectrum, which can distort the signal.

Oxic: An environment having normal oxygen concentrations. In lakes this commonly implies dissolved O_2 concentrations in the range of 2.0–7.0 ml l^{-1} (cf. anoxic, dysoxic).

Oxidation: The process whereby an atom or ion loses electrons, causing its oxidation number to increase.

Oxidation number: The charge that an atom or ion would have if all the electrons in its bonds belonged to the more electronegative atoms or ions.

Oxygen hole centre (OHC ESR) signal: A complex ESR signal arising from a missing electron associated with oxygen in quartz and related minerals. Measured at room temperature, the OHC is used to date quartz, flint, and fault gouge.

Oxygen Isotope Stages: A sequence of sedimentation events recognized in deep oceanic cores by the oxygen isotope ($^{18}O/^{16}O$) ratios in the benthic and planktonic foraminifera therein.

P, p, $\mathscr{P}$: 1) Percentage error. 2) Mean annual flux of fallout ^{210}Pb from the atmosphere and to the sediment record.

P: Paleodose.

^{231}Pa/^{235}U (dating): A radiometric geochronological method often used to date Quaternary speleothem, travertine, calcrete deposits, corals, teeth, and other fossils <200 ka old. A U series disequilibrium dating method, it uses the ratios between ^{231}Pa/^{235}U, two ^{235}U daughter isotopes, to determine the age (also known as Pa/U dating).

Pa: Pascal.

Pack-rat middens: The eclectic collections of bones, shells, other debris, and bodily waste products deposited by pack-rats (*Neotoma*) in their nests, often cemented by calcite.

515

PAGES: Past Global Changes, a core project of IGBP.

Palaeomagnetic Secular Variation (PSV): Regional, short lasting fluctuations (10^1 to 10^4 years), of the geomagnetic field of smaller amplitude than those involved in complete reversals of the Earth's geomagnetic field.

Paleodiet: The diets for fossil animals, including humans.

Paleodietary analysis: Any analytical method that can establish the diets for fossil animals, including humans. Common methods include oxygen, carbon, and nitrogen isotopic analyses, sometimes coupled with amino acid racemization analyses.

Paleodose: This term is often used as an alternative to equivalent dose; since it is not the same as the past radiation dose, the term paleodose equivalent would be more appropriate.

Paleoindian: An archaeological cultural complex thought to be the first in North America defined by complex lancolate points, including fluted points, and other bifacially flaked tools. Ranging in age from 12.5–8 ka , its range recently has been extended back as early as 20–30 ka by some archaeologists to include sites that predate the appearance of fluted points in North America.

Paleomagnetic: Records that indicate past variations in the Earth's magnetic field.

Paleosol: A buried soil dating from some previous period of subaerial exposure.

Paleotemperature: A past temperature.

Paleowind: A wind of the geologic past.

Parabolic kinetics: A model used to approximate the kinetics for amino acid racemization in some molluscs and other fossils. A racemization reaction is assumed to be parabolic if $[D/L]$ versus $t^{0.5}$ gives a linear regression.

Partial-bleach method: There are a number of different electron traps in a mineral, and the time it takes sunlight to empty electrons from them varies. A normal thermoluminescence measurement makes no distinction among them. The partial bleach method is designed to extract that component of the thermoluminescence that arises from those traps that are most easily emptied by sunlight. This is accomplished by utilizing the difference in thermoluminescence between aliquots that have had, and aliquots that have not had, a short laboratory light exposure.

Particulate organic carbon (POC): Carbon contained in smaller organic particles.

Pascal: SI unit of pressure equal to 1 newton/m^2.

^{210}Pb: Radioactive isotope of lead with a 22.26 year half-life produced by decay of ^{226}Ra.

^{214}Pb: Short-lived decay product of ^{226}Ra used to determine ^{226}Ra in gamma assay. In sealed samples it reaches equilibrium with ^{226}Ra in about 28 days.

PDB: Pee Dee Belemnite: a belemnite fossil from Pee Dee Formation, South Carolina, USA.

Peat: An unconsolidated combustible deposit of partially carbonized plant remains, common in bogs, fens, and lakes in tundra or periglacial regions.

PeeDee Belemnite (PDB): Oxygen and carbon isotope ratios are measured as relative deviations ($\delta^{18}O$ per mil; $\delta^{13}C$ per mil) from a laboratory standard value. The standard normally employed for the analysis of carbonates is a PDB limestone (belemnite shell from the Cretaceous Peedee Formation of South Carolina).

PEP: Pole-Equator-Pole transects (a PAGES projects).

Peptide: 1) Any string of amino acids that itself is not a viable protein, also called a polypeptide. 2) The bond commonly found in proteins (see peptide bond).

Peptide bond: A chemical bond common in proteins between successive amino acids, in which the carboxyl group (-COOH) and amine group ($-NH_2$) lose a water molecule to become a -C(O)NH- link.

Peroxy (P_1 ESR) signal: An ESR signal that may prove useful for dating pre-Quaternary rock and sediment.

PFPA (pentafluoroproprionic anhydride): A volatile compound with the formula $C_2F_5CO_2H$, that smells like sweaty socks, used in esterifying amino acid solutions prior to gas chromatography.

Phe: Phenylalanine (an amino acid).

Phenylalanine (Phe): A uniasymmetric amino acid occurring abundantly in natural proteins with the formula $C_6H_5CH_2CH(NH_2)CO_2H$, also known as α-aminohydrocinnamic acid, β-phenyl-α-proprionic acid.

Phi grade scale (also phi scale; phi unit): A logarithmic transformation of the Wentworth grade scale in which the particle diameter becomes the negative logarithm to the base 2 of the diameter measured in millimeters.

Photoluminescence (PL): The photons of light emitted in response to an incident beam of photons, the emitted photons usually having lower energies (longer wavelengths) than those of the incident photons. If the emitted photons have higher energies the term optically-stimulated luminescence is normally used.

Photon: A quantum (particle) of electromagnetic radiation. The energy of a photon is equal to hc/λ, where h is Planck's constant, c is the speed of light, and λ is the wavelength of the light. The light emitted by an ordinary light bulb, light-emitting diode, the sun etc. can be thought of as a stream of photons. Radio waves, microwaves, X-rays, and gamma rays can also be thought of as photons, although they are not visible to the eye.

Phytolith: 1) Paleontology: A microscopic biologically precipitated silica or calcium oxalate nodule found in the stems and other tissues in plants. 2) Sedimentology: A rock or partially cemented sediment formed by plant activity.

Pisolite: 1) A coarsely grained oolitic limestone deposit. 2) A small round or ellipsoidal accretionary body in a sedimentary deposit. A grain in a pisolitic deposit; the term is synonymous with pisolith.

Piston: Cylindrical disk moving inside a pipe.

Piston corer: A coring device with a piston. A piston corer can be lowered in the sediment in a closed position until the required coring depth is reached without being contaminated by sediment on its way down. On the way up the piston acts as a sucking device and hampers the sediment from falling out of the core barrel.

Plagioclase: A variety of feldspar; a group of triclinic silicates with the general formula: $(Na, Ca) Al(Al, Si)Si_2O_8$.

Plateau technique or test: A protocol in ESR dating for testing the reliability of accumulated dose estimates and checking for signal interference in which accumulated doses are calculated at several different g values.

Playa lake: A lake, typically shallow, that exists on an intermittent basis, such as in wet seasons or in especially wet years.

Plerospheres: Hollow inorganic ash spheres (cenospheres) containing encapsulated smaller spheres

Plinian: Volcanic eruption of great violence, named after Pliny the Younger who recorded the AD 79 eruption of Mt. Vesuvius.

Pliocene: 1) Time: The last epoch in the Tertiary Period. The Pliocene is estimated to have begun at roughly 5.3 Ma, but its end is currently controversial, either having lasted until the Gauss-Matuyama at approximately 2.48 Ma (assuming the long Quaternary chronology) or until about 1.78 Ma (according to the short and traditionally defined Quaternary chronology; see also Quaternary). 2) Rocks: The sediment and rocks formed during the Pliocene Epoch.

Pneumatic corer: A corer whose main operating force is generated by a compressed gas.

^{210}Po: Radioactive isotope of polonium with a 138.4 day half-life produced by decay of ^{210}Pb.

POC: See Particulate Organic Carbon.

Pollen: Grains formed by seed-producing plants. They are widely spread by a variety of means (wind, water, insects, birds, animals) and are usually well preserved in lake sediments and peat bogs. Such deposits have extensively been investigated by palynologists. Changing pollen assemblages in superimposed layers are analysed to interpret temporal changes in vegetation cover, environment and climate.

518

Pollen assemblage zones (paz): Analysis of the pollen content of a specific sediment horizon will reveal a mixture of pollen types, that is named pollen assemblage. Samples from different horizons, but with a similar pollen assemblage composition are called pollen assemblage zones.

Polymer: A compound having multiple repeated structural units.

Polypeptide: Any string of amino acids that itself is not a viable protein, also called a peptide.

Pond: 1) A natural or artificial standing freshwater body occupying a small surface depression, often synonymous with lake or pool. 2) A water body formed by stream ponding.

ppb: Part per billion, a unit of concentration.

ppm: Part per million, a unit of concentration.

Precision: 1) The degree of uniformity between repeated successive measurements for a quantity or an operational performance (also known as quality control; see also accuracy). 2) The deviation of several measurements about their mean.

Premolar: A mammalian tooth between the canines and the molars, often multiply rooted with bi- or multiple cusped occlussal surfaces, also called bicuspids in humans.

Pro: Proline (an amino acid).

Proline (Pro): A uniasymmetric amino acid occurring abundantly in natural proteins with the formula $HN(CH_2)_3CHCO_2H$, also known as 2-pyrrolidine carboxylic acid.

Propagation constant: Describes how EM energy with travel through a substance. The propagation constant is a function of the magnetic and electric properties of the material and the frequency of the energy.

Propagation velocity: The rate at which a pulse of EM energy moves through a medium.

Proportional U uptake or leaching: U uptake or leaching in which all tooth subsamples gain (for primary or secondary uptake) or lose (for leaching) a constant percentage, $\varphi\%$, of their initial concentrations in the enamel, dentine, cementum, and attached bone.

Protein: Any complex organic molecule containing predominantly amino acids bonded by peptide bonds.

Provenance: The area or geological unit from which a sediment or fossil originates, also called provenience.

Proxy data: Data obtained by analytical techniques used to reconstruct past conditions (usually environmental) that are used as surrogates (proxies) for other variables, such as climate. The word has the same root as "approximation".

Proxy records: Materials that accumulate and retain information about past environments or environmental processes (e.g., water chemistry, climate).

PSV: Palaeomagnetic secular variation.

Pulsed ESR: An ESR measurement system in which a short intense microwave pulse instead of a continuous wave is used to induce the spin resonance, which is followed by a Fourier signal transformation to allow the entire ESR spectrum to be measured in microseconds rather than over 2–30 minutes in standard ESR spectrometers.

Pumice: Highly vesicular siliceous glass.

PVC: Polyvinyl chloride.

P-wave velocity: Travel speed of ultrasonic compressional waves (primary or P-waves) through a sediment core. Data improve the understanding of seismic profiles and can be applied to characterize sediment properties such as bulk density, carbonate content, or grain size.

Pyroclast: Fragmented rocks ejected by a volcanic eruption.

Pyroxene: Silicate rock-forming mineral.

Polarity transitions: Large amplitude, low frequency variations in the Earth's magnetic field. The duration of a polarity transition is approximately 10^4 years. The direction of the field changes to that of the opposite polarity during the transition and remains in the opposite polarity after the transition.

Pyrolysis: Heating of organic compounds to very high temperatures.

Q, Q_C: Inventories of fallout radionuclides in a lake and its catchment.

Q-band ESR: ESR spectrometry using microwave frequencies near 35 GHz under a 100 kHz field modulation with a field center at 1250 mT for $g \sim 2$.

Quaternary: 1) Time: The second and last period in the Cenozoic Era. Normally, it is subdivided into the Holocene (Recent) and Pleistocene Epochs. The short chronology places its beginning at the traditionally defined boundary at roughly 1.78 Ma, while the long chronology places it at the start of the Matuyama Magnetochron at approximately 2.48 Ma. 2) Rocks: The sediment and rocks formed during the Quaternary Period.

r: Dry mass sedimentation rate.

$\bar{r}_b$: Estimated mean dry mass sedimentation rate at the base of the unsupported ^{210}Pb record.

R: 1) Reflection coefficient. 2) Reflectance.

R-group(s): The ion(s), atom(s), or ligand(s) attached to an amino acid that defines the amino acid.

Racimers: 1) *Sensu stricto*: Stereoisomers whose molecule structures mirror each other at one or more asymmetrical carbon atoms. 2) *Sensu lato*: Stereoisomers whose molecule structures mirror each other at all asymmetrical carbon atoms (also called isomers, stereoisomers; see also enantiomers, epimers).

Racemic concentration ratio: The concentration ratio at which the D to L enantiomers are in geochemical equilibrium. For uniasymmetric amino acids at room temperature, the racemic D/L ratio is usually 1.0.

Racemic mixture: A mixture in which the D and L enantiomers are in geochemical equilibrium.

Racemization: 1) *Sensu lato*: Any process by which an asymmetric carbon atom in optically active organic molecule changes its optical configuration. For amino acids in living tissues, the process usually involves converting L-amino acids to a mixture of D and L-amino acids (see also enantiomerization, epimerization). 2) *Sensu lato*: Any process by which an optically active stereoisomer is converted into a mixture of two isomers having no optical activity. 3) *Sensu stricto*: The process in which a molecule with one optically active (asymmetric) carbon atom changes in its optical configuration.

Rad: An ionizing radiation unit corresponding to the absorption of 100 ergs/g energy in any medium (1 rad = 10 Gray).

Radiation-sensitive signal: An ESR or luminescent signal that grows with added radiation dose, often useful for dating.

Radioactive decay: Radioactive decay (atomic transformation) is time dependent and exponential and considered in terms of the half-life. If the half-life is known and the extent of decay of a radioactive element can be measured, the age of rocks, sediments or fossils can be established. Radioactive decay processes are governed by atomic constants; the number of transformations per unit time is proportional to the number of atoms present, and for each decay scheme there is a decay constant (λ), which represents the probability that an atom will decay in a given period of time.

Radioactive disequilibrium: A radioactive decay chain (sequence of decaying nuclei) is in equilibrium if the activity (number of decays per second) is the same for all nuclei of the chain except the last one (which does not decay). This is a dynamic equilibrium situation. If the activities are not all the same then the chain is said to be in disequilibrium.

Radiocarbon dating: The application of the ^{14}C isotope and its radioactive decay for age determinations.

Radiocarbon plateau: Time period of constant radiocarbon ages.

Radiogenic isotope: Any isotope produced by radioactive decay.

Radioisotope: A radioactive isotope.

Radioluminescence (RL) dating: A new radiochemically based luminescent or trapped-charge geochronological technique that measures the light emitted by electrons trapped in crystal defects formed by ionizing radiation upon their release from their traps when stimulated by radiowaves (see electron spin resonance dating, thermoluminescence dating optically stimulated luminescence dating, luminescent dating).

Radiolysis: The process in which ionizing radiation cleaves chemical bonds within organic molecules.

Radiometric dating: Any of the dating methods that use the radioactive decay of one or more radioactive isotopes to calculate the ages. These include ^{14}C, $^{40}Ar/^{39}Ar$, $^{230}Th/^{234}U$, and many other methods.

Raman-scattered photons: A photon incident on a crystal can gain or lose energy by interaction with vibrations of the crystal lattice. When this occurs, the resultant photon will have either more or less energy than the original one and is said to be Raman scattered.

Ramping technique: A protocol for the additive dose method for ESR dating that uses only 3–4 sample aliquots, in which two or three are successively irradiated to ever higher added doses, and one or two are repeatedly used to calibrate the spectrometer for each measurement set.

Random U uptake or leaching: U uptake or leaching in which no pattern exists between the subsamples' U concentrations before and after the uptake or leaching event. This will not produce viable ESR isochrons.

Ratite (birds): Large ground-dwelling flightless birds having a flat (keelless) sternum, including the ostrich *(Struthio)*, the emu, rhea, and several extinct species, such as *Genyornis.*

Reaction force: A force acting in the opposite direction from that of a directed force. Examples being the recoil of a pistol or rifle shot, or the backward motion of a garden hose as water leaves the nozzle.

Recent: See Holocene.

Recent (U) uptake (RU): 1) An U uptake model used for calculating ESR ages for tooth enamel, mollusk shells, and other fossils tissues that can absorb U secondarily. It assumes U uptake very late in the sample's burial history, which reduces its internally generated dose, A_{int}, to a minor contribution compared to A_Σ. This gives the maximum possible age. 2) The equivalent U uptake model used calculating external radiation dose rates in sediment with components that absorb secondary U.

Recovery: The ratio, usually in percentages, between the actual amount of sediment obtained in one drive and the amount that should have been obtained. For example, if a drilling tool is lowered by 100 cm, it is expected that 100 cm of sediment will have been sampled. If so, the recovery is 100%. In many cases, the recovery is less than 100%.

Re-deposition: Deposition of reworked or eroded minerogenic and/or organic material.

Reduction: The process whereby an atom or ion gains electrons, causing its oxidation number to decrease.

Reference dates and accumulation rates: Independently determined dates and accumulation rates in lake sediment cores used in calculating corrections to CRS model ^{210}Pb dates.

Reflection coefficient: A measure of the proportion of energy reflected from a given interface.

Regression analysis: A statistical technique used to determine the degree of mutual association between an independent variable and one or more dependent variables in a paired data set.

Remanent magnetization or remanence: The magnetization remaining (in a sample) after exposure to an external magnetic field.

Reservoir effect: If the ^{14}C age (^{14}C/^{12}C ratio) of the carbon, from which aquatic plants and animals built up their tissues, was older (lower) than the ^{14}C age (^{14}C/^{12}C ratio) of the CO_2 of the contemporaneous atmosphere, the dating of such organic remains will be affected by the so-called "lake reservoir effect". See also marine reservoir effect.

Residual signal: An ESR signal remaining after some exposure to intense heat, light, or pressure, common in archaeologically heated flints and cherts (see zeroing, unbleachable component).

Reversals: A complete and stable change in the polarity of the Earth's magnetic field.

RGB: Red-green-blue.

R-gamma, R-beta: The original name for the partial bleach method. "R-gamma" if laboratory gamma doses are used, and "R-beta" if beta doses are used to construct the dose responses.

RH: Relative humidity.

Rhyolite: A volcanic rock with $>68\%$ SiO_2.

Rift lake: Lake formed in a basin produced by continental rifting. Typical modern examples are Lake Baikal and most of the large East African lakes.

Rift valley: A long narrow valley bounded by one or two normal or listric faults, often filled with deep lakes and/or thick lake sediment deposits. Often associated with volcanoes or flood basalts, they form during nascent continental separation or along the mid-ocean ridge or in failed arms of triple junctions.

RL: Radioluminescence, usually referring to dating unless otherwise specified.

Rod string: The assembled rods, sometimes called extension rods, that are fastened to the sampler in order to advance it into the sediment and recover it.

Rotation system: Drilling technique where the core barrel and (or) drill string is rotated.

rpm: Rotations or revolutions per minute.

RR: Sediment recycling ratio (dimensionless).

RU: Recent U uptake model (ESR and luminescent dating).

S: Siemens (a measure of conductivity).

Sandwich modular raft: (model used by the LRC) The raft is a twin-hulled vessel with a moon pool and central gap for handling core barrels. The 2 floats are 2.44 m open Carolina Skiff unsinkable fiberglass. The deck is 4.87×6.60 m with a catwalk to 8.53 m, made of interlocking 1.22×2.44 m, light-weight plates of hard styrofoam glued to thin diamond plate aluminum sheets. These sandwiches provide tremendous rigidity and strength, combined with the safety flotation needed if the boats filled with water. A light aluminum frame bolts the floats to the deckplates, and a rubber gasket around the gunwales forms a water-tight seal for the boats.

Sangamonian: 1) Time: The time period or physical events associated with the penultimate glacial advance(s) that occurred in North America before the Last Interglacial, and considered equivalent to the Eemian or Ipswichian. Some controversy surrounds the exact period covered by this interval, due to the problem in correlating pre-Wisconsinan glacial events, the time period is often cited as approximately 190–128 ka (Oxygen Isotope Stage 6), 300–128 ka (Oxygen Isotope Stages 8–6), or even occasionally 360–128 ka (Oxygen Isotopes Stages 10–6). 2) Rocks: The sediment or rocks deposited during the Sangamonian.

Scoria: Highly vesicular pyroclastic material. Typically andesitic or basaltic in composition.

SCP: See spheroidal carbonaceous particle.

Scanning electron microscope (SEM): An electron microscope from which an image of a sample is produced by scanning an electron beam in a television-like raster over the sample and displaying the resultant signal from an electron detector on a screen.

SEC gain: Spherical and exponentially compensation gain.

Secondary mineral: A mineral formed later than the material enclosing it.

Secondary mineralization: A common form of diagenetic alteration in which original minerals in fossils and sediment are replaced by or their interstices are infilled by a new (suite of) mineral(s) that may be identical to the original minerals.

Secular Varition Patterns: Radiometrically dated stratigraphies of geomagnetic inclination, declination, and paleointensity that may be used for regional stratigraphic correlation and dating.

Sediment core: A sample of sediment retrieved from a lake, pond, or other environment and typically stored in a pipe.

Sediment focussing: 1) Process causing sediments to accumulate at different rates in different parts of a basin through time; dependent on the lake's shape, water depth, total basin depth and wind exposure. 2) Process by which lake sediments deposited in marginal zones are remobilized and redeposited in deeper water zones.

Sedimentation rate: The amount of sediment deposited during a certain time span.

Sedimentology: The physical and chemical study of sediments, which includes clastic particles and chemical precipitates; particle composition, size, shape, sorting, orientation, and sedimentary structures reveal details regarding the environment in which the sediments were deposited.

Seismogram: Record of elastic waves generated by earthquakes or generated artificially by explosions. A means of geophysical prospecting providing information about the internal structure of *in situ* sediments or rocks.

Self-absorption: Absorption of gamma photons within the sample before they can interact with the detector.

SEM: Scanning electron microscope or microscopy.

Ser: Serine (an amino acid).

Serine: A uniasymmetric amino acid occurring in natural proteins with the formula $HOCH_2CH(NH_2)CO_2H$, also known as 2-amino-3-hydroxypropanoic acid.

Shallow trap: See deep trap.

Shear strength: The strength of material resisting forces that tend to make one part or piece to slide over another. Forcing a core tube into the sediment is an example in which shear forces predominate.

Sheet flood (also sheetflood): 1) A broad expanse of moving, often storm-generated, water that spreads as a shallow, continuous uniform film over a large area. 2) A flow or the process of water movement in which the fluid is not concentrated in a discrete channel or defined river banks.

Shelby sampling technique: The Shelby tube consists of a thin-walled stainless steel tube ($i\phi$ 75 mm). The leading edge of the tube is beveled and crimped such that the entry ϕ is fractionally smaller than the body ϕ. The tube is usually 50 cm-long with the top end designed to fit into an adapter. The adapter has a one-way valve built onto it to allow water to escape as to prevent compression of the sample. The Shelby tube sampler is attached to the drill string in place of a core barrel and is lowered to the base of the borehole and pressed onto the soft sediment using the drill rig hydraulics. This sampling technique is employed to obtain undisturbed material from soft and very soft cohesive sediments with a drill rig.

Shine plateau: The optically-stimulated luminescence decreases with time after the incident photon beam is switched on, because the traps are being emptied of electrons. It is normal to evaluate the equivalent dose as a function of this time. If it is constant, as it should be, then there is said to be a shine plateau.

Shoreline preservation index: The sum of the lengths of mappable preserved segments of a relict shoreline divided by the reconstructed original length of the once-laterally continuous shoreline.

Side-wall sampling: Any type of sampling that selectively removes and recovers material adjacent to the bored or cored hole, rather then the bottom, as is the case for most coring and drilling types of sampling.

Siemens (S): The measure of the electrical conductivity of a material.

Signal fading: The diminution of an ESR or luminescent signal's intensity reflecting the loss of trapped electrons related to the signal lifetime.

Signal height or intensity (I): A direct measure for the number of trapped charges at a given crystal lattice site, which form an ESR or luminescent spectral peak. Also called peak height or intensity.

Signal lifetime, τ: The time required for the number of ESR-active defects to drop to $1/e$ of the original number. Caused by thermal annealing, this is a function of the trap's or hole's thermal stability. τ must exceed the desired dating range by at least 2–3 orders of magnitude to ensure reliable ages.

Signal reflection: The redirection of a pulse of energy as a result of the change in the physical properties of the material it has encountered. When a pulse of EM energy travelling through the ground comes upon an interface between different sediment types a portion of the energy is reflected back to the surface.

Signal saturation, $A_{\Sigma \cdot sat}$: The accumulated dose at which an ESR or luminescent signal no longer grows despite exposure to continued radiation doses. All the charge defects contain trapped electrons or all the potential holes have been created.

Signal subtraction: A mathematical protocol for correcting signal interference in spectral peaks, especially in ESR, the luminescent, some radiometric dating methods and other analytical techniques. After generating a pure interference signal, it is calibrated to, and then subtracted from, the peak with the interference.

Single-channel seismic reflection profiling: Seismic reflection method that uses a single source and receiver configuration.

Single-entry system: A coring system wherein only one single penetration of the sediment is possible from the sediment surface downwards

Silcrete: A conglomeratic surficial sedimentary crust cemented by secondary silica, often formed under strongly evaporative conditions or in deforested tropical soils (see also caliche, calcrete, gypcrete, ferricrete, laterite).

Smooth units: A well-sorted mono-mineralogical sedimentary unit (see also lumpy unit).

Soft water lakes: Lakes with low pH (<6.5) and alkalinity. They are usually situated in carbonate-free or carbonate poor areas, without input of carbonate or bicarbonate ions.

SOL: See alkali-soluble.

Sonar: A geophysical technique used to measure water depth and map bottom and sub-bottom structures. Acoustical energy is radiated into water, such as a lake, and return time and strength of reflections from the lake bottom and interfaces between differing materials below the lake is recorded.

Sound velocity: Travel speed of ultrasonic compressional waves (primary or P-waves) through a sediment core.

Spatial filtering: The application of numeric algorithms (e.g., averaging) to data values recorded at the same position on adjacent traces. This enables the enhancement or elimination of specific returns from a given depth.

Speciation: 1) The process whereby new species evolve from pre-existing species. 2) The taxonomic identification of organisms.

Species effects: Any phenomenon that has different expressions in different taxonomic species. For example, molluscs have different amino acid racemization rates depending on their species.

Spectroscopy: The production and analysis of a spectrum.

Spheres of influence: The sphere around an ESR, TL, OSL, or RL dating sample or sediment loci from which can come a radiation dose affecting the ESR or luminescent signal in the sample or sediment. For α radiation, the sphere is $\sim 20\,\mu$m in diameter, for β, ~ 2–3 mm, and for terrestrially produced γ, ~ 30 cm.

Spherical and exponentially compensation gain: A gain function that compensates for spherical divergence of energy in the subsurface.

Spheroidal carbonaceous particles: Microscopic spheroids of elemental carbon produced from the incomplete combustion of fossil fuels at industrial temperatures. A component of fly-ash.

Spike (solution or compound): A solution or compound added to a chemical analysis that serves as the basis for the measurement standard.

Spit: A wave- and current-constructed coastal depositional landform attached to land at only one end; the free end may be recurved or hooked back towards land.

Split-core logging: Measurement of a physical property (e.g., magnetic susceptibility) performed on the split surface of an opened core section (working half or archive half).

Split-spoon drilling and recovery: Drilling of unconsolidated material, where a double tube is inserted into the drill string and then pushed into the sediment. The inner of the two tubes is split in half longitudinally before deployment, and when brought to the surface and removed from the outer tube, the two halves are separated to provide access to the sediment.

Stalactite: A conical or cylindrical speleothem that grows down from a cave ceiling.

Stalagmite: A conical or cylindrical speleothem that grows up from a cave floor.

Standard error: Uncertainty in a measured quantity, usually measured by standard deviation of repeated determinations.

Steady state limit or level ($A_{\Sigma,ss}$): An ESR or luminescent signal intensity below the saturation level, where the number of electrons gained through natural or artificial irradiation equals the number lost by trap evacuation or retrapping.

Stereoscopic: Visually three-dimensional; the overlap zone on adjacent air photos can be seen as a three-dimensional image with the use of a simple optical device called a stereoscope.

Stick-slip motion: A type of intermittent linear motion resulting from the accumulation and periodic release of energy in the sheared material. This can occur in some material regardless of the uniform and constant, applied force.

Strain: A change in shape or volume for a body responding to stress.

Stratal relationships: Diagnostic stratigraphic geometries observed on seismic reflection profiles, in outcrop, or in a series of cores and wells that characterize discrete depositional and erosional events within a sedimentary basin. Examples include erosional truncation, onlap, downlap, and toplap.

Stratigraphy: The study of stratified rocks and sediments especially for the purpose of correlation. Important considerations include each unit's physical, biological, and chemical characteristics, boundary conditions, spatial distribution, relative position with respect to other strata, and age.

Stratosphere: The layer of the atmosphere lying above the troposphere, in which the temperature ceases to fall with height.

Stratotype: A locality where a particular stratigraphic unit is clearly and fully recorded, and where the upper and lower boundaries are securely defined.

Stress: In a solid, the force per unit area action on any surface within it.

Stromatolite: An internally layered mound or tabular concretion produced by sediment trapping, binding, and/or precipitating on successive generations' growing surfaces on microorganisms, principally cyanobacteria (blue-green algae). They often form as circular to semi-circular domes up to a metre across, but also form reefs and hardgrounds. They are the oldest known fossils.

528

Sub-bottom profiling: A process that uses acoustic energy to generate a vertical profile of the subsurface beneath the sea floor or the bottom of a lake.

Subsample: A portion of a larger solution or other sample that need not be geochemically identical (see also aliquot).

Substage: Subdivision of a stage (i.e., a geological time period).

Supported ^{210}Pb: Component of ^{210}Pb activity in a sample that derives from decay of the *in situ* ^{226}Ra.

Surface sediment: Upper layer of sediment currently being deposited at the interface between the water and the sediment. It can yield proxy data that can be directly linked to known modern environmental and climatic conditions, allowing calibration of earlier records. Generally refers to the upper 1–5 cm of sediment.

SV: Geomagnetic secular variation.

Swedish Time Scale: Annual chronology based upon the successive matching of overlapping clay varve-thickness diagrams. The older varved clays are glacio-lacustrine sediments (formed in a glacial lake), which were deposited in the Baltic Ice Lake (a glacial precursor of the Baltic Sea) during the last deglaciation due to the seasonal melting of the Scandinavian inland ice. The younger varved clays of Holocene age were deposited as delta sediments at the river mouth of River Ångermanälven. Varve thickness diagrams have been constructed and connected all along the Swedish east coast by measuring the annual thickness of each varve in open sections or on sediment cores and correlating diagrams from different areas.

T: 1) Temperature. 2) Tesla, a unit of magnetic flux density: $1\,T = 1\,kg/(s^2A) = 1\,Vs/m^2$. 3) a ton or 1000 kilograms; 4) Transmittance; 5) Relative transmittance of X-rays.

^{230}Th/^{234}U (dating): A radiometric geochronological method often used to date Middle and Late Quaternary speleothem, travertine, calcrete deposits, corals, teeth, and other fossils. A U series disequilibrium dating method, it uses the ratios between ^{230}Th/^{234}U, two daughter ^{238}U isotopes, and ^{234}U/^{238}U to determine the age (also known as U/Th dating, and by the outdated term ionium dating).

T_d: The symbol commonly used for the diagenetic temperature (amino acid racemization analysis).

TDIE: Thermal diffusion isotopic enrichment.

T_C: Curie temperature.

T_W, T_S, T_L: Residence times of water, sediments and radionuclides in a lake.

T_{Rn}, T_{Pb}: Radioactivity half-lives of ^{222}Rn and ^{210}Pb.

Teeth: Complex biologically produced mineralized tissues. In vertebrates, they are usually mixtures of the mineralized tissues, enamel, dentine, dental cementum, and purely organic tissues, such as pulp and blood vessels.

Tektite: A glassy nodule of frozen molten rock formed by meteorite impacts.

Telmatic: Zone in a lake, which is situated between the high and low water level, also called swamp zone.

Temperature sensitivity: The propensity for an ESR or luminescent signal to partially zero when exposed to temperatures above 25 °C (see also thermal stability).

Tephra: Volcanic ash particles that are wide spread during a volcanic eruption and can form secure, time-synchronous marker horizons in sediments.

Tephrochronology: Geochronological tool for dating past events by their association with tephra.

Ternary plot (or diagram): A graphical data representation in which three axes are used to show the relationships between three interdependent variables, often three compositional components or trace elements.

Terrigenous: Used synonymously with minerogenic.

Tesla (T): Unit of magnetic induction.

Thermal diffusion isotopic enrichment (TDIE): Method used for ^{14}C dating old samples, older than 45,000 years BP; the amount of ^{14}C in a sample is enhanced so that the frequency of the decay can be more accurately measured by gas or liquid scintillation counts. This approach uses either thermal diffusion or photodissociation by means of a laser beam.

Thermal Ionisation Mass Spectrometry (TIMS): Used for U/Th dating; direct counting of individual atoms as opposed to the monitoring of α particles emitted during radioactive decay (see also accelerator mass spectrometry).

Thermal stability: The propensity for an ESR, or luminescent signal to partially zero when exposed to temperatures above 25 °C (see also temperature sensitivity).

Thermal transfer correction: If the optically-stimulated luminescence of a well-bleached mineral is measured it is (ideally) found to be zero. If, however, the mineral is first given the usual preheat for an equivalent dose determination, the measured luminescence is not zero; this is believed to be due to the heating causing the transfer of electrons from traps that are not sampled during the measurement to traps that are measured; hence the name thermal transfer. This process occurs during an equivalent dose determination and thus must be corrected for.

Thermocline: A marked thermal gradient, which separates warm surface waters from deeper, cooler water in lakes and in the ocean.

Thermoluminescence: The light emitted by a mineral when it is heated and which results from a prior radiation dose.

Thermoluminescence (TL) dating: A radiochemically based luminescent or trapped-charge geochronological technique that measures the light emitted by electrons trapped in crystal defects formed by ionizing radiation upon their escape from their traps when stimulated by a ramped heating protocol (see electron spin resonance dating, optically stimulated luminescence dating, radioluminescence dating, luminescent dating).

Thin section: Sample that is mounted on a glass slide and ground to near-transparency. In practice, most samples will be 30 to $100\,\mu$m thick. Soft materials found in recent sediments can be hardened by embedding the samples with a hard material like epoxy resin or wax.

Thixotropic liquid: A liquid whose viscosity is dependent on the degree of mixing (velocity) and the length of time the motion has been applied. The faster the liquid is moved, the less viscous that the thixotropic liquid becomes.

Thr: Threonine (an amino acid).

Threonine (Thr): A multiasymmetric amino acid occurring in natural proteins with the formula $CH_3CH(OH)CH(NH_2)CO_2H$, also known 2-amino-3-hydroxybutyric acid.

Threshold control: A lake that is discharging over its topographic threshold adjusts to net changes in its water supply primarily by adjusting outflow velocity while holding lake level approximately constant so that the threshold elevation acts as a control on the lake level.

Till: Term to denote poorly sorted sediment deposited by an active glacier or ice sheet or in connection with melting of stagnant ice from a decaying glacier or ice sheet.

Time-averaged dose rate: A radiation dose rate that has been averaged over the time that it has affected the sample, used in ESR and luminescent dating.

Time-integrated dose rate: A radiation dose rate that has been integrated over the time that it has affected the sample, used in ESR and luminescent dating, particularly the isochron protocols for these methods.

Time window: The duration over which a geophysical instrument, such as a seismograph or a GPR, records returned energy after a given energy pulse is transmitted into the ground.

TIMS: Thermal ionisation mass spectrometry.

Titanium (Ti ESR) signal: A complex ESR signal arising from $(TiO_4/H^+)^0$, $(TiO_4/Li^+)^0$, or $(TiO_4/Na^+)^0$ defects, seen mainly in quartz and related minerals measured at $70\,°K$.

TL: Thermoluminescence, usually referring to dating unless otherwise specified.

TL dosimeter: A devise emplaced within the sediment at a site to measure the current external dose rate, from sedimentary γ and cosmic sources over 0.5–2.0 years. It usually contains a mineral with a strongly radiation sensitive themoluminescent signal, such as $CaSO_4$, in a sealed copper tube that is zeroed by intense heating before emplacement, and measured by standard TL techniques to assess.

Tombolo: A wave- and current-constructed coastal depositional landform connected at both ends to islands or to an island at one end and the mainland at the other.

Tomographic images: Non-destructive 3-dimensional examinations of the internal structures of sediments using X-ray computed tomography.

Toplap: Initially inclined strata terminate against an upper boundary as a consequence of non-deposition.

Topographic threshold: The lowest point on the perimeter of a lake basin; if the lake level reaches the threshold elevation, surface outflow will become established over that point on the perimeter.

Torque: A force or system of forces producing rotation. A twisting motion producing force about an axis.

Total ^{210}Pb: Total ^{210}Pb activity in a sample including both fallout ^{210}Pb and supported ^{210}Pb.

Total-bleach method: An extreme case of the partial-bleach method in which all the traps that can be emptied by sunlight, or other defined light, exposure in a day or so are emptied.

Trace: Time versus amplitude graph of energy received. Traces from a number of positions along a GPR or seismic survey line are plotted side by side to show the spatial character of subsurface interfaces.

Transducer: Electronic components that change electrical signals into ultrasonic compressional waves and back again.

Transponder: Instrument for generating and detecting acoustic pulses in water.

Transport rate: Rate of transfer of a substance between two separate environmental compartments, e.g., from water column of a lake to the bottom sediments.

Trapezium rule: Method for calculating the integral of a function (area under its graph) in which the area is divided into a series of narrow trapezia bounded by successive ordinates.

Trapped-charge dating techniques: Any dating technique that measures electrons trapped in crystal defects in response to low levels of ionizing radiation from the environment (see electron spin resonance dating, thermoluminescence dating, optically stimulated luminescence dating, radioluminescence dating, luminescent dating).

Travertine: 1) Any finely crystalline massive concretionary limestone formed by surface or groundwater evaporation. 2) Any flowstone deposit, hence the usage travertine implying stalagmitic flowstone. 3) Any carbonate deposit formed within a cave or karst system by secondary precipitation. 4) Massive, hard, dense carbonate deposits formed in association with open-air springs, streams, lakes, or marshes that precipitate carbonate.

Tree rings: See dendrochronology.

Trepan: Tool to make holes in hard layers.

Tricone: A conical-shaped trepan with three toothed wheels.

Trona: A white or yellow-white monoclinic mineral: $Na_3(CO_3)(CO_2OH) \cdot 2H_2O$.

Troposphere: The layer of the atmosphere extending about 10 km above the Earth's surface in which the temperature falls with height and in which "weather phenomena" occur.

Tufa: 1) A chemical sedimentary rock composed of calcium carbonate, which form in nonmarine settings. 2) Any subaerial travertine deposit formed near a spring mouth, along a stream, in a marsh or lake. The term often refers to thin, porous, or soft deposits, as opposed to travertine used for massive, hard, dense deposits.

Tusk: A modified incisor or canine tooth, which grows outside the lips or facial tissues, found in elephantids, walrus, narwhales, and boars.

Type section: The originally described stratigraphic sequence that forms the basis for a stratigraphic unit's definition.

u-channel: Type of continuous sub-core that is sampled by an u-shaped plastic profile, normally 2H2 cm in size and 100 B 150 cm in length, that is pressed into the split surface of a core section half. After extracting the u-channel it is closed by lid. Used in paleomagnetic studies.

(initial) U isotopic ratio $[(^{234}U/^{238}U)_0]$: A parameter used in calculating ages for $^{230}Th/^{234}U$, $^{234}U/^{238}U$, ESR, and several other relating dating methods that rely on the U isotopic ratio in some mineral phase or groundwater associated with the mineral in order to calculate the age.

U series (disequilibrium) dating: A family of some 40 radiometric geochronological methods, including $^{230}Th/^{234}U$, $^{231}Pa/^{235}U$, ^{210}Pb, $^{231}Pa/^{230}Th$, and $^{234}U/^{238}U$, often used to date Quaternary (and occasionally Pliocene) sediment, rocks, and fossils. All the methods rely on the disequilibrium produced by differential solubilities of the ^{235}U, ^{238}U, and ^{232}Th decay series daughter isotopes to determine the ages.

Unbleachable component: The part of an ESR, TL, OSL, or RL signal remaining after prolonged, intense heating, light exposure, strain, or comminution, common in archaeologically heated flints and cherts (see zeroing, residual signal).

Uniasymmetric amino acids: Amino acids having only one asymmetric carbon atom.

Uniform U uptake or leaching: U uptake or leaching in which all tooth subsamples gain (for primary or secondary uptake) or lose (for leaching) x ppm in the enamel, y ppm in the dentine, z ppm in the dental cementum, and w ppm in any attached bone, despite their initial U concentrations, also called equal U uptake or leaching.

Unsupported ^{210}Pb: Component of ^{210}Pb activity in a sample that derives from fallout from the atmosphere, usually measured as the difference between total and supported ^{210}Pb.

Uranium (U) uptake model: A mathematical construct used in ESR dating and dosimetry to account for U uptake in materials that absorb U after deposition.

USGS: United States Geological Survey.

U/Pb: Any or all of the family of radiometric geochronological techniques that include ^{235}U/^{207}Pb, ^{238}U/^{208}Pb, ^{232}Th/^{206}Pb, or combinations thereof. Capable of dating events back to 4.5 Ga, they are also called uranium or lead dating.

U/Th calibration curve: Radiocarbon calibration curve based on ^{234}U/^{230}Th and ^{14}C dated corals. Since ^{234}U/^{230}Th measurements are assumed to equal calendar years, the paired age determinations can be used to assess the difference between ^{14}C and absolute age at a given time interval. See also calibration curve.

$(^{234}$U/^{238}U$)_0$: The symbol used to designate the initial U isotopic ratio.

UV (light or radiation): Ultraviolet light (or radiation).

V: Propagation velocity.

V_p: Velocity of P-waves.

Val: Valine (an amino acid).

Valence band: The energy levels at which an electron is bound to an atom and cannot move freely about within a given mineral phase, which includes its ground state and some excited states (see also conduction band, ground state, excited state).

Valine (Val): A uniasymmetric amino acid occurring abundantly in natural proteins with the formula $(CH_3)_2CHCH(NH_2)CO_2H$, also known as α-aminoisovaleric acid, 2-amino-3-methyl-butanoic acid.

Varve: A sedimentary structure (couplet) representing a single year of deposition. Generally found in environments with strong seasonal changes in environmental conditions.

Volume susceptibility: Magnetic susceptibility related to the volume of a test sample, generally $10 \, cm^3$; dimensionless.

Volumetrically averaged or integrated dose rate: A dose rate that has been modeled to account for inhomogeneously distributed radiation sources within the sphere of influence (α, β, or γ), used in ESR and luminescent dating methods.

W: Watt, a unit of power, $1\,W = m^2kg/s^3 = 1\,J/s$.

Wavelength dispersive spectrometry (WDS): Electron microprobe-type, generating sequential elemental acquisition.

Well-logs: A measure of rock or sediment properties as a function of depth in a well or borehole.

Whiting: The rapid precipitation of fine-grained inorganic carbonates in an epilimnion caused by the rapid photosynthetic uptake of dissolved CO_2 by algae and accompanying supersaturation of the carbonate mineral in water.

Whole-core logging: Measurement of a physical property (e.g., magnetic susceptibility) performed on unopened core sections.

Wiggle matching method: Matching of a contiguous series of ^{14}C measurements spanning a relatively short time interval to time-equivalent short-term fluctuations (wiggles) in the ^{14}C calibration curve. This procedure can be used to accurately determine the calendar age of the ^{14}C measurements.

Wisconsinan: 1) Time: The time period or physical events associated with the last glacial advance(s) that occurred in North America between the Last Interglacial and the Holocene, and considered equivalent to the Würm, Weichselian, or Devensian. Although some controversy surrounds the exact period covered by this interval, depending on where one places the end of the Last Interglacial and the start of the Holocene, the time range is often cited as approximately 115–10.8 ka (Oxygen Isotope Stages 5d–2 for the long, or 74–10.8 ka (Oxygen Isotope Stages 4–2) for the short Wisconsinan chronology. 2) Rocks: The rocks or sediment formed during the Wisconsinan.

wt%: Weight percent, which is equal to: (weight of component/total weight)100.

X-band ESR: ESR spectrometry usually at 1–20 mW power using microwave frequencies near 8–10 GHz under a 100 kHz field modulation with a field center near 336.0 mT for $g\sim 2$.

X-radiograph: An image produced by exposure of a sample material and photographic paper to an x-ray source. Samples with higher density will attenuate the x-ray more than lower density samples, thus decrease the exposure of the photographic paper.

X-radiography: X-rays that pass through heterogeneous media such as sediments are absorbed by certain components. Passing X-rays are captured on an X-ray sensitive film where they produce radiographs of the sediment structure.

X-rays: Electromagnetic radiation with a wavelength shorter than UV and longer than gamma radiation.

X-ray diffractometry (X-ray diffraction; XRD): The diffraction of a beam of X-rays by the three-dimensional array of atoms in a crystal.

X-ray fluorescence (XRF): 1) Method for quantitative determination of the geochemical composition (major and minor elements) of sediments. 2) Analytical technique in which samples are exposed to an X-ray beam. The beam generates secondary X-rays (fluorescence) that have wavelengths characteristic of elements in the sample.

X-ray fluorescence analysis: An elemental analysis method based on x-ray fluorescence.

X-ray images: Comparable to X-radiographs, X-ray images capture the sediment structure by sediment-penetrating X-rays. They differ because the X-ray source is coupled with a digital camera providing digital images that allow a direct image processing.

XRD: X-ray diffraction, diffractometer, or diffractometry.

XRF: X-ray fluorescence.

XRF-ED: X-ray fluorescence analysis using energy dispersive detection.

XRF-scanner: Automated sediment profiling instrument for qualitative determination of major elements such as K, Sr, Ca, Mg, Mn, Fe, Ti.

Yield: Probability that an atom will emit a gamma photon during radioactive decay.

Younger Dryas: A well-known and much discussed cold period during the last deglacial period, which terminates the Late Weichselian, or Late Glacial time. It occurred between 10,700–10,000 ^{14}C years BP or between 12,650–11,550 GRIP ice core years BP.

Zeroing: Any process that can reduce an ESR signal's intensity to a level indistinguishable from background levels. Most newly formed minerals have ESR signals with zero intensities. Strong heating at >250–$500\,°C$ will zero most signals in all minerals, while remineralization and diagenesis effectively zero signals. For some minerals, exposure to intense sunlight, high pressure or strain will cause zeroing.

INDEX

540

548

CPSIA information can be obtained at www.ICGtesting.com
Printed in the USA
LVOW011047160613

338773LV00002B/79/A